Lecture Notes in Con

4019

T0230292

Commenced Publication in 1973
Founding and Former Series Editor
Gerhard Goos, Juris Hartmanis, anc

Michael Johnson Varmo Vene (Eds.)

Algebraic Methodology and Software Technology

11th International Conference, AMAST 2006
Kuressaare, Estonia, July 5-8, 2006
Proceedings

Springer

Volume Editors

Michael Johnson
Macquarie University
Information and Communication Sciences
2109, Australia
E-mail: mike@ics.mq.edu.au

Varmo Vene
University of Tartu
Liivi 2, EE-50409, Estonia
E-mail: varmo@cs.ut.ee

Library of Congress Control Number: 2006928059

CR Subject Classification (1998): F.3-4, D.2, C.3, D.1.6, I.2.3, I.1.3

LNCS Sublibrary: SL 2 – Programming and Software Engineering

ISSN 0302-9743
ISBN-10 3-540-35633-9 Springer Berlin Heidelberg New York
ISBN-13 978-3-540-35633-2 Springer Berlin Heidelberg New York

Springer is a part of Springer Science+Business Media

springer.com

Printed in Germany

Typesetting: Camera-ready by author, data conversion by Scientific Publishing Services, Chennai, India
Printed on acid-free paper SPIN: 11784180 06/3142 5 4 3 2 1 0

Preface

This is the proceedings of the 11th edition of the Algebraic Methodology and Software Technology (AMAST) conference series. The first conference was held in the USA in 1989, and since then AMAST conferences have been held on (or near) five different continents and have been hosted by many of the most prominent people and organizations in the field.

The AMAST initiative has always sought to have practical effects by developing the science of software and basing it on a firm mathematical foundation. AMAST has interpreted software technology broadly, and has, for example, held AMAST workshops in areas as diverse as real-time systems and (natural) language processing. Similarly, algebraic methodology is interpreted broadly and includes abstract algebra, category theory, logic, and a range of other mathematical subdisciplines. The truly distinguishing feature of AMAST is that it seeks rigorous mathematical developments, but always strives to link them to real technological applications. Our meetings frequently include industry-based participants and are a rare opportunity for mathematicians and mathematically minded academics to interact technically with industry-based technologists. Over the years AMAST has included industrial participants from organizations specializing in safety-critical (including medical) systems, transport (including aerospace), and security-critical systems, amongst others.

AMAST has continued to grow and change. Much of the work that was the subject of early meetings is now established and used. A good deal of it has been presented in the eight monographs that have so far appeared as part of Springer's LNCS series. Many of the issues that the AMAST community was concerned with academically have now become part of major industrial organizations' research and development as security, correctness, and safety-critical performance become more and more important in the systems we use daily. Other issues remain unresolved, and new questions continually arise. What is certain is that in the future the fundamental character of AMAST—serious mathematics developed for real technology—will remain important.

The 11th edition of the conference was held in Kuressaare in Estonia, hosted by the Institute of Cybernetics at Tallinn University of Technology. Among the 55 full submissions, the Programme Committee selected 24 regular papers and 3 system demonstrations. All submissions were reviewed by three PC members with the help of external reviewers. In addition to the accepted papers, the conference also featured invited talks by three distinguished speakers: Ralph Back (Åbo Akademi University, Finland), Larry Moss (Indiana University, USA) and Till Mossakowski (Universität Bremen, Germany).

After the successful dual meeting in Stirling in 2004, the conference was co-located with Mathematics of Program Construction (MPC) for the second time. We thank the MPC organizers for suggesting this co-location. It is also worth noting that AMAST enjoys the cooperation and overlapping organizational participation with other like-minded conferences including CALCO, CMCS and WADT.

AMAST 2006 was the result of a considerable effort by a number of people. It is our pleasure to express our gratitude to the AMAST Programme Committee and additional referees for their expertise and diligence in reviewing the submitted papers, and to the AMAST Steering Committee for its general guidance. Our special thanks go to Tarmo Uustalu and his colleagues from the Institute of Cybernetics for taking care of practical matters in the local organization. We are also grateful to Andrei Voronkov for providing the EasyChair system, which was used to manage the electronic submissions, the review process, the electronic PC meeting, and to assemble the proceedings. Finally, we would like to express our thanks to Springer for its continued support in the publication of the proceedings in the *Lecture Notes in Computer Science* series.

April 2006

Michael Johnson
Varmo Vene

Conference Organization

Programme Chairs

Michael Johnson (Macquarie University, Australia)
Varmo Vene (University of Tartu, Estonia)

Programme Committee

Luís Soares Barbosa (Universidade do Minho, Portugal)
Gilles Barthe (INRIA Sophia-Antipolis, France)
Michel Bidoit (École Normale Supérieure de Cachan, France)
Gregor v. Bochmann (University of Ottawa, Canada)
Manfred Broy (Technische Universität München, Germany)
Cristian Calude (University of Auckland, New Zealand)
Christine Choppy (Université Paris Nord, France)
Arthur Fleck (University of Iowa, USA)
Marcelo Fabián Frias (Universidad de Buenos Aires, Argentina)
Nicolas Halbwachs (Université Grenoble I / CNRS, France)
Anne Haxthausen (Technical University of Denmark, Denmark)
Antónia Lopes (Universidade de Lisboa, Portugal)
Michael Mislove (Tulane University, USA)
Peter Mosses (University of Wales Swansea, UK)
Monica Nesi (Università degli Studi di L'Aquila, Italy)
Rocco De Nicola (Università degli Studi di Firenze, Italy)
Anton Nijholt (Universiteit Twente, The Netherlands)
Dusko Pavlovic (Kestrel Institute, USA)
Jaco van de Pol (CWI, The Netherlands)
Charles Rattray (University of Stirling, UK)
Teodor Rus (University of Iowa, USA)
Giuseppe Scollo (Università degli Studi di Catania, Italy)
Carolyn Talcott (SRI International, USA)
Andrzej Tarlecki (Warsaw University, Poland)
Ken Turner (University of Stirling, UK)
Irek Ulidowski (University of Leicester, UK)
Martin Wirsing (Ludwig-Maximilians-Universität München, Germany)

Organizing Committee

Tarmo Uustalu (Chair), Juhan Ernits, Monika Perkmann, Ando Saabas, Olha Shkaravska, Kristi Uustalu (Institute of Cybernetics, Estonia)

Additional Referees

Daniel Amyot
Matthias Anlauff
Eugene Asarin
Christel Baier
Hubert Baumeister
Marek Bednarczyk
Francesco Bellomi
Marcin Benke
Michele Boreale
Jewgenij Botaschanjan
Domenico Cantone
Maura Cerioli
Jacek Chrząszcz
Horatiu Cirstea
Robert Clark
Agostino Cortesi
Olivier Danvy
Amy Felty
Antoine Girard
Guillem Godoy Balil
Laure Gonnord
Benjamin Gregoire
Eric Gregoire
Will Harwood
Reiko Heckel
Peter Homeier
Neil Jones
Nicolás Kicillof
Bartek Klin
Alexander Knapp
Leonid Kof
Piotr Kosiuczenko
Steve Kremer
Moez Krichen
Peeter Laud
Martin Leucker
Luigi Logrippo
Michele Loreti
Etienne Lozes
Savi Maharaj
Mieke Massink
Laurent Mazare
Greg Meredith
Härmel Nestra
Isabel Nunes
Joel Ouaknine
Anne Pacalet
Olga Pacheco
Giuseppe Pappalardo
Jan Storbank Pedersen
Anna Philippou
David Pichardie
Jorge Sousa Pinto
Markus Roggenbach
Jan Romberg
Judi Romijn
Fernando Schapachnik
Carron Shankland
Élodie-Jane Sims
Doug Smith
Maria Spichkova
Fausto Spoto
Xavier Urbain
Tarmo Uustalu
Arnaud Venet
Herna L. Viktor
Joost Visser
Vesal Vojdani-Ghamsari
Stefan Wagner
James Worrell
Jinzhi Xia

Sponsoring Institutions

National Centers of Excellence Programme of the Estonian Ministry of Education and Research
Institute of Cybernetics at Tallinn University of Technology, Estonia

Table of Contents

Invited Talks

Contributed Papers

System Descriptions

Incremental Software Construction with Refinement Diagrams

Ralph-Johan Back

Åbo Akademi University, Department of Computer Science
Lemminkainenkatu 14 A, SF-20520 Turku, Finland
backrj@abo.fi

Abstract. We propose here a mathematical framework for *incremental software construction* and *controlled software evolution*. The framework allows incremental changes of a software system to be described on a high architecture level, but still with mathematical precision so that we can reason about the correctness of the changes. The framework introduces *refinement diagrams* as a visual way of presenting the architecture of large software systems. Refinement diagrams are based on lattice theory and allow reasoning about lattice elements to be carried out directly in terms of diagrams. A refinement diagram proof will be equivalent to a Hilbert like proof in lattice theory. We show how to apply refinement diagrams and refinement calculus to the incremental construction of large software system. We concentrate on three topics: (i) *modularization* of software systems with component specifications and the role of information hiding in this approach, (ii) *layered extension* of software by adding new features one-by-one and the role of inheritance and dynamic binding in this approach, and (iii) *evolution of software* over time and the control of successive versions of software.

M. Johnson and V. Vene (Eds.): AMAST 2006, LNCS 4019, p. 1, 2006.

Recursive Program Schemes: Past, Present, and Future

Lawrence S. Moss

Department of Mathematics, Indiana University
831 East Third Street, Bloomington, IN 47405-7106 USA
lsm@cs.indiana.edu

Abstract. This talk describes work on one of the first applications of algebra to theoretical computer science, the study of recursive program schemes. I would like to put a lot of the past work in perspective and then to describe recent work by Stefan Milius and myself which reworks the classical theory of uninterpreted and interpreted recursive program schemes using tools from coalgebraic recursion theory. Finally, I hope to speculate on whether the new work could be of interest to those pursuing AMAST's goal of "setting of software technology on a firm, mathematical basis."

M. Johnson and V. Vene (Eds.): AMAST 2006, LNCS 4019, p. 2, 2006.

Monad-Based Logics for Computational Effects

Till Mossakowski

DFKI Laboratory, Bremen, and
Department of Computer Science, University of Bremen
till@tzi.de

Abstract. The presence of computational effects, such as state, store, exceptions, input, output, non-determinism, backtracking etc., complicates the reasoning about programs. In particular, usually for each effect (or each combination of these), an own logic needs to be designed.

Monads are a well-known tool from category theory that originally has been invented for studying algebraic structures. Monads have been used very successfully by Moggi [1] to model computational effects (in particular, all of those mentioned above) in an elegent way. This has been applied both to the semantics of programming languages (e.g. [2, 3, 4, 5]) and to the encapsulation of effects in pure functional languages such as Haskell [6].

Several logics for reasoning about monadic programs have been introduced, such as evaluation logic [7, 8], Hoare logic [9] and dynamic logic [10, 11]. Some of these logics have a semantics and proof calculus given in a completely monad independent (and hence, effect independent) way. We give an overview of these logics, discuss completeness of their calculi, as well as some application of these logics to the reasoning about Haskell and Java programs, and a coding in the theorem prover Isabelle [12].

References

[1] Moggi, E.: Notions of computation and monads. Inform. and Comput. **93** (1991) 55–92

[2] Moggi, E.: An abstract view of programming languages. Technical Report ECS-LFCS-90-113, Dept. of Computer Science, Edinburgh Univ. (90)

[3] Wadler, P.: Comprehending monads. In: LFP '90: Proceedings of the 1990 ACM conference on LISP and functional programming, New York, NY, USA, ACM Press (1990) 61–78

[4] Jacobs, B., Poll, E.: Coalgebras and Monads in the Semantics of Java. Theoret. Comput. Sci. **291** (2003) 329–349

[5] Shinwell, M.R., Pitts, A.M.: On a monadic semantics for freshness. Theoret. Comput. Sci. **342** (2005) 28–55

[6] Wadler, P.: How to declare an imperative. ACM Computing Surveys **29** (1997) 240–263

[7] Pitts, A.: Evaluation logic. In: Higher Order Workshop. Workshops in Computing, Springer (1991) 162–189

[8] Moggi, E.: A semantics for evaluation logic. Fund. Inform. **22** (1995) 117–152

M. Johnson and V. Vene (Eds.): AMAST 2006, LNCS 4019, pp. 3–4, 2006.

[9] Schröder, L., Mossakowski, T.: Monad-independent Hoare logic in HasCasl. In: Fundamental Aspects of Software Engineering. Volume 2621 of LNCS. (2003) 261–277

[10] Schröder, L., Mossakowski, T.: Monad-independent dynamic logic in HasCasl. J. Logic Comput. **14** (2004) 571–619

[11] Mossakowski, T., Schröder, L., Goncharov, S.: Completeness of monad-based dynamic logic. Technical report, University of Bremen (2006)

[12] Walter, D.: Monadic dynamic logic: Application and implementation. Master's thesis, University of Bremen (2005)

State Space Representation for Verification of Open Systems

Irem Aktug and Dilian Gurov

KTH Computer Science and Communication
Osquars Backe 2, 100 44
Stockholm, Sweden
{irem, dilian}@nada.kth.se

Abstract. When designing an open system, there might be no implementation available for certain components at verification time. For such systems, verification has to be based on assumptions on the underspecified components. When component assumptions are expressed in Hennessy-Milner logic (HML), the state space of open systems can be naturally represented with modal transition systems (MTS), a graphical specification language equiexpressive with HML. Having an explicit state space representation supports state space exploration based verification techniques. Besides, it enables proof reuse and facilitates visualization for the user guiding the verification process in interactive verification. As an intuitive representation of system behavior, it aids debugging when proof generation fails in automatic verification.

However, HML is not expressive enough to capture temporal assumptions. For this purpose, we extend MTSs to represent the state space of open systems where component assumptions are specified in modal μ-calculus. We present a two-phase construction from process algebraic open system descriptions to such state space representations. The first phase deals with component assumptions, and is essentially a maximal model construction for the modal μ-calculus. In the second phase, the models obtained are combined according to the structure of the open system to form the complete state space. The construction is sound and complete for systems with a single unknown component and sound for those without dynamic process creation. For establishing open system properties based on the representation, we present a proof system which is sound and complete for prime formulae.

1 Introduction

In an *open system*, certain components can join the system after it has been put in operation. For example, applications can be loaded on a smart card after the card has been issued (see e.g. [SGH04]). Since the implementations of certain components are not yet available, the verification of the system has to be based on behavioural assumptions on such components. Security protocols can be verified in this manner, for instance by treating an unpredictable attacker as an unknown component of the system [1].

M. Johnson and V. Vene (Eds.): AMAST 2006, LNCS 4019, pp. 5–20, 2006.

Modal transition systems (MTS) were introduced by Larsen as a graphical specification language [2]. Certain kinds of properties are easier to express graphically than in temporal logics. Each MTS specifies a set of processes as an interval determined by necessary and admissable transitions. MTSs are equiexpressive with Hennessy-Milner logic (HML), i.e. an HML formula can be characterized by an MTS and vice versa. As a result, MTSs provide a natural representation of the state space of open systems when assumptions on the behavior of the not-yet-available components are specified in HML. When the assumptions are temporal properties, however, MTSs are not expressive enough for this purpose. In [3], we extend MTSs to represent the state space of open systems when the component assumptions are written in the modal μ-calculus [4]. This logic adds the expressive power of least and greatest fixed point recursion to HML. Besides the *must* (necessary) and *may* (admissable) transitions of MTS, our notion, *extended modal transition system* (EMTS) has sets of states (instead of single states) as targets to transitions - an extension which is needed for dealing with disjunctive assumptions, and well-foundedness constraints to handle least fixed point assumptions.

Having a way to capture the state space of an open system explicitly can be useful in various phases of the development of open systems. In *the modeling phase*, this formalism can be used as an alternative means of graphical specification of open system behavior. In *interactive verification*, an explicit state space representation facilitates visualization of the system behaviour, assisting the user in guiding the proof. This visualization facility is beneficial in *automatic verification* when the automatic proof construction fails and an understanding of the open system behaviour becomes necessary for debugging. Furthermore, computing the whole state space enables proof reuse when the same system is to be checked for several properties.

In this paper, we address the problem of constructing an explicit state space representation from an open system description and verifying open system properties based on this representation. In a process algebraic setting, the behaviour of an open system can be specified by an *open process term with assumptions* (OTA). An OTA consists of a process term equipped with a list of behavioral assumptions on the free variables of the term. We offer a two-phase construction that, under given restrictions, automatically extracts an EMTS from an OTA. The first phase in the construction corresponds to a maximal model construction for each component assumption. For the fixed point cases, a powerset construction is used that is similar to the one used in the Büchi automata constructions of [5] and [6]. In the second phase, the maximal models are composed according to the structure of the open system. The construction is *sound* (resp. *complete*) if the set of systems denoted by the OTA is a subset (resp. superset) of the denotation of the resulting EMTS. We show soundness of the construction for systems without dynamic process creation, and soundness and completeness for systems with a single unknown component. Finally, we present a proof system for showing open system properties based on EMTSs. The proof system is sound and complete for *prime* formulae, a prime formula being one that logically implies one of the disjuncts whenever it logically implies a disjunction. The relative simplicity of the proof system and its use is an indication of the adequateness of EMTSs for open system state space representation.

Related Work. In this strand of research, our work follows earlier work on using maximal model constructions for modular verification for various fragments of the μ-calculus: for ACTL by Grumberg and Long [7], ACTL* by Kupferman and Vardi [8], and the fragment without least fixed points and diamond modalities by Sprenger et al [9]. In automata based approaches (see for instance [10, 6, 11]), various structures like alternating tree automata, Büchi and Rabin automata have been employed for capturing temporal properties. Although expressively powerful, we argue that these structures do not provide an intuitive representation of the state space for branching-time logics.

Proof system based methods have previously been suggested for the interactive verification of open systems [12, 13] where modal μ-calculus is used to express the temporal assumptions on components as well as the desired property of the system. These interactive methods explore the state space implicitly as much as it is necessary for the particular verification task. In contrast to these methods, we separate the tasks of constructing a finite representation of the state space of an open system from the task of verifying its properties. This separation provides a state visualization facility to the user guiding the interactive proof, and offers greater possibilities for proof reuse.

Organization. The paper is organized as follows. In section 2, we make the syntax of OTAs precise by a brief account of the logic used in behavioral assumptions and the process algebra used to define the process term. Section 3 is a summary of important definitions related to the notion of EMTS. We present the translation from OTA to EMTS in Section 4, and provide correctness results. In Section 5, we give a proof system for showing open system properties of EMTSs. The last section presents conclusions and identifies directions for future work.

2 Specifying Open Systems Behaviour

A system, the behaviour of which is parameterized on the behaviour of certain components, is conveniently represented as a pair $\Gamma \rhd E$, where E is an open process-algebraic term, and Γ is a list of assertions of the shape $X : \Phi$ where X is a process variable free in E and Φ is a closed formula in a process logic.

In the present study, we work with the class of Basic Parallel Processes (BPP)[14]. The terms of BPP are generated by:

$$E ::= \mathbf{0} \mid X \mid a.E \mid E + E \mid E \parallel E \mid \mathit{fix}X.E$$

where X ranges over a set of process variables *ProcVar* and a over a finite set of actions A. We assume that *ProcVar* is partitioned into assumption process variables *AssProcVar* used in assertions, and recursion process variables *RecProcVar* bound by *fix*. A term E is called *linear* if every assumption process variable occurs in E at most once. The operational semantics of closed process terms (called processes and ranged over by t) is standard, where the operator $\|$ signifies *merge composition*. Closed process terms give rise to *labeled transition systems* (LTS) through this standard semantics.

As a process logic for specifying behavioural assumptions of components, as well as for specifying system properties to be verified, we consider the modal μ-calculus [4]. Its formulas are generated by:

$$\Phi ::= \mathrm{tt} \mid \mathrm{ff} \mid Z \mid \Phi_1 \wedge \Phi_2 \mid \Phi_1 \vee \Phi_2 \mid [a]\,\Phi \mid \langle a \rangle\,\Phi \mid \nu Z.\Phi \mid \mu Z.\Phi$$

where Z ranges over a set of propositional variables *PropVar*. The semantics of the μ-calculus is standard and given in terms of its denotation on some LTS $\mathcal{T}$=($S_\mathcal{T}$, A, $\longrightarrow_\mathcal{T}$). The denotation of a modal μ-calculus formula Φ, written $||\Phi||^\mathcal{T}_\mathrm{V}$, is a subset of the set of states of $\mathcal{T}$, where $\mathrm{V} : PropVar \rightarrow S_\mathcal{T}$ is a valuation that maps propositional variables to states of $\mathcal{T}$. As usual, we write $t \models^\mathcal{T}_\mathrm{V} \Phi$ whenever $t \in ||\Phi||^\mathcal{T}_\mathrm{V}$. In the sequel, we omit the subscript V when Φ is a closed formula.

We say that an OTA $\Gamma \rhd E$ is *guarded* when the term E and all modal μ-calculus formula Φ in Γ are guarded. Similarly, we say an OTA is linear when the term it contains is linear.

The behaviour specified by an open term with assumptions is given with respect to a LTS $\mathcal{T}$ that is closed under the transition rules and is closed under substitution of processes for assumption process variables in subterms of the OTA. The denotation of an OTA is then the set of all processes obtained by substituting each assumption process variable in the term by a process from $\mathcal{T}$ satisfying the respective assumptions.

Definition 1 (OTA Denotation). *Let $\Gamma \rhd E$ be an OTA, $\mathcal{T}$ be an LTS, and ρ_R : RecProcVar $\rightarrow S_\mathcal{T}$ be a recursion environment. The* denotation *of $\Gamma \rhd E$ relative to $\mathcal{T}$ and ρ_R is defined as:*

$$[\![\Gamma \rhd E]\!]_{\rho_R} \triangleq \{E\rho_R\rho_A \mid \forall (X : \Phi) \in \Gamma.\ \rho_A(X) \models^\mathcal{T} \Phi\}$$

where ρ_A : AssProcVar $\rightarrow S_\mathcal{T}$ ranges over assumption environments.

Example. Consider an operating system in the form of a concurrent server that spawns off *Handler* processes each time it receives a request. These processes run system calls for handling the given requests to produce a result (modeled by the action $\overline{out}$). *Handler* is defined as $Handler \stackrel{def}{=} In \,\|\, \overline{out}.\mathbf{0}$ where $In \stackrel{def}{=} in.In$. Although it is possible to communicate with request handlers through the attached channel (modeled by the action *in*), they do not react to further input. A property one would like to prove of such a server is that it stabilizes whenever it stops receiving new requests. Eventual stabilization can be formalized in the modal μ-calculus as $\mathbf{stab} \triangleq \nu X.\mu Y.\,[in]\,X \wedge [\overline{out}]\,Y$. We can reduce this verification task to proving that the open system modeled by the OTA

$$X : \mathbf{stab} \rhd X \,\|\, Handler$$

which consists of *Handler* and any stabilizing process X, eventually stabilizes.

3 Extended Modal Transition Systems

In [3], we proposed Extended Modal Transition Systems (EMTS) as an explicit state space representation for open systems with temporal assumptions, with an extensional representation for the well-foundedness constraints. In this section, we summarize the main definitions, and propose a concrete representation of well-foundedness constraints. The notion of EMTS is based on Larsen's Modal Transition Systems [2]. Kripke Modal Transition Systems (KMTS) have been first introduced by Huth et. al. [15], and later refined by Grumberg and Shoham [16] for representing state space abstractions in an abstraction refinement framework. EMTS is similar to KMTS with a constraint added to deal with termination assumptions.

In addition to may and must transitions for dealing with modalities, EMTSs include sets of states (instead of single states) as targets to transitions to capture disjunctive assumptions, and a set of prohibited infinite runs defined through a coloring function to represent termination assumptions.

Definition 2 (EMTS). *An* extended modal transition system *is a structure*

$$\mathcal{E} = (S_\mathcal{E}, A, \longrightarrow^\Diamond_\mathcal{E}, \longrightarrow^\Box_\mathcal{E}, c)$$

where (i) $S_\mathcal{E}$ is a set of abstract states, *(ii) A is a set of* actions, *(iii) $\longrightarrow^\Diamond_\mathcal{E}, \longrightarrow^\Box_\mathcal{E} \subseteq S_\mathcal{E} \times A \times 2^{S_\mathcal{E}}$ are* may *and* must transition relations, *and (iv) $c : S_\mathcal{E} \to \mathbb{N}^k$ is a* coloring function *for some $k \in \mathbb{N}$.*

May transitions of an EMTS show possible behaviours of the closed systems represented, while must transitions specify behaviour shared by all these closed systems. A *run* (or may–run) of $\mathcal{E}$ is a possibly infinite sequence of transitions $\rho_\mathcal{E} = s_0 \xrightarrow{a_0}_\mathcal{E} s_1 \xrightarrow{a_1}_\mathcal{E} s_2 \xrightarrow{a_2}_\mathcal{E} \ldots$ where for every $i \geq 0$, $s_i \xrightarrow{a_i}^\Diamond_\mathcal{E} S$ for some S such that $s_{i+1} \in S$. Must–runs are defined similarly. We distinguish between two kinds of *a-derivatives* of a state s: $\partial^\Diamond_a(s) \triangleq \{S \mid s \xrightarrow{a}^\Diamond_\mathcal{E} S\}$ and $\partial^\Box_a(s) \triangleq \{S \mid s \xrightarrow{a}^\Box_\mathcal{E} S\}$.

The coloring function c specifies a set $W_\mathcal{E}$ of prohibited infinite runs, which plays a similar role to fairness constraints of e.g. [7], by means of a *parity acceptance condition* (cf. [17, 10]). The function c is extended to infinite runs so that $c(\rho_\mathcal{E}) = (c(s_0)(1) \cdot c(s_1)(1) \ldots, \ldots, c(s_0)(k) \cdot c(s_1)(k) \ldots)$ is a k-tuple of infinite words where $c(s)(j)$ denotes the jth component of $c(s)$. Let $inf(c(\rho_\mathcal{E})(i))$ denote the set of infinitely occurring colors in the ith word of this tuple. Then the run $\rho_\mathcal{E}$ is prohibited, $\rho_\mathcal{E} \in W_\mathcal{E}$, if and only if $max\,(inf\,(c(\rho_\mathcal{E})(i)))$ is odd for some $1 \leq i \leq k$, i.e. the greatest number that occurs infinitely often in one of these k infinite words is odd.

Next, we define a simulation relation between the states of an EMTS as a form of mixed fair simulation (cf. e.g. [7, 18]).

Definition 3 (Simulation). *$R \subseteq S_\mathcal{E} \times S_\mathcal{E}$ is a* simulation relation *between the states of $\mathcal{E}$ if whenever $s_1 R s_2$ and $a \in A$:*

1. *if* $s_1 \xrightarrow{a}^{\Diamond}_{\mathcal{E}} S_1$*, then there is a* S_2 *such that* $s_2 \xrightarrow{a}^{\Diamond}_{\mathcal{E}} S_2$ *and for each* $s_1' \in S_1$*, there exists a* $s_2' \in S_2$ *such that* $s_1' R s_2'$*;*
2. *if* $s_2 \xrightarrow{a}^{\Box}_{\mathcal{E}} S_2$*, then there is a* S_1 *such that* $s_1 \xrightarrow{a}^{\Box}_{\mathcal{E}} S_1$ *and for each* $s_1' \in S_1$*, there exists a* $s_2' \in S_2$ *such that* $s_1' R s_2'$*;*
3. *if the run* $\rho_{s_2} = s_2 \xrightarrow{a_1}_{\mathcal{E}} s_2^1 \xrightarrow{a_2}_{\mathcal{E}} s_2^2 \xrightarrow{a_3}_{\mathcal{E}} \dots$ *is in* $W_{\mathcal{E}}$ *then every infinite run* $\rho_{s_1} = s_1 \xrightarrow{a_1}_{\mathcal{E}} s_1^1 \xrightarrow{a_2}_{\mathcal{E}} s_1^2 \xrightarrow{a_3}_{\mathcal{E}} \dots$ *such that* $s_1^i R\, s_2^i$ *for all* $i \geq 1$ *is also in* $W_{\mathcal{E}}$*.*

We say that abstract state s_2 *simulates* abstract state s_1, denoted $s_1 \preceq s_2$, if there is a simulation relation R such that $s_1 R s_2$. Simulation can be generalized to two different EMTSs $\mathcal{E}_1$ and $\mathcal{E}_2$ in the natural way.

Labeled transition systems can be viewed as a special kind of EMTS, where: $\longrightarrow^{\Box}_{\mathcal{E}} = \longrightarrow^{\Diamond}_{\mathcal{E}}$, the target sets of the transition relation are singleton sets of states, and the set of prohibited runs W is empty. We give the meaning of an abstract state relative to a given LTS, as the set of concrete LTS states simulated by the abstract state.

Definition 4 (Denotation). *Let* $\mathcal{E}$ *be an EMTS, and let* $\mathcal{T}$ *be an LTS. The* denotation *of abstract state* $s \in S_{\mathcal{E}}$ *is the set* $[\![s]\!]_{\mathcal{T}} \triangleq \{t \in S_{\mathcal{T}} \mid t \preceq s\}$*. This notion is lifted to sets of abstract states* $S' \subseteq S_{\mathcal{E}}$ *in the natural way:* $[\![S']\!]_{\mathcal{T}} \triangleq \bigcup\{[\![s]\!]_{\mathcal{T}} \mid s \in S'\}$*.*

In the rest of the paper, we shall assume that EMTSs obey the following *consistency* restrictions: $\longrightarrow^{\Box}_{\mathcal{E}} \subseteq \longrightarrow^{\Diamond}_{\mathcal{E}}$, $s \xrightarrow{a}^{\Box}_{\mathcal{E}} S$ implies S is non-empty, and W does not contain runs corresponding to infinite must–runs of the EMTS. The meaning of abstract states would not be altered if the targets of may transitions were restricted to singletons, but we prefer the targets of both kinds of transitions to be sets of states for reasons of uniformity.

In section 5, we present a proof system for proving properties of abstract states. For this purpose, we define when an abstract state s satisfies a modal μ-calculus formula Φ. The global nature of the set W in EMTSs makes it cumbersome to define the denotation of a fixed point formula compositionally as a set of abstract states. We therefore give an indirect definition of satisfaction, by means of the denotation $[\![s]\!]_{\mathcal{T}}$ of a state s.

Definition 5 (Satisfaction). *Let* $\mathcal{E}$ *be an EMTS,* $s \in S_{\mathcal{E}}$ *be an abstract state of* $\mathcal{E}$ *and* Φ *be a modal* μ*-calculus property. Then* s *satisfies* Φ *under valuation* $\mathcal{V} : PropVar \to 2^{S_{\mathcal{E}}}$*, denoted* $s \models^{\mathcal{E}}_{\mathcal{V}} \Phi$*, if and only if for any LTS* $\mathcal{T}$ $[\![s]\!]_{\mathcal{T}} \models^{\mathcal{T}}_{\mathrm{V}} \Phi$ *where valuation* $\mathrm{V} : PropVar \to 2^{S_{\mathcal{T}}}$ *is induced by* $\mathcal{V}$ *as* $\mathrm{V}(Z) \triangleq \bigcup\{[\![s]\!]_{\mathcal{T}} \mid s \in \mathcal{V}(Z)\}$*.*

Example. The state space of the open system introduced in the previous section is captured by the EMTS in Figure 1. For any labeled transition system $\mathcal{T}$, the processes simulated by the state s_1 are those denoted by the open term $X : \textbf{stab} \rhd X \parallel Handler$. The EMTS consists of six abstract states, each state

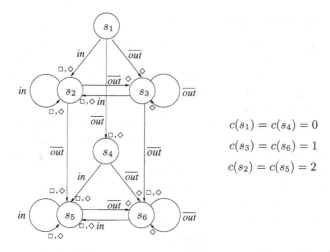

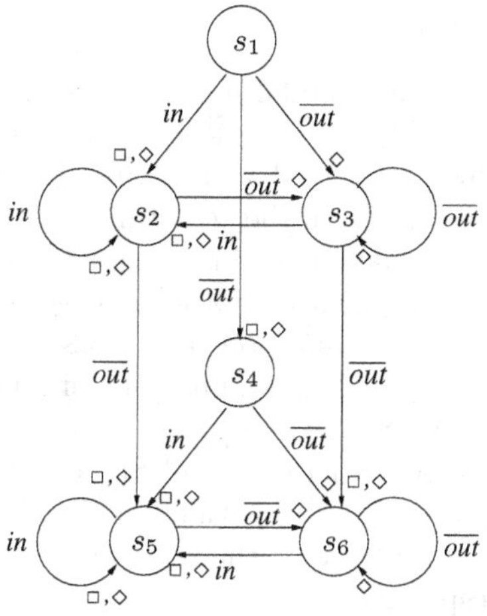

$$c(s_1) = c(s_4) = 0$$
$$c(s_3) = c(s_6) = 1$$
$$c(s_2) = c(s_5) = 2$$

Fig. 1. EMTS for $X : \boldsymbol{stab} \rhd X \parallel Handler$

denoting the set of processes which it simulates. For instance, states s_5 and s_6 in the example denote all processes which can engage in arbitrary interleavings of *in* and $\overline{out}$ actions, but so that *in* has to be enabled throughout while $\overline{out}$ has not. Infinite runs stabilizing on $\overline{out}$ actions are prohibited by the coloring of s_3 and s_6. A proof of eventual stabilization of the system using this representation can be found in [19].

4 From OTA to EMTS

In this section, we address the problem of providing an explicit state space representation for a given open term $\Gamma \rhd E$, by means of an EMTS $\mathcal{E}$. While it is tempting to define $\longrightarrow^{\Diamond}_{\mathcal{E}}$ and $\longrightarrow^{\Box}_{\mathcal{E}}$ through transition rules, the global nature of the well-foundedness constraints suggests that a direct construction would be more convenient for automatic construction. We propose a two-phase construction ε that translates an open term $\Gamma \rhd E$ to an EMTS, denoted $\varepsilon(\Gamma \rhd E)$. In the first phase, an EMTS is constructed for each underspecified component. This part is essentially a maximal model construction as developed by Grumberg and Long for ACTL [7], extended to ACTL* by Kupferman and Vardi [8], and applied by Sprenger et al to the fragment of the modal μ-calculus without least fixed points and diamond modalities [9]. For the construction of the fixed point cases, we adapt a powerset construction used earlier to convert fragments of the modal μ-calculus to Büchi automata which was introduced by Dam [5] for linear time μ-calculus and extended by Kaivola [6] to the Π_2 fragment. The second phase consists of combining the EMTSs produced in the first step according to the structure of the term E. We then show the correctness of the construction by relating the set of states simulated by the constructed EMTS to the denotation of the given OTA.

4.1 Maximal Model Construction

We define the function ε which maps modal μ-calculus formulas to triples of the shape $(\mathcal{E}, S, \lambda)$, where $\mathcal{E} = (S_{\mathcal{E}}, A, \longrightarrow^{\Diamond}_{\mathcal{E}}, \longrightarrow^{\Box}_{\mathcal{E}}, c)$ is an *EMTS*, $S \subseteq S_{\mathcal{E}}$ is a set of *start states* of $\mathcal{E}$, and $\lambda : S_{\mathcal{E}} \rightarrow 2^{PropVar}$ is a *labeling function.*

The function is defined inductively on the structure of Φ as shown in Table 1. The meaning of open formulae that arises in intermediate steps are given by the valuation which assigns the whole set of processes $S_{\mathcal{T}}$ to each propositional variable. Essentially, the particular valuation used does not play a role in the final EMTS, since the properties used as assumptions of an OTA are closed.

In the definition, let $\varepsilon(\Phi_1)$ be $((S_{\mathcal{E}_1}, A, \longrightarrow^{\Diamond}_{\mathcal{E}_1}, \longrightarrow^{\Box}_{\mathcal{E}_1}, c_1), S_1, \lambda_1)$ and $\varepsilon(\Phi_2)$ be $((S_{\mathcal{E}_2}, A, \longrightarrow^{\Diamond}_{\mathcal{E}_2}, \longrightarrow^{\Box}_{\mathcal{E}_2}, c_2), S_2, \lambda_2)$ where $S_{\mathcal{E}_1}$ and $S_{\mathcal{E}_2}$ are disjoint sets. The new state s_{new} is not in $S_{\mathcal{E}_1}$ and a and a' are actions in A. The coloring functions $c_1 : S_{\mathcal{E}_1} \rightarrow \mathbb{N}^{k_1}$ and $c_2 : S_{\mathcal{E}_2} \rightarrow \mathbb{N}^{k_2}$ color the states of $\mathcal{E}_1$ and $\mathcal{E}_2$ with integer tuples of length k_1 and k_2 respectively.

For a set S, $S\mid_{\Box}$ denotes the largest transition-closed set contained in S such that there is no element $s \in S\mid_{\Box}$ with the empty set as the target to a must transition, that is, there is no s such that for some $a \in A$, $s \overset{a}{\longrightarrow}^{\Box}_{\mathcal{E}} \emptyset$ and each state s is reachable from some start state.

In what follows, we explain the various cases of the construction. The EMTS for formula tt consists of the single state s_{tt} with may transitions to itself for every action, while the EMTS for ff is the empty EMTS. The EMTS for a propositional variable consists of a single start state with may transitions to s_{tt} for each action.

The states of the EMTS for the conjunction of two formulas is the cross product of the states of the EMTSs constructed for each conjunct, excluding pairs with incompatible capabilities. If a state s_1, which has a must transition for an action a to some set S_1, is producted with a state s_2 that has multiple may transitions for a, then the product state has a must a-transition to the product of S_1 with the set of all may-successors of s_2. The color of a state of $\varepsilon(\Phi_1 \wedge \Phi_2)$ is the concatenation of the colors of the paired states. In the case of disjunction, the set of start states of $\varepsilon(\Phi_1 \vee \Phi_2)$ is the union of the start states of $\varepsilon(\Phi_1)$ and $\varepsilon(\Phi_2)$ which reflects the union of their denotation. The color of a state is given by padding with 0's from either the left or right.

For the modal cases, a new state s_{new} is set as the start state. The EMTS for $\varepsilon([a]\,\Phi)$ has a single may transition for a, which is to the set of initial states of $\varepsilon(\Phi)$. This is to ensure all simulated processes satisfy Φ after engaging in an a. Additionally, there is a may transition to s_{tt} for all other actions. The EMTS for $\varepsilon(\langle a\rangle\,\Phi)$ includes a must transition for a from this start state to the start states of $\varepsilon(\Phi)$, along with may transitions for all actions to s_{tt} forcing the simulated processes to have an a transition to some process satisfying Φ and allowing any other transitions besides.

The construction for fixed point formulae is a powerset construction, which is similar to the constructions given in [5] and [6] for the purpose of constructing Büchi Automata for linear time and the alternation-depth class Π_2 fragments of the μ-calculus, respectively. The states of $\varepsilon(\sigma Z.\Phi)$ consist of sets of states of

Table 1. Maximal Model Construction

- $\varepsilon(\mathrm{tt}) \triangleq ((\{s_{\mathrm{tt}}\},\ A,\ \longrightarrow^{\diamond}_{\mathcal{E}},\ \emptyset,\ \{s_{\mathrm{tt}} \mapsto 0\}),\ \{s_{\mathrm{tt}}\},\ \{s_{\mathrm{tt}} \mapsto \emptyset\})$
  where $s_{\mathrm{tt}} \xrightarrow{a}{}^{\diamond}_{\mathcal{E}} \{s_{\mathrm{tt}}\}$ for all $a \in A$.
- $\varepsilon(\mathrm{ff}) \triangleq ((\emptyset,\ A,\ \emptyset,\ \emptyset,\ \emptyset),\ \emptyset,\ \emptyset)$
- $\varepsilon(Z) \triangleq ((\{s_{new}, s_{\mathrm{tt}}\},\ A,\ \longrightarrow^{\diamond}_{\mathcal{E}},\ \emptyset,\ \{s_{\mathrm{tt}} \mapsto 0, s_{new} \mapsto 0\}),\ \{s_{new}\},\ \{s_{new} \mapsto \{Z\}, s_{\mathrm{tt}} \mapsto \emptyset\})$
  where $s_{new} \xrightarrow{a}{}^{\diamond}_{\mathcal{E}} \{s_{\mathrm{tt}}\}$ and $s_{\mathrm{tt}} \xrightarrow{a}{}^{\diamond}_{\mathcal{E}} \{s_{\mathrm{tt}}\}$ for all $a \in A$.
- $\varepsilon(\Phi_1 \wedge \Phi_2) \triangleq (((S_{\mathcal{E}_1} \times S_{\mathcal{E}_2})|_{\Box},\ A,\ \longrightarrow^{\diamond}_{\mathcal{E}},\ \longrightarrow^{\Box}_{\mathcal{E}},\ W),\ (S_{\mathcal{E}_1} \times S_{\mathcal{E}_2})|_{\Box} \cap (S_1 \times S_2),\ \lambda)$ where

  $\longrightarrow^{\diamond}_{\mathcal{E}} \triangleq \{(s,r) \xrightarrow{a}{}^{\diamond}_{\mathcal{E}} S' \times \cup\partial^{\diamond}_{a}(r) \mid s \xrightarrow{a}{}^{\Box}_{\mathcal{E}_1} S'\}$
  $\cup\ \{(s,r) \xrightarrow{a}{}^{\diamond}_{\mathcal{E}} \cup\partial^{\diamond}_{a}(s) \times R' \mid r \xrightarrow{a}{}^{\Box}_{\mathcal{E}_2} R'\}$
  $\cup\ \{(s,r) \xrightarrow{a}{}^{\diamond}_{\mathcal{E}} S' \times R' \mid s \xrightarrow{a}{}^{\diamond}_{\mathcal{E}_1} S' \wedge r \xrightarrow{a}{}^{\diamond}_{\mathcal{E}_2} R' \wedge S' \notin \partial^{\Box}_{a}(s) \wedge R' \notin \partial^{\Box}_{a}(r)\}$
  $\longrightarrow^{\Box}_{\mathcal{E}} \triangleq \{(s,r) \xrightarrow{a}{}^{\Box}_{\mathcal{E}} (S' \times \cup\partial^{\diamond}_{a}(r)) \mid s \xrightarrow{a}{}^{\Box}_{\mathcal{E}_1} S'\}$
  $\cup\ \{(s,r) \xrightarrow{a}{}^{\Box}_{\mathcal{E}} (\cup\partial^{\diamond}_{a}(s) \times R') \mid r \xrightarrow{a}{}^{\Box}_{\mathcal{E}_2} R'\}$
  $c \triangleq \{(s,r) \mapsto c_1(s) \cdot c_2(r) \mid s \in S_{\mathcal{E}_1} \wedge r \in S_{\mathcal{E}_2}\}$
  $\lambda \triangleq \{(s,r) \mapsto \lambda_1(s) \cup \lambda_2(r) \mid s \in S_{\mathcal{E}_1} \wedge r \in S_{\mathcal{E}_2}\}$
- $\varepsilon(\Phi_1 \vee \Phi_2) \triangleq ((S_{\mathcal{E}_1} \cup S_{\mathcal{E}_2},\ A,\ \longrightarrow^{\diamond}_{\mathcal{E}},\ \longrightarrow^{\Box}_{\mathcal{E}},\ c),\ S_1 \cup S_2,\ \lambda_1 \cup \lambda_2)$ with:

  $\longrightarrow^{\diamond}_{\mathcal{E}} \triangleq \longrightarrow^{\diamond}_{\mathcal{E}_1} \cup \longrightarrow^{\diamond}_{\mathcal{E}_2}$
  $\longrightarrow^{\Box}_{\mathcal{E}} \triangleq \longrightarrow^{\Box}_{\mathcal{E}_1} \cup \longrightarrow^{\Box}_{\mathcal{E}_2}$
  $c \triangleq \{s \mapsto c_1(s) \cdot 0^{k_2} \mid s \in S_{\mathcal{E}_1}\} \cup \{s \mapsto 0^{k_1} \cdot c_2(s) \mid s \in S_{\mathcal{E}_2}\}$
- $\varepsilon([a]\,\Phi_1) \triangleq ((S_{\mathcal{E}_1} \cup \{s_{new}, s_{\mathrm{tt}}\},\ A,\ \longrightarrow^{\diamond}_{\mathcal{E}},\ \longrightarrow^{\Box}_{\mathcal{E}_1},\ c),\ \{s_{new}\},\ \lambda)$ with:

  $\longrightarrow^{\diamond}_{\mathcal{E}} \triangleq \longrightarrow^{\diamond}_{\mathcal{E}_1} \cup \{s_{\mathrm{tt}} \xrightarrow{a'}{}^{\diamond}_{\mathcal{E}} \{s_{\mathrm{tt}}\} \mid a' \in A\} \cup \{s_{new} \xrightarrow{a}{}^{\diamond}_{\mathcal{E}} S_1\}$
  $\cup\ \{s_{new} \xrightarrow{a'}{}^{\diamond}_{\mathcal{E}} \{s_{\mathrm{tt}}\} \mid a' \neq a \wedge a' \in A\}$
  $c \triangleq c_1 \cup \{s_{new} \mapsto 0^{k_1}\} \cup \{s_{\mathrm{tt}} \mapsto 0^{k_1}\}$
  $\lambda \triangleq \lambda_1 \cup \{s_{new} \mapsto \emptyset\} \cup \{s_{\mathrm{tt}} \mapsto \emptyset\}$
- $\varepsilon(\langle a\rangle\,\Phi_1) \triangleq \varepsilon(\mathrm{ff})$ if $S_1 = \emptyset$
  $\varepsilon(\langle a\rangle\,\Phi_1) \triangleq ((S_{\mathcal{E}_1} \cup \{s_{new}, s_{\mathrm{tt}}\},\ A,\ \longrightarrow^{\diamond}_{\mathcal{E}},\ \longrightarrow^{\Box}_{\mathcal{E}},\ c),\ \{s_{new}\},\ \lambda)$ otherwise, with:

  $\longrightarrow^{\diamond}_{\mathcal{E}} \triangleq \longrightarrow^{\diamond}_{\mathcal{E}_1} \cup \{s_{new} \xrightarrow{a}{}^{\diamond}_{\mathcal{E}} S_1\} \cup \{s_{new} \xrightarrow{a'}{}^{\diamond}_{\mathcal{E}} \{s_{\mathrm{tt}}\} \mid a' \in A\} \cup \{s_{\mathrm{tt}} \xrightarrow{a'}{}^{\diamond}_{\mathcal{E}} \{s_{\mathrm{tt}}\} \mid a' \in A\}$
  $\longrightarrow^{\Box}_{\mathcal{E}} \triangleq \longrightarrow^{\Box}_{\mathcal{E}_1} \cup \{s_{new} \xrightarrow{a}{}^{\Box}_{\mathcal{E}} S_1\}$
  $c \triangleq c_1 \cup \{s_{new} \mapsto 0^{k_1}\} \cup \{s_{\mathrm{tt}} \mapsto 0^{k_1}\}$
  $\lambda \triangleq \lambda_1 \cup \{s_{new} \mapsto \emptyset\} \cup \{s_{\mathrm{tt}} \mapsto \emptyset\}$
- $\varepsilon(\sigma Z.\Phi_1)\ ((2^{S_{\mathcal{E}_1}}|_{\Box}, A, \longrightarrow^{\diamond}_{\mathcal{E}}, \longrightarrow^{\Box}_{\mathcal{E}}, c_\sigma), 2^{S_{\mathcal{E}_1}}|_{\Box} \cap \{\{s\} \mid s \in S_1\}, \lambda)$ where $\sigma \in \{\nu, \mu\}$ with:

  $\longrightarrow^{\diamond}_{\mathcal{E}} \triangleq \{\{s_1, \ldots, s_n\} \xrightarrow{a}{}^{\diamond}_{\mathcal{E}} S \mid \exists i.\exists S'_i.s_i \xrightarrow{a}{}^{\Box}_{\mathcal{E}_1} S'_i \wedge$
  $S = \partial_{\mathcal{P}}((\cup\partial^{\diamond}_{a}(s_1), \ldots, S'_i, \ldots, \cup\partial^{\diamond}_{a}(s_n)), S_1, \lambda_1, Z)\}$
  $\cup\ \{\{s_1, \ldots, s_n\} \xrightarrow{a}{}^{\diamond}_{\mathcal{E}} S \mid \forall j.\exists S'_j.s_j \xrightarrow{a}{}^{\diamond}_{\mathcal{E}_1} S'_j \wedge S'_j \notin \partial^{\Box}_{a}(s_j) \wedge$
  $S = \partial_{\mathcal{P}}((S'_1, \ldots, S'_n), S_1, \lambda_1, Z)\}$
  $\longrightarrow^{\Box}_{\mathcal{E}} \triangleq \{\ \{s_1, \ldots, s_n\} \xrightarrow{a}{}^{\Box}_{\mathcal{E}} S \mid \exists i.\exists S'_i.s_i \xrightarrow{a}{}^{\Box}_{\mathcal{E}_1} S'_i \wedge$
  $S = \partial_{\mathcal{P}}((\cup\partial^{\diamond}_{a}(s_1), \ldots, S'_i, \ldots, \cup\partial^{\diamond}_{a}(s_n)), S_1, \lambda_1, Z)\}$

  $$c_\nu(\{s_1, \ldots, s_n\})(j) \triangleq \begin{cases} \underset{1 \leq i \leq n}{maxodd}(c_1(s_i)(j)) & \text{if } \forall i.Z \notin \lambda_1(s_i) \\ \overset{even}{\underset{s \in S_{\mathcal{E}_1}}{\bigsqcap}} c_1(s)(j) & \text{if } \exists i.Z \in \lambda_1(s_i) \end{cases}$$

  $$c_\mu(\{s_1, \ldots, s_n\})(j) \triangleq \begin{cases} \underset{1 \leq i \leq n}{maxodd}(c_1(s_i)(j)) & \text{if } \forall i.Z \notin \lambda_1(s_i) \\ \overset{odd}{\underset{s \in S_{\mathcal{E}_1}}{\bigsqcap}} c_1(s)(j) & \text{if } \exists i.Z \in \lambda_1(s_i) \end{cases}$$

  $$\lambda \triangleq \{\{s_1, \ldots, s_n\} \mapsto \bigcup_{1 \leq i \leq n} \lambda_1(s_i) - \{Z\} \mid \{s_1, \ldots, s_n\} \in 2^{S_{\mathcal{E}_1}}\}$$

$\varepsilon(\Phi)$ and its start states are singletons containing some start state of $\varepsilon(\Phi)$. An invariant of the maximal model construction is that start states do not have incoming transitions. (The case for $\varepsilon(\mathrm{tt})$ is the only exception and can be easily adapted to satisfy the invariant.) For a transition of state $q = \{s_1, \ldots, s_n\}$ of $\varepsilon(\sigma Z.\Phi)$, each state s_i has a transition in $\varepsilon(\Phi)$. A member state of the target of this transition, then, contains a derivative for each s_i. A member of the target state additionally contains an initial state of $\varepsilon(\Phi)$ if one of the derivatives included is labeled by Z. The definition of Table 1 makes use of the target set function $\partial_{\mathcal{P}}$ defined below.

Definition 6 (Target Set Function $\partial_{\mathcal{P}}$). *Let Φ be a modal μ-calculus formula, σ be either μ or ν, $\varepsilon(\Phi)$ be $(\mathcal{E}_1,S,\lambda)$ where $\mathcal{E}_1 = (S_{\mathcal{E}_1}, A, \longrightarrow^{\Diamond}_{\mathcal{E}}, \longrightarrow^{\Box}_{\mathcal{E}}, c)$ is an EMTS, $S \subseteq S_{\mathcal{E}_1}$ is a set of start states, $\lambda : S_{\mathcal{E}_1} \rightarrow 2^{PropVar}$ is a function that maps states of $\mathcal{E}$ to propositional variables, $c : S_{\mathcal{E}} \rightarrow N^k$ is a coloring function that maps states of $\mathcal{E}$ to k-tuples, and let $Z \in PropVar$ be a propositional variable. Given a tuple consisting of a target set for each element of a state of $\varepsilon(\sigma Z.\Phi)$, the function $\partial_{\mathcal{P}} : (2^{S_{\mathcal{E}_1}} \times \ldots \times 2^{S_{\mathcal{E}_1}}) \times 2^{S_{\mathcal{E}_1}} \times (S_{\mathcal{E}_1} \rightarrow 2^{PropVar}) \times PropVar \rightarrow 2^{2^{S_{\mathcal{E}_1}}}$ defines the target set of a transition of $\varepsilon(\sigma Z.\Phi)$ for this state as follows:*

$$\partial_{\mathcal{P}}((S_1,\ldots,S_n),S,\lambda,Z) \triangleq \{\{s_1,\ldots,s_n\} \mid \forall i.s_i \in S_i \wedge \nexists j.Z \in \lambda(s_j)\} \cup \{\{s_1,\ldots,s_n,s_0\} \mid \forall i.s_i \in S_i \wedge \exists j.Z \in \lambda(s_j) \wedge s_0 \in S\}$$

Each component of the color of state q is determined by comparing the corresponding entries of the member states s_i. When, for at least one s_i, this entry is odd, the greatest of the corresponding odd entries is selected as the entry of q, otherwise the maximum entry is selected for the same purpose. In Table 1, the function *maxodd* selects the greater of two numbers if both of them are odd or both of them are even, and the odd one otherwise. The color of q is further updated if it contains a state s_i labeled by Z. When Z identifies a greatest fixed point formula, each entry of the constructed tuple is defined to be the least even upper bound of the integers used in this entry of $\varepsilon(\Phi)$. Whereas, when Z identifies a least fixed point formula, the least odd upper bound of the integers is the entry for the color of q. In Table 1, least even and least odd upper bounds are denoted by the operators $\overset{even}{\sqcap}$ and $\overset{odd}{\sqcap}$, respectively.

4.2 Composing EMTSs

We extend the function ε to the domain of OTAs so that $\varepsilon(\Gamma \rhd E) = (\mathcal{E},S,\lambda)$, where $\mathcal{E} = (S_{\mathcal{E}}, A, \longrightarrow^{\Diamond}_{\mathcal{E}}, \longrightarrow^{\Box}_{\mathcal{E}}, c)$ is an EMTS, $S \subseteq S_{\mathcal{E}}$ is the set of *start states* of $\mathcal{E}$, and $\lambda : S_{\mathcal{E}} \rightarrow 2^{RecProcVar}$ is a *labeling function.*

The function ε is defined inductively on the structure of E as shown in Table 2. In the definition, we let $\varepsilon(\Gamma \rhd E_1)$ be $((S_{\mathcal{E}_1}, A, \longrightarrow^{\Diamond}_{\mathcal{E}_1}, \longrightarrow^{\Box}_{\mathcal{E}_1}, c_1), S_1, \lambda_1)$ and $\varepsilon(\Gamma \rhd E_2)$ be $((S_{\mathcal{E}_2}, A, \longrightarrow^{\Diamond}_{\mathcal{E}_2}, \longrightarrow^{\Box}_{\mathcal{E}_2}, c_2), S_2, \lambda_2)$, where $S_{\mathcal{E}_1}$ and $S_{\mathcal{E}_2}$ are disjoint sets. The new state s_{new} is not in $S_{\mathcal{E}_1}$. The coloring functions $c_1 : S_{\mathcal{E}_1} \rightarrow \mathbb{N}^{k_1}$ and $c_2 : S_{\mathcal{E}_2} \rightarrow \mathbb{N}^{k_2}$ color the states of $\mathcal{E}_1$ and $\mathcal{E}_2$ with integer tuples of length k_1 and k_2 respectively.

Table 2. EMTS Construction for Process Algebra Terms

- $\varepsilon(\Gamma \rhd \mathbf{0}) \triangleq ((\{s_{new}\},\ A,\ \emptyset,\ \emptyset,\ \{s_{new} \mapsto 0\}),\ \{s_{new}\},\ \{s_{new} \mapsto \emptyset\})$
- $\varepsilon(\Gamma \rhd X) \triangleq \varepsilon(\Phi)$ if $X \in AssProcVar$
  where $\Phi = \bigwedge_{X:\Psi \in \Gamma} \Psi$ (defaults to tt when Γ contains no assumption on X).
- $\varepsilon(\Gamma \rhd X) \triangleq ((\{s_{new}\},\ A,\ \emptyset,\ \emptyset,\ \{s_{new} \mapsto 0\}),\ \{s_{new}\},\ \{s_{new} \mapsto \{X\}\})$ if $X \in RecProcVar$
- $\varepsilon(\Gamma \rhd a.E_1) \triangleq ((S_{\mathcal{E}_1} \cup \{s_{new}\},\ A,\ \longrightarrow^{\diamond}_{\mathcal{E}},\ \longrightarrow^{\Box}_{\mathcal{E}},\ c),\ \{s_{new}\},\ \lambda_1 \cup \{s_{new} \mapsto \emptyset\})$ with:
$$\longrightarrow^{\diamond}_{\mathcal{E}} \triangleq \longrightarrow^{\diamond}_{\mathcal{E}_1} \cup \{s_{new} \overset{a}{\longrightarrow}{}^{\diamond}_{\mathcal{E}} S_1\}$$
$$\longrightarrow^{\Box}_{\mathcal{E}} \triangleq \longrightarrow^{\Box}_{\mathcal{E}_1} \cup \{s_{new} \overset{a}{\longrightarrow}{}^{\Box}_{\mathcal{E}} S_1\}$$
$$c \triangleq c_1 \cup \{s_{new} \mapsto 0^{k_1}\}$$
- $\varepsilon(\Gamma \rhd E_1 + E_2) \triangleq ((S_{\mathcal{E}_1} \cup S_{\mathcal{E}_2} \cup (S_1 \times S_2),\ A,\ \longrightarrow^{\diamond}_{\mathcal{E}},\ \longrightarrow^{\Box}_{\mathcal{E}},\ c),\ S_1 \times S_2,\ \lambda)$
$$\longrightarrow^{\diamond}_{\mathcal{E}} \triangleq \longrightarrow^{\diamond}_{\mathcal{E}_1} \cup \longrightarrow^{\diamond}_{\mathcal{E}_2} \cup \{(s,r) \overset{a}{\longrightarrow}{}^{\diamond}_{\mathcal{E}} S' \mid s \in S_1 \wedge r \in S_2 \wedge (s \overset{a}{\longrightarrow}{}^{\diamond}_{\mathcal{E}_1} S' \vee r \overset{a}{\longrightarrow}{}^{\diamond}_{\mathcal{E}_2} S')\}$$
$$\longrightarrow^{\Box}_{\mathcal{E}} \triangleq \longrightarrow^{\Box}_{\mathcal{E}_1} \cup \longrightarrow^{\Box}_{\mathcal{E}_2} \cup \{(s,r) \overset{a}{\longrightarrow}{}^{\Box}_{\mathcal{E}} S' \mid s \in S_1 \wedge r \in S_2 \wedge (s \overset{a}{\longrightarrow}{}^{\Box}_{\mathcal{E}_1} S' \vee r \overset{a}{\longrightarrow}{}^{\Box}_{\mathcal{E}_2} S')\}$$
$$c \triangleq \{s \mapsto c_1(s) \cdot 0^{k_2} \mid s \in S_{\mathcal{E}_1}\} \cup \{r \mapsto 0^{k_1} \cdot c_2(r) \mid r \in S_{\mathcal{E}_2}\} \cup \{(s,r) \mapsto c_1(s) \cdot c_2(r) \mid (s,r) \in S_1 \times S_2\}$$
$$\lambda \triangleq \lambda_1 \cup \lambda_2 \cup \{(s,r) \mapsto \lambda_1(s) \cup \lambda_2(r) \mid s \in S_1 \wedge r \in S_2\}$$
- $\varepsilon(\Gamma \rhd \mathit{fix}X.E_1) \triangleq ((S_{\mathcal{E}_1},\ A,\ \longrightarrow^{\diamond}_{\mathcal{E}},\ \longrightarrow^{\Box}_{\mathcal{E}},\ c_1),\ S_1,\ \lambda)$ with:
$$\longrightarrow^{\diamond}_{\mathcal{E}} \triangleq \{s \overset{a}{\longrightarrow}{}^{\diamond}_{\mathcal{E}} S \mid (s \overset{a}{\longrightarrow}{}^{\diamond}_{\mathcal{E}_1} S) \vee (\exists s_1 \in S_1 . s_1 \overset{a}{\longrightarrow}{}^{\diamond}_{\mathcal{E}_1} S \wedge X \in \lambda_1(s) \wedge s \text{ is reachable from } s_1)\}$$
$$\longrightarrow^{\Box}_{\mathcal{E}} \triangleq \{s \overset{a}{\longrightarrow}{}^{\Box}_{\mathcal{E}} S \mid (s \overset{a}{\longrightarrow}{}^{\Box}_{\mathcal{E}_1} S) \vee (\exists s_1 \in S_1 . s_1 \overset{a}{\longrightarrow}{}^{\Box}_{\mathcal{E}_1} S \wedge X \in \lambda_1(s) \wedge s \text{ is reachable from } s_1)\}$$
$$\lambda \triangleq \{s \mapsto (\lambda_1(s) - \{X\}) \mid s \in S_{\mathcal{E}_1}\}$$
- $\varepsilon(\Gamma \rhd E_1 \parallel E_2) \triangleq ((S_{\mathcal{E}_1} \times S_{\mathcal{E}_2} \times \{1,2\},\ A,\ \longrightarrow^{\diamond}_{\mathcal{E}},\ \longrightarrow^{\Box}_{\mathcal{E}},\ c),\ S_1 \times S_2 \times \{1,2\},\ \lambda)$
$$\longrightarrow^{\diamond}_{\mathcal{E}} \triangleq \{(s,r,x) \overset{a}{\longrightarrow}{}^{\diamond}_{\mathcal{E}} S' \times \{r\} \times \{1\} \mid s \overset{a}{\longrightarrow}{}^{\diamond}_{\mathcal{E}_1} S'\} \cup \{(s,r,x) \overset{a}{\longrightarrow}{}^{\diamond}_{\mathcal{E}} \{s\} \times R' \times \{2\} \mid r \overset{a}{\longrightarrow}{}^{\diamond}_{\mathcal{E}_2} R'\}$$
$$\longrightarrow^{\Box}_{\mathcal{E}} \triangleq \{(s,r,x) \overset{a}{\longrightarrow}{}^{\Box}_{\mathcal{E}} S' \times \{r\} \times \{1\} \mid s \overset{a}{\longrightarrow}{}^{\Box}_{\mathcal{E}_1} S'\} \cup \{(s,r,x) \overset{a}{\longrightarrow}{}^{\Box}_{\mathcal{E}} \{s\} \times R' \times \{2\} \mid r \overset{a}{\longrightarrow}{}^{\Box}_{\mathcal{E}_2} R'\}$$
$$c \triangleq \{(s,r,1) \mapsto c_1(s) \cdot 0^{k_2} \mid s \in S_{\mathcal{E}_1} \wedge r \in S_{\mathcal{E}_2}\} \cup \{(s,r,2) \mapsto 0^{k_1} \cdot c_2(r) \mid s \in S_{\mathcal{E}_1} \wedge r \in S_{\mathcal{E}_2}\}$$
$$\lambda \triangleq \{(s,r,x) \mapsto \emptyset \mid s \in S_{\mathcal{E}_1} \wedge r \in S_{\mathcal{E}_2} \wedge x \in \{1,2\}\}$$

The EMTS corresponding to the nil process **0** consists of an abstract state without outgoing transitions, indicating that no transition is allowed for processes simulated by this state. If a process variable X in the term E stands for an underspecified component of the system, that is if X is an assumption process variable, then the EMTS for X is a maximal model for the conjunction of the properties specified for this component in the assumption list Γ.

The EMTS for a recursion process variable X is a single state without outgoing transitions, since the capabilities of the processes simulated are determined by the binding *fix*-expression. The function λ labels the state X. Given the EMTS for the term of the *fix*-expression where X is free, the transitions of the start states are transferred to the states labeled by X.

The EMTS for a subterm prefixed by an action a is given by a start state with a must a-transition to the set of start states of the EMTS for the subterm. The EMTS for the sum operator consists of an EMTS where the start states are the cross product of the start states of the EMTSs for the subterms. It is assumed for this case that there are no incoming transitions to the start states of the EMTSs being combined. This is an invariant of the construction, except the case for tt which can be trivially converted to an equivalent EMTS to satisfy the property.

Finally, the states of the EMTS for a parallel composition of two components consist of a state from each component. Each state has transitions such that one of the components make a transition while the other stays in the same state. Each state is further marked by 1 or 2 to keep track of which component has performed the last transition; this is necessary to enable a run of the composition if the interleaved runs are enabled.

4.3 Correctness Results

The aim of the above construction is to capture, by means of an EMTS, exactly those behaviors denoted by the given OTA. The construction is *sound* (resp. *complete*) if the denotation of the OTA is a subset (resp. superset) of the denotation of the resulting EMTS. Our first result establishes that the first part of the construction is a maximal model construction for the modal μ-calculus.

Theorem 1. *Let $\mathcal{T}$ be a transition-closed LTS, Φ be a closed and guarded modal μ-calculus formula and $\varepsilon(\Phi) = (\mathcal{E}, S, \lambda)$. Then $[\![S]\!]_{\mathcal{T}} = ||\Phi||^{\mathcal{T}}$.*

Our next result shows that the construction is sound and complete when assumptions exist on only one of the components that are running in parallel and the rest of the system is fully determined.

Theorem 2. *Let $\mathcal{T}$ be a transition-closed LTS, $\Gamma \rhd E \| t$ be a guarded linear OTA where E does not contain parallel composition and t is closed, and let $\varepsilon(\Gamma \rhd E \| t) = (\mathcal{E}, S, \lambda)$. Then $[\![S]\!]_{\mathcal{T}}$ is equal to the set $[\![\Gamma \rhd E \| t]\!]_{\rho_0}$ up to bisimulation, where ρ_0 maps each recursion process variable X to $\mathbf{0}$.*

Theorems 1 and 2 are proved by induction on the structure of the logical formula and the process term, respectively. The proofs can be found in [19].

In the general case, when multiple underspecified components run in parallel, we only have soundness: our construction is sound for systems without dynamic process creation. For systems with dynamic process creation, the construction does not terminate.

Theorem 3. *Let $\mathcal{T}$ be a transition-closed LTS, $\Gamma \rhd E$ be a guarded linear OTA where every recursion process variable in the scope of parallel composition is bound by a* fix *operator in the same scope, and let $\varepsilon(\Gamma \rhd E) = (\mathcal{E}, S, \lambda)$. Then the set $[\![S]\!]_{\mathcal{T}}$ includes $[\![\Gamma \rhd E]\!]_{\rho_0}$ up to bisimulation.*

The proof of the theorem is as the proof of Theorem 2, but includes a more general case for parallel composition and can be found in [19].

Our last result reflects the fact that verification of open systems in the presence of parallel composition is undecidable for the modal μ-calculus in general. Completeness results can, however, be obtained for various fragments of the μ-calculus, such as ACTL, ACTL* and the simulation logic of [9]. In our approach, the tasks of constructing a finite representation of the state space in the form of an EMTS and the task of verifying properties of this representation are separated. This allows different logics to be employed for expressing assumptions on components and for specifying system properties, giving rise to more refined completeness results.

5 A Proof System for EMTS

In [3], we presented a proof system for verifying that an abstract state s of an EMTS $\mathcal{E}$ satisfies a modal μ-calculus formula Φ. In this section, we give a summary of this proof system and provide an alternative termination condition that uses the coloring function c instead of the earlier condition that assumed an extensional definition of the set of prohibited runs $W_{\mathcal{E}}$. The system is a specialization of a proof system by Bradfield and Stirling [20, 21] for showing μ-calculus properties for sets of LTS states. The relationship between the two proof systems is clear when one considers that each EMTS state denotes a set of LTS states.

A proof tree is constructed using the rules below, where σ ranges over μ and ν. The construction starts with the goal and progresses in a goal-directed fashion, checking at each step if a terminal node was reached.

$$\frac{s \vdash^{\mathcal{E}}_{\mathcal{V}} \Phi \wedge \Psi}{s \vdash^{\mathcal{E}}_{\mathcal{V}} \Phi \quad s \vdash^{\mathcal{E}}_{\mathcal{V}} \Psi} \qquad \frac{s \vdash^{\mathcal{E}}_{\mathcal{V}} \Phi \vee \Psi}{s \vdash^{\mathcal{E}}_{\mathcal{V}} \Phi} \qquad \frac{s \vdash^{\mathcal{E}}_{\mathcal{V}} \Phi \vee \Psi}{s \vdash^{\mathcal{E}}_{\mathcal{V}} \Psi}$$

$$\frac{s \vdash^{\mathcal{E}}_{\mathcal{V}} \sigma Z.\Phi}{s \vdash^{\mathcal{E}}_{\mathcal{V}} Z} \qquad \frac{s \vdash^{\mathcal{E}}_{\mathcal{V}} [a]\,\Phi}{s_1 \vdash^{\mathcal{E}}_{\mathcal{V}} \Phi \ \ldots\ s_n \vdash^{\mathcal{E}}_{\mathcal{V}} \Phi}\ \{s_1, \ldots, s_n\} = \cup\, \partial^{\Diamond}_{a}(s)$$

$$\frac{s \vdash^{\mathcal{E}}_{\mathcal{V}} Z}{s \vdash^{\mathcal{E}}_{\mathcal{V}} \Phi}\ Z \text{ identifies } \sigma Z.\Phi \qquad \frac{s \vdash^{\mathcal{E}}_{\mathcal{V}} \langle a\rangle\,\Phi}{s_1 \vdash^{\mathcal{E}}_{\mathcal{V}} \Phi \ \ldots\ s_n \vdash^{\mathcal{E}}_{\mathcal{V}} \Phi}\ \{s_1, \ldots, s_n\} \in \partial^{\Box}_{a}(s)$$

A successful tableau (or proof) is a finite proof tree having successful terminals as leaves. If $n : r \vdash^{\mathcal{E}}_{\mathcal{V}} Z$ is a node where Z identifies a fixed point formula, and there is an identical ancestor node of n, $n' : r \vdash^{\mathcal{E}}_{\mathcal{V}} Z$ and for any other fixed point variable Y on this path, Z subsumes Y, then node n is called a σ-terminal. So no further rules are applied to it. The most recent node making n a σ-terminal is named n's companion. The conditions for a leaf node $r \vdash^{\mathcal{E}}_{\mathcal{V}} \Psi$ of a proof tree to be a successful terminal are listed below.

Successful Terminals

1. $\Psi = \mathrm{tt}$, or else $\Psi = Z$, Z is free in the initial formula, and $r \in \mathcal{V}(Z)$
2. $\Psi = [a]\,\Phi$ and $\cup\partial^{\Diamond}_{a}(r) = \emptyset$

3. $\Psi = Z$ where Z identifies a fixed point formula $\sigma Z.\Phi$, and the sequent is a σ-terminal with companion node $n : r \vdash^{\mathcal{E}}_{\mathcal{V}} \Psi$, then
   (a) If $\sigma = \nu$, then the terminal is successful.
   (b) If $\sigma = \mu$, then the terminal is successful if every infinite run of the EMTS that corresponds to an infinite sequence of trails of the companion node n_0 is in $W_{\mathcal{E}}$. (The notion of trail is explained below.) When the set $W_{\mathcal{E}}$ is encoded using a coloring function c, the condition is that for any set S_T of trails of n_0, there should exist $1 \leq j \leq k$, so that $max(\bigcup_{T \in S_T} c(\alpha(T))(j))$ is odd. This ensures, for an infinite run $w_{n_0} = \alpha(T_1) \circ \alpha(T_2) \circ \alpha(T_3) \ldots$ where for all $i \geq 1$, T_i is a trail of n_0, that there exists some $1 \leq j' \leq k$ such that $max\,(inf\,(c(w_{n_0})(j')))$ is odd.

Unsuccessful Terminals

1. $\Psi = \mathrm{ff}$, or else $\Psi = Z$, Z is free in the initial formula, and $r \notin \mathcal{V}(Z)$
2. $\Psi = \langle a \rangle\, \Phi$ and $\cup \partial^{\Box}_{a}(r) = \emptyset$
3. $\Psi = Z$ where Z identifies the least fixed point formula $\mu Z.\Phi$, and the sequent is a σ-terminal with companion node n_0, then the terminal is unsuccessful if there exists a set S_T of trails of n_0 such that for every $1 \leq j \leq k$, $max\,(\bigcup_{T \in S_T} c(\alpha(T))(j))$ is even. This means that some infinite run w_{n_0} of the EMTS, which corresponds to an infinite sequence of trails of the companion node n_0, is not in $W_{\mathcal{E}}$.

Trails and corresponding runs are defined as follows. Assume that node $n_k{:}r \vdash^{\mathcal{E}}_{\mathcal{V}} Z$ is a μ-terminal and node $n_0{:}r \vdash^{\mathcal{E}}_{\mathcal{V}} Z$ is its companion. A trail T of the companion node n_0 is a sequence of state–node pairs $(r, n_0), \ldots, (r, n_k)$ such that for all $0 \leq i < k$, one of the following holds:

1. $n_{i+1} : r_{i+1} \vdash^{\mathcal{E}}_{\mathcal{V}} \Psi_{i+1}$ is an immediate successor of $n_i : r_i \vdash^{\mathcal{E}}_{\mathcal{V}} \Psi_i$, or
2. n_i is the immediate predecessor of a σ-terminal node $n' : r' \vdash^{\mathcal{E}}_{\mathcal{V}} Z'$ where $n' \neq n_k$ whose companion is $n_j : r' \vdash^{\mathcal{E}}_{\mathcal{V}} Z'$ for some $j : 0 \leq j \leq i$, $n_{i+1} = n_j$, and $r_{i+1} = r'$.

In order to convert a trail to a corresponding run, we use the function α, which returns the empty string when the trail contains only one pair, and is defined for longer trails as follows:

$$\alpha((r_1, n_1) \cdot (r_2, n_2) \cdot T) \triangleq \begin{cases} (r_1 \xrightarrow{a}_{\mathcal{E}} r_2) \cdot \alpha((r_2, n_2) \cdot T) & \Box_a \text{ or } \Diamond_a\text{-rule is applied to } n_1 \\ \alpha((r_2, n_2) \cdot T) & \text{otherwise.} \end{cases}$$

A formula is prime if whenever it logically implies a disjunction then it also implies one of the disjuncts. As we show in [3], the proof system is sound and complete for all formulas with only prime subformulas. An example proof is given in [19].

6 Conclusion

In this paper we investigate a state space representation for open systems specified as open process terms with behavioural assumptions written in the modal μ-calculus. This representation can serve both as a graphical specification formalism and as a basis for verification, supporting state space exploration based techniques and state visualization for interactive methods. We present a two-phase construction of such a representation from an open term with assumptions, and show it sound for terms without dynamic process creation and complete for systems with a single underspecified component. Finally, we adapt an existing proof system for the task of proving behavioural properties of open systems based on the given state space representation. The relative simplicity of the proof system and its use is an indication of the adequateness of EMTSs for open system state space representation.

Future work is required to characterize more precisely the construction and the μ-calculus fragments for which it is complete, taking into account that the fragment for specifying component assumptions need not be the same as the fragment chosen for specifying system properties. In addition to automatic state space construction, interactive state space exploration will be considered, allowing a wider class of open systems to be handled. Finally, we plan to demonstrate the utility of the proposed approach by means of tool support and case studies.

References

1. Martinelli, F.: Analysis of security protocols as open systems. Theoretical Computer Science **290** (2003) 1057–1106
2. Larsen, K.G.: Modal specifications. Automatic Verification Methods for Finite State Systems (1989) 232–246
3. Aktug, I., Gurov, D.: Verification of open systems based on explicit state space representation. In: AVIS'05: Proceedings of Automated Verification of Infinite Systems. To appear (2005)
4. Kozen, D.: Results on the propositional mu-calculus. Theoretical Computer Science **27** (1983) 333–354
5. Dam, M.: Fixed points of Büchi automata. In: FSTTCS '92: Proceedings of 12th Conference on Foundations of Software Technology and Theoretical Computer Science. Volume 652 of Lecture Notes in Computer Science. (1992) 39–50
6. Kaivola, R.: On modal mu-calculus and Büchi tree automata. Information Processing Letters **54** (1995) 17–22
7. Grumberg, O., Long, D.: Model checking and modular verification. ACM Transactions on Programming Languages and Systems **16(3)** (1994) 843–871
8. Kupferman, O., Vardi, M.: An automata-theoretic approach to modular model checking. ACM Transactions on Programming Languages and Systems **22** (2000) 87–128
9. Sprenger, C., Gurov, D., Huisman, M.: Compositional verification for secure loading of smart card applets. In Heitmeyer, C., Talpin, J.P., eds.: Proc. MEMOCODE'04, IEEE (2004) 211–222

10. Emerson, E.A., Jutla, C.S.: Tree automata, mu-calculus and determinacy (extended abstract). In: Proceedings of 32nd Annual Symposium on Foundations of Computer Science. IEEE, Computer Society Press (1991) 368–377
11. Kupferman, O., Vardi, M.Y., Wolper, P.: An automata-theoretic approach to branching-time model checking. Journal of ACM **47** (2000) 312–360
12. Dam, M., Fredlund, L., Gurov, D.: Toward parametric verification of open distributed systems. In Langmaack, H., Pnueli, A., de Roever, W.P., eds.: Compositionality: the Significant Difference. Volume 1536 of Lecture Notes in Computer Science. Springer-Verlag (1998) 150–185
13. Dam, M., Gurov, D.: Compositional verification of CCS processes. In: PSI '99: Proceedings of the Third International Andrei Ershov Memorial Conference on Perspectives of System Informatics, Springer-Verlag (2000) 247–256
14. Christensen, S.: Decidability and Decomposition in Process Algebras. PhD thesis, University of Edinburgh (1993)
15. Huth, M., Jagadeesan, R., Schmidt, D.A.: Modal transition systems: A foundation for three-valued program analysis. In: ESOP '01: Proceedings of the 10th European Symposium on Programming Languages and Systems. Volume 2028., London, UK, Springer-Verlag (2001) 155–169
16. Grumberg, O., Shoham, S.: Monotonic abstraction-refinement for CTL. In: TACAS'04: Proceedings of the 10th International Conference on Tools and Algorithms for the Construction and Analysis of Systems. Volume 2988 of Lecture Notes in Computer Science., Springer-Verlag (2004) 546–560
17. Mostowski, A.W.: Regular expressions for infinite trees and a standard form of automata. Computation Theory **208** (1984) pages 157–168
18. Bustan, D., Grumberg, O.: Applicability of fair simulation. In: TACAS '02: Proceedings of the 8th International Conference on Tools and Algorithms for the Construction and Analysis of Systems. Volume 2280 of Lecture Notes in Computer Science., Springer-Verlag (2002) 401–414
19. Aktug, I., Gurov, D.: State space representation for verification of open systems. Technical report, KTH CSC, `http://www.nada.kth.se/~irem/sefros/techrep06.ps` (2006)
20. Bradfield, J., Stirling, C.: Local model checking for infinite state spaces. Theoretical Computer Science **96** (1992) 157–174
21. Stirling, C.: Modal and Temporal Properties of Processes. Texts in Computer Science. Springer-Verlag (2001)

Data Movement Optimisation in Point-Free Form

Brad Alexander and Andrew Wendelborn

School of Computer Science
University of Adelaide
5005, Adelaide, SA, Australia
{brad, andrew}@cs.adelaide.edu.au

Abstract. Programs written in point-free form express computation purely in terms of functions. Such programs are especially amenable to local transformation. In this paper, we describe a process for optimising the transport of data through point-free programs. This process systematically applies local transformations to achieve effective global optimisation. We describe the strategies we employ to ensure this process is tractable. This process has been implemented as an intermediate stage of a compiler. The optimiser is shown to be highly effective, producing code of comparable efficiency to hand-written code.

1 Introduction

Transformation is the key to any program improvement process. By using highly transformable programming notations we pave the way for the application of deep and pervasive transformation techniques. Programs written in point-free form are particularly amenable to transformation[5]. In point-free code all computation is expressed purely in terms of functions. Point-free code contains no variables to store values generated during program execution. As a consequence, functions cannot rely on variables as agents to transmit data through the program. Instead, the functions themselves are responsible for routing data through the code. This exposed data-transport maps well to distributed-memory platforms and there have been a number of experiments mapping functions found in point-free form to such platforms[18, 8, 6].

As well as providing a path to distributed implementation, exposing the transport of data also provides scope for direct optimisation of this transport. This avenue of research is less well explored. In this paper we outline an automated process to reduce the volume of data movement through point-free code through the systematic use of local transformations. We show that this process is highly effective and describe the techniques we found useful.

1.1 Outline of This Paper

The next section outlines the context in which our optimisation process takes place. Section 3 defines the broad strategies we applied in all parts of our optimisation process. Section 4 focuses on one part of the optimisation process -

M. Johnson and V. Vene (Eds.): AMAST 2006, LNCS 4019, pp. 21–35, 2006.

the vector optimisation of map functions. Section 5 presents some of our results. Section 6 outlines related work and we present our conclusions in Sect. 7.

2 Context

The work described in this paper is part of a of a prototype implementation of a compiler mapping a simple functional language, Adl, to point-free code[2]. The focus of this paper is the optimisation phase, which reduces the flow of data through point-free programs. The optimiser was developed in CENTAUR[4], using rules expressed in Natural Semantics[11]. A simple translator from Adl to point-free form provides the input to the optimisation process. To provide a context for this paper, we outline salient features of Adl, point-free code, and the translation process next.

2.1 Adl

Adl is a small, strict, vector-oriented, functional language. Adl can be described as *point-wise* because, like most languages, it supports the use of variables as a means of storing data. Adl also provides standard operations for iteration - using `while`, conditional execution - using `if`, and scoping - using `let`. Adl supports implicit parallelism through second-order data-parallel operations over nestable single-dimensional arrays, called vectors, including `map`, `reduce` and `scan`. Other vector operations include a length operator (#), an index operator (!) and an iota operation to dynamically allocate a contiguous vector of indices. Adl also supports tuples of arbitrary arity and these are manipulated through pattern-matching. Adl places no restrictions on the nesting of datatypes and operations. Recursion is not supported in its initial implementation. Figure 1 shows an Adl program that adds corresponding elements of two input vectors.

We have built a number of applications in Adl and found it to be a simple and expressive language to use.

2.2 Point-Free Form

Our dialect of point-free form is derived from a point-free expression of the theory of concatenate-lists in the Bird-Meertens-Formalism (BMF)[7]. In this paper, we restrict ourselves to the functions required to express the translation and optimisation of Adl, omitting point-free equivalents of `reduce` and `scan`, which are beyond the immediate scope of this paper but discussed in[2].

```
main (a: vof int, b: vof int)
  := let
       f x := a!x + b!x
     in
       map(f,iota #a)
     endlet
```

Fig. 1. An Adl program to add corresponding elements of two vectors

Table 1. A selection of functions in point-free form

Description	Symbol(s)	Semantics
Function Composition	$\cdot$	$(f \cdot g)\, x = f(g\, x)$
Vector map	$*$	$f * [x_0, \ldots, x_{n-1}] = [f\, x_0, \ldots, f\, x_{n-1}]$
All applied to for tuples	$(f_1, \ldots, f_n)^\circ$	$(f_1, \ldots, f_n)^\circ\, x = (f_1\, x, \ldots, f_n\, x)$
Identity function	id	$\mathsf{id}\, x = x$
Tuple access	${}^n\pi_i$	${}^n\pi_i\, (a_1, \ldots, a_n) = a_i$
Constant functions	K	$\mathsf{K}\, x = K$
Arithmetic operators	$+, -, \div, \times, \ldots$	$+\, (x, y) = x + y$ *etc.*
Left distribute	distl	$\mathsf{distl}\, (a, [x_0, \ldots, x_{n-1}]) = [(a, x_0), \ldots, (a, x_{n-1})]$
Zip	zip	$\mathsf{zip}\, ([x_0, \ldots, x_{n-1}], [y_0, \ldots, y_{n-1}]) = [(x_0, y_0), \ldots, (x_{n-1}, y_{n-1})]$
Value repetition	repeat	$\mathsf{repeat}\, (a, n) = [\overbrace{a, \ldots, a}^{n\ times}]$
Vector transpose	$\mathsf{transpose}$	$\mathsf{transpose}\, a = b : \begin{array}{l} \forall (i,j) \in indices(a), a(i,j) = b(j,i) \wedge \\ \forall (i,j) \notin indices(a), a(i,j) = b(j,i) = \bot \end{array}$
Vector enumeration	iota	$\mathsf{iota}\, n = [0, 1, \ldots, n-1]$
Vector length	$\#$	$\#\, [x_0, \ldots, x_{n-1}] = n$
Vector indexing	$!$	$!\, ([x_0, \ldots, x_{n-1}], i) = x_i$
Vector selection	select	$\mathsf{select}\, (v, [x_0, \ldots, x_{n-1}]) = [v!x_0, \ldots, v!x_{n-1}]$

Most point-free programs produced by our compiler consist of sequences of composed functions:

$$f_n \cdot f_{n-1} \cdot \ldots \cdot f_1$$

where input data enters at f_1 and flows toward f_n at the end of the program. In the remainder of this paper, we refer to the beginning of the program (f_1) as the *upstream* end of the program and we refer to the end of the program (f_n) as the *downstream* end of the program.

Translation. Translation from point-wise Adl to point-free form strips all variable references from Adl code and replaces these with code to transport values between operations. A detailed description of the translation process is given in [2]. Similar translation processes have been defined in[5, 16, 13].

The translation process is conservative. It transports every value in the scope of an operation to the doorstep of that operation. This approach, though simple and robust, results in large surplus transport of data through translator-code. This can be seen in the translation of the Adl code from Fig. 1:

$$(+ \cdot (! \cdot (\pi_1 \cdot \pi_1, \pi_2)^\circ, ! \cdot (\pi_2 \cdot \pi_1, \pi_2)^\circ)^\circ) * \mathsf{distl} \cdot (\mathsf{id}, \mathsf{iota} \cdot \#\pi_1)^\circ$$

where the distl operation distributes a copy of both input vectors to each instance of the map function downstream. The aim of the optimiser is to transform programs in order to reduce this volume of copying. We outline the general strategy of our optimiser next.

3 Optimisation Strategy

The optimiser works through the application of simple, semantics-preserving, rules. Taken alone this set of rules is neither confluent or terminating[1]. Moreover, the large number of steps typically required for optimisation, coupled with multiple rule and site choices at each step, leads to a case-explosion. To control these factors we must apply our rules strategically.

Our main strategy is to propagate the optimisation process, on a localised front[2], from the downstream end of the program to the upstream end. The front moves a step upstream by specialising functions immediately upstream of the front with respect to the needs of the optimised code immediately downstream. The front leaves a trail of optimised code in its wake as it steps upstream. The specialisation that takes place in in each step consists of three phases:

Pre-processing: applies sets of normalisation rules to code at the front. These rules standardise the form of the code to make the application of key optimisation rules easier.

Key-Rule-Application: applies rules that substantially increase efficiency by either eliminating functions responsible for making redundant copies of data, or facilitating the removal of such functions further upstream.

Post-processing: re-factors code to eliminate some small inefficiencies introduced by pre-processing. and exposes functions for optimisation further upstream.

The phases in each step are applied iteratively until the localised front of optimisation reaches the start of the program.

The broad pattern of processing we have just described applies to all optimisation stages of our implementation. A full description of these stages is beyond the scope of this paper. Instead, we illustrate the process by describing a key part of optimisation, the vector optimisation of map functions.

4 Vector Optimisation of Map Functions

The map operator is a second-order function that applies a parameter function to an input vector. In point-free programs, copies of all data required by the parameter function must be explicitly routed to that function. Vector optimisation of map functions reduces the amount of data that must be copied by changing code so that it selectively routes *vector* data to its destination[3]. Specifically, we seek to replace multiple applications of indexing operations on copies of vectors

[1] Most rules could be applied in both directions which, trivially, leads to loops. It has been observed [9] that confluence seems an implausible objective in the context of program optimisation.

[2] In our implementation, we delineate this front by associating function compositions to make the functions on the front appear as an outermost term.

[3] The impact of this optimisation is most strongly felt in a distributed parallel context where explicit routing of data is required and this routing incurs a cost.

with a single bulk selection operator to direct data to where it is needed. Two general rules are used to achive this aim. For the purposes of explanation, we first introduce two specialised versions of these rules:

$$(!) * \cdot \mathsf{distl} \cdot (\mathsf{id}, R)^\circ \Rightarrow \mathsf{select} \cdot (\mathsf{id}, R)^\circ \tag{1}$$

$$\begin{array}{c}\underline{(! \cdot (\pi_1 \cdot \pi_1, \pi_2 \cdot \pi_1)^\circ)} * \cdot \mathsf{distl} \cdot (\mathsf{id}, R)^\circ \Rightarrow \\ \mathsf{repeat} \cdot (! \cdot (\pi_1 \cdot \pi_1, \pi_2 \cdot \pi_1)^\circ, \# \cdot \pi_2)^\circ \cdot (\mathsf{id}, R)^\circ\end{array} \tag{2}$$

The code $(\mathsf{id}, R)^\circ$, though not modified by these rules, is a product of the translation of all calls to `map` in Adl, and provides important context for later discussions. The code R can take various forms but, given an input value, a, always generates some vector of values $[x_0, \ldots, x_{n-1}]$.

Rule 1, above, fuses the vector indexing function and a corresponding distl function into a select. If we factor out $(\mathsf{id}, R)^\circ$ from both sides, the equivalence underlying rule 1 can be informally stated:

$$\begin{aligned}(!) * \cdot \mathsf{distl}\ (a, [x_0, \ldots, x_{n-1}]) &= \\ \mathsf{select}\ (a, [x_0, \ldots, x_{n-1}]) &= \\ [a!x_0, \ldots, a!x_{n-1}]\end{aligned}$$

The important difference between the two sides is that distl creates the, often large, intermediate structure: $[(a, x_0), \ldots, (a, x_{n-1})]$ whereas select avoids this.

Rule 2 applies where the code that carries out the the indexing (underlined) accesses only the first element in each of the tuples in the vector produced by distl[4]. Again, factoring out $(\mathsf{id}, R)^\circ$, the equivalence underlying rule 2 can be informally stated:

$$\begin{aligned}(! \cdot (\pi_1 \cdot \pi_1, \pi_2 \cdot \pi_1)^\circ) * \cdot \mathsf{distl}\ ((a, x), [y_0, \ldots, y_{n-1}]) &= \\ \mathsf{repeat} \cdot (! \cdot (\pi_1 \cdot \pi_1, \pi_2 \cdot \pi_1)^\circ, \# \cdot \pi_2)^\circ\ ((a, x), [y_0, \ldots, y_{n-1}]) &= \\ [a!x, \ldots, a!x]\end{aligned}$$

where $\#\ [a!x, \ldots, a!x] = \#\ [y_0, \ldots, y_{n-1}] = n$. It should be noted that while rule 2 reduces the size of intermediate structures, it also moves the indexing function further upstream where it can be accessed by subsequent optimisation steps.

We emphasise that rules 1 and 2 are specialisations of corresponding, more general, rules in our vector optimiser implementation. We present these rules shortly, but first we describe the pre-processing steps that allow such rules to be applied effectively.

4.1 Pre-processing

The pre-processing of code for vector optimisation of map functions consists of three main stages:

[4] We know this because both $\pi_1 \cdot \pi_1$ and $\pi_2 \cdot \pi_1$ first execute π_1 which accesses the first element of a tuple.

1. **Compaction:** composed functions are coalesced to help reveal true data dependencies and to bring functions of interest as far upstream as possible.
2. **Isolation:** functions are isolated from each other to allow them to be processed individually.
3. **Reorientation:** transpose functions are added, where necessary, to reorient the vectors to aid further processing.

We describe each of these stages in turn.

Compaction. Compaction transforms code to minimise the number of composed functions between functions of interest, in this case, indexing functions and the code further upstream. When compaction is complete, index functions are "on the surface" and exposed for further processing. As an example - prior to compaction, the code:

$$\underline{(!.(\pi_2, \pi_1)^\circ \cdot (\pi_2, \pi_1)^\circ)} * \cdot \mathsf{distl} \cdot (\mathsf{id}, R)^\circ$$

is not amenable to immediate optimisation because the contents of the map function, underlined, is recognisable to neither rule 1 or 2. However, after compacting: $(\pi_2, \pi_1)^\circ \cdot (\pi_2, \pi_1)$ to $(\pi_1, \pi_2)^\circ$ the code takes the form:

$$(! \cdot (\pi_1, \pi_2)^\circ) * \cdot \mathsf{distl} \cdot (\mathsf{id}, R)^\circ$$

and rule 1 can be applied after eliminating the redundant identity $(\pi_1, \pi_2)^\circ$.

Isolation. Often, the combination of functions in code confounds the matching of optimisation rules. For example the code:

$$(+ \cdot (! \cdot (\pi_1 \cdot \pi_1, \pi_2)^\circ, ! \cdot (\pi_1 \cdot \pi_1, \pi_2 \cdot \pi_1)^\circ)^\circ) * \cdot \mathsf{distl} \cdot (\mathsf{id}, R)^\circ$$

matches neither rule 1 or 2. However, if the indexing components are isolated from each other to produce the equivalent code:

$$\begin{aligned}(+) * \cdot \mathsf{zip} \cdot (\,&(! \cdot (\pi_1 \cdot \pi_1, \pi_2)^\circ) * \cdot \mathsf{distl} \cdot (\mathsf{id}, R)^\circ,\\ &(! \cdot (\pi_1 \cdot \pi_1, \pi_2 \cdot \pi_1)^\circ) * \cdot \mathsf{distl} \cdot (\mathsf{id}, R)^\circ)^\circ\end{aligned}$$

to which rule 2 can be applied immediately, and to which a more general form of rule 1 can be applied.

Reorientation. In code where indexing functions are nested it is sometimes the case that the dimensions of the input vector are accessed in an order that defeats immediate optimisation. For example, it is not instantly clear how to optimise:

$$(! \cdot (! \cdot (\pi_1 \cdot \pi_1, \pi_2)^\circ, \pi_1 \cdot \pi_2)^\circ) * \cdot \mathsf{distl} \cdot (\mathsf{id}, R)^\circ$$

However, if we transpose the vector we can switch the functions used to create the indices giving:

$$(! \cdot (! \cdot (\mathsf{transpose} \cdot \pi_1 \cdot \pi_1, \pi_2 \cdot \pi_1)^\circ, \pi_2)^\circ) * \cdot \mathsf{distl} \cdot (\mathsf{id}, R)^\circ$$

which *almost* matches the specialised rule 1 and actually *does* match the corresponding rule (rule 3) described next.

4.2 Key-Rule Application

The key vector optimisation rules rules are shown in Fig. 2. Rules 3 and 4 correspond to the archetype rules 1 and 2 respectively.

Select introduction

$$\frac{\begin{array}{c} Not\_In\_oexp(f_1, \pi_2) \\ In\_oexp(f_2, \pi_2) \\ Opt(f_2 \Rightarrow f_2') \end{array}}{(! \cdot (f_1, f_2)^{\circ}) * \cdot \mathsf{distl} \cdot (\mathsf{id}, R)^{\circ} \Rightarrow \mathsf{select} \cdot (f_1, f_2')^{\circ} \cdot (\mathsf{id}, R)^{\circ}} \quad (3)$$

Repeat introduction

$$\frac{\begin{array}{c} Not\_In\_oexp(f_1, \pi_2) \\ Not\_In\_oexp(f_2, \pi_2) \end{array}}{(! \cdot (f_1, f_2)^{\circ}) * \cdot \mathsf{distl} \cdot (\mathsf{id}, R)^{\circ} \Rightarrow \mathsf{repeat} \cdot (! \cdot (f_1, f_2)^{\circ}, \# \cdot \pi_2)^{\circ} \cdot (\mathsf{id}, R)^{\circ}} \quad (4)$$

Fig. 2. Two key rules of the vector optimiser

Rules 3 and 4 are expressed in Natural Semantics. The parts above the line in each rule are premises that must be true in order to apply the transformation specified below the line. These rules capture a wider variety of code than their respective archetypes and are thus more useful in an actual implementation.

Both rules hinge on calls to the predicates *In_oexp* and *Not_In_oexp* which test for the presence, and absence, respectively, of π_2 as a most-upstream function in f_1 and f_2. The presence of a π_2 function as the most upstream functions indicates a reference to the output value of R.

During the application of rules 3 and 4, the fate of the f_2 function in each rule differs. In rule 3, the truth of $In\_oexp(f_2, \pi_2)$ implies that f_2 references R. This referencing means that at least some code in f_2 cannot be carried upstream of R for further processing. In light of this constraint, the recursive call to the vector optimiser, $Opt(f_2 \Rightarrow f_2')$, is made to exploit a last opportunity to, locally, optimise f_2 before the process moves upstream. In rule 4, f_2 does not reference the output of R and thus *can* be carried upstream of R for further processing.

On a related note, some thought about the premises of both rules reveals that code such as:

$$(! \cdot (! \cdot (\pi_1, \pi_2)^{\circ}, \pi_2)^{\circ}) * \cdot \mathsf{distl} \cdot (\mathsf{id}, R)^{\circ}$$

will match neither rule 3 or 4. In these cases we apply a default rule, not shown here, leaves outer index function intact. The post-processing phase is then left to salvage what it can from the code that generates its parameters for optimisation upstream.

4.3 Post-processing

After the application of the key rules in the last section, code is often in no fit state for immediate processing further upstream. It is the task of post-processing to *compact* and *combine* optimisable code fragments to prepare them for the next optimisation step. As an example of compaction, after applying rule 4 to the code:

$$(! \cdot (\pi_1 \cdot \pi_1, \pi_2 \cdot \pi_1)^\circ) * \cdot \mathsf{distl} \cdot (\mathsf{id}, R)^\circ$$

we have:

$$\mathsf{repeat} \cdot (! \cdot (\pi_1 \cdot \pi_1, \pi_2 \cdot \pi_1)^\circ, \# \cdot \pi_2)^\circ \cdot (\mathsf{id}, R)^\circ$$

where the functions of interest: $! \cdot (\pi_1 \cdot \pi_1, \pi_2 \cdot \pi_1)^\circ$, and $\# \cdot \pi_2$ are not in the most-upstream section of code, ready for further processing. Post-processing compacts these functions into $(\mathsf{id}, R)^\circ$ producing the, more accessible, code:

$$\mathsf{repeat} \cdot (! \cdot (\pi_1, \pi_2)^\circ, \# \cdot R)^\circ$$

As an example of combination, the application of rules 3 and 4, plus compaction, to the code:

$$\begin{array}{l}(+) * \cdot \mathsf{zip} \cdot (((! \cdot (\pi_1 \cdot \pi_1, \pi_2 \cdot \pi_1)^\circ) * \cdot \mathsf{distl} \cdot (\mathsf{id}, R)^\circ, \\ \quad ! \cdot (\pi_1 \cdot \pi_1, \pi_2)^\circ) * \cdot \mathsf{distl} \cdot (\mathsf{id}, R)^\circ)^\circ\end{array}$$

produces:

$$\begin{array}{l}(+) * \cdot \mathsf{zip} \cdot (\mathsf{repeat} \cdot (! \cdot (\pi_1, \pi_2)^\circ, \# \cdot R)^\circ, \\ \quad \mathsf{select} \cdot (\pi_1, R)^\circ)^\circ)^\circ\end{array}$$

Subsequently, further post-processing *combines* the two zipped sections of code to produce the more easily processed:

$$(+) * \cdot \mathsf{distl} \cdot (! \cdot (\pi_1, \pi_2)^\circ, \mathsf{select} \cdot (\pi_1, R)^\circ)^\circ$$

Note that the re-introduced distl function now transports just the required values to downstream code rather than broadcasting copies of whole vectors.

This concludes our description of the vector optimisation of map. The code resulting from vector optimisation has a significantly reduced flow of surplus vector elements. However, surplus flows of other values remain. Our implementation reduces these flows during, much-simpler, subsequent passes of optimisation. A discussion of these other passes is beyond the scope of this paper but their effects are evident in the performance of code produced by the optimiser in its entirety. We examine this performance next.

5 Results

We now examine the impact of the optimisation process on the performance of point-free program code. After this, we briefly discuss the influence that point-free form has on the design of the optimiser.

5.1 Performance Model

To measure the effect of data movement optimisation we created an instrumented model for measuring execution time and space consumption on point-free code. To keep the design of the model simple and consistent we implemented the following basic strategy for memory allocation and deallocation:

- memory for each data element is *a*llocated just prior to when it is needed.
- memory for each data element is *de*-allocated just after its last use.

Our model assigned unit costs to all scalar operations and unit costs to allocating and to copying scalars. Vectors and tuples were treated as collections of scalars.

```
main (a: vof vof int, b: vof vof int)
 := let
        f x :=
           let
                g y := a!x!y + b!x!y
           in
                map(g,iota #(a!x))
           endlet
    in
       map(f,iota (# a))
    endlet
```

(*a*)

$$
\begin{aligned}
&((+ \cdot (! \cdot (! \cdot (\pi_1 \cdot \pi_1 \cdot \pi_1, \pi_2 \cdot \pi_1)^\circ, \pi_2)^\circ,\\
&\quad\; ! \cdot (! \cdot (\pi_2 \cdot \pi_1 \cdot \pi_1, \pi_2 \cdot \pi_1)^\circ, \pi_2)^\circ)^\circ) * \cdot\\
&\;\mathsf{distl} \cdot (\mathsf{id}, \mathsf{iota} \cdot \# \cdot ! \cdot (\pi_1 \cdot \pi_1, \pi_2)^\circ)^\circ \cdot \mathsf{id}) * \cdot\\
&\mathsf{distl} \cdot (\mathsf{id}, \mathsf{iota} \cdot \# \cdot \pi_1)^\circ \cdot \mathsf{id}
\end{aligned}
$$

(*b*)

$$
\begin{aligned}
&((+) * \cdot \mathsf{zip}\\
&\quad \cdot(\mathsf{select} \cdot ({}^3\pi_1, {}^3\pi_2)^\circ, \mathsf{select} \cdot ({}^3\pi_3, {}^3\pi_2)^\circ)^\circ \cdot\\
&\quad (\pi_1 \cdot \pi_1, \mathsf{iota} \cdot \pi_2, \pi_2 \cdot \pi_1)^\circ) * \cdot\\
&\mathsf{zip} \cdot (\, ((\pi_2, \pi_1)^\circ) * \cdot \mathsf{zip} \cdot\\
&\quad\; (\mathsf{select} \cdot ({}^3\pi_1, {}^3\pi_2)^\circ, \mathsf{select} \cdot ({}^3\pi_3, {}^3\pi_2)^\circ)^\circ,\\
&\quad\; (\#) * \cdot \mathsf{select} \cdot ({}^3\pi_3, {}^3\pi_2)^\circ)^\circ \cdot\\
&(\pi_1, \mathsf{iota} \cdot \# \cdot \pi_2, \pi_2)^\circ \cdot (\pi_2, \pi_1)^\circ
\end{aligned}
$$

(*c*)

$$
\begin{aligned}
&(\,(+) * \cdot \mathsf{zip} \cdot\\
&\;\; (\mathsf{select} \cdot (\pi_1 \cdot \pi_1, \pi_2)^\circ\\
&\;\; \mathsf{select} \cdot (\pi_2 \cdot \pi_1, \pi_2)^\circ)^\circ \cdot\\
&\;\; (\mathsf{id}, \mathsf{iota} \cdot \# \cdot \pi_1)^\circ) * \cdot\\
&\mathsf{zip} \cdot\\
&(\mathsf{select} \cdot (\pi_1 \cdot \pi_1, \pi_2)^\circ\\
&\mathsf{select} \cdot (\pi_2 \cdot \pi_1, \pi_2)^\circ)^\circ \cdot\\
&(\mathsf{id}, \mathsf{iota} \cdot \# \cdot \pi_1)^\circ
\end{aligned}
$$

(*d*)

Fig. 3. Source code - part (a), translator code - part (b), optimiser code - part (c), and hand-crafted code - part(d) for `map_map_addpairs.Adl`

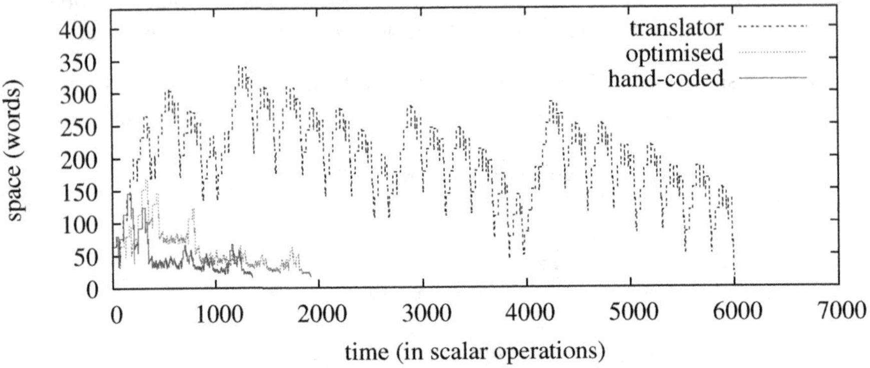

Fig. 4. Performance plots for map_map_addpair applied to the pair of ragged vectors: ([[1], [2, 3, 4], [5, 6], [], [7, 8, 9, 10]], [[1], [2, 3, 4], [5, 6], [], [7, 8, 9, 10]])

It must be noted that, when mapping point free code to imperative sequential code there are optimisations that could be applied that our model doesn't reflect. However, in a distributed context, we have found that high data-transport costs in this model map to high communications costs[1].

5.2 Experiments

We ran the translator and then the optimiser over a series of Adl programs and used an implementation of the model above to compare the performance of translator and optimiser code. As a benchmark, we hand-coded efficient solutions in point-free form and ran these against the model too. The results of two of these experiments are presented next.

Adding Corresponding Elements of Nested Vectors. The source code for `map_map_addpairs.Adl` is shown in Fig. 3(a). This program uses a nested map operation to add corresponding elements of two nested input vectors. The translator code, the optimiser code and the handed-coded point-free version are shown in parts (b), (c) and (d) respectively. The translator code distributes large amounts of data to the inner map function to be accessed by index functions. The optimiser code in part (c) has replaced all of the indexing operations by select operations. The hand-coded version in part (d) has the same basic structure as the code in part (c) but forms fewer intermediate tuples.

The performance of the three versions of point-free code, applied to a pair of nested input vectors, is shown in Fig. 4. The translator code fares the worst due to the cost of distributing aggregate values. The optimiser code exhibits substantially better performance. The hand optimised version performs even better, with a similar pattern of data allocation on a slightly smaller scale. Close inspection of the code reveals that the optimiser has been more aggressive than necessary in moving the # (length) function out of the inner-map function. This resulted in extra code to transmit the output of the # that has now been moved

```
main a: vof int :=
  let
    stencil x
      := a!x + a!(x-1) + a!(x+1);
    addone x := x + 1;
    element_index
      := map(addone,iota ((# a)-2))
  in
    map (stencil, element_index)
  endlet
```

(*a*)

$$
\begin{array}{l}
(+\cdot(+\cdot(!\cdot(\pi_1\cdot\pi_1,\pi_2)^\circ,\\
\qquad\qquad !\cdot(\pi_1\cdot\pi_1,-\cdot(\pi_2,1)^\circ)^\circ)^\circ,\\
\qquad !\cdot(\pi_1\cdot\pi_1,+\cdot(\pi_2,1)^\circ)^\circ)^\circ)*\cdot\\
\mathsf{distl}\cdot(\mathsf{id},\pi_2)^\circ\cdot\mathsf{id}\cdot\\
(\mathsf{id},(+\cdot(\pi_2,1)^\circ)*\mathsf{distl}\cdot(\mathsf{id},\mathsf{iota}\cdot-\cdot(\#\cdot\mathsf{id},2)^\circ)^\circ)^\circ
\end{array}
$$

(*b*)

$$
\begin{array}{l}
(+\cdot(+\cdot\pi_1,\pi_2)^\circ)*\cdot\mathsf{zip}\cdot\\
(\mathsf{zip}\cdot\\
(\,\mathsf{select},\\
\;\mathsf{select}\cdot(\pi_1,(-)*\cdot\mathsf{zip}\cdot\\
\qquad(\mathsf{id},\mathsf{repeat}\cdot(1,\#)^\circ)^\circ\cdot\pi_2)^\circ)^\circ,\\
\;\mathsf{select}\cdot(\pi_1,(+)*\cdot\mathsf{zip}\cdot(\mathsf{id},\mathsf{repeat}\cdot(1,\#)^\circ)^\circ\cdot\pi_2)^\circ)^\circ\cdot\\
(\mathsf{id},(+\cdot(\mathsf{id},1)^\circ)*\cdot\mathsf{iota}\cdot-\cdot(\#,2)^\circ)^\circ
\end{array}
$$

(*c*)

$$
\begin{array}{l}
(+\cdot(\pi_1\cdot\pi_1,+\cdot(\pi_2\cdot\pi_1,\pi_2)^\circ)^\circ)*\cdot\\
\mathsf{zip}\cdot(\mathsf{zip}\cdot(\pi_1\cdot\pi_1,\pi_2\cdot\pi_1)^\circ,\pi_2)^\circ\cdot\\
((\,\mathsf{select},\\
\quad\mathsf{select}\cdot(\pi_1,(+\cdot(\mathsf{id},1)^\circ)*\cdot\pi_2)^\circ)^\circ,\\
\;\mathsf{select}\cdot(\pi_1,(-\cdot(\mathsf{id},1)^\circ)*\cdot\pi_2)^\circ)^\circ\cdot\\
(\mathsf{id},(+\cdot(\mathsf{id},1)^\circ)*\cdot\mathsf{iota}\cdot-\cdot(\#,2)^\circ)^\circ
\end{array}
$$

(*d*)

Fig. 5. Source code - part (a), translator code - part (b), optimiser code - part (c), and hand-crafted code - part(d) for `finite_diff.Adl`

upstream. The movement wasn't warranted in this case because the input vector to this invocation of # had to be transmitted to the inner map function anyway. This result indicates that there is scope for tempering the aggression of the vector optimiser in certain cases.

A Simple Stencil Operation. The source code for `finite_diff.Adl` is shown in Fig. 3(a). This program applies a very simple stencil operation to a one-

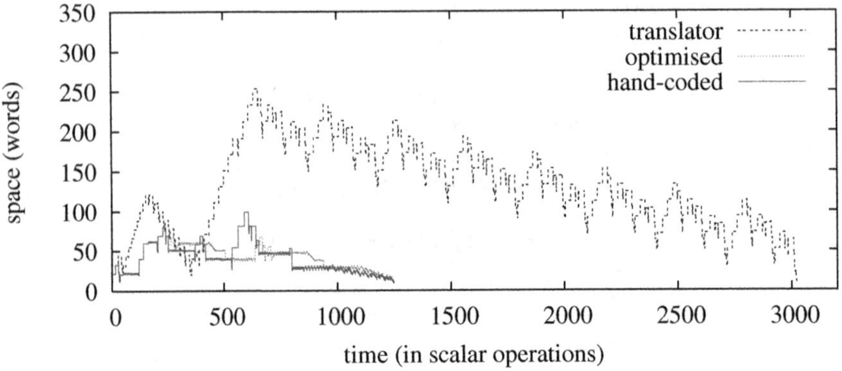

Fig. 6. Performance plots for `finite_diff` applied to a short one-dimensional vector

dimensional vector. It is a good example of the use of multiple indexing operations into a single vector and the use of arithmetic operators on vector indices. The translator code, the optimiser code and the handed-coded point-free version are shown in parts (b), (c) and (d) respectively. Again the translator code distributes surplus data through the distl operations. The optimiser code in part (c) has removed this distribution. The hand-coded solution in part (d) is similar to the hand-coded version but avoids the use of repeat.

Figure 6 shows performance of the three versions of point-free code. Again, the translator code is the worst-performing. The efficiencies of the optimised and hand-coded versions are very similar with the optimiser code very slightly ahead on time and space performance. Close inspection of the code shows that the optimiser code in part (c) has been more thorough in eliminating transport of tuple elements in downstream parts of the code.

5.3 Point-Free Form and Optimiser Design

The performance of code produced by the optimiser in the experiments presented here, and in other experiments we have carried out, is encouraging. In addition to these results we make the following general observations regarding the influence of point-free form on optimiser design.

First, pre-processing, using normalisation rules, is an essential part of automatically processing point-free form. It is infeasible to write enough rules to match code as-is, so code needs to be processed to match the rules. We applied, and reused, normalisation rules quite frequently and in a variety of circumstances to maximise the effect of other transformation rules.

Second, it pays to have interfaces. If there is a mismatch between the output of one transformation component and the expected input of another, the cause of the resulting error can be difficult to find. The use of interfaces, even on an informal, ad-hoc, basis makes matching, and constructing compilation components much less error-prone.

Last, it is difficult to write intelligent rules. It is not easy to trace dependencies through, and then alter, at a distance, point-free code. It is easier to propagate dependencies by local transformations to the code itself.

6 Related Work

There is a large body of work exploring program transformation in functional notations. [17] summarises many of the important issues that arise during transformation. The use and transformation of point-free programs was espoused in [3]. A broad range of rules and techniques applicable to both point-free and point-wise programs were developed with Bird-Meertens Formalism[7] we used a number of the algebraic identities developed in this work in our optimiser.

The idea of making programs more efficient by reducing the intermediate data produced is strongly related to concept of program fusion. Automated systems have been developed to perform fusion[14, 10] in point-wise recursive definitions. Our work, while using some of the rules derived by research into fusion, is more specialised in its purpose. However, there remains broad scope for further applications of techniques from the work above in refinements of our optimiser.

Point-free notation has been used in a few experiments and implementations including FP*[21] and EL*[16]. Some early experiments with transforming point-free code are described in[12]. More recently a complete translation process from recursive functions in Haskell to point-free form has been defined[5]. However, none of these implementations perform data movement optimisation to the extent of the implementation described here[5].

7 Conclusions and Future Work

We have described an optimisation process to reduce data movement through point-free code. We have shown that this process is effective in significantly reducing the flow of data through such programs. This process is incremental with the program itself serving as the sole repository of the state of the transformation process.

We envisage three major improvements to the implementation as defined so far. First, the system could be made more extensible by the formal definition of syntax interfaces for transformation components as in[19]. Second, the volume of transformation rules may be significantly reduced by separating rule application strategies from the rewrite rules themselves[20]. Last, an efficient mapping of point free code to imperative code on sequential architectures needs to be defined, work toward this goal is underway[15].

[5] FP* performs some optimisation as code is being mapped to an imperative language but explicit data transfer information, useful for mapping to a distributed machine, is lost in the process.

To conclude, optimisation in point free form is a highly effective process that requires different strategies to more traditional approaches. Like all optimisation, this process is open-ended and much interesting work remains to be done.

References

1. Alexander, B. , Wendelborn, A. L. 2004, *Automated Transformation of BMF Programs*, The First International Workshop on Object Systems and Software Architectures., pp. 133-141,
   URL: `http://www.cs.adelaide.edu.au/~wossa2004/HTML/19-brad-2.pdf`
2. Alexander, B. 2006, *Compilation of Parallel Applications via Automated Transformation of BMF Programs*, Phd. Thesis, University of Adelaide,
   URL: `http://www.cs.adelaide.edu.au/~brad/thesis/main.pdf`
3. Backus, J. 1978, *Can Programming Be Liberated from the von Neumann Style? A functional Style and Its Algebra of Programs*, 'Communications of the ACM', Vol. 21, No. 8, pp. 613 – 641.
4. Borras, P., Clément, D., Despeyroux, Th., Incerpi, J., 1989 , Kahn, G., *CENTAUR: the system*, SIGSOFT software engineering notes / SIGPLAN: SIGPLAN notices, Vol. 24, No. 2, The ACM.
5. Cunha, A., Pinto, S. P., , Proenca, J., 2005, *Down with Variables*, Technical report, No. DI-PURe-05.06.01.
6. Crooke, D. C., 1999, *Practical Structured Parallelism Using BMF*, Thesis, University of Edinburgh.
7. Gibbons, J., 1994, *An introduction to the Bird-Meertens Formalism*, 'New Zealand Formal Program Development Colloquium', Hamilton, NZ.
8. Hamdan, M., 2000 , *A Combinational Framework for Parallel Programming Using Algorithmic Skeletons*, Thesis, Department of Computing and Electrical Engineering. Heriot-Watt University.
9. Jones, S., Hoare, T. and T. Hoare., Tolmach, A., 2001,*Playing by the rules: rewriting as a practical optimisation technique*, ACM SIGPLAN Haskell Workshop.
10. Johann, P., Visser, E., 1997 , *Warm fusion in Stratego: A case study in the generation of program transformation systems*, Technical report, Department of Computer Science, Universiteit Utrecht, 1999.
11. Kahn, G., 1987 , *Natural Semantics*, Fourth Annual Symposium on Theoretical Aspects of Computer Science, Lecture Notes in Computer Science, Vol. 247, Springer-Verlag.
12. Martin, U. , Nipkow, T., 1990 , *Automating Squiggol*, Programming Concepts and Methods, pp. 233–247, Eds. Broy, M., Jones, C. D., North-Holland.
13. Mauny, M. , Ascander Suarez, 1986 , *Implementing functional languages in the Categorical Abstract Machine*,
    LFP '86: Proceedings of the 1986 ACM conference on LISP and functional programming, pp. 266–278, ACM Press.
14. Onoue, Y., Hu, Z., Takeichi, M., , Iwasaki H. 1997 , *A calculational fusion system HYLO*, 'Proceedings of the IFIP TC 2 WG 2.1 international workshop on Algorithmic languages and calculi', pp. 76–106, Chapman & Hall, Ltd.
15. Pettge, S. 2005, *A Fast Code Generator for Point-Free Form*, Hons. Thesis, University of Adelaide,
    URL: `http://www.cs.adelaide.edu.au/~brad/students/seanp.pdf`

16. Rao, P. , Walinsky, C., 1993, *An equational language for data-parallelism*, Proceedings of the fourth ACM SIGPLAN symposium on Principles and practice of parallel programming, pp. 112–118, ACM Press.
17. Sands, D., 1996, *Total correctness by local improvement in the transformation of functional programs*, 'ACM Trans. Program. Lang. Syst.', Vol. 18, No. 2.
18. Skillicorn, D. B. , Cai, W., 1994, *Equational code generation: Implementing categorical data types for data parallelism*, TENCON '94, Singapore, IEEE.
19. de Jonge, M. , Visser, J., 2001, *Grammars as Contracts*, Lecture Notes in Computer Science, Vol. 2177, Springer-Verlag.
20. Visser, E., 2001, *Stratego: A Language for Program Transformation Based on Rewriting Strategies*,
RTA '01: Proceedings of the 12th International Conference on Rewriting Techniques and Applications, pp. 357–362, Springer-Verlag.
21. Walinksy, C., Banerjee, D., 1994, *A Data-Parallel FP Compiler*, Journal of Parallel and Distributed Computing, Vol. 22, pp. 138–153.

Measuring the Speed of Information Leakage in Mobile Processes

Benjamin Aziz

Department of Computing
Imperial College London
180 Queen's Gate, London SW7 2RH, U.K.
baziz@doc.ic.ac.uk

Abstract. This paper presents a syntax-directed non-uniform static analysis of the stochastic π-calculus to safely approximate the amount of time required before name substitutions occur in a process. Name substitutions form the basis for defining security properties, like information leakage. The presence of the quantitative and qualitative information in the results of the analysis allows us to reason about the speed at which sensitive information is leaked in a computing environment with malicious mobile code. We demonstrate the applicability of the analysis through a simple example of firewall breaches.

1 Introduction

Measuring the rate of propagation of malicious and intrusive mobile code is a key factor in preventing and reducing the harmful effects of such code particularly, given that over the past few years, the rate of propagation of viruses and worms have increased dramatically. In one of CERT's incident reports [1], it mentions that *"The speed at which viruses are spreading is increasing ... Beginning with the Code Red worm (CA-2001-19, CA-2001-23) in 2001 up through the Slammer worm (CA-2003-04) earlier this year [2003], we have seen worm propagation times drop from hours to minutes"*. In the case of the Sapphire/Slammer worm, the main novelty of the worm was its speed of infection. According to [2], the worm *"achieved its full scanning rate (over 55 million scans per second) after approximately three minutes"*. Clearly, the moral behind such and other examples is that the earlier the malicious code is discovered the higher the possibility of protecting against its harmful effects.

Formal models of mobility, in particular message-passing process algebra such as the π-calculus [3] and the ambient calculus [4], provide an attractive basis for reasoning about qualitative and quantitative features of mobile systems due to their inherent expressive power and simplicity. In fact, there have been several extensions of these models targeted towards quantitative features of mobile systems, for example stochasticity [5, 6, 7], probability [8], semiring-based cost [9, 10, 11] and time [12]. Nonetheless, most of these models only provide pure quantitative reasoning; they do not state, for example, how the security of mobile systems is affected by changes in the quantitative features of such systems.

M. Johnson and V. Vene (Eds.): AMAST 2006, LNCS 4019, pp. 36–50, 2006.

The motivation for the work presented in this paper is to construct a static analysis that provides approximate reasoning about the speed of information leaks in processes modelled by the stochastic π-calculus [6]. More specifically, we are interested in the amount of time, starting from the initial state of the system, before a particular communication takes place. Such communications are characterised by name substitutions in message-passing calculi and have significant security implications. For the sake of brevity, we only deal with simple properties like privacy breaches resulting from high-level names substituting low-level input parameters. Other properties like authenticity and denial of service may also be incorporated in the future.

The work presented here stems from previous works dealing with the static analysis of mobile processes. Name substitution-based security properties were defined for the π-calculus in [13] and for a semiring-based extension of the π-calculus in [9]. Aside from these works, there are a few other related works. In [14], guessing attacks in security protocols are modelled in an extension of the π-calculus which adopts a computational security model with random sampling of new names. Also some works deal with simpler languages without mobility, for example [15], in which a bisimulation-based confinement property is analysed for a probabilistic version of CCS [16]. Similar approach is followed in [17] in defining a non-interference-based information flow property for the probabilistic concurrent constraint programming language [18]. Other frameworks include PEPA [19] and EMPA [20], several analysis tools for which have been defined. In general, frameworks like PEPA and EMPA do not cater for mobility aspects like fresh name creation and message-passing.

The structure of the paper is as follows. In Section 2, we review the syntax of the language of stochastic π-calculus. In Section 3, we define a domain-theoretic model of processes and define a syntax-directed semantics of the language. In Section 4, we extend the standard semantics to be able to capture name substitutions and the elapsed time from the initial state of the process. In Section 5, we introduce a couple of abstraction functions to limit the number of bound names generated in the interpretation and to produce an abstract view of time. In Section 6, we define the timed version of the information leakage property and finally, in Section 7, we demonstrate the applicability of the analysis in a simple example of firewall breaches.

2 The Stochastic π-Calculus

We recall here the syntax of the stochastic π-calculus.

Definition 1 (Stochastic π-calculus). *Let $\mathcal{N}$ be the infinite countable set of names, $\{a, b, c, x, y, z \ldots\}$, then processes, $P, Q, R, \ldots \in \mathcal{P}$, are built as follows:*

$$P ::= \mathbf{0} \mid (\pi, r).P \mid (\nu x)P \mid [x = y]P \mid (P \mid Q) \mid P + Q \mid\ !P$$

Where $r \in \mathbb{R}^+$ is a positive real number and π is defined as $\pi ::= x(y) \mid \overline{x}\langle y \rangle$.

The syntax is quite similar to the standard π-calculus syntax [3], with which we assume the reader to be familiar with, along with the notions of free/bound names and name substitutions. The only syntactic difference is in $(\pi, r).P$, which is a process capable of firing action, π, with rate, r. These actions are either input, $x(y)$, or output, $\overline{x}\langle y\rangle$. From now on, we only deal with *normal* processes.

Definition 2 (Normal Processes). *A process, P, is said to be normal if it has no occurrences of homonymous bound names*[1] *and $bn(P) \cap fn(P) = \{\}$.*

3 A Domain-Theoretic Model

We follow the approach of [21, 13] (which was originally inspired by [22]) in defining a domain-theoretic model of the stochastic π-calculus. We start by defining the following predomain equations, which describe the basic actions of processes:

$$Pi = 1 + \mathbb{P}(Tau + In + Out) \tag{1}$$
$$Tau = R^+ \times Pi_\perp \tag{2}$$
$$In = N \times R^+ \times (N \to Pi_\perp) \tag{3}$$
$$Out = N \times R^+ \times (N \times Pi_\perp + N \to Pi_\perp) \tag{4}$$

These equations are explained as follows: Pi is the predomain of processes constructed from Plotkin's powerdomain operation [23] applied to the sum of the predomains Tau, In and Out and adjoined to the single-element domain, 1, as in [24, Def. 3.4] to express terminated or deadlocked processes. The domain of processes is formed by lifting Pi to $Pi_\perp$, where $\perp_{Pi_\perp}$ represents the divergent or undefined process. The predomain, Tau, of silent actions is defined as a pair: the first element, R^+, represents the rate of synchronisation (since silent actions in our case occur solely as a result of synchronisations) whereas the second element, $Pi_\perp$, represents the residual process. The predomain, In, of input actions consists of a triple: the first two elements, $N \times R^+$, are the channel of communication and the execution rate whereas the third element is a function, $N \to Pi_\perp$, which takes a name and yields the residual process. Finally, the predomain, Out, of free and bound output actions consists of a triple: the first two elements, $N \times R^+$, are the channel and the execution rate whereas the third element represents the sum of free, $N \times Pi_\perp$, and bound, $N \to Pi_\perp$, outputs.

Finding a solution to equations (1)–(4) consists in finding a definition of the concrete elements of each of the (pre)domains involved. One such solution is shown in Figure 1, where $\mathcal{K}$ is the set underlying any (pre)domain. The definition of N is trivial: N is a flat predomain, therefore its structure is similar to $\mathcal{N}$. The same applies to R^+. In fact, in what follows, we abuse the set membership operator and write $x \in N$ and $r \in R^+$ to mean $x \in \mathcal{K}(N)$ and $r \in \mathcal{K}(R^+)$, respectively. On the other hand, $Pi_\perp$ is defined as a multiset of semantic elements, where $\{\!|\perp|\!\}$ is the bottom element representing the undefined process and $\emptyset$ is

[1] This implies that no two bound names are the same, a property that can be achieved by applying α-conversion intintially.

$$
\begin{array}{lcl}
\textit{Elements of } N: & & \\
x \in \mathcal{N} & \Rightarrow & x \in \mathcal{K}(N) \\
 & & \\
\textit{Elements of } R^+: & & \\
x \in \mathbb{R}^+ & \Rightarrow & x \in \mathcal{K}(R^+) \\
 & & \\
\textit{Elements of } Pi_\perp: & & \\
\{\!|\perp|\!\} \in \mathcal{K}(Pi_\perp) & & \\
\emptyset \in \mathcal{K}(Pi_\perp) & & \\
p, q \in \mathcal{K}(Pi_\perp) & \Rightarrow & p \uplus q \in \mathcal{K}(Pi_\perp) \\
p \in \mathcal{K}(Pi_\perp), r \in R^+ & \Rightarrow & \{\!|tau(r,p)|\!\} \in \mathcal{K}(Pi_\perp) \\
x, y \in \mathcal{K}(N), p \in \mathcal{K}(Pi_\perp), r \in R^+ & \Rightarrow & \{\!|in(x,r,\lambda y.p)|\!\} \in \mathcal{K}(Pi_\perp) \\
x, y \in \mathcal{K}(N), p \in \mathcal{K}(Pi_\perp), r \in R^+ & \Rightarrow & \{\!|out(x,r,y,p)|\!\} \in \mathcal{K}(Pi_\perp) \\
x, y \in \mathcal{K}(N), p \in \mathcal{K}(Pi_\perp), r \in R^+ & \Rightarrow & \{\!|out(x,r,\lambda y.p)|\!\} \in \mathcal{K}(Pi_\perp) \\
x \in \mathcal{K}(N), p \in \mathcal{K}(Pi_\perp) & \Rightarrow & new(x,p) \in \mathcal{K}(Pi_\perp) \\
 & & \\
\textit{Definition of new}: & & \\
new(x, \emptyset) & = & \emptyset \\
new(x, \{\!|\perp|\!\}) & = & \{\!|\perp|\!\} \\
new(x, \{\!|in(y,r,\lambda z.p)|\!\}) & = & \begin{cases} \emptyset, & \text{if } x = y \\ \{\!|in(y,r,\lambda z.new(x,p))|\!\}, & \text{otherwise} \end{cases} \\
new(x, \{\!|out(y,r,z,p)|\!\}) & = & \begin{cases} \emptyset, & \text{if } x = y \\ \{\!|out(y,r,\lambda z.p)|\!\}, & \text{if } x = z \neq y \\ \{\!|out(y,r,z,new(x,p))|\!\}, & \text{otherwise} \end{cases} \\
new(x, \{\!|out(y,r,\lambda z.p)|\!\}) & = & \begin{cases} \emptyset, & \text{if } x = y \\ \{\!|out(y,r,\lambda z.new(x,p))|\!\}, & \text{otherwise} \end{cases} \\
new(x, \{\!|tau(r,p)|\!\}) & = & \{\!|tau(r,new(x,p))|\!\} \\
new(x, (p_1 \uplus p_2)) & = & new(x,p_1) \uplus new(x,p_2)
\end{array}
$$

Fig. 1. Elements of N, R^+ and $Pi_\perp$

the empty multiset representing terminated or deadlocked processes[2]. Other elements are defined as follows: $\uplus$ is the standard multiset union of two elements. The singleton map, $\{\!|\ |\!\}$, takes tuples representing input, output and silent actions and creates a singleton multiset of each tuple. These tuples are $in(x, r, \lambda y.p)$ (input action), $out(x, r, y, p)$ (free output action), $out(x, r, \lambda y.p)$ (bound output action) and $tau(r, p)$ (silent action). In these tuples, x is the channel of communication, y is the message or input parameter, r is the rate of execution (synchronisation) and p is the residual process. The use of λ-abstraction to model the binding effect in input and bound output actions implies that the actual residual process is obtained only once its function is instantiated.

The effects of restriction are modelled by the *new* operator. In general, *new* captures deadlocked situations arising from the attempt to communicate over restricted non-extruded channels. It also turns a free output into a bound output once a restricted message is directly sent over a channel (scope extrusion). In

[2] Following [22], we take $\{\!|\perp|\!\} \sqsubseteq \emptyset$ and $\emptyset$ is incomparable otherwise.

$$
\begin{array}{|l|}
\hline
(\mathcal{S}1)\ \mathcal{S}(\![\mathbf{0}]\!)\ \rho\ \phi_S \qquad = \emptyset \\
(\mathcal{S}2)\ \mathcal{S}(\![(x(y),r).P]\!)\ \rho\ \phi_S = \{\!|in(\phi_S(x),r,\lambda y.\mathcal{R}(\![\{\!|P|\!\} \uplus \rho]\!)\ \phi_S)|\!\} \\
(\mathcal{S}3)\ \mathcal{S}(\![(\overline{x}\langle y\rangle,r).P]\!)\ \rho\ \phi_S = \{\!|out(\phi_S(x),r,\phi_S(y),\mathcal{R}(\![\{\!|P|\!\} \uplus \rho]\!)\ \phi_S)|\!\} \uplus \\
\quad (\biguplus_{(x'(z),r_z).P' \in \rho:\ \phi_S(x)=\phi_S(x')} \{\!|tau(r_{syn},\mathcal{R}(\![\{\!|P|\!\} \uplus \rho[P'/x'(z).P']]\!)\ \phi_S[z \mapsto \phi_S(y)])|\!\}) \\
\quad \text{where, } r_{syn} = (r \times (r + \sum_{(\overline{x}\langle y\rangle,r_y).P \in \rho} r_y)^{-1}) \times (r' \times (\sum_{(x'(z),r_z).P \in \rho} r_z)^{-1}) \times \\
\qquad\qquad min((\sum_{(\overline{x}\langle y\rangle,r_y).P \in \rho} r_y + r), \sum_{(x'(z),r_z).P \in \rho} r_z) \\
(\mathcal{S}4)\ \mathcal{S}(\![(\nu x)P]\!)\ \rho\ \phi_S \qquad = new(x,\mathcal{R}(\![\{\!|P|\!\} \uplus \rho]\!)\ \phi_S) \\
(\mathcal{S}5)\ \mathcal{S}(\![[x=y]P]\!)\ \rho\ \phi_S \qquad = \begin{cases} \mathcal{R}(\![\{\!|P|\!\} \uplus \rho]\!)\ \phi_S, & \text{if } \phi_S(x) = \phi_S(y) \\ \emptyset, & \text{otherwise} \end{cases} \\
(\mathcal{S}6)\ \mathcal{S}(\![P \mid Q]\!)\ \rho\ \phi_S \qquad = \mathcal{R}(\![\{\!|P|\!\} \uplus \{\!|Q|\!\} \uplus \rho]\!)\ \phi_S \\
(\mathcal{S}7)\ \mathcal{S}(\![P + Q]\!)\ \rho\ \phi_S \qquad = \mathcal{R}(\![\{\!|P|\!\} \uplus \rho]\!)\ \phi_S \ \uplus\ \mathcal{R}(\![\{\!|Q|\!\} \uplus \rho]\!)\ \phi_S \\
(\mathcal{S}8)\ \mathcal{S}(\![!P]\!)\ \rho\ \phi_S \qquad = snd(fix\ \mathcal{F}\ (0,\{\!|\bot|\!\})) \\
\quad \text{where, } \mathcal{F} = \lambda f \lambda(j,p).f\ (if\quad p = \mathcal{R}(\![(\biguplus_{i=0}^{j} \{\!|(P)\sigma|\!\})\ \uplus\ \rho]\!)\ \phi_S \\
\qquad\qquad then\ (j,p) \\
\qquad\qquad else\ ((j+1),(\mathcal{R}(\![(\biguplus_{i=0}^{j} \{\!|(P)\sigma|\!\})\ \uplus\ \rho]\!)\ \phi_S))) \\
\quad \text{and, } \sigma = [bn_i(P)/bn(P)], bn_i(P) = \{x_i \mid x \in bn(P)\} \\
(\mathcal{R}0)\ \mathcal{R}(\![\rho]\!)\ \phi_S \qquad = \biguplus_{P \in \rho} \mathcal{S}(\![P]\!)\ (\rho \backslash \{\!|P|\!\})\ \phi_S \\
\hline
\end{array}
$$

Fig. 2. The syntax-directed semantics of the stochastic π-calculus

all other cases, restriction has no effect and it is simply passed to the residue or distributed over multiset union.

Using the semantic elements of Figure 1, it is possible to give a syntax-directed semantics for the stochastic π-calculus as a function, $\mathcal{S}(\![P]\!)\ \rho\ \phi_S \in Pi_\bot$, defined over the syntactic structure of P, as shown in Figure 2. In this semantics there are two environments, $\rho : \wp(\mathcal{P})$, which is a multiset of all the processes in parallel with the current process being interpreted where rule $(\mathcal{R}0)$ is used to interpret the contents of this multiset, and $\phi_S : N \to N$, which maps names to their substitutions, where initially, $\forall x \in N : \phi_{S0}(x) = x$. The rules are described as follows. In rule $(\mathcal{S}1)$, a null process is interpreted as the empty multiset. In rule $(\mathcal{S}2)$, input actions are interpreted as a singleton multiset of the corresponding *in* semantic tuple. Note that no communications are considered in this rule and are simply interpreted in the following rule, $(\mathcal{S}3)$, for the case of output actions. In this rule, the meaning of a process guarded by an output action is interpreted by either the corresponding *out* tuple, which is necessary to express the case of no communications, and *tau* elements to express the case of communications with matching input actions in ρ, where ϕ_S is updated accordingly. In the latter case, the rate of synchronisation, r_{syn}, is computed according to [6, Eq. 1] but adapted to suit the structure of the ρ multiset. In rule $(\mathcal{S}4)$, restriction is interpreted directly using the new operator and in rule $(\mathcal{S}5)$, name matching is resolved according to the equality of ϕ_S-values of the matched names. Rule $(\mathcal{S}6)$ interprets parallel composition by adding the parallel processes to the rest in ρ. On the other hand, non-deterministic choice is interpreted in rule ($\mathcal{S}7$) as the standard multiset union of the individual meaning of each process. Rule ($\mathcal{S}8$) deals with the interpretation of replication in terms of the higher-order non-recursive functional, $\mathcal{F}$, whose fixed point meaning, $\mathit{fix}\,\mathcal{F} = \mathcal{F}\mathit{fix}\,\mathcal{F}$, is a pair of elements, (j, p). The meaning of the replicated process is then taken as p. The rule also attaches a subscript label to the bound names of each copy of the replicated process signifying the number of that copy. This is needed in order to maintain the normality of processes. It is interesting to mention that since the semantic domain, $Pi_{\perp}$, has an infinite size, then the fixed point calculation is not guaranteed to terminate.

4 Extended Semantics

The semantics of the previous section was designed in order to capture the standard operational meaning of a process as an element of the domain, $Pi_{\perp}$. In this section, we extend the semantics to capture the *time* meaning of each name substitution. More specifically, we capture the quantity, t, used to compute the time duration a substitution induces according to Gillespie's algorithm [25]:

$$t \times ln(n^{-1})$$

where n is some random number ranged over $[0, 1]$. This quantity t, which we call *the exponential time factor*, in fact turns out to be equal to r_{syn}^{-1} as defined in the standard semantics of the previous section (see [26, Def. 12]).

First, we define the special environment, $\phi_{\mathcal{E}} : N \times N \rightarrow R_{\perp}^{+}$, such that $\phi_{\mathcal{E}}(x/y) = t$ means that the name substitution, x/y, occurred at some time, t, from the beginning of the interpretation of the process. This time is in fact the accumulation of the exponential time factors encountered through each flow of control, over each side of $\uplus$. For example, in the process:

$$(\overline{x}\langle a\rangle, r_1).(\overline{x}\langle b\rangle, r_2).P \mid (x(y), r_1').(x(w), r_2').Q$$

and assuming that the current time is t_0, the first synchronisation of $\overline{x}$ and x has a rate of r_{syn1} (as a parameter of r_1 and r_1') and the second has a rate r_{syn2} (as a parameter of r_2 and r_2'), then the name substitutions a/y and b/w will occur at times $(t_0 + (r_{syn1}^{-1})) \times ln(n^{-1})$ and $(t_0 + (r_{syn1}^{-1}) + (r_{syn2}^{-1})) \times ln(n^{-1})$, parameterised by n, a random number[3]. In what follows, we ignore the random element, $ln(n^{-1})$, and capture only the accumulated exponential time factors.

Initially, the $\phi_{\mathcal{E}}$ environment is defined such that $\phi_{\mathcal{E}0}(x/y) = 0$ if $x = y$, otherwise $\phi_{\mathcal{E}0}(x/y) = \perp$. In fact, since the extended semantics is a precise semantics and we only deal with normal processes, then each input parameter can be substituted with one name at most in the domain of $\phi_{\mathcal{E}}$, i.e. $\forall x, y, z \in N$:

[3] In fact, n is a simplification of the random numbers, n_1, n_2 and n_3, since the actual Gillespie's algorithm would have computed times $(t \times ln(n_1^{-1})) + (r_{syn1}^{-1} \times ln(n_2^{-1}))$ and $(t \times ln(n_1^{-1})) + (r_{syn1}^{-1} \times ln(n_2^{-1})) + (r_{syn2}^{-1} \times ln(n_3^{-1}))$.

$x/z, y/z \in dom(\phi_\mathcal{E}) \Rightarrow x = y$. The semantic domain of $\phi_\mathcal{E}$ environments is then defined as $D_\perp = N \times N \rightarrow R^+_\perp$ with the following ordering:

$$\forall \phi_{\mathcal{E}1}, \phi_{\mathcal{E}2} \in D_\perp : \phi_{\mathcal{E}1} \sqsubseteq_{D_\perp} \phi_{\mathcal{E}2} \Leftrightarrow dom(\phi_{\mathcal{E}1}) \subseteq dom(\phi_{\mathcal{E}2})$$

where the bottom element is, $\perp_{D_\perp} = \phi_{\mathcal{E}0}$. We also define the union of $\phi_\mathcal{E}$ environments as follows:

$$\forall x, y \in N, \phi_{\mathcal{E}1}, \phi_{\mathcal{E}2} \in D_\perp : (\phi_{\mathcal{E}1} \cup_\phi \phi_{\mathcal{E}2})(x/y) = \begin{cases} \phi_{\mathcal{E}1}(x/y), \text{ if } x/y \in dom(\phi_{\mathcal{E}1}) \\ \phi_{\mathcal{E}2}(x/y), \text{ if } x/y \in dom(\phi_{\mathcal{E}2}) \\ \perp, \quad \text{otherwise} \end{cases}$$

Using $D_\perp$, we can define an extended semantics for the stochastic π-calculus through the semantic function, $\mathcal{E}([P])\ \rho\ t\ \phi_\mathcal{E} \in Pi_\perp \times D_\perp$, defined by the rules of Figure 3, where t is the total time expressed by accumulated exponential time factors resulting from synchronisations in each flow of control. We

$$
\begin{array}{ll}
(\mathcal{E}1) & \mathcal{E}([\mathbf{0}])\ \rho\ t\ \phi_\mathcal{E} = (\emptyset, \phi_\mathcal{E}) \\
(\mathcal{E}2) & \mathcal{E}([(x(y), r).P])\ \rho\ t\ \phi_\mathcal{E} = (\{|in(x', r, \lambda y.fst(\mathcal{R}([\{|P|\} \uplus \rho])\ t\ \phi_\mathcal{E}))|\}, \phi_\mathcal{E}) \\
& \text{where, } x'/x \in dom(\phi_\mathcal{E}) \\
(\mathcal{E}3) & \mathcal{E}([(\overline{x}\langle y \rangle, r).P])\ \rho\ t\ \phi_\mathcal{E} = (\{|out(w, r, u, fst(\mathcal{R}([\{|P|\} \uplus \rho])\ t\ \phi_\mathcal{E}))|\} \uplus \\
& \quad (\biguplus_{\forall (x'(z), r_z).P' \in \rho, \exists x'':\ x''/x', x''/x \in dom(\phi_\mathcal{E})} \{|tau(r_{syn}, p')|\})), \\
& \quad (\phi_\mathcal{E} \cup_\phi (\bigcup_{\phi\ \forall (x'(z), r').P' \in \rho, \exists x'':\ x''/x', x''/x \in dom(\phi_\mathcal{E})} \phi'_\mathcal{E})) \\
& \text{where, } w/x \in dom(\phi_\mathcal{E}), u/y \in dom(\phi_\mathcal{E}) \\
& (p', \phi'_\mathcal{E}) = \mathcal{R}([\{|P|\} \uplus \rho[P'/x'(z).P']])\ t'\ \phi_\mathcal{E}[u/z \mapsto t + r_{syn}^{-1}] \\
& \text{and } r_{syn} = (r \times (r + \sum_{(\overline{x}\langle y \rangle, r_y).P \in \rho} r_y)^{-1}) \times (r' \times (\sum_{(x'(z), r_z).P \in \rho} r_z)^{-1}) \times \\
& \qquad min((\sum_{(\overline{x}\langle y \rangle, r_y).P \in \rho} r_y + r), \sum_{(x'(z), r_z).P \in \rho} r_z) \\
(\mathcal{E}4) & \mathcal{E}([(\nu x)P])\ \rho\ t\ \phi_\mathcal{E} = new(x, fst(\mathcal{R}([\{|P|\} \uplus \rho])\ t\ \phi_\mathcal{E})), snd(\mathcal{R}([\{|P|\} \uplus \rho])\ t\ \phi_\mathcal{E}) \\
(\mathcal{E}5) & \mathcal{E}([[x = y]P])\ \rho\ t\ \phi_\mathcal{E} = \begin{cases} \mathcal{R}([\{|P|\} \uplus \rho])\ t\ \phi_\mathcal{E}, \text{ if } \exists z:\ z/x, z/y \in dom(\phi_\mathcal{E}) \\ (\emptyset, \phi_\mathcal{E}), \quad \text{otherwise} \end{cases} \\
(\mathcal{E}6) & \mathcal{E}([P \mid Q])\ \rho\ t\ \phi_\mathcal{E} = \mathcal{R}([\{|P|\} \uplus \{|Q|\} \uplus \rho])\ t\ \phi_\mathcal{E} \\
(\mathcal{E}7) & \mathcal{E}([P + Q])\ \rho\ t\ \phi_\mathcal{E} = (p_1 \uplus p_2), (\phi_1 \cup_\phi \phi_2) \\
& \text{where, } (p_1, \phi_1) = \mathcal{R}([\{|P|\} \uplus \rho])\ t\ \phi_\mathcal{E} \text{ and } (p_2, \phi_2) = \mathcal{R}([\{|Q|\} \uplus \rho])\ t\ \phi_\mathcal{E} \\
(\mathcal{E}8) & \mathcal{E}([!P])\ \rho\ t\ \phi_\mathcal{E} = snd(fix\ \mathcal{F}\ (0, (\{|\perp|\}, \perp_{D_\perp}))) \\
& \text{where, } \mathcal{F} = \lambda f \lambda(j, e).f\ (if\quad e = \mathcal{R}([(\biguplus_{i=0}^{j} \{|(P)\sigma|\})\ \uplus\ \rho])\ t\ \phi_\mathcal{E} \\
& \qquad then\ (j, e) \\
& \qquad else\ ((j+1), (\mathcal{R}([(\biguplus_{i=0}^{j} \{|(P)\sigma|\})\ \uplus\ \rho])\ t\ \phi_\mathcal{E}))) \\
& \text{and, } \sigma = [bn_i(P)/bn(P)], bn_i(P) = \{x_i \mid x \in bn(P)\} \\
(\mathcal{R}0) & \mathcal{R}([\rho])\ t\ \phi_\mathcal{E} = \biguplus_{P \in \rho} fst(\mathcal{E}([P])\ (\rho \backslash \{|P|\})\ t\ \phi_\mathcal{E}), \\
& \qquad \bigcup_{\phi\ P \in \rho} snd(\mathcal{E}([P])\ (\rho \backslash \{|P|\})\ t\ \phi_\mathcal{E})
\end{array}
$$

Fig. 3. The non-standard semantics of the stochastic π-calculus

explain here a few interesting rules. In rule ($\mathcal{E}2$), and since communications are only considered in the following rule ($\mathcal{E}3$), the $\phi'_{\mathcal{E}}$ environment resulting from the interpretation of the residual process, P, is neglected because the input action guarding P cannot be fired in this rule. In rule ($\mathcal{E}3$), communications are dealt with by introducing the *tau* element and updating the $\phi_{\mathcal{E}}$ and time information. The rate of synchronisation, r_{syn}, is computed exactly as in the standard semantics of the previous section. On the other hand, the exponential time factor, r_{syn}^{-1}, which induces a time delay of $(r_{syn}^{-1} \times ln(n^{-1}))$ according to Gillespie's algorithm, is added to the current value, t. The equality of the names of the synchronising channels is tested by finding a common name that substitutes both channel names in the domain of $\phi_{\mathcal{E}}$. Finally, rule ($\mathcal{E}8$) deals with the case of replication using the subscript labelling of bound names and a fixed point calculation over the higher-order non-recursive functional, $\mathcal{F}$. As in the case of the standard semantics, this calculation is not guaranteed to terminate due to the infinite size of the semantic domain, $Pi_{\perp} \times D_{\perp}$.

In the following theorem, we show that the extended semantics is *correct* with respect to the standard semantics of the previous section by proving that the standard element of the former is equal to the latter and that any name substitutions (excluding their time value) captured in the former are also captured by the latter.

Theorem 1 (Correctness of the Extended Semantics).
$\forall P \in \mathcal{P}, \rho, t, \phi_{\mathcal{E}}, \phi_{\mathcal{S}}, \mathcal{E}(\![P]\!)\ \rho\ t\ \phi_{\mathcal{E}} = (p', \phi'_{\mathcal{E}}), \mathcal{S}(\![P]\!)\ \rho\ \phi_{\mathcal{S}} = p : \quad p = p' \ \wedge$
$\exists P', \rho', \phi'_{\mathcal{S}} : \mathcal{S}(\![P']\!)\ \rho'\ \phi'_{\mathcal{S}} \in p \ \wedge \ (\forall x, y \in N : x/y \in dom(\phi_{\mathcal{E}}) \Rightarrow \phi'_{\mathcal{S}}(x) = y)$

Proof sketch. The proof is by induction over the rules of both semantics. In particular, we take care in noting that each time a substitution is recorded in $\phi_{\mathcal{E}}$, then the same substitution is recorded in the corresponding $\phi_{\mathcal{S}}$. □

5 Abstract Semantics

Despite the fact that the extended semantics of the previous section captures the property of interest, i.e. name substitutions and the exponential time factor which can be used to determine the point in time in which a substitution takes place, the semantics is non-computable due to the infinite size of $Pi_{\perp} \times D_{\perp}$. Therefore, it is necessary to introduce abstraction functions in order to obtain a termination result. The abstraction we adopt is based on two ideas: First, limiting the number of copies each bound name is allowed to have and second, considering time as an abstract quantity instead of the infinite time span of R^{+}. Moreover, we remove elements of $Pi_{\perp}$ from the abstract meaning of a process since these are not required anymore in the security properties defined later.

Before introducing our abstraction, we define the flat predomain of tags, *Tag*, ranged over by l, l'. These tags will be used to mark each message of an output action in the specification. For example, $\overline{x}\langle y^{l}\rangle.\overline{z}\langle y^{l'}\rangle.\overline{u}\langle a^{l''}\rangle.\mathbf{0}$. This tagging is necessary so that homonymous output messages are distinguished, for example, the dual occurrence of y as a message above[4]. We also define the functions:

$$\mathit{value\_of}(l) = y$$
$$\mathit{tags\_of}(P) = \{l_1, \ldots, l_n\}$$

where $\mathit{value\_of} : \mathit{Tag} \to N$ gives the name value corresponding to a tag and $\mathit{tags\_of} : \mathcal{P} \to \wp(\mathit{Tag})$ gives the set of tags used in a process. So, in the above example, we have that $\mathit{value\_of}(l) = \mathit{value\_of}(l') = y$ and $\mathit{value\_of}(l'') = a$. Also, $\mathit{tags\_of}(\overline{x}\langle y^{l}\rangle.\overline{z}\langle y^{l'}\rangle.\overline{u}\langle a^{l''}\rangle.\mathbf{0}) = \{l, l', l''\}$. Now, we define our name and tag abstraction function.

Definition 3 (Name and Tag Abstraction Function). *Define the abstract function,* $\alpha_k : (N \cup \mathit{Tag}) \to (N^\sharp \cup \mathit{Tag}^\sharp)$*, as follows:*

$$\forall u \in (N \cup \mathit{Tag}) : \alpha_k(u) = \begin{cases} u_k, & \mathit{if}\ u = u_i\ \wedge\ i > k \\ u, & \mathit{otherwise} \end{cases}$$

The definition of α_k, curried with respect to the positive natural number, k, ensures that all copies of names and tags beyond k are abstracted to the kth copy. As a result, the abstract flat predomains, $N^\sharp$, $\mathit{Tag}^\sharp$, are defined as,

$$\mathcal{K}(N^\sharp) = \mathcal{K}(N)\backslash\{x_i \mid i > k\}$$
$$\mathcal{K}(\mathit{Tag}^\sharp) = \mathcal{K}(\mathit{Tag})\backslash\{l_i \mid i > k\}$$

In the case where $k = 1$, the analysis becomes *uniform*. Otherwise, it is *non-uniform* and the choice of k will determine the precision of the results. This choice is dependent on the properties the analysis is designed for.

Next, we define the notion of abstract time durations as follows.

Definition 4 (Abstract Time Durations). *We define abstract time durations as the set,* $T^\sharp = \{\mathit{long}, \mathit{medium}, \mathit{short}\}$*, ordered as follows:*

$$\mathit{short} \sqsubseteq_{T^\sharp} \mathit{medium} \sqsubseteq_{T^\sharp} \mathit{long}$$

We also define a function that estimates abstract time durations from concrete exponential time factors as follows.

Definition 5 (Time Abstraction Function). *Given two exponential time factors,* t_1 *and* t_2*, we define the time abstraction function,* $\beta_{t_1,t_2} : R^+ \to T^\sharp$*, as:*

$$\forall t \in R^+ : \beta_{t_1,t_2}(t) = \begin{cases} \mathit{short}, & \mathit{if}\ t \le t_1 \\ \mathit{medium}, & \mathit{if}\ t_1 < t \le t_2 \\ \mathit{long}, & \mathit{if}\ t_2 < t \end{cases}$$

[4] Such a distinction was not required in the concrete semantics of the previous sections since a message was always going to instantiate at most one input parameter. This is not the case in the abstract semantics.

The time abstraction function, β_{t_1,t_2}, is parameterised by two times, t_1 and t_2, which define the boundaries of *short*, *medium* and *long* time periods.

Based on the predomains $N^\sharp$, $Tag^\sharp$ and $T^\sharp$, we can define the abstract environment, $\phi_{\mathcal{A}} : Tag^\sharp \times N^\sharp \rightarrow \wp(T^\sharp)_\perp$, to denote the fact that a particular substitution of an abstract name by an abstract tag may occur at any point in abstract time among a set of such points. Initially, $\forall x \in N^\sharp, l \in Tag^\sharp : \phi_{\mathcal{A}0}(l/x) = \{\}$ if $value\_of(l) = x$, otherwise, $\phi_{\mathcal{A}0}(l/x) = \perp$. Furthermore, we can define the abstract domain, $D^\sharp_\perp = Tag^\sharp \times N^\sharp \rightarrow \wp(T^\sharp)_\perp$, with the following ordering:

$$\forall \phi_{\mathcal{A}1}, \phi_{\mathcal{A}2} \in D^\sharp_\perp : \phi_{\mathcal{A}1} \sqsubseteq_{D^\sharp_\perp} \phi_{\mathcal{A}2} \Leftrightarrow dom(\phi_{\mathcal{A}1}) \subseteq dom(\phi_{\mathcal{A}1})$$

where the bottom element is $\perp_{D^\sharp_\perp} = \phi_{\mathcal{A}0}$. We also redefine the union operation over abstract environments as follows:

$$\forall \phi_{\mathcal{A}1}, \phi_{\mathcal{A}2} \in D^\sharp_\perp, x \in N^\sharp, l \in Tag^\sharp : (\phi_{\mathcal{A}1} \cup_\phi \phi_{\mathcal{A}2})(l/x) = \phi_{\mathcal{A}1}(l/x) \cup \phi_{\mathcal{A}2}(l/x)$$

Given the abstract domain, $D^\sharp_\perp$, we define the abstract semantics of the stochastic π-calculus as an element, $\mathcal{A}(\![P]\!)\ \rho\ t^\sharp\ \phi_{\mathcal{A}} \in D^\sharp_\perp$, inductively over the structure of processes as shown in Figure 4. The rules of the abstract semantics are explained as follows. Rules $(\mathcal{A}1)$ and $(\mathcal{A}2)$ do not change the $\phi_{\mathcal{A}}$ environment since they do not induce any communications. Instead, communications are dealt with in rule $(\mathcal{A}3)$, where the $\phi_{\mathcal{A}}$ environment is updated with the abstract time value corresponding to a particular tag/name substitution and then added to the original environment representing the no-communications case. Note that since the semantics is approximate, a substitution may have a set (not just one element) of abstract time values associated with it. The new time, $t'^\sharp$, resulting from the synchronisation of the matching input/output channels will be joined to the current time using the least upper bound operator. This is justified by the fact that slower synchronisations have a stronger effect on the overall time in any flow of control. The rest of the rules are straightforward, except for rule $(\mathcal{A}8)$, where the fixed point calculation, unlike the case of concrete semantics, is guaranteed to terminated in this semantics, as shown by the following result.

Theorem 2 (Termination of the Abstract Semantics). *The calculation of rule* $(\mathcal{A}8)$ *terminates.*

Proof sketch. The proof relies on satisfying two requirements in the abstract semantics: The first is that the abstract domain is finite, this can be shown from the definition of $D^\sharp_\perp$. The second is that the meaning of a process is monotonic with respect to increments in the number of copies of P, i.e.

$$\mathcal{R}(\!|(\biguplus_{i=0}^{j} \{\!|P|\!\}) \uplus \rho|\!)\ t^\sharp\ \phi_{\mathcal{A}} \sqsubseteq_{D^\sharp_\perp} \mathcal{R}(\!|(\biguplus_{i=0}^{j+1} \{\!|P|\!\}) \uplus \rho|\!)\ t^\sharp\ \phi_{\mathcal{A}}$$

This latter property can be proven by showing that the extra copy of P can only induce more communications, not less. □

$$
\begin{array}{ll}
(\mathcal{A}1)\ \mathcal{A}(\![\mathbf{0}]\!)\ \rho\ t^\sharp\ \phi_\mathcal{A} & = \phi_\mathcal{A}\\
(\mathcal{A}2)\ \mathcal{A}(\![(x(y),r).P]\!)\ \rho\ t^\sharp\ \phi_\mathcal{A} & = \phi_\mathcal{A}\\
(\mathcal{A}3)\ \mathcal{A}(\![(\overline{x}\langle y^l\rangle,r).P]\!)\ \rho\ t^\sharp\ \phi_\mathcal{A} & = \phi_\mathcal{A} \cup_\phi\\
\quad \Big(\bigcup\nolimits_{\phi\ \forall(x'(z),r').P'\in\rho,\exists l:\ l/x',l/x\in dom(\phi_\mathcal{A})} \mathcal{R}(\![\{\![P]\!\} \uplus \rho[P'/x'(z).P']]\!)\ t'^\sharp\ \phi'_\mathcal{A} & \\
\quad \text{where, } \phi'_\mathcal{A} = \phi_\mathcal{A}[\alpha_k(l)/\alpha_k(z) \mapsto \{t'^\sharp\} \cup \phi_\mathcal{A}(\alpha_k(l)/\alpha_k(z))]\Big) & \\
\quad t'^\sharp = t^\sharp \sqcup \beta_{t_1,t_2}(r_{syn}^{-1}) & \\
\quad \text{and } r_{syn} = (r \times (r + \sum\limits_{(\overline{x}\langle y\rangle,r_y).P\in\rho} r_y)^{-1}) \times (r' \times (\sum\limits_{(x'(z),r_z).P\in\rho} r_z)^{-1}) \times & \\
\qquad\qquad min((\sum\limits_{(\overline{x}\langle y\rangle,r_y).P\in\rho} r_y + r), \sum\limits_{(x'(z),r_z).P\in\rho} r_z) & \\
(\mathcal{A}4)\ \mathcal{A}(\![(\nu x)P]\!)\ \rho\ t^\sharp\ \phi_\mathcal{A} & = \mathcal{R}(\![\{\![P]\!\} \uplus \rho]\!)\ t^\sharp\ \phi_\mathcal{A}\\
(\mathcal{A}5)\ \mathcal{A}(\![[x=y]P]\!)\ \rho\ t^\sharp\ \phi_\mathcal{A} & = \begin{cases} \mathcal{R}(\![\{\![P]\!\} \uplus \rho]\!)\ t^\sharp\ \phi_\mathcal{A}, \\ \qquad \text{if } \exists l : l/\alpha_k(x), l/\alpha_k(y) \in dom(\phi_\mathcal{A}) \\ \phi_\mathcal{A},\ \text{otherwise} \end{cases}\\
(\mathcal{A}6)\ \mathcal{A}(\![P \mid Q]\!)\ \rho\ t^\sharp\ \phi_\mathcal{A} & = \mathcal{R}(\![\{\![P]\!\} \uplus \{\![Q]\!\} \uplus \rho]\!)\ t^\sharp\ \phi_\mathcal{A}\\
(\mathcal{A}7)\ \mathcal{A}(\![P + Q]\!)\ \rho\ t^\sharp\ \phi_\mathcal{A} & = \mathcal{R}(\![\{\![P]\!\} \uplus \rho]\!)\ t^\sharp\ \phi_\mathcal{A} \cup_\phi \mathcal{R}(\![\{\![Q]\!\} \uplus \rho]\!)\ t^\sharp\ \phi_\mathcal{A}\\
(\mathcal{A}8)\ \mathcal{A}(\![!P]\!)\ \rho\ t^\sharp\ \phi_\mathcal{A} & = snd(\mathit{fix}\ \mathcal{F}\ (0, \bot_{D^\sharp_\bot}))\\
\quad \text{where, } \mathcal{F} = \lambda f \lambda(j,\phi).f\ (\mathit{if}\quad \phi = \mathcal{R}(\![(\biguplus_{i=0}^{j} \{\![(P)\sigma]\!\})\ \uplus\ \rho]\!)\ t^\sharp\ \phi_\mathcal{A} & \\
\qquad\qquad\qquad\qquad \mathit{then}\ j, \phi & \\
\qquad\qquad\qquad\qquad \mathit{else}\ (j+1), (\mathcal{R}(\![(\biguplus_{i=0}^{j} \{\![(P)\sigma]\!\})\ \uplus\ \rho]\!)\ t^\sharp\ \phi_\mathcal{A})) & \\
\quad \sigma = [bn_i(P)/bn(P)][tags\_of_i(P)/tags\_of(P)], bn_i(P) = \{x_i \mid x \in bn(P)\} & \\
\quad \text{and } tags\_of_i(P) = \{x_i \mid x \in tags\_of(P)\} & \\
(\mathcal{R}0)\ \mathcal{R}(\![\rho]\!)\ t^\sharp\ \phi_\mathcal{A} & = \bigcup\nolimits_{\phi\ P\in\rho} \mathcal{A}(\![P]\!)\ (\rho \backslash \{\![P]\!\})\ t^\sharp\ \phi_\mathcal{A}
\end{array}
$$

Fig. 4. The abstract semantics of the stochastic π-calculus

The following safety result states that the abstract semantics always captures, in an approximate manner, the same information captured in the concrete extended semantics and therefore, the former is a safe abstraction of the latter.

Theorem 3 (Safety of the Abstract Semantics).
$\forall P, \rho, \phi_\mathcal{E}, \phi_\mathcal{A}, k, t_1, t_2, t, \mathcal{E}(\![P]\!)\ \rho\ t\ \phi_\mathcal{E} = (p, \phi'_\mathcal{E}), \mathcal{A}(\![P]\!)\ \rho\ \beta_{t_1,t_2}(t)\ \phi_\mathcal{A} = \phi'_\mathcal{A}$:

$$
\begin{array}{l}
(\exists x, y \in N : x/y \in dom(\phi_\mathcal{E}) \Rightarrow \\
\quad \exists l \in Tag^\sharp, y \in N^\sharp : l/y^\sharp \in dom(\phi_\mathcal{A}) \wedge value\_of(l) = \alpha_k(x) \wedge y^\sharp = \alpha_k(y) \wedge \\
\quad \beta_{t_1,t_2}(\phi_\mathcal{E}(x/y)) \in \phi_\mathcal{A}(l/y)) \\
\Rightarrow \\
(\exists x, y \in N : x/y \in dom(\phi'_\mathcal{E}) \Rightarrow \\
\quad \exists l \in Tag^\sharp, y \in N^\sharp : l/y^\sharp \in dom(\phi'_\mathcal{A}) \wedge value\_of(l) = \alpha_k(x) \wedge y^\sharp = \alpha_k(y) \wedge \\
\quad \beta_{t_1,t_2}(\phi'_\mathcal{E}(x/y)) \in \phi'_\mathcal{A}(l/y))
\end{array}
$$

Proof sketch. The proof of the theorem is by induction over the rules of the abstract semantics and the extended semantics and relies on a lemma showing that the $\cup_\phi$ operation preserves a similar safety property. □

6 Timed Information Leakage

Using the results of the abstract semantics of the previous section, we define in this section the notion of *timed information leakage* as a property to measure how quickly information may be leaked in a system.

6.1 Security Policies

We first define the notion of a *security policy*, written as ξ, to refer to a classification of the different data and input parameters using the well-known lattice structures.

Definition 6 (Security Policy). *Assume that* $S = (S_L, \sqsubseteq_S, \sqcap_S, \sqcup_S, \top_S, \bot_S)$ *is a finite lattice of security levels ranged over by* $\mathcal{L}, \mathcal{L}' \in S_L$*, then a security policy is a function,* $\xi : N \rightarrow S$*, such that* $\xi(x) = \mathcal{L}$ *means that name* x *is classified at security level* $\mathcal{L}$*.*

According to this definition, a security policy assigns to each name in a process specification the security level of that name. For names occurring as messages, this level is a reflection of the sensitivity of that message. On the other hand, for names occurring as input parameters, the level reflects the sensitivity of the process doing the input.

6.2 Information Leakage

An essential threat that arises once data/input parameters have been classified in some security policy is the threat of information leakage; namely, a high-level piece of data may be caught by some low-level input parameter. In our context, we define the information leakage threat as follows.

Definition 7 (Information Leakage). *Given a security policy,* ξ*, and some abstract environment,* $\phi_{\mathcal{A}}$*, then a name,* x*, with security level,* $\xi(x) = \mathcal{L}_x$*, is leaked to another name,* y*, with security level,* $\xi(y) = \mathcal{L}_y$*, if* $\mathcal{L}_y \sqsubseteq_S \mathcal{L}_x$ *and the following holds true:*

$$\exists l^\sharp \in Tag^\sharp, y^\sharp \in N^\sharp, k \in \mathbb{N} : \ l^\sharp/y^\sharp \in dom(\phi_{\mathcal{A}}) \wedge value\_of(l^\sharp) = \alpha_k(x) \wedge y^\sharp = \alpha_k(y)$$

The property relies on name substitutions captured by a $\phi_{\mathcal{A}}$ environment, which results from the analysis of some process. We shall write, *leaked*$(x, y, \xi, \phi_{\mathcal{A}})$, to say that the message, x, is leaked to the input parameter, y, given particular definitions of $\phi_{\mathcal{A}}$ and ξ.

6.3 Timed Information Leakage

The presence of time information in $\phi_{\mathcal{A}}$ allows us to define a timed version of the information leakage threat, as given in the following definition.

Definition 8 (Timed Information Leakage). *Given a security policy, ξ, and some abstract environment, $\phi_{\mathcal{A}}$, the names, x and y, then we say that x is leaked to y in abstract time, $t^\sharp$, if:*

$$leaked(x, y, \xi, \phi_{\mathcal{A}}) \ \wedge \ t^\sharp = \bigsqcap_{t \in \phi_{\mathcal{A}}(l^\sharp/y^\sharp)} t$$

This definition adds the timed condition ($t^\sharp = \bigsqcap_{t \in \phi_{\mathcal{A}}(l^\sharp/y^\sharp)} t$) to the information leakage property. The abstract time, $t^\sharp$, is the shortest time among the times estimated by $\phi_{\mathcal{A}}$ for the substation, x/y to take place. Taking the shortest time (or the greatest lower bound of all times) is necessary in order to consider the worst possible scenario.

7 Example: Firewall Breach

We discuss in this section a simple example to demonstrate the applicability of the timed information leakage property as defined in the previous section. The example consists of a *Firewall* with two secret names of gateways, *gate* and *gate'*. Behind the firewall we have a private *LAN* containing some sensitive *data* and outside the firewall there is a malicious *Intruder*, which is assumed to be able to discover any channel's name, use that name to inject a *leak* name and finally, use that name to obtain sensitive date. The *Intruder*, *Firewall* and *LAN* processes are given the following specifications:

$$\begin{aligned}
Intruder &\stackrel{\text{def}}{=} \ !(obtain(ch), 5000).(\overline{ch}\langle leak\rangle, 5000).(leak(mine), 5000)\\
Firewall &\stackrel{\text{def}}{=} \ !((gate(x), 7).(\overline{pgm}\langle x\rangle, 5000) \mid (gate'(x'), 5000).(\overline{pgm'}\langle x'\rangle, 5000))\\
LAN &\stackrel{\text{def}}{=} \ !((pgm(y), 5000).(\overline{y}\langle data\rangle, 5000) \mid (pgm'(y'), 5000).(\overline{y'}\langle data\rangle, 5000))
\end{aligned}$$

In the first case, assuming that the intruder is only capable of discovering the first gate's name, *gate*, then the definition of the overall system would be:

$$\begin{aligned}
System \stackrel{\text{def}}{=} \ & Intrduer \mid (\nu data)(\nu gate)(\nu gate')(\nu pgm)(\nu pgm')(\\
& \quad Firewall \mid LAN \mid (\overline{obtain}\langle gate\rangle, 5000))
\end{aligned}$$

Also, we assume a time abstraction function, $\beta_{100,500}$. Performing a uniform abstract interpretation, $\mathcal{A}([\![System]\!])\ \{\![\,]\!\}\ short\ \phi_{\mathcal{A}0}$, we obtain the following fixed point value for $\phi_{\mathcal{A}}$:

$$gate/ch_1 \mapsto \{short\},\ leak/x_1 \mapsto \{long\},\ leak/y_1 \mapsto \{long\},\ data/mine \mapsto \{long\}$$

In the second case, we modify the definition of the system so that now the intruder discovers both gateway names, *gate* and *gate'*:

$$\begin{aligned}
System \stackrel{\text{def}}{=} \ & Intrduer \mid (\nu data)(\nu gate)(\nu gate')(\nu pgm)(\nu pgm')(\\
& \quad Firewall \mid LAN \mid (\overline{obtain}\langle gate\rangle, 5000) \mid (\overline{obtain}\langle gate'\rangle, 5000))
\end{aligned}$$

Then, by re-performing the same abstract interpretation as above, we obtain the following fixed point value for $\phi_{\mathcal{A}}$:

$$gate/ch_1 \mapsto \{short\},\ leak/x_1 \mapsto \{long\},\ leak/y_1 \mapsto \{long\},$$
$$leak/x_1' \mapsto \{short\},\ leak/y_1' \mapsto \{short\},\ data/mine \mapsto \{long, short\}$$

Assuming that we have a security policy, ξ, such that $\xi(leak) \sqsubseteq_S \xi(data)$, then we find according to Definition (8) that in the first definition of our system, the leakage of *data* to *mine* requires a long amount of time due to the fact that the intruder can only interact with the slow action, $gate(x)$. However, in the second case, the same leakage now requires only a *short* amount of time since there is a faster alternative to $gate(x)$, with which the intruder can interact, namely $gate'(x')$. Therefore, the worst case scenario for leaking *data* to *mine* will require a *short* amount of time (equivalent to $long \sqcap short$).

Acknowledgements. Thanks for the useful comments of the anonymous referees on this paper. The work was supported by EPSRC.

References

1. CERT: CERT© incident note in-2003-01. Technical report, Software Engineering Institute (2003)
2. Moore, D., Paxson, V., Savage, S., Shannon, C., Staniford, S., Weaver, N.: Inside the slammer worm. IEEE Security and Privacy **1(4)** (2003) 33–39
3. Milner, R., Parrow, J., Walker, D.: A calculus of mobile processes (parts I & II). Information and Computation **100(1)** (1992) 1–77
4. Cardelli, L., Gordon, A.: Mobile ambients. In: Proceedings of the 1st International Conference on the Foundations of Software Science and Computation Structures. Volume 1378 of Lecture Notes in Computer Science., Lisbon, Portugal, Springer Verlag (1998) 140–155
5. Hillston, J., Ribaudo, M.: Modelling mobility with pepa nets. In Aykanat, C., Dayar, T., Korpeoglu, I., eds.: Proceedings of the 19th International Conference on Computer and Information Sciences. Volume 3280 of Lecture Notes in Computer Science., Antalya, Turkey, Springer Verlag (2004) 513–522
6. Priami, C.: Stochastic π-calculus. The Computer Journal **38**(7) (1995) 578–589
7. Priami, C.: Stochastic π-calculus with general distributions. In: Proceedings of the 4th International Workshop on Process Algebra and Performance Modelling, Torino, Italy, GLUT Press (1996) 41–57
8. Herescu, O.M., Palamidessi, C.: Probabilistic asynchronous pi-calculus. In Tiuryn, J., ed.: Proceedings of the 3rd International Conference on the Foundations of Software Science and Computation Structures. Volume 1784 of Lecture Notes in Computer Science., Berlin, Germany, Springer Verlag (2000) 146–160
9. Aziz, B.: A semiring-based quantitative analysis of mobile systems. In: Proceedings of the 3rd International Workshop on Software Verification and Validation. Electronic Notes in Theoretical Computer Science, Manchester, UK, Elsevier (2005) to appear.

10. Buchholz, P., Kemper, P.: Quantifying the dynamic behavior of process algebras. In: Proceedings of the Joint International Workshop on Process Algebra and Probabilistic Methods, Performance Modeling and Verification. Volume 2165 of Lecture Notes in Computer Science., Aachen, Germany, Springer Verlag (2001) 184–199
11. Hirsch, D., Tuosto, E.: Shreq: Coordinating application level qos. In: Proceedings of the 3rd IEEE International Conference on Software Engineering and Formal Methods, Koblenz, Germany, IEEE Computer Society Press (to appear) (2005)
12. Rounds, W.C., Song, H.: The phi-calculus: A language for distributed control of reconfigurable embedded systems. In Maler, O., Pnueli, A., eds.: Proceedings of the 6th International Workshop on Hybrid Systems: Computation and Control. Volume 2623 of Lecture Notes in Computer Science., Prague, Czech Republic, Springer Verlag (2003) 435–449
13. Aziz, B., Hamilton, G.: A privacy analysis for the π-calculus: The denotational approach. In: Proceedings of the 2nd Workshop on the Specification, Analysis and Validation for Emerging Technologies. Number 94 in Datalogiske Skrifter, Copenhagen, Denmark, Roskilde University (2002)
14. Chothia, T.: Guessing attacks in the pi-calculus with computational justification. `http://www.lix.polytechnique.fr/ tomc/Papers/pigPaperLong.pdf` (2005)
15. Pierro, A.D., Hankin, C., Wiklicky, H.: Measuring the confinement of probabilistic systems. Theoretical Computer Science **340(1)** (2005) 3–56
16. Milner, R.: A calculus of communicating systems. Lecture Notes in Computer Science **92** (1980)
17. Pierro, A.D., Hankin, C., Wiklicky, H.: Approximate non-interference. In: Proceedings of the 15th IEEE Computer Security Foundations Workshop, Cape Breton, Nova Scotia, Canada, IEEE Press (2002) 3–17
18. Pierro, A.D., Wiklicky, H.: An operational semantics for probabilistic concurrent constraint programming. In: Proceedings of the 1998 International Conference on Computer Languages, Chicago, USA, IEEE Computer Society (1998) 174–183
19. Gilmore, S., Hillston, J.: The pepa workbench: A tool to support a process algebra-based approach to performance modelling. In Thomas, W., ed.: Proceedings of the 7th International Conference on Modelling Techniques and Tools for Computer Performance Evaluation. Volume 794 of Lecture Notes in Computer Science., Vienna, Austria, Springer Verlag (1994) 353–368
20. Bernardo, M., Gorrieri, R.: Extended markovian process algebra. In: Proceedings of the 7th International Conference on Concurrency Theory. Volume 1119 of Lecture Notes in Computer Science., Pisa, Italy, Springer Verlag (1996) 315–330
21. Aziz, B., Hamilton, G.: A denotational semantics for the π-calculus. In: Proceedings of the 5th Irish Workshop in Formal Methods. Electronic Workshops in Computing, Dublin, Ireland, British Computing Society Publishing (2001)
22. Stark, I.: A fully abstract domain model for the π-calculus. In: Proceedings of the 11th Annual IEEE Symposium on Logic in Computer Science, New Brunswick, New Jersey, USA, IEEE Computer Society (1996) 36–42
23. Plotkin, G.: A powerdomain construction. SIAM Journal on Computing **5**(3) (1976) 452–487
24. Abramsky, S.: A domain equation for bisimulation. Information and Computation **92**(2) (1991) 161–218
25. Gillespie, D.T.: Exact stochastic simulation of coupled chemical reactions. Journal of Physical Chemistry **81**(25) (1977) 2340–2361
26. Phillips, A., Cardelli, L.: A correct abstract machine for the stochastic pi-calculus. In: In Proceedings of the 2nd Workshop on Concurrent Models in Molecular Biology, London, U.K. (2004)

Formal Islands

Emilie Balland, Claude Kirchner, and Pierre-Etienne Moreau

UHP & LORIA, INRIA & LORIA, and INRIA & LORIA,
BP 101, 54602 Villers-lès-Nancy Cedex France
{Emilie.Balland, Claude.Kirchner, Pierre-Etienne.Moreau}@loria.fr

Abstract. Motivated by the proliferation and usefulness of Domain Specific Languages as well as the demand for enriching well established languages by high level capabilities like pattern matching or invariant checking, we introduce the *Formal Islands* framework.

The main idea consists to integrate, in existing programs, formally defined parts called islands, on which proofs and tests can be meaningfully developed. Then, *Formal Islands* could be safely dissolved into their hosting language to be transparently integrated in the existing development environment.

The paper presents this generic framework and shows that the properties valid on the Formal Islands are also valid on the corresponding dissolved host codes. Formal Islands can be used as a general methodology to develop new DSLs and we show that language extensions like SQLJ—embedding SQL capabilities in Java—, or Tom—a Java language extension allowing for pattern matching and rewriting—are indeed *islands*.

1 Introduction

At all the levels of our social and scientific organizations, the development of formal proofs of program properties is recognized as a priority of fundamental interest. But this faces at least three important difficulties. First is the lack of formal environments for existing widely used programming languages like Java, C or ML. Second is the scalability to allow for the proof of properties of large programs. Third is the fact that on the enormous corpus of active software, maintenance and adaptation should be conducted without having to rewrite or deeply transform the existing running code. Therefore we are in need of having language extensions, formally defined, adaptable to existing widely used programming languages and that do not induce dependence on a new language.

To contribute to solve these problems given the above constraints, we propose the concept of *Formal Islands* and show how it could be implemented and used. Indeed, taking the geography metaphor as well as a terminology already used for island grammars [11], we call *Ocean* the language of interest, typically C or Java, and *Island* the language extension that we would like to define.

As shown in Figure 1, the island cycle of life is composed of 4 phases:

- *anchor* which relates the grammars and the semantics of the two languages,
- *construction* which inserts some island code in an ocean program,

M. Johnson and V. Vene (Eds.): AMAST 2006, LNCS 4019, pp. 51–65, 2006.

In pictures

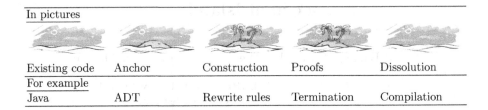

Existing code	Anchor	Construction	Proofs	Dissolution
For example				
Java	ADT	Rewrite rules	Termination	Compilation

Fig. 1. Formal Islands in picture

- *proofs* or program transformations on islands,
- *dissolution* of the islands in the ocean language.

The anchoring step consists in defining the grammar and semantics of the island language and in relating it to the existing ocean one. This step should in particular take care of the data representation correspondence between the island and ocean constructions.

The construction phase consists in writing a program in the combined island and ocean languages. For example, we could consider, as it is already possible in Tom[1], to define functions using matching constructs (of the form `%match pattern -> JavaCode`) or using term rewrite rules (of the form `%rule term -> term`). What is quite appealing at this level is the possibility to mix both language constructions to ease either the expressivity or the references to the existing ocean structures or functionalities.

Then comes the proof phase. It is not necessarily used, but defining formally such a framework enables developers of language extensions to formally check their well-formedness and properties. For example, defining in Tom a set of rewrite rules on top of Java, one could check at that step the termination of the rewrite system, therefore ensuring a better confidence in the program behavior.

Last, the island should be dissolved. This means that the framework should provide a compilation of island built programs (that may embed ocean subparts) into pure ocean ones. For example again, in Tom, a set of rewrite rules will be compiled into a Java program implementing the normalization process for these rules. Of course the framework setting should ensure that the properties proved at the island level are still valid after dissolution into the concerned ocean code.

To achieve these goals, after introducing the basic notations in Section 2, we present in Section 3 the anchoring mechanism, in Section 4 the dissolution one and in Section 5 the *island framework*, making precise the properties that the island should fulfill to be *formal* and to preserve proofs. From these definitions and results, we illustrate in Section 6 how domain specific languages [14] can be implemented in this Formal Islands framework. Finally, we present a main application in the context of the Tom project.

[1] `http://tom.loria.fr`

2 Preliminaries

When considering the problem of combining two different languages, we have to understand the relationship that exists between the grammars of these languages, the programs that can be written in these grammars, their semantics, and the objects that are manipulated by these programs.

We assume the reader to be familiar with the basic definitions of languages constructions and first order term rewriting as given for example in [2]. We briefly recall or introduce notations for the main concepts that will be used along this paper.

A *grammar* is a tuple $\mathcal{G} = (A, N, T, R)$ where A denotes the axiom, N and T, disjoint finite sets of respectively, non-terminal and terminal symbols, and R a finite set of production rules of the form $N \rightarrow (N \cup T)^*$. $\texttt{left}(R)$ is the set of left-hand sides of R. We note $\mathcal{L}(\mathcal{G})$ the *language* recognized by the grammar G. When a grammar G is not ambiguous, to each valid program we can associate *abstract syntax tree* representations. Assuming given such a representation, we note $\mathsf{AST}(\mathcal{G})$ the set of all abstract syntax trees, and their subtrees.

In the following, we only consider unambiguous grammars. Therefore, we make no distinction between the notions of grammars and valid programs p, and the notions of signature and abstract syntax trees. We note p_{ast} the abstract syntax tree that represents p. Given a term $t = p_{\mathsf{ast}}$, $t \in \mathsf{AST}(\mathcal{G})$, we note $\texttt{getSort}(t)$ its sort, which corresponds to the non-terminal generating p.

In addition to the definition of grammars, we use a big-step reduction relation *à la* Kahn, written $\mapsto_{bs}$, to characterize the semantics of the ocean and the islands languages. Given a set $\mathcal{O}$ of objects manipulated by a program, corresponding to all possible instances of the data-model, an *environment* is a function from $\mathcal{X}$ to $\mathcal{O}$, where $\mathcal{X}$ is a set of variables. $\mathcal{E}nv$ denotes the set of all environments. The reduction relation $\mapsto_{bs}$ is defined using a set of inference rules of the form: $\langle \epsilon, i \rangle \mapsto_{bs} \epsilon'$ with $i \in \mathsf{AST}(\mathcal{G})$, and $\epsilon, \epsilon' \in \mathcal{E}nv$

In the following, we consider two languages il and ol, the island and the ocean languages, to which correspond a grammar $\mathcal{G}_{\mathsf{il}}$ (resp. $\mathcal{G}_{\mathsf{ol}}$), a set of variables $\mathcal{X}_{\mathsf{il}}$ (resp. $\mathcal{X}_{\mathsf{ol}}$), a semantics bs_{il} (resp. bs_{ol}) based on a set of objects $\mathcal{O}_{\mathsf{il}}$ (resp. $\mathcal{O}_{\mathsf{ol}}$), and a set of inference rules R_{il} (resp. R_{ol}).

3 Anchor

Given two languages il and ol, we introduce the notions of *syntactic anchor* and *representation function*, which make a connection between il and ol in syntactic and semantic ways.

3.1 Syntax

The *syntactic anchor* consists in associating ol non-terminals to il non-terminals, to obtain ol programs with il parts. In Definition 1, we introduce two types of anchors corresponding to two types of islands. One called *simple island*, corresponds to pure il constructs and the other called *islands with lakes*, corresponds to islands which can recursively contain ol constructs.

Definition 1. *Given two grammars* $\mathcal{G}_{ol} = (A_{ol}, N_{ol}, T_{ol}, R_{ol})$ *and* $\mathcal{G}_{il} = (A_{il}, N_{il}, T_{il}, R_{il})$*, we define two kinds of syntactic anchors:*

- *A* simple syntactic anchor *is a function* $\mathit{anch}(\mathcal{G}_{ol}, \mathcal{G}_{il}) \in N_{ol} \rightarrow N_{il}$ *where we assume that* $(T_{ol} \cap T_{il}) = \emptyset \wedge (N_{ol} \cap N_{il}) = \emptyset$*,*
- *A* syntactic anchor with lakes *is a function* $\mathit{anch}(\mathcal{G}_{ol}, \mathcal{G}_{il}) \in N_{ol} \rightarrow N_{il}$ *where we assume that* $(T_{ol} \cap T_{il}) = \emptyset \wedge (N_{ol} \cap \mathtt{left}(R_{il})) = \emptyset$*.*

From this definition, the grammar $\mathcal{G}_{oil}$ *resulting from the combination of* ol *and* il*, is defined by:* $\mathcal{G}_{oil} = (A_{ol}, N_{ol} \cup N_{il}, T_{ol} \cup T_{il}, R_{ol} \cup R_{il} \cup \mathit{anch}(\mathcal{G}_{ol}, \mathcal{G}_{il}))$*.*

Therefore, the syntax of the language oil, combination of ol and il is function of the grammars $\mathcal{G}_{ol}$ and $\mathcal{G}_{il}$, and of the syntactic anchor noted anch.

Example 1. As a first example, let us consider the two grammars, $\mathcal{G}_{ol} = (\{A\}, \{A\}, \{a\}, \{(A ::= a), (A ::= Aa)\})$ and $\mathcal{G}_{il} = (\{B\}, \{B\}, \{b\}, \{(B ::= b)\})$. The language $\mathcal{L}(\mathcal{G}_{ol})$ is the set of sequences a, aa, aaa, ... The language $\mathcal{L}(\mathcal{G}_{il})$ contains only b. By considering the *simple syntactic anchor* $\mathtt{anch}(\mathcal{G}_{ol}, \mathcal{G}_{il}) = \{(A ::= B)\}$ we define the language $\mathcal{L}(\mathcal{G}_{oil})$ which consists of words like a, b, aa, ba and more generally of any sequence of a or b ended by a.

For *simple syntactic anchors*, the condition $T_{ol} \cap T_{il} = \emptyset \wedge N_{ol} \cap N_{il} = \emptyset$ ensures that there is no conflict between the two grammars. But in some cases, it is interesting to allow the embedding of ocean constructs inside island code. We call *lakes* such constructs that are not modified by the dissolution phase. In term of syntactic anchor, this means that the il grammar can use non-terminals from $\mathcal{G}_{ol}$. For this notion of *syntactic anchor with lakes*, the non-conflict condition becomes $T_{ol} \cap T_{il} = \emptyset \wedge N_{ol} \cap \mathtt{left}(R_{il}) = \emptyset$.

Example 2. To illustrate the notion of *anchor with lakes*, we now consider an *ocean language* ol which allows to manipulate arrays of integers. The considered island language il allows to manipulate lists of integers, where the notion of integers comes from the ocean language: this is why it is considered as a lake. The grammars of both languages are given in Figure 2.

In ol, an array can be allocated and filled with 0 using the construction $array(n)$. Given an array t and an integer n, $t[n]$ allows to read the contents of t. Similarly, $t[n] = i$, with $i \in \mathbb{N}$, allows to modify the contents of t.

In il, data structures are lists, which are classically defined by two constructors *nil* and *cons*. This language defines islands with lakes since the non-terminal $\langle int \rangle$ comes from $\mathcal{G}_{ol}$. To interconnect the two languages, we define the anchor $\mathtt{anch} = \{(\langle instr \rangle ::= \langle instruction \rangle), (\langle array \rangle ::= \langle list \rangle), (\langle int \rangle ::= \langle expr \rangle)\}$.

Using the grammar defined in Figure 2, the following program is valid in the ocean language: `t=array(5); t[0]=3; t[1]=7`. This program can be extended by `l←cons(t[1],cons(t[2],nil)); x=l[1]; y=head(l)`. This shows that a list of the island language can be considered as an array by the ocean language (`l[1]`). The integer `t[1]` and `t[2]` are lakes in the island `l←cons(t[1],cons(t[2],nil))`.

The ocean language

$$
\begin{array}{lcl}
\langle instr \rangle & ::= & \langle instr \rangle ; \langle instr \rangle \\
 & | & \langle vararray \rangle = \langle array \rangle \\
 & | & \langle varint \rangle = \langle int \rangle \\
 & | & \langle array \rangle [\langle int \rangle] = \langle int \rangle \\
\langle array \rangle & ::= & array(\langle int \rangle) \\
 & | & \langle vararray \rangle \\
\langle vararray \rangle & ::= & x \in \mathcal{X} \\
\langle varint \rangle & ::= & x \in \mathcal{X} \\
\langle int \rangle & ::= & i \in \mathbb{N} \\
 & | & \langle varint \rangle \\
 & | & size(\langle vararray \rangle) \\
 & | & \langle array \rangle [\langle int \rangle]
\end{array}
$$

The island language

$$
\begin{array}{lcl}
\langle instruction \rangle & ::= & \langle varlist \rangle \leftarrow \langle list \rangle \\
\langle list \rangle & ::= & nil \\
 & | & cons(\langle expr \rangle, \langle list \rangle) \\
 & | & tail(\langle list \rangle) \\
 & | & \langle varlist \rangle \\
\langle varlist \rangle & ::= & x \in \mathcal{X} \\
\langle expr \rangle & ::= & \langle int \rangle \\
 & | & head(\langle list \rangle)
\end{array}
$$

The syntactic anchor relation

$$
\begin{array}{lcl}
\langle instr \rangle & ::= & \langle instruction \rangle \\
\langle array \rangle & ::= & \langle list \rangle \\
\langle int \rangle & ::= & \langle expr \rangle
\end{array}
$$

Fig. 2. Syntax of the combination of the tool languages

3.2 Semantics

As for the syntax, we assume given a semantics definition for each language. In the most general case, the objects manipulated by these two languages are not of the same nature. For example, the ocean language can manipulate tuples and the island language, algebraic terms. Before giving a semantics to the extended language, we have to make precise the data-structure relation between island and ocean objects (the *representation and abstraction functions*) and how the data-structure properties in il are mapped to data-structure properties in ol (the *predicate mapping*). The notion of *representation and abstraction functions* is originally from *data refinement theory* [6, 1], used to convert an abstract data model (such as lists) into implementable data structures (such as arrays). In our framework, islands are considered as abstract comparing to ocean.

Definition 2. *Given a set of island objects $\mathcal{O}_{il}$ and a set of ocean objects $\mathcal{O}_{ol}$, a* representation function $\lceil\ \rceil$ *is an injective total function from $\mathcal{O}_{il}$ to $\mathcal{O}_{ol}$ and an* abstraction function $\lfloor\ \rfloor$ *is a surjective function (potentially partial) from $\mathcal{O}_{ol}$ to $\mathcal{O}_{il}$ such that $\lfloor\ \rfloor.\lceil\ \rceil = Id_{\mathcal{O}_{il}}$.*

Example 3. Suppose that we manipulate sets in the island and these sets are represented by lists in the ocean. $\lceil\ \rceil$ can associate to every set the list containing the same elements in a determined order and $\lfloor\ \rfloor$ can associate to every list from ocean the set of its elements, which is not the inverse of $\lceil\ \rceil$.

Example 4 (from example 2). Every list from the island language can be represented by an array in the ocean language which contains exactly the same integers in the same order. We note this representation function map_1. The function map_1^{-1} can be simply chosen as an abstraction. The second kind of objects manipulated by the ocean language is the integers whose representation is the same in the two languages.

In Definition 2, representation and abstraction functions have been introduced to establish a correspondence between data structures in the island and in the ocean language. However, we did not put any constraint on the representation of objects. In particular, the function $\lceil\ \rceil$ does not necessarily preserve structural properties of island objects. In practice, we need to consider mappings such that properties are preserved. Therefore, for each language we consider a *set of predicates* noted $\mathcal{P}_{\mathsf{ol}}$ and $\mathcal{P}_{\mathsf{il}}$ corresponding to structural properties, and we introduce the notion of *predicate mapping.*

Definition 3. *Given a set of island predicates $\mathcal{P}_{il}$ and a set of ocean predicates $\mathcal{P}_{ol}$, a* predicate mapping *ϕ is an injective mapping from $\mathcal{P}_{il}$ to $\mathcal{P}_{ol}$ such that $\forall p \in \mathcal{P}_{il}, arity(p) = arity(\phi(p))$. This mapping is extended by morphism on first order formulae, using the representation mapping:*

$$\forall p \in \mathcal{P}_{il}, \forall t_1, \ldots, t_n, \phi(p(t_1, \ldots, t_n)) = \phi(p)(t'_1, \ldots, t'_n)$$
where $t'_i = \lceil t_i \rceil$ if $t_i \in \mathcal{O}_{il}$ and t_i otherwise (i.e. when t_i is a variable),

$$\begin{array}{l|l} \phi(\forall x\, P) = \forall x\, \phi(P), & \phi(\exists x\, P) = \exists x\, \phi(P), \\ \phi(P_1 \vee P_2) = \phi(P_1) \vee \phi(P_2), & \phi(P_1 \wedge P_2) = \phi(P_1) \wedge \phi(P_2), \\ \phi(\neg P) = \neg\phi(P), & \phi(P_1 \rightarrow P_2) = \phi(P_1) \rightarrow \phi(P_2). \end{array}$$

Definition 4. *Given a predicate mapping ϕ, a representation function $\lceil\ \rceil$ is said ϕ-*formal *if $\forall p \in \mathcal{P}_{il}, \forall o_1, \ldots, o_n \in \mathcal{O}_{il}$ with $n = arity(p)$*

$$p(o_1, \ldots, o_n) \Leftrightarrow \phi(p)(\lceil o_1 \rceil, \ldots, \lceil o_n \rceil)$$

Example 5. Consider the relations of equality $=_{\mathsf{il}}$ and $=_{\mathsf{ol}}$ as an example of predicates respectively defined on lists and arrays, and the predicate mapping $\phi_1 = \{(=_{\mathsf{il}}, =_{\mathsf{ol}})\}$. The representation function map_1, introduced in example 4, is ϕ_1*-formal* because two lists are equal with $=_{\mathsf{il}}$ (composed by the same integers) if and only if their representations are equal with $=_{\mathsf{ol}}$. As a counterexample we consider the representation function map_2 that associates a list to an array, but whose elements are in reverse order.

- $eq_{head}(l, l') \equiv (head(l) = head(l'))$,
- $eq_{elt}(t, t') \equiv (t[0] = t'[0])$

When considering the predicate mapping $\phi_2 = \{(eq_{head}, eq_{elt})\}$, the representation function map_2 is not ϕ_2*-formal* because we can construct two lists $l_1 = (1, 2), l_2 = (1, 3)$ such that $eq_{head}(l_1, l_2)$ is true but $eq_{elt}(map_2(l_1), map_2(l_2))$, which is equal to $eq_{elt}([2, 1], [3, 1])$ is false.

Given a representation function $\lceil\ \rceil$ and an abstraction function $\lfloor\ \rfloor$, we can simulate the behavior of il programs in the ol environment. Suppose we have a big-step semantics for each language (with their respective reduction relation bs_{ol} and bs_{il} in their respective set of environments $\mathcal{E}nv_{\mathsf{ol}}$ and $\mathcal{E}nv_{\mathsf{il}}$).

To define the evaluation of il programs in an ol environment $\epsilon_{\mathsf{ol}} \in \mathcal{E}nv_{\mathsf{ol}}$, we need to translate ϵ_{ol} in an il environment $\epsilon_{\mathsf{il}} \in \mathcal{E}nv_{\mathsf{il}}$. Therefore, we extend the representation function to environments.

Definition 5. *The extension of the representation (resp. abstraction) function to environments also noted* $\lceil\ \rceil \in \mathcal{E}nv_{il} \to \mathcal{E}nv_{ol}$ *(resp.* $\lfloor\ \rfloor \in \mathcal{E}nv_{ol} \to \mathcal{E}nv_{il}$*) is such that:*

$$\forall \epsilon_{il} \in \mathcal{E}nv_{il}, \forall x \in \mathcal{X}_{il}, \forall v \in \mathcal{O}_{il},\ \langle x, v\rangle \in \epsilon_{il} \Leftrightarrow \langle x, \lceil v\rceil\rangle \in \lceil \epsilon_{il}\rceil$$

Even if $\lceil\ \rceil$ is total and injective, $\lfloor\ \rfloor \in \mathcal{E}nv_{\mathsf{ol}} \to \mathcal{E}nv_{\mathsf{il}}$ can be partial. To obtain a total function, we extend it with the empty environment for ol environments that are not in its domain.

We simulate the reduction of il programs in an ol environment with the reduction relation of il by translating the ol environment with $\lfloor\ \rfloor$. The semantics rules of ol, extended with mapped il rules, give a semantics to the extended language.

Definition 6. *Given two semantics* bs_{il} *and* bs_{ol} *respectively defined by sets of inference rules* $\mathcal{R}_{il}$ *and* $\mathcal{R}_{ol}$*, we define the semantics of oil as:*

- *the reduction relation* $bs_{oil} = bs_{ol}$*,*
- *the set of inference rules* $\mathcal{R}_{oil} = \mathcal{R}_{ol} \cup \mathcal{R}'_{il} \cup \{r_1, r_2\}$ *where*
  - $\mathcal{R}'_{il} = \mathcal{R}_{il}$ *where* $\langle \epsilon, i\rangle \mapsto_{bs_{il}} \epsilon'$ *is replaced by* $\langle \epsilon, \delta, i\rangle \mapsto_{bs_{il}} \langle \epsilon', \delta\rangle$ $(\delta \in \mathcal{E}nv_{ol}$ *and* $\epsilon, \epsilon' \in \mathcal{E}nv_{il})$*,*
  - *the inference rules* r_1 *and* r_2*:*

$$\frac{\langle \lfloor\epsilon\rfloor, \gamma(\epsilon), i\rangle \mapsto_{bs_{il}} \langle \epsilon', \delta\rangle}{\langle \epsilon, i\rangle \mapsto_{bs_{ol}} \lceil\epsilon'\rceil \cup \delta}\ r_1 \qquad \frac{\langle \lceil\epsilon\rceil \cup \delta, i\rangle \mapsto_{bs_{ol}} \epsilon'}{\langle \epsilon, \delta, i\rangle \mapsto_{bs_{il}} \langle \lfloor\epsilon'\rfloor, \gamma(\epsilon')\rangle}\ r_2$$

  where $\gamma(x) = x - \lceil\lfloor x\rfloor\rceil$ *denotes the elements of ocean environment* x *that do not represent an island object.*

The function $\lfloor\ \rfloor$ gives the corresponding il environment restricted to ol objects which are island object's representations, then the il construction is evaluated in il semantics, we obtain a new environment that can be mapped by $\lceil\ \rceil$ to give the target environment in ol semantics. The objects in ϵ that cannot be represented in the island $(\epsilon' - \lceil\lfloor\epsilon'\rfloor\rceil)$ are given as parameter to island evaluation in case of lakes. This is why ϵ' corresponds to the union of part of ϵ not represented in the island but that can be modified by lakes (which corresponds to δ) and the representation in the ocean of the evaluation of il instructions (which corresponds to $\lceil\epsilon'\rceil$).

The introduction of δ in the rules of the il semantics is required for the reduction of lakes. Indeed, we need to keep track of ol environment when we evaluate an island. Otherwise the lakes would be kept separated from ocean constructs, and no reference to ocean variables would be possible. The inference rule r_1 and r_2 link the two semantics: r_1 is the bridge from ol to il semantics for island evaluation and r_2 is the bridge from il to ol semantics for the lakes evaluation.

We notice that in r_1, $\lceil\epsilon'\rceil \cup \delta$ is a function only if $\mathtt{dom}(\lceil\epsilon'\rceil) \cap \mathtt{dom}(\delta) = \emptyset$. This condition means that a variable cannot represent both an island and ocean objects in the same environment. The rule r_2 introduces a similar condition.

From now on, we consider semantics that verify these two conditions. In practice, it means that islands and lakes have to introduce fresh variables with respect to both ol and il environments.

Finally, the semantics of the language oil, combination of ol and il, is function of big-step semantics bs_{ol} and bs_{il}, the representation function $\lceil\ \rceil$ and the abstraction function $\lfloor\ \rfloor$.

We can now use the island formalism to extend the ol language by new constructs. In the following, we will see how to implement this idea in practice: for this new language, instead of building a new compiler from scratch, we consider a *dissolution* phase which replaces islands constructs by ocean constructs. With such an approach, an existing ocean compiler could be reused.

4 Dissolution

At the syntax level, the dissolution step consists of replacing all il constructs that appear in the ol AST by ol constructs, in order to obtain a complete ol AST.

Definition 7. *Given two grammars $\mathcal{G}_{il}$ and $\mathcal{G}_{ol}$, we call* dissolution *a function* $\mathtt{diss} : \mathsf{AST}(\mathcal{G}_{il}) \rightarrow \mathsf{AST}(\mathcal{G}_{ol})$.

Such a function is said lake preserving *when* $\forall i \in \mathsf{AST}(\mathcal{G}_{il}), \forall l \in \mathtt{lakes}(\mathtt{diss}(i))$, *we have* $l \in \mathtt{lakes}(i)$, *where* $\mathtt{lakes}$ *is a function that gives the set of lakes contained in an AST (i.e an il construct).*

The condition of *lake-preserving* authorizes dead-code elimination but ensures that the remaining lakes have not been modified. In practice, it is verified by constructing with the same strategy (for example top-down) a list of lakes in the source and target program and the condition consists simply to compare the two lists. Finding lakes in a dissolved program can be realized by marking generated code during dissolution in order to distinguish lakes from generated code in the target program.

Example 6. Considering again the previously introduced program:
`t=array(5); t[0]=3; t[1]=7;`
`l←cons(t[1],cons(t[2],nil)); x=l[1]; y=head(l),`
we can distinguish three ol islands:
`l←cons(t[1],cons(t[2],nil))`, `l` (from `l[1]`), and `head(l)`.

The dissolution of these islands could (depending on the implementation) result in the following program:
`t=array(5); t[0]=3; t[1]=7; x=t[0];`
`l=array(2); l[0]=t[1]; l[1]=t[2]; x=l[1]; y=l[0];`.

In term of semantics, the ol constructs that are generated must have the same evaluation as the il constructs that they replace.

Definition 8. *Given representation and abstraction functions* $\lceil\ \rceil, \lfloor\ \rfloor$, *a* dissolution *function* $\mathtt{diss}$ *is* well-formed *if:*

- *for every* $i \in \mathsf{AST}(\mathcal{G}_{il})$, *for every environment* $\epsilon \in \mathcal{E}nv_{il}$, *we have:* $\langle \epsilon, i \rangle \mapsto_{bs_{oil}} \epsilon' \Leftrightarrow \langle \lceil \epsilon \rceil, \boldsymbol{diss}(i) \rangle \mapsto_{bs_{ol}} \epsilon'$,
- *the* ol *program resulting from dissolution is syntactically correct. More formally,* $\forall i \in \mathsf{AST}(\mathcal{G}_{il})$, $\mathtt{getSort}(i) \in \boldsymbol{anch}(\mathtt{getSort}(\boldsymbol{diss}(i)))$,
- *the dissolution function is lake preserving.*

Figure 3 shows the link between the evaluation of an il instruction with il semantics and the execution of the corresponding ol instruction (by dissolution). The states after evaluation are the same.

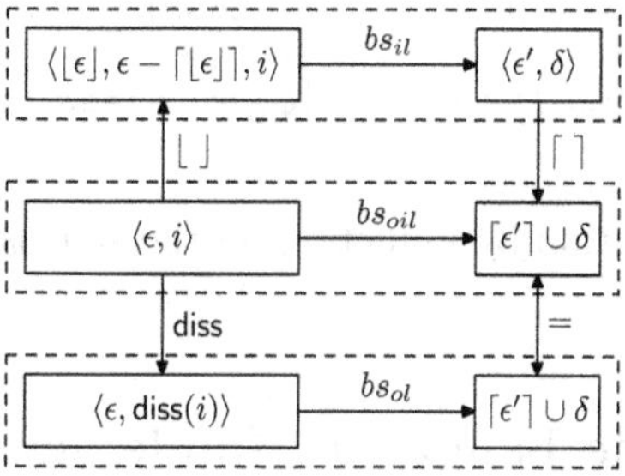

Fig. 3. Reduction of an il dissolution in ol semantics

To summarize, an island language should fulfill the following requirements:

Definition 9. *Given two languages* ol *and* il *described by a grammar, a big-step semantics, a set of objects, and a set of predicates in these objects,* il is an island language for ol *if there exist:*

- *a syntactic anchor* $\boldsymbol{anch}$, *either simple or with lakes (Definition 1),*
- *representation and abstraction functions* $\lceil\ \rceil, \lfloor\ \rfloor$ *(Definition 2),*
- *a dissolution function* $\boldsymbol{diss}$ *(Definition 7).*

5 Formal Islands

Definition 10. *Given two languages* ol *and* il, il *is a* Formal Island *over* ol *if:*

1. il *is an island for* ol *(Definition 9),*
2. *there exists a predicate mapping* ϕ *for objects (Definition 3) such that the representation function* $\lceil\ \rceil$ *is* ϕ*-formal (Definition 4),*
3. *the dissolution function* $\boldsymbol{diss}$ *is well-formed (Definition 8).*

Condition 1 is purely syntactic and simple to verify. Condition 3 is similar to the correctness of a compilation process and condition 2 is more specific to the island formalism. This definition of Formal Islands allows us to ensure the preservation of properties.

The environment is extended by morphism on first order formulae:

$$\forall p \in \mathcal{P}_{\mathsf{il}}, \forall t_1, \ldots, t_n, \epsilon(p(t_1, \ldots, t_n)) = \epsilon(p)(t'_1, \ldots, t'_n)$$
where $t'_i = \epsilon(t_i)$ if $t_i \in \mathcal{X}_{\mathsf{il}}$ and t_i otherwise (*i.e.* t_i is an object),

$$\epsilon(\forall x\ P) = \forall x\ \epsilon(P), \qquad \epsilon(\exists x\ P) = \exists x\ \epsilon(P),$$
$$\epsilon(P_1 \vee P_2) = \epsilon(P_1) \vee \epsilon(P_2), \qquad \epsilon(P_1 \wedge P_2) = \epsilon(P_1) \wedge \epsilon(P_2),$$
$$\epsilon(\neg P) = \neg\epsilon(P), \qquad \epsilon(P_1 \rightarrow P_2) = \epsilon(P_1) \rightarrow \epsilon(P_2).$$

From this definition, we note $\epsilon \models pre \Leftrightarrow \epsilon(pre)$.

Proposition 1. *Given a Formal Island* il *over* ol *and* $pre, post$ *two first order formulae built over* $\mathcal{P}_{\mathsf{il}}$ *predicates,* $\forall i \in \mathbf{dom}(\mathbf{diss}), \epsilon \in \mathcal{E}nv_{\mathsf{il}}$*, we have:*

$$\epsilon \models \{pre\}i\{post\} \Leftrightarrow \lceil\epsilon\rceil \models \{\phi(pre)\}\mathbf{diss}(i)\{\phi(post)\}$$

The proof of this proposition by induction on the structure of the formulae *pre* and *post* is given in [3].

6 Domain Specific Languages Implemented by Formal Island

A *Domain Specific Language* (DSL) is a programming language designed for a very specific task or domain, contrary to general programming languages like Java. A few papers give an overview on the Domain Specific Languages implementation methodology [13, 10, 14]. Summarizing the main ideas, this can be achieved by language specialization (removing features of an existing language), language extension (adding new features to an existing language), language invention (designed from scratch with no commonality with existing languages) or piggyback (using partially an existing language). Some works on modular and extensible semantics [7] are well-tailored for DSL specifications.

In [13], Spinellis proposes eight recurring patterns to classify DSL design and implementation. One of this pattern is the *piggyback pattern* which corresponds informally to the design of island languages. The piggyback structural pattern uses the capabilities of an existing language to be a hosting base for a new DSL. Thus, the DSL shares common elements with an existing language and is compiled in the host language.

In the classification of [10], the patterns correspond to different phases of DSL development: decision, analysis, design and implementation. Formal Island gets involved in two related phases of DSL development: the design and implementation phases. Formal Islands correspond in terms of design patterns to the *Language exploitation* (i.e. based on an existing language) and in terms of implementation pattern to the *Embedding* and *Preprocessor* patterns.

MetaBorg [4] proposes a general method to provide concrete syntax for domain abstractions to application programmers and thus promotes APIs to the language level (for example SQLJ would have been easily implemented in Metaborg).

This method is based on SDF for the syntax and Stratego language for the assimilation. The concept is very similar to Formal Islands. However, the data-structures manipulated by the domain-specific-language and the host language are identical. *MetaBorg* can therefore be seen as an implementation technique for purely syntactic Formal Islands.

To illustrate the link between DSL and Formal Islands, we will detail the language SQLJ [5, 9], an example of DSL implemented using the piggyback pattern. SQLJ is an interface to the JDBC domain-specific-library hiding the complexity of the API.

To avoid grammar conflicts and make identification of constructs easier, each SQLJ constructs starts with `#sql` token (which is not a legal Java identifier). The simplest SQLJ executable clauses consist of the token `#sql` followed by a SQL statement enclosed in *curly braces*. For example, the following SQLJ clause may appear instead of a Java instruction.

```
public void honors(float limit) {
  #sql{
    SELECT STUDENT AS "name", SCORE AS "grade"
    FROM   GRADE_REPORTS
    WHERE  SCORE >= :limit
  }
}
```

The SQL statements can contain variable names that correspond to Java variables (the variable `limit` for example). These variables are prefixed by a colon and they correspond to the notion of *lake* introduced previously.

Syntactic anchor. If we have a grammar of Java where $\langle Statement \rangle$ and $\langle Instruction \rangle$ are Java non-terminals, we can define the following simple syntactic anchor **anch** $= \{(\langle Statement \rangle ::= \langle Declaration \rangle), (\langle Instruction \rangle ::= \langle ExecutableStatement \rangle)\}$. The non-terminal $\langle Declaration \rangle$ corresponds to SQLJ declarations used to initialize a JDBC connection. As illustrated above, $\langle ExecutableStatement \rangle$ corresponds to embedded SQL queries.

Data-structure anchor. In SQLJ, the representation and abstraction functions between SQL objects and Java objects are given by conversions from SQL types to Java types. For example, the SQL `CHAR` type is converted into a Java `String`. Therefore, the results of a SQL query have to be translated into Java objects before being stored in Java variables.

Dissolution. In the SQLJ formalism, the SQL language is not really the embedded language because this is not the SQL requests which are dissolved in Java, but rather the SQLJ instructions which contain SQL requests. The SQLJ preprocessor provides type-checking and schema-object-checking to detect syntax errors and missing or misspelled object errors in SQL statements at translation time rather than at runtime (like in JDBC). Programs written in SQLJ are, therefore, more robust than JDBC programs. We just give the intuition of the translation step by giving the dissolution in Java of the program given previously:

```
public void honors(float limit) {
  java.sql.PreparedStatement ps = recs.prepareStatement(
      "SELECT STUDENT, SCORE "
    + "FROM   GRADE_REPORTS "
    + "WHERE  SCORE >= ? ");
  ps.setFloat(1, limit);
  ps.executeQuery();
}
```

The object `recs` is a JDBC connection of type `java.sql.Connection`. The SQLJ translator verifies that in the SQLJ statement, `limit` is of type `float` in order to be compared with `SCORE`, whose SQL type is `REAL`.

In the case of SQLJ, there is no formal property given for the mapping between types or for the compilation of the SQLJ instructions. We cannot ensure that the Java compiled code is consistent. Our framework is a base to formalize DSL implemented using the piggyback pattern. It gives conditions to ensure that properties established at the DSL level are preserved by compilation.

7 Tom: A Formal Island for Pattern-Matching

An other example of island language is Tom, which adds pattern matching facilities to imperative languages such as C and Java. Indeed, it is in this context that we identified the need to have a notion of *Formal Island framework*. This helps us to understand how properties of Tom can be preserved by compilation.

As presented in [12], a Tom program is written on top of a host language and extended by several new constructs. It is out of the scope of this paper to present the language in detail and it is sufficient here to consider that Tom provides three main constructs:

- `%op` allows to define an algebraic signature (i.e. names of constructors with their profile),
- `%match` corresponds to an extension of `switch/case`, well known in functional programming languages,
- ` allows to build an algebraic term from the host language.

Therefore, a program can be seen as a list of Tom constructs (the islands) interleaved with some sequences of characters (the ocean). During the compilation process, all Tom constructs are dissolved and replaced by instructions of the host language, as it is usually done by a preprocessor. From this point, we consider that the ocean language is Java and we call JTom this specialized version of Tom.

The following example shows how a simple symbolic computation (addition) over Peano integers can be defined. This supposes the existence of a data-structure and a mapping (defined using `%op`) where Peano integers are represented by *zero* and *successor*; for instance the integer 3 is denoted *suc(suc(suc(zero)))*.

```
public class PeanoExample {
  %op Term zero() { ... }
  %op Term suc(Term) { ... }
  ...
  Term plus(Term t1, Term t2) {
    %match(t1, t2) {
      x,zero   -> { return `x; }
      x,suc(y) -> { return `suc(plus(x,y)); }
    }
  }
  void run() {
   System.out.println("plus(1,2) = " + plus(`suc(zero),`suc(suc(zero))));
  }
}
```

In this example, given two terms t_1 and t_2 (that represent Peano integers), the evaluation of `plus` returns the sum of t_1 and t_2. This is implemented by pattern matching: t_1 is matched by x, t_2 is possibly matched by the two patterns *zero* and $suc(y)$. When *zero* matches t_2, the result of the addition is x (with $x = t_1$, as instantiated by matching). When $suc(y)$ matches t_2, this means that t_2 is rooted by a *suc* symbol: the subterm y is added to x and the successor of this number is returned, using the ` construct. The definition of `plus` is given in a functional programming style, but the `plus` function can be used in Java to perform computations. This example illustrates how the `%match` construct can be used in conjunction with the considered native language. We can notice that JTom programs contain lakes (the right part of a rule is a Java statement). Note also that lakes can contains islands, introduced by ` for example.

From the definition of Formal Islands (Definition 10), we define for JTom the syntactic anchor, the representation function, the predicate mapping, and gives the intuition of the dissolution function which corresponds to the Tom compiler task.

7.1 Syntactic Anchor

In the case of JTom, the syntactic anchor `anch` is defined as follow:

$$\texttt{anch} = \left\{ \begin{array}{c} (\langle Statement\rangle ::= \langle OpConstruct\rangle), \\ (\langle Instruction\rangle ::= \langle MatchConstruct\rangle), \\ (\langle Expression\rangle ::= \langle BackQuoteConstruct\rangle) \end{array} \right\}$$

7.2 Data-Representation Anchor

In JTom, the notion of *term* can be implemented by any data-structure. Once given such an implementation, the data-representation anchor can be defined. Let us consider that terms are implemented using a record `(sym:integer, sub:array of term)`, where the first slot (`sym`) denotes the top symbol, and the second slot (`sub`) corresponds to the subterms. It is easy to check that the following definition of the predicate mapping provides a *formal anchor*:

$$\mathtt{eq}(\mathtt{t}_1,\mathtt{t}_2) \triangleq \lceil \mathtt{t}_1 \rceil.\mathtt{sym} = \lceil \mathtt{t}_2 \rceil.\mathtt{sym} \wedge \forall i \in [1..\mathrm{ar}(\lceil \mathtt{t}_1 \rceil.\mathtt{sym})], \mathtt{eq}(\mathtt{t}_1.\mathtt{sub}[i], \mathtt{t}_2.\mathtt{sub}[i])$$

$$\mathtt{is\_fsym}(\mathtt{t},\mathtt{f}) \triangleq \lceil \mathtt{t} \rceil.\mathtt{sym} = \lceil \mathtt{f} \rceil$$

The first definition says that two terms are equal if the representations of their root symbol are equal and the subterms are respectively equal. The second definition says that a term t is rooted by f if the representation of t (which is a record) has the representation of f as first element.

7.3 Dissolution

Due to lack of space, we cannot give in detail the complete definition of the dissolution function which corresponds to the compilation phase. Therefore, we just give the intuition of the translation step by illustrating the dissolution of the `PeanoExample` program given previously.

```
public Term plus(Term t1, Term t2) {
  if (is_fsym_zero(t2)) {
    return t1;
  } else if (is_fsym_suc(t2)) {
    return make_suc(plus(t1,subterm_suc(t2,1)));
  }
}
```

With these definitions, Tom is an island for Java. We have proved that the anchor is formal. The last condition to obtain a Formal Island is the proof that the dissolution is well-formed. As shown in [8], a first step in this direction is the development of a certifying compiler which proves, for each compilation, that the dissolution preserves the semantics of the pattern matching.

8 Conclusion and Future Work

We have defined the notion of Formal Island to provide a formal framework allowing language designers to base their languages extensions. For this framework to back-up properties proofs, e.g. about safety or security, we have shown that under sufficient conditions, properties established at the island level are preserved once dissolved into the host language. We have then shown application of this framework to DSL like SQLJ and to Tom.

Amongst the many applications that we envision, the safe treatment of XML transformations, via appropriate Java based islands, is particularly promising and is currently under development. Of course such a framework should be closely linked to proving tools adapted to the properties to be checked: another direction that we are also investigating.

Acknowledgments. We sincerely thank the anonymous referees for their valuable and detailed comments that led to a substantial improvement of the paper.

References

[1] T. Alves, P. Silva, J. Visser, and J. N. Oliveira. Strategic Term Rewriting And Its Application To A VDM-SL to SQL Conversion. In *Formal Methods 2005*, volume 3582 of *LNCS*, pages 399–414, 2005.

[2] F. Baader and T. Nipkow. *Term Rewriting and all That*. Cambridge University Press, 1998.

[3] E. Balland, C. Kirchner, and P.-E. Moreau. Formal Islands. Technical report, LORIA, 2006. available at http://hal.inria.fr/inria-00001146.

[4] M. Bravenboer, R. de Groot, and E. Visser. Metaborg in action: Examples of domain-specific language embedding and assimilation using Stratego/XT. In *Participants Proceedings of the Summer School on Generative and Transformational Techniques in Software Engineering (GTTSE'05)*, Braga, Portugal, July 2005.

[5] G. Clossman, P. Shaw, M. Hapner, J. Klein, R. Pledereder, and B. Becker. Java and relational databases (tutorial): SQLJ. In *Proceedings of the ACM SIGMOD International Conference on Management of Data*, page 500. ACM, 1998.

[6] W.-P. de Roever and K. Engelhardt. *Data Refinement: Model-Oriented Proof Methods and their Comparison.* Number 47 in Cambridge Tracts in Theoretical Computer Science. Cambridge University Press, Cambridge, UK, November 1998.

[7] K.-G. Doh and P. D. Mosses. Composing programming languages by combining action-semantics modules. *Science of Computer Programming*, 47(1):3–36, 2003.

[8] C. Kirchner, P.-E. Moreau, and A. Reilles. Formal validation of pattern matching code. In P. Barahona and A. Felty, editors, *Proceedings of the 7th ACM SIGPLAN International Conference on Principles and Practice of Declarative Programming*, pages 187–197. ACM, 2005.

[9] J. Melton and A. Eisenberg. *Understanding SQL and Java Together: A Guide to SQLJ, JDBC, and Related Technologies.* Morgan-Kaufmann, 2000.

[10] M. Mernik, J. Heering, and A. M. Sloane. When and how to develop domain-specific languages. Technical report, CWI, 2003.

[11] L. Moonen. Generating robust parsers using island grammars. In *Proceedings of the 8th Working Conference on Reverse Engineering*, pages 13–22. IEEE Computer Society Press, 2001.

[12] P.-E. Moreau, C. Ringeissen, and M. Vittek. A Pattern Matching Compiler for Multiple Target Languages. In G. Hedin, editor, *12th Conference on Compiler Construction, Warsaw (Poland)*, volume 2622 of *LNCS*, pages 61–76. Springer-Verlag, 2003.

[13] D. Spinellis. Notable design patterns for domain-specific languages. *Journal of Systems and Software*, 56(1):91–99, 2001.

[14] A. van Deursen, P. Klint, and J. Visser. Domain-specific languages: An annotated bibliography. *SIGPLAN Notices*, 35(6):26–36, 2000.

Some Programming Languages for LOGSPACE and PTIME

Guillaume Bonfante

Loria, Calligramme project, B.P. 239, 54506 Vandœuvre-lès-Nancy Cedex, France
and École Nationale Supérieure des Mines de Nancy, INPL, France
Guillaume.Bonfante@loria.fr

Abstract. We propose two characterizations of complexity classes by means of programming languages. The first concerns LOGSPACE while the second leads to PTIME. This latter characterization shows that adding a choice command to a PTIME language (the language WHILE of Jones [1]) may not necessarily provide NPTIME computations. The result is close to Cook in [2] who used "auxiliary push-down automata". LOGSPACE is obtained through a decidable mechanism of tiering. It is based on an analysis of deforestation due to Wadler in [3]. We get also a characterization of NLOGSPACE.

We propose a contribution to the program of Jones [1]: "We maintain that Computability and Complexity theory, and Programming Language and Semantics [...] have much to offer each other, in both directions." ; we give characterizations of complexity classes by means (of restrictions) of programming languages.

The present contribution belongs to a largely wider program (see [4, 5, 6]) where we have shown the interest of that kind of characterization. Let us recall it briefly. From a practical point of view, a static analysis allows an evaluation of the bounds on the resources before computations are effectively performed. It can be used by an operating system to manage processes. In particular, it avoids the monitoring of memory usage. Maybe more interestingly, it can be used by the compiler to deal with memory management, and so, to optimize the complexity of programs. Such analyses are the theoretical core of projects like Amadio's CRISS project[1] whose objective is to control resources—time and space—for synchronous systems.

We propose a characterization of LOGSPACE. It is obtained with respect to a kind of tiering discipline. This fruitful approach has been initially considered by Bellantoni and Cook in [7] and Leivant and Marion [8] who characterized PTIME. Leivant and Marion showed that such a stratification could be used for other complexity classes, see [9, 10]. Following Bellantoni-Cook, Neergaard [11] has shown what restrictions of the language B leads to LOGSPACE. The current proposition differs from the preceding ones in the following way. The role of tiering is not to control recursion but rather to restrict the width of the call graph. It is essentially based on the work of Wadler [3] and Jones [1].

[1] http://www.pps.jussieu.fr/~amadio/Criss/criss.html

M. Johnson and V. Vene (Eds.): AMAST 2006, LNCS 4019, pp. 66–80, 2006.

Among space characterizations, we mention the work of Hofmann [12, 13] who, in particular, showed how to compile functional programs to `malloc()`-free C. Applications of these techniques are to be found in the Embounded Project[2] whose aim is to quantify and certify resources for real-time embedded systems. An other approach, based on linear logic, was carried on by Baillot and Terui, see [14]. In the vein of Leivant, Oitavem proposed an interesting characterization of small complexity classes in [15, 16].

The second characterization we propose deals with non determinism. We show, and the result is surprising, that adding some choice command in the language `WHILE` of Jones does not change the class of computed functions. Naively, one would have expected to characterize NPTIME. Indeed, if one considers — as does Jones — the class of functions computed in polynomial time in `WHILE`, one gets PTIME. Adding the choice command, one gets NPTIME. Since `WHILE`-cons-free programs characterize PTIME, adding the choice command "should" have resulted in NPTIME. It is not the case, and we show that such a system characterize PTIME. The result is all the more surprising that for the corresponding space characterization, that is of LOGSPACE, adding the choice command leads to the corresponding non-deterministic complexity class NLOGSPACE. We mention here the work of Cook [2] whose characterization of PTIME by means of auxiliary pushdown automata is in essence close to us. The main difference lies in the fact that we have an implicit call stack (for recursion) where Cook has an explicit one.

The upper bound on the complexity of functions computed in LOGSPACE is obtained through a mechanism of compilation. The syntactical restrictions make the method sound. Two points. First, we do not explicitly compute time or space bounds. We know that they exist as a consequence of Jone's analysis of life without `cons`. Second, the proposition enters the field of implicit complexity as one has to compile the program in order to stay within LOGSPACE. Computing in the original framework leads to polynomial time computations.

The structure of the paper is as follows. In Section 1, we define the two programming languages we will consider, namely the language `WHILE` of Jones and a functional language. We define also some syntactical restrictions on them, in particular, we present our tiering discipline. In Section 2, we show the LOGSPACE characterization. At that point, we show how tiering can be used for the evaluation of programs. Section 3 deals with non-determinism, choice command and the corresponding complexity classes. We give a polynomial time procedure for computing non-deterministic `WHILE`-cons-free programs.

1 Programming Languages

We introduce two programming languages, `WHILE` programs, and `FOFP` programs. We suppose from now on, that we are given a signature *Cns*, that is, some symbols with their arity. These symbols are called the constructor symbols. The signature defines a term algebra $\mathcal{T}(Cns)$ on which computations will be

[2] `http://www.embounded.org/`

performed. We suppose that among these symbols one is `nil` of arity 0. The expression `nil` serves as "false". Any other value is "true".

For each constructor symbol $\mathbf{c}$, we define a set of destructor functions $\mathbf{d}_{\mathbf{c},k}$ which map $\mathbf{c}(t_1, \cdots, t_n) \mapsto t_k$. Finally, we suppose, for any constructor symbol $\mathbf{c}$, we are given a pattern matching function $\mathbf{p}_\mathbf{c}$ that tells whether a term has the form $\mathbf{c}(t_1, \cdots, t_n)$ or not.

1.1 The WHILE Language

Let us begin with the syntax, which is due to Jones [1], except that we authorize more than one input. This is only for convenience.

Definition 1. *A* WHILE *program is given by the following grammar:*

$\mathtt{P : Program}$	$::= \mathtt{read\ X1, ..., Xn; C; write\ Y}$
$\mathtt{C : Command}$	$::= \mathtt{Z := E}$
	$\mid \mathtt{C_1; C_2}$
	$\mid \mathtt{if\ E\ then\ C_1\ else\ C_2}$
	$\mid \mathtt{while\ E\ do\ C\ done}$
$\mathtt{E : Expression}$	$::= \mathtt{Z}$
	$\mid \mathtt{D}$
	$\mid \mathbf{c}(\mathtt{E_1, E_2, \ldots, E_k})$
	$\mid \mathbf{d}_{\mathbf{c},k}\ \mathtt{E}$
	$\mid \mathbf{p}_\mathbf{c}\ \mathtt{E}$
$\mathtt{X, Y, Z : Variable}$	$::= \mathtt{X0} \mid \mathtt{X1} \mid \ldots$
$\mathtt{D : Data - value}$	$::= \mathcal{T}(\mathit{Cns})$

We note Var(p) *the variables appearing in a program* p.

The semantics are given by Jones. We recall it informally. A store for a program p is a function $\sigma^\mathtt{p} : \mathit{Var}(\mathtt{p}) \to \mathcal{T}(\mathit{Cns})$. The initial store given the input data $\mathtt{d_1, d_2, \ldots, d_n}$ is the store $\sigma_0^\mathtt{p}(\mathtt{d_1, \ldots, d_n}) = [\mathtt{X_1} \mapsto \mathtt{d_1}, \ldots, \mathtt{X_n} \mapsto \mathtt{d_n}, \mathtt{Z} \mapsto \mathtt{nil}, \ldots, \mathtt{Y} \mapsto \mathtt{nil}]$. Commands have the intuitive meaning. For instance, in an expression $\mathtt{if\ E\ then\ C_1\ else\ C_2}$, one tests if $\mathtt{E} = \mathtt{nil}$, in which case one executes $\mathtt{C_1}$. Otherwise, one executes $\mathtt{C_2}$. Assignments modify the store.

Definition 2. *Given a program* p*, its execution induces a partial function* $[\![\mathtt{p}]\!] : \mathcal{T}(\mathit{Cns})^\mathtt{n} \to \mathcal{T}(\mathit{Cns})$ *which maps* $\mathtt{d_1, d_2, \ldots, d_n}$ *to* $\sigma(\mathtt{Y})$ *if the program terminates and where* σ *is the last store of the computation, otherwise it is undefined.*

Definition 3. *A program is called cons-free, if it does not use an expression of the form* $\mathbf{c}(\mathtt{E_1, \ldots, E_k})$*. We note* $\mathtt{WHILE}^{\mathit{cons\text{-}free}}$ *the set of cons-free programs.*

Theorem 1 (Jones [1]). *The set of decision problems computed by cons-free programs is exactely* LOGSPACE.

Definition 4. *The recursive extension of* WHILE *is described as follows. To* WHILE*, we add the instruction* ***call*** *that calls some sub-procedure. A program is given by*

```
globalvariable U_1, ..., U_u;
procedure P1; localvariable P11, ..., P1v;
C1;
procedure P2; localvariable P21, ..., P2w;
C2;
.......
read U1; call P1; write U1
```

Variables appearing in `Ci` *belong to the local variables of the procedure or to the global variables. The semantics are briefly as follows. Each time one calls a new procedure, one stacks some fresh local variables. Then, one executes the instructions until one reaches the end of the procedure (modifying the fresh local variables and the global ones). At this point, just forget the local variables. We note* $\texttt{WHILE}^{rec-cons-free}$ *the set of such programs.*

Theorem 2 (Jones [1]). *The set of decision problems computed by* $\texttt{WHILE}^{rec\text{-}cons\text{-}free}$ *programs is exactly the set of* PTIME *decision problems.*

1.2 FOFP

We define a generic first order functional programming language. The vocabulary $\Sigma = \langle Cns, Op, Fct\rangle$ is composed of three disjoint domains of symbols. The set of programs is defined by the following grammar.

$$
\begin{array}{lll}
\texttt{Programs} \ni \mathbf{p} & ::= & d_1, \cdots, d_m \\
\texttt{Definitions} \ni d & ::= & \texttt{f}(x_1, \cdots, x_n) = e^{\texttt{f}} \\
\texttt{Expression} \ni e & ::= & x \mid \mathbf{op}(e_1, \cdots, e_n) \mid \texttt{f}(e_1, \cdots, e_n) \\
 & & \mid \mathbf{c}(e_1, \cdots, e_n) \\
 & & \mid \textbf{if } e_1 \textbf{ then } e_2 \textbf{ else } e_3 \\
 & & \mid \textbf{let } x = e_1 \textbf{ in } e_2 \\
 & & \mid \textbf{case } x_1, \cdots, x_n \textbf{ of } \overline{p_1} \rightarrow e_1 \ldots \overline{p_\ell} \rightarrow e_\ell \\
\texttt{Patterns} \ni p & ::= & x \mid \mathbf{c}(p_1, \cdots, p_n)
\end{array}
$$

where $x \in Var$ is a variable, $\mathbf{c} \in Cns$ is a constructor, $\mathbf{op} \in Op$ is an operator, $\texttt{f} \in Fct$ is a function symbol, and $\overline{p_i}$ is a sequence of n patterns. Throughout, we generalize this notation to expressions and we write $\overline{e}$ to express a sequence of expressions, that is $\overline{e} = e_1, \ldots, e_n$, for some n clearly determined by the context.

Throughout the proofs which follows, we make no distinction between operators and function symbols. We have introduced operators only for convenience when writing the examples.

The set of variables Var is disjoint from Σ. In a definition, $e^{\texttt{f}}$ is called the body of `f`. A variable of $e^{\texttt{f}}$ is either a variable in the parameter list $x_1, \cdots, x_n$ of the definition of `f` or a variable which occurs in a pattern of a case definition. In a case expression, patterns are supposed to be non overlapping. We will come back to this Hypothesis in the Section on non determinism.

Given a function symbol $\mathtt{f}$, we say that an expression e is $\mathtt{f}$-free if there is no occurrences of $\mathtt{f}$ in e. We call functional an expression of the form $\mathtt{g}(e_1, \cdots, e_n)$.

Lastly, it is convenient, because it avoids tedious details, to restrict our attention to programs without nested **case** , **let** , **if** expressions within functional expressions. This is not a severe restriction as one can easily transform programs to avoid this nesting. For instance, one transforms $\mathtt{f}(\ldots, \textbf{if } e_1 \textbf{ then } e_2 \textbf{ else } e_3, \ldots)$ into $\textbf{if } e_1 \textbf{ then } \mathtt{f}(\ldots, e_2, \ldots) \textbf{ else } \mathtt{f}(\ldots, e_3, \ldots)$.

Definition 5. *Rules for evaluation are given by Fig. 1. A function $f : \mathcal{T}(Cns)^k \to \mathcal{T}(Cns)$ is computed by a program* **p** *if there is a function $f \in$* **p** *such that $\forall \bar{t} \in \mathcal{T}(Cns)^k : f(\bar{t}) \downarrow f(\bar{t})$.*

From now on, we suppose that the programs that we consider are terminating. Any method for proving their termination can be considered, for instance Recursive Path Orderings, Dependency Pairs, and so on.

$$\frac{t \in \mathcal{T}(Cns)}{t \downarrow t} \qquad \frac{e_1 \downarrow v_1 \ldots e_n \downarrow v_n \quad \mathtt{f}(x_1, \cdots, x_n) = \mathtt{e}^{\mathtt{f}} \quad \mathtt{e}^{\mathtt{f}}[x_i \leftarrow v_i] \downarrow v}{\mathtt{f}(e_1, \cdots, e_n) \downarrow v}$$

$$\frac{e_1 \downarrow \mathtt{tt} \quad e_2 \downarrow v}{\textbf{if } e_1 \textbf{ then } e_2 \textbf{ else } e_3 \downarrow v} \qquad \frac{e_1 \downarrow \mathtt{ff} \quad e_3 \downarrow v}{\textbf{if } e_1 \textbf{ then } e_2 \textbf{ else } e_3 \downarrow v}$$

$$\frac{e_1 \downarrow u \quad e_2[x \leftarrow u] \downarrow v}{\textbf{let } x = e_1 \textbf{ in } e_2 \downarrow v} \qquad \frac{e_k \downarrow u_k \quad \exists \sigma,\ i\ :\ \overline{p_i}\sigma = \overline{u} \quad e_i\sigma \downarrow v}{\textbf{case } t_1, \cdots, t_n \textbf{ of } \overline{p_1} \to e^1 \ldots \overline{p_\ell} \to e^\ell \downarrow v}$$

Fig. 1. Call by value semantics of a program **p**

Definition 6. *We say that a program* **p** *is cons-free if the definitions do not use the rule $\mathbf{c}(e_1, \cdots, e_n)$ of the grammar. In other words, there are only constructors in patterns. The set of such cons-free programs is noted $\mathtt{FOFP}^{cons\text{-}free}$.*

Definition 7. *A definition $f(x_1, \cdots, x_n) = e^f$ induces a relation on function symbols. Say that f calls g if g appears in the body of f. We note this relation $\to$. The reflexive-transitive closure of this relation induces a pre-order on function symbols, noted $\xrightarrow{*}$. The corresponding equivalence relation $\simeq$ is defined by $f \simeq g \Leftrightarrow (f \xrightarrow{*} g \wedge g \xrightarrow{*} f)$. The corresponding strict partial order is noted $\prec$. We have $g \prec f \Leftrightarrow (f \xrightarrow{*} g \wedge \neg(f \xrightarrow{*} g))$.*

Definition 8. *We say that an expression e is tail-recursive w.r.t. a function symbol f if*

1. *$e = x$,*
2. *$e = g(e_1, \cdots, e_n)$ where for all $h \in e$, $h \prec f$,*

3. $e = f(x_1, \cdots, x_n)$,
4. $e =$ **if** e_1 **then** e_2 **else** e_3 *and* e_1 *is* f*-free and both* e_2 *and* e_3 *are tail recursive wrt* f,
5. $e =$ **case** $\overline{x}$ **of** $\overline{p_1} \to e_1 \ldots \overline{p_\ell} \to e_\ell$ *and for all* $i \leq \ell$ *the expression* e^k *is tail recursive.*

A definition $f(x_1, \cdots, x_n) = e^f$ *is tail recursive if* e^f *is tail recursive wrt* f. *A program is tail recursive if any definition is tail recursive. The set of such tail recursive programs is noted* $\mathtt{FOFP}^{tr}$. *We note* $\mathtt{FOFP}^{tr\text{-}cons\text{-}free}$ *the set of programs that are both tail recursive and cons-free.*

The following is due to Jones [17].

Theorem 3

1. *The set of decision problems computed by* $\mathtt{FOFP}^{cons-free}$ *programs is exactly the set of* PTIME *decision problems.*
2. *The set of decision problems computed by* $\mathtt{FOFP}^{tr-cons-free}$ *programs is exactly the set of* LOGSPACE *decision problems.*

In the following, we reinforce Definition 8 to allow nesting of functions. We propose a finer discipline on programs that stays within LOGSPACE.

Example 1.

$$x_1 < x_2 = \textbf{case } x_1, x_2 \textbf{ of} \quad x'_1, \mathbf{0} \to \mathbf{ff} \quad \mathbf{0}, \mathtt{s}(x'_2) \to \mathbf{tt} \quad \mathtt{s}(x'_1), \mathtt{s}(x'_2) \to x'_1 < x'_2$$

$$x_1 - x_2 = \textbf{case } x_1, x_2 \textbf{ of} \quad x'_1, \mathbf{0} \to x_1 \quad \mathbf{0}, x'_2 \to \mathbf{0} \quad \mathtt{s}(x'_1), \mathtt{s}(x'_2) \to x'_1 - x'_2$$

$$\begin{aligned} \mathtt{pgcd}(x_1, x_2) = \textbf{case } & x_2 \textbf{ of} \\ & \mathbf{0} \to x_1 \\ & \mathtt{s}(x'_2) \to \textbf{if } x_1 < x_2 \\ & \quad \textbf{then } \mathtt{pgcd}(x_2, x_1) \\ & \quad \textbf{else } \mathtt{pgcd}(x_1 - x_2, x_2) \end{aligned}$$

Here the first two definitions are tail-recursive. This is not the case of the third expression. Note that it cannot be directly handled by Wadler's approach, see [3], as there is some composition of function symbols. Note also that there is more than one occurrence of pgcd in the right hand side of the second rule. In the following, we show how to compute pgcd in LOGSPACE.

Definition 9. *An expression is strongly tail-recursive if it follows Definition 8 except that clauses (2) and (3) are replaced by clauses*

2'. $e = g(e_1, \cdots, e_n)$ *and for all the* $i \leq n$, e_i *is* f*-free. Here,* g *may be equal to* f ;
6'. $e =$ **let** $x = e_1$ **in** e_2 *where* e_1 *is* f*-free and* e_2 *is strongly tail-recursive.*

This extends to the definition of function symbols and to programs. We note the set of such programs $\texttt{FOFP}^{s-tr}$.

One may observe that the above definition of the pgcd respects the strong-tail recursiveness condition. The next Section shows that programs in $\texttt{FOFP}^{s\text{-}tr\text{-}cons\text{-}free}$ can be computed within LOGSPACE.

Theorem 4. *The set of decision problems computed by* $\texttt{FOFP}^{s-tr-cons-free}$ *programs is exactly the set of* LOGSPACE *decision problems.*

Finally, we propose a notion that goes beyond strong tail-recursion.

Definition 10 (Linear programs). *Given a function symbol* f*, the level of an expression is given by the inductive rules:*

- $lvl_f(x) = 0$,
- $lvl_f(g(e_1, \cdots, e_n)) = 1 + \sum_{k \leq n} lvl_f(e_k)$ *where* $g \simeq f$,
- $lvl_f(g(e_1, \cdots, e_n)) = \sum_{k \leq n} lvl_f(e_k)$ *where* $g \prec f$,
- $lvl_f(\textbf{let } x = e_1 \textbf{ in } e_2) = lvl_f(e_1) + lvl_f(e_2)$,
- $lvl_f(\textbf{if } e_1 \textbf{ then } e_2 \textbf{ else } e_3) = lvl_f(e_1) + \max(lvl_f(e_2), lvl_f(e_3))$,
- $lvl_f(\textbf{case } \overline{x} \textbf{ of } p_1 \to e_1, \ldots, p_k \to e_k) = \max(lvl_f(e_1), \ldots, lvl(e_k))$.

We say that a definition $f(\overline{x}) = e^f$ *is linear if* $lvl_f(e^f) = 1$. *A program is linear if any definition has level* 1. *The set of such programs is noted* $\texttt{FOFP}^{lin}$.

Theorem 5. *Decision problems decided by linear cons-free programs are exactly* LOGSPACE *decision problems.*

Example 2. The following program is not strongly-tail-recursive but linear.

$$
\begin{aligned}
\texttt{pred}(x) &= \textbf{case } x \textbf{ of} \\
&\quad \mathbf{0} \to \mathbf{0} \\
&\quad \texttt{s}(x') \to x' \\
\texttt{incr}(x, y) &= y - \texttt{pred}(y - x) \\
\texttt{log}(x) &= \textbf{case } x \textbf{ of} \\
&\quad \mathbf{0} \to \mathbf{0} \\
&\quad \texttt{s}(x') \to \texttt{incr}(\texttt{log}(\texttt{half}(x)), x)
\end{aligned}
\qquad
\begin{aligned}
\texttt{half}(x) &= \textbf{case } x \textbf{ of} \\
&\quad \mathbf{0} \to \mathbf{0} \\
&\quad \texttt{s}(\mathbf{0}) \to \mathbf{0} \\
&\quad \texttt{s}(\texttt{s}(x')) \to \texttt{incr}(\texttt{half}(x), x)
\end{aligned}
$$

2 Compiling FOFP Programs

The proofs of the Theorems 4, 5 involve the same argument. We compile programs in $\texttt{FOFP}^{s-tr-cons-free}$ and in $\texttt{FOFP}^{lin-cons-free}$ to WHILE-cons-free. As a consequence, function computable in these two languages can be computed within LOGSPACE due to Theorem 3. The converse part, that is to show that all LOGSPACE decidable problems can be computed by strongly-tail-recursive programs or linear-cons-free programs is a direct consequence of the fact that $\texttt{FOFP}^{tr-cons-free} \subseteq \texttt{FOFP}^{s-tr-cons-free} \subseteq \texttt{FOFP}^{lin-cons-free}$.

We first begin with a few observations.

Proposition 1. *Given a program in* $\mathtt{FOFP}^{cons-free}$*, for any constructor terms* $\bar{t}$*, and any expression* e*, suppose that* $e[x_i \leftarrow t_i] \downarrow v$*, then,* v *is a subterm of one of the* t_i*.*

Corollary 1. *For a* $\mathtt{FOFP}^{cons-free}$ *program, the height of the evaluation tree (cf. Fig. 1) is bounded by a polynomial in the input.*

Proof. Given a term t, the number of subterms of t is linear in the size of t. Suppose we are given a constant d. Consider the set $C_{\bar{t}} = \{s_1, \cdots, s_n \mid n \leq d \wedge (\forall i \leq n : \exists j \leq n : s_i \trianglelefteq t_j)\}$. Then, this set has polynomial size in the size of $\bar{t}$. Now, take d to be the maximal arity of a symbol, the polynomial bound together the property of termination of programs and the preceding proposition gives the result.

Proposition 2. *Suppose that* $f_1, \ldots, f_k$ *are* FOFP *programs which are* LOGSPACE *computable. Then any function* $f(x_1, \ldots, x_n) = e^f$ *with* e^f *a composition of the functions* $f_1, \ldots, f_k$ *is computed by a* WHILE*-cons free program, and so, is* LOGSPACE*. Given an expression* e*, we note the corresponding code* $\mathtt{C_e}$*.*

Proof. The proof is by induction on the expression $e^{\mathtt{f}}$. Suppose that we are given for each function $\mathtt{f}_i$ a WHILE-cons-free program **read** Xfi1, Xfi2, ..., Xfik; $\mathtt{C_{fi}}$; **write** Y; that computes it. We suppose w.l.o.g that these programs do not change de values of the input variables. Suppose that $e^{\mathtt{f}} = x_k$, it is computed by:

```
read X1, X2, ..., Xn;
Y := Xk;
write Xk;
```

Suppose now that $e^{\mathtt{f}} = \mathtt{g}(\overline{e})$. Then, by induction we can suppose that e_i is computed by $\mathtt{C_{e_i}}$. In that case, the following WHILE-cons-free program computes f.

```
read X1, ..., Xn;
Ce1;
X1g := Y;
Ce2;
X2g := Y;
.
.
.
Cek;
Xkg := Y;
Cg;
write Y;
```

Remark 1. It is well known that if f and g are LOGSPACE, so is $f \circ g$. Since the output of functions are subterms of the inputs, we have a much easier proof of the composition. Furthermore, we use the construction throughout the paper. This is why we give an explicit construction.

2.1 Strongly-Tail-Recursive Programs

We prove now Theorem 4. First of all, let us eliminate the **let** construction.

Proposition 3. *Consider the program transformation that maps* **let** $x = e_1$ **in** e_2 *to* $e_2[x \leftarrow e_1]$*. It preserves the semantics of the program. Furthermore, if a program is strongly-tail recursive, so is its transform.*

As a consequence, we may consider w.l.o.g only programs without the **let** construction.

For each definition, we build a WHILE-cons-free program that computes it. We proceed by induction on the ordering $\prec$. For the sake of the proof, to avoid a tedious case analysis, we suppose that for all symbols $\mathtt{g} \in e^{\mathtt{f}}$, either $\mathtt{g} = \mathtt{f}$ or $\mathtt{g} \prec \mathtt{f}$. In other words, we avoid mutual definitions.

For minimal elements, observe that their definitions are tail-recursive. So, one applies Theorem 3 to get a WHILE-cons-free programs that computes them.

Now, we suppose that we are given an expression $e^{\mathtt{f}}$ that computes f. We suppose by induction that we have a program computing any function $\mathtt{g} \neq \mathtt{f}$ involved in the definition of f. As a consequence, applying Proposition 2, one gets for each composition of such functions some program that computes it.

We now perform an induction on the structure of the definition of f. In the following compiling procedure, we suppose (by induction) that we are given for any sub-expression e some program **read** $\mathtt{X_{e1}}, \ldots, \mathtt{X_{en}}; \mathtt{C_e};$ **write**$\mathtt{Y}$; where the $\mathtt{X}_{\mathtt{e}i}$ are the variables of e_i .

By compositions of pattern expression $\mathbf{p_c}$ as well as destructors $\mathbf{d_c}$, we build for each pattern p a code $P_p \in \mathtt{WHILE}^{cons-free}$ that returns tt if the inputs verify the pattern. We suppose given the code for the unification of variables in patterns. So, after $\overline{X'_p} :=_p \overline{X}$, the variables in the patterns are supposed to have their value after pattern matching.

Given an expression e, Table 1 gives rules to build the code $\mathtt{D_e}$.

Table 1. Compilation rules

$\mathtt{D}_{\mathtt{case}\ \overline{\mathtt{x}}\ \mathtt{of}\ \mathtt{p_1 \to e_1, \ldots, p_k \to e_k}}$	$\mathtt{D}_{\mathtt{if}\ \mathtt{e_1}\ \mathtt{then}\ \mathtt{e_2}\ \mathtt{else}\ \mathtt{e_3}}$	$\mathtt{D}_{\mathtt{f(e_1, \cdots, e_n)}}$	$\mathtt{D}_{\mathtt{g}(\overline{\mathtt{e}})}$
P_{p_1}; if Y then $\overline{\mathtt{X'_{p_1}}} :=_{\mathtt{p_1}} \overline{\mathtt{X}}; \mathtt{D_{e_1}}$; else $\mathtt{P_{p_2}}$; if Y then $\overline{\mathtt{X'_{p_2}}} :=_{\mathtt{p_2}} \overline{\mathtt{X}}; \mathtt{D_{e_2}}$; ⋮ $\overline{X'_{p_k}} :=_{p_3} \overline{X}; \mathtt{D_{e_k}}$;	$\mathtt{C_{e_1}}$; if Y then $\mathtt{D_{e_2}}$; else $\mathtt{D_{e_3}}$;	$\mathtt{C_{e_1}}$; $\mathtt{X_1} := \mathtt{Y}$; $\mathtt{C_{e_2}}$; $\mathtt{X_2} := \mathtt{Y}$; ⋮ $\mathtt{C_{e_k}}$; $\mathtt{X_k} := \mathtt{Y}$;	$\mathtt{C_{g(\overline{e})}}$; R := ff;

What do these programs do? If the expression is f-free, that is if we reached the end of the recursion, Y is assigned the value of the computation. In that case, the flag variable R is turned to tt. Otherwise, it computes the arguments of the next call of f, and continues the process.

So, the function f is computed by the following program:

```
read X1, ··· , Xn;
R := tt;
while R do
De^f;
write Y;
```

2.2 Linear Programs

The rationale behind the compilation of linear programs is the following. In the first part of the computation, we just compute the arguments of the intermediate calls of f and forget the context in which they appear. At the end of the process, one knows two crucial points.

First, one knows the exact number of nested calls of f, moreover, due to Corollary 1, this number is polynomial in the size of the input. As a consequence, it is representable in log-space. Second, one knows the value of the function f on its terminating call.

In the second part, you just redo what has been said above except that at each step you compute one less nesting of calls of f and reuse the last result of the loop to compute the value of f in its full context.

Contrarily to what happened for strong-tail-recursion, we cannot get rid off the **let** construction. Indeed, the transform does not preserve linearity. A counter-example is the definition $\mathtt{f}(x) = \textbf{case } x \textbf{ of } (\mathbf{0} \to \mathbf{0}, \mathtt{s}(x') \to \textbf{let } y = \mathtt{f}(x') \textbf{ in } \mathtt{g}(y, y))$ where g is an already defined binary function. So, don't forget we have to cope with **let** .

Proof. As above, we proceed by induction on the $\prec$ order. For minimal elements, the definition is tail-recursive. For those, we have already seen that we have a procedure. Suppose now that we are trying to compute function f whose definition is $\mathtt{f}(x_1, \cdots, x_n) = e^{\mathtt{f}}$. As above, we suppose that there is no symbol equivalent to f in $e^{\mathtt{f}}$ (except for f of course!). We suppose that we have built for any function $\mathtt{g} \prec \mathtt{f}$ some WHILE-cons-free code that computes it. Moreover, using Proposition 2, we suppose that we are able to compute any expression composed of such symbols.

Now, suppose we are given a functional term $\mathtt{h}(\overline{e})$ of level 1. It contains one occurrence of f. It can be seen as $C[\mathtt{f}(\overline{e'})]$ where the context C can be seen as an expression over the variables of $\mathtt{h}(\overline{e})$ plus an extra variable F that corresponds to the call of f. We can suppose that for any of these expressions, we have some code that computes them. We use a similar notation to that of the proof above.

To compute the value of the last call, we build a code analogous to what we have done for strong-tail recursion. The rules are somewhat different.

- For the **case** construction, we use the construction of Table 1.
- The **if** case splits into two sub-cases. When $\mathtt{lvl}_{\mathtt{f}}(e_1) = 0$, we use the rule of Table 1. The other case, is shown below.
- The **let** construction also splits into two parts that are considered below.
- The last case correspond to the functional expressions. When the functional expression has level 0, we use the rule of Table 1. The other case is presented below.

$\mathtt{D}_{\mathbf{if}\ \mathtt{e}_1\ \mathbf{then}\ \mathtt{e}_2\ \mathbf{else}\ \mathtt{e}_3}$ when $\mathtt{lvl}_{\mathtt{f}_1}(e_1) = 1$	$\mathtt{D}_{\mathbf{let}\ \mathtt{v}=\mathtt{e}_1\ \mathbf{in}\ \mathtt{e}_2}$ when $\mathtt{lvl}_{\mathtt{f}}(e_1) = 0$	$\mathtt{D}_{\mathbf{let}\ \mathtt{v}=\mathtt{e}_1\ \mathbf{in}\ \mathtt{e}_2}$ when $\mathtt{lvl}_{\mathtt{f}}(e_1) = 1$	$\mathtt{D}_{\mathtt{C}[\mathtt{f}(\overline{\mathtt{e}'})]}$
$\mathtt{D}_{\mathtt{e}_1}$	$\mathtt{C}_{\mathtt{e}_1}$; $\mathtt{V} := \mathtt{Y}; \mathtt{D}_{\mathtt{e}_2}$	$\mathtt{D}_{\mathtt{e}_1}$	$\mathtt{C}_{\mathtt{e}'_1}$; $\mathtt{X}_1 := \mathtt{Y}$; $\mathtt{C}_{\mathtt{e}'_2}$; $\mathtt{X}_2 := \mathtt{Y}$; $\vdots$ $\mathtt{C}_{\mathtt{e}'_k}$; $\mathtt{X}_k := \mathtt{Y}$;

Given an expression e, we define now (by induction) a code $\mathtt{E}_{\mathtt{e}}$ that computes the value of $\mathtt{f}$ given that $\mathtt{F}$ contains the value of the sub-call of $\mathtt{f}$.

$\mathtt{E}_{\mathbf{case}\ \overline{\mathtt{x}}\ \mathbf{of}\ \mathtt{p}_1 \to \mathtt{e}_1, \ldots, \mathtt{p}_k \to \mathtt{e}_k}$	$\mathtt{E}_{\mathbf{if}\ \mathtt{e}_1\ \mathbf{then}\ \mathtt{e}_2\ \mathbf{else}\ \mathtt{e}_3}$ when $\mathtt{lvl}_{\mathtt{f}}(e_1) = 0$	$\mathtt{E}_{\mathbf{let}\ \mathtt{v}=\mathtt{e}_1\ \mathbf{in}\ \mathtt{e}_2}$ when $\mathtt{lvl}_{\mathtt{f}}(e_1) = 0$

P_{p_1};
if Y
then $\overline{\mathtt{X}'_{\mathtt{p}_1}} :=_{\mathtt{p}_1} \overline{\mathtt{X}}; \mathtt{E}_{\mathtt{e}_1}$;
else $\mathtt{P}_{\mathtt{p}_2}$;
    if Y
    then $\overline{\mathtt{X}'_{\mathtt{p}_2}} :=_{\mathtt{p}_2} \overline{\mathtt{X}}; \mathtt{E}_{\mathtt{e}_2}$;
        $\vdots$
        $\overline{X'_{p_k}} :=_{p_k} \overline{X}; \mathtt{E}_{\mathtt{e}_k}$;
	$\mathtt{C}_{\mathtt{e}_1}$;
	if Y
	then $\mathtt{E}_{\mathtt{e}_2}$;
	else $\mathtt{E}_{\mathtt{e}_3}$;
		$\mathtt{C}_{\mathtt{e}_1}$;
		$\mathtt{V} := \mathtt{Y}$;
		$\mathtt{E}_{\mathtt{e}_2}$;

$\mathtt{E}_{\mathtt{C}[\mathtt{f}(\overline{\mathtt{e}'})]}$	$\mathtt{E}_{\mathbf{if}\ \mathtt{e}_1\ \mathbf{then}\ \mathtt{e}_2\ \mathbf{else}\ \mathtt{e}_3}$ when $\mathtt{lvl}_{\mathtt{f}}(e_1) = 1$	$\mathtt{E}_{\mathbf{let}\ \mathtt{v}=\mathtt{e}_1\ \mathbf{in}\ \mathtt{e}_2}$ when $\mathtt{lvl}_{\mathtt{f}}(e_1) = 1$
$\mathtt{C}_{\mathtt{C}}$;	$\mathtt{E}_{\mathtt{e}_1}$; if Y then $\mathtt{C}_{\mathtt{e}_2}$; else $\mathtt{C}_{\mathtt{e}_3}$;	$\mathtt{E}_{\mathtt{e}_1}$; $\mathtt{V} := \mathtt{Y}$; $\mathtt{C}_{\mathtt{e}_2}$;

We can now compute f by the following code:

```
read X_{1,0}, ..., X_{n,0};
X̄_i := X̄_{i,0}; //a copy of the inputs
R := tt;
while R do
   D_{e^f}; incr N;
done; N := pred N;
while N ≠ 0 do
X̄_i := X̄_{i,0};
   M = N;
   while M ≠ 0 do
      D_{e^f}; M := pred M;
   done;
   N := pred N;
   E_{e^f}; F := Y;
done;
Y := F;
writeY;
```

Some last words about this code. Our management of the counters N and M is licit, even the incrementing, since we have a polynomial bound due to Proposition 1 on the two counters. We refer to Jones [17] who extensively discusses how to carry this out.

3   Non-determinism

This part of the paper introduces some "non-determinism" to the languages. To WHILE, we add a new command **choose**. We propose non-confluence as a functional correspondence of this instruction.

Definition 11. *Following Jones, to WHILEwe add the expression* **choose** C_1 C_2 *whose operational semantics is to evaluate either* C_1 *or* C_2. *A program induces now a relation between inputs and outputs. We say that a decision problem f is computed by a program f if for all inputs $\bar{t}$, the value of $f(\bar{t})$ is true iff one execution of f on $\bar{t}$ reaches* **tt**. *We note* WHILE^n *the set of programs with this extra instruction.* WHILE$^{n\text{-}ptime}$ *denotes the set of (non deterministic) programs working in polynomial time, etc.*

Theorem 6 (Jones [17]).

1. WHILE$^{n\text{-}ptime}$ = NPTIME;
2. WHILE$^{n-log-space}$ = NLOGSPACE.

Definition 12. *We consider here some FOFP programs without the confluence property, that is, patterns may overlap each other. A normal form is one possible result of the computation. Following Grädel and Gurevich [18], the value of any* *term is the maximal normal form of the term (for a given order on terms). Notice that this includes the usual definition for decision problem by choosing* ***true*** $>$ ***false****. We add the superscript n to denote the fact that we include non-deterministic programs.*

Theorem 7.

1. $\mathtt{WHILE}^{n-cons-free} = \mathtt{FOFP}^{n-lin-cons-free} = \text{NLOGSPACE}$*;*
2. $\mathtt{WHILE}^{n-rec-cons-free} = \mathtt{FOFP}^{n\text{-}cons\text{-}free} = \text{PTIME}$.

This latter fact is surprising as it breaks a similarity (similarity that holds for logspace):

$$\begin{array}{ccccc}
\mathtt{WHILE}^{n-rec-cons-free} \neq & & \mathtt{WHILE}^{n-ptime} & = & \text{NPTIME} \\
\downarrow = & & \downarrow & & \downarrow \\
\mathtt{WHILE}^{rec-cons-free} & = & \mathtt{WHILE}^{ptime} & = & \text{PTIME}
\end{array}$$

This result is analogous to that of Cook [2] Th.2 p7. He gives a characterization of PTIME by means of auxiliary pushdown automata working in logspace, that is a Turing Machine working in logspace plus an extra (unbounded) stack. It is also the case that the result holds whether or not the auxiliary pushdown automata is deterministic.

3.1 Bound on FOFP^n-Cons-Free

We propose a proof for FOFP-programs. The case of WHILE-programs is similar. The key observation is that Proposition 1 remains true in the context of non-confluent programs. As a consequence, following a call-by-value semantics, any arguments in subcomputations are some subterms of the initial inputs. From that, it is possible to use memoization, see [17]. The original point is that we have to manage non-determinism.

So, the crucial point is that the arguments of functions are subterms of the input and moreover, that the cardinality of this set is polynomial as was shown in Proposition 1. A second point is that normal forms are also subterms of the input. It means that, for each defined symbol, the induced relation can be stored in polynomial space. This leads to a procedure where we remember the normal forms of each (already computed) function on arguments and reuse it when necessary.

Suppose we are given a program f which is n-cons-free. Given input $t_1, \cdots, t_n$, let us note $I = \{t \trianglelefteq t_i \mid i \leq n\}$. We have $\sharp I \leq O(|t_1, \cdots, t_n|)$.

We consider a 3D table. The first dimension corresponds to $\mathcal{F}$, the second to I^A (where A is the maximal arity of a symbol), that is the arguments of functions. The third to I, the possible values of the relation. The entries of the table are boolean, and T[g][t][v] is (intended to be) true iff $\mathtt{g}(t) \xrightarrow{+} v$. This table has a polynomial size w.r.t. the inputs.

Consider the following algorithm (at the beginning, the entries of table `T[g][t][v]` are false):

```
var r : Term;
for i := 1 to |F|  * O(|t1,..., tn|^A) * O(|t1,...,tn|) do
for g in F do
for t in I^A do
for v in I do
        r1, ..., rn := find(g,t);
        for l := 1 to n do
        T[g][t][v] := T[g][t][v] || compute(ri,t,v);
        end
end;
end;
end;
end;
```

`find(g,t)` is charged to give the list of all rules that can be applied on t. There are finitely many of these. This can be done in linear time.

`compute(ri,t,v)` is charged to see if the rule `ri` given by `find` may lead to value v. That is, to see if the subcalls (with the corresponding inputs) in `r` have been already computed, choose for all of them the already computed values and finally turns the table cell to true if one of these choices leads to the value `v`. This process is easily proved polynomial by a simple induction on the construction of the rule `ri`. So, the instructions inside the loop take polynomial time.

For each loop on `i`, one will fulfil some of the $T[\mathtt{g}][t][v]$. As a consequence, the bound on the exterior loop is enough to get the result. So, the fixpoint is reached within a polynomial in the number of entries in the table. This algorithm works in polynomial time, hence we obtain the following corollary:

Corollary 2. $\mathtt{FOFP}^{n-cons-free} \subseteq$ PTIME.

Concerning the counterpart of the proof. In the case of PTIME, one may note that $\mathtt{FOFP}^{cons-free} \subseteq \mathtt{FOFP}^{n-cons-free}$. As a consequence, w.r.t. Theorem 3, it is PTIME complete.

3.2 $\mathtt{FOFP}^{\textbf{n-Lin-Cons-Free}}$

First, proving that $\mathtt{WHILE}^{nlogspace} \simeq \mathtt{WHILE}^{n-cons-free}$ can be achieved following Jones's proof that $\mathtt{WHILE}^{logspace} \simeq \mathtt{WHILE}^{cons-free}$. Here, non-determinism plays no special role.

For the case of non-deterministic linear programs, one may note that the rules for the case analysis can be transformed to take into account the fact that more than one pattern applies. At this point, use the choice operator to decide which pattern to take.

As a consequence, the analysis of Section 2 can be used here. So, by the remark at the beginning of the subsection, the global process can be performed within NLOGSPACE.

Acknowledgment. This work has been largely inspired by "life without `cons`" of Jones.

References

1. Jones, N.D.: LOGSPACE and PTIME characterized by programming languages. Theoretical Computer Science **228** (1999) 151–174
2. Cook, S.: Characterizations of pushdown machines in terms of time-bounded computers. Journal of the ACM **18(1)** (1971) 4–18
3. Wadler, P.: Deforestation: Transforming programs to eliminate trees. In: ESOP '88. European Symposium on Programming, Nancy, France, 1988 (Lecture Notes in Computer Science, vol. 300), Berlin: Springer-Verlag (1988) 344–358
4. Marion, J.Y., Moyen, J.Y.: Efficient first order functional program interpreter with time bound certifications. In: LPAR 2000. Volume 1955 of Lecture Notes in Computer Science., Springer (2000) 25–42
5. Bonfante, G., Marion, J.Y., Moyen, J.Y.: On lexicographic termination ordering with space bound certifications. In: PSI 2001, Ershov Memorial Conference. Volume 2244 of Lecture Notes in Computer Science., Springer (2001)
6. Bonfante, G., Marion, J.Y., Moyen, J.Y.: Quasi-Interpretations and Small Space Bounds. In Giesl, J., ed.: Rewrite Techniques and Applications. Volume 3467 of Lecture Notes in Computer Science., Springer (2005) 150–164
7. Bellantoni, S., Cook, S.: A new recursion-theoretic characterization of the polytime functions. Computational Complexity **2** (1992) 97–110
8. Leivant, D., Marion, J.Y.: Lambda calculus characterizations of poly-time. Fundamenta Informaticae **19** (1993) 167,184
9. Leivant, D., Marion, J.Y.: Predicative functional recurrence and poly-space. In Bidoit, M., Dauchet, M., eds.: TAPSOFT'97, Theory and Practice of Software Development. Volume 1214 of Lecture Notes in Computer Science., Springer (1997) 369–380
10. Leivant, D., Marion, J.Y.: A characterization of alternating log time by ramified recurrence. Theoretical Computer Science **236** (2000) 192–208
11. Neergaard, P.: A functional language for logarithmic space. In Springer-Verlag, L., ed.: In Proc. 2nd Asian Symp. on Prog. Lang. and Systems (APLAS 2004). (2004)
12. Hofmann, M.: Linear types and non-size-increasing polynomial time computation. In: Proceedings of the Fourteenth IEEE Symposium on Logic in Computer Science (LICS'99). (1999) 464–473
13. Hofmann, M.: The strength of non size-increasing computation. In Notices, A.S., ed.: POPL'02. Volume 37. (2002) 260 – 269
14. Baillot, P., Terui, K.: Light types for polynomial time computation in lambda-calculus. In Press, I.C.S., ed.: Proceedings. (2004)
15. Oitavem, I.: Characterizing nc with tier 0 pointers. Archive for Mathematical Logic **41** (2002) 35–47
16. Oitavem, I.: A term rewriting characterization of the functions computable in polynomial space. Archive for Mathematical Logic **41** (2002) 35–47
17. Jones, N.D.: Computability and complexity, from a programming perspective. MIT press (1997)
18. Grädel, E., Gurevich, Y.: Tailoring recursion for complexity. Journal of Symbolic Logic **60** (1995) 952–969

Opaque Predicates Detection by Abstract Interpretation

Mila Dalla Preda[1], Matias Madou[2], Koen De Bosschere[2], and Roberto Giacobazzi[1]

[1] Department of Computer Science
University of Verona, Italy
dallapre@sci.univr.it,
roberto.giacobazzi@.univr.it

[2] Electronics and Information Systems Department
Ghent University, Belgium
{mmadou, kdb}@elis.UGent.be

Abstract. Code obfuscation and software watermarking are well known techniques designed to prevent the illegal reuse of software. Code obfuscation prevents malicious reverse engineering, while software watermarking protects code from piracy. An interesting class of algorithms for code obfuscation and software watermarking relies on the insertion of opaque predicates. It turns out that attackers based on a dynamic or an hybrid static-dynamic approach are either not precise or time consuming in eliminating opaque predicates. We present an abstract interpretation-based methodology for removing opaque predicates from programs. Abstract interpretation provides the right framework for proving the correctness of our approach, together with a general methodology for designing efficient attackers for a relevant class of opaque predicates. Experimental evaluations show that abstract interpretation based attacks significantly reduce the time needed to eliminate opaque predicates.

1 Introduction

The aim of *malicious reverse engineering* of software is to understand the inner workings of programs in order to identify vulnerabilities, to make unauthorized modifications or to steal the intellectual property of software. *Code obfuscation* is a well-known low cost approach to prevent malicious reverse engineering of software [2, 3]. The basic idea of code obfuscation is to transform programs so that the obfuscated programs are so difficult to understand that reverse engineering becomes too expensive in terms of resources or time. *Software piracy* refers to the illegal reproduction and distribution of software applications, whether for business or personal use. The aim of *software watermarking* is to dissuade illegal copying and reseal of programs. Software watermarking is a program transformation technique that embeds a signature into the software in order to encode some identifying information about it [4, 22].

1.1 The Problem

A predicate is opaque if its value is known a priori to a program transformation, while it is difficult for attackers to deduce it [2]. Opaque predicates can be used both for obfus-

M. Johnson and V. Vene (Eds.): AMAST 2006, LNCS 4019, pp. 81–95, 2006.

cating and watermarking programs. In the case of code obfuscation, a class of obfuscating transformations known as control code obfuscators act by masking software control flow. Control code obfuscators often rely on inserting opaque predicates. Consider for example the insertion of a branch instruction controlled by an opaque predicate that always evaluates $true$, i.e., the $true$ path is always followed. Attackers are not aware of the constantly $true$ value of the opaque predicate, and have to take into account both $true$ and $false$ paths. On the other side, Monden et al. [22] store the watermark in a piece of dead code and then they make the watermark potentially reachable by inserting a true opaque predicate whose false branch transfers the control to the dead code containing the watermark. Therefore, a static analysis-based dead code removal does not eliminate the watermark, while the dead code itself is never executed. A different approach by Myles and Collberg [23] instead encodes the watermark in the constants used in opaque predicates. The resilience of an opaque predicate to attacks measures the resilience of the corresponding obfuscating/watermarking transformation. Here, we consider opaque predicates from number theory [1, 5, 23] such as $\forall x \in \mathbb{Z} : n|f(x)$, i.e., the function f always returns a multiple of n. More in general, we consider opaque predicates $\forall x \in \mathbb{Z} : f(x) \subseteq P$, i.e., the result of the function f always satisfies the property P. An attacker is a malicious user that wants to reverse engineer or copy a program for unlawful purposes, thus to succeed it has to defeat expected software protection techniques such as opaque predicate insertion. Once an opaque predicate is inserted in a program, it is possible to further protect the code using transformations meant to mask the opaque predicate itself. For example, hiding constant values by use of address computations or using bit-level operations to hide arithmetic manipulations are obfuscating transformations that mask the inserted opaque predicates. The de-obfuscation of these additional transformations and the opaque predicates detection are problems that can be studied independently. In the following we study a general and efficient methodology for disclosing opaque predicates, assuming that potential additional transformations have already been handled. We introduce a novel and efficient methodology of attack, based on Cousot and Cousot's abstract interpretation technique [7, 9], for eliminating opaque predicates. The present approach builds over the semantics-based view to code obfuscation introduced in [10, 11].

1.2 Main Results

We analyze two different approaches to opaque predicates detection. The first one is based on purely dynamic information, while the second one is based on hybrid static/dynamic information [16]. Experimental evaluations on a limited set of inputs show that a dynamic attack removes any opaque predicate, but it has the drawback of classifying many predicates as opaque, while they are not. Thus, dynamic attacks do not provide a trustful solution. Randomized algorithm may be used to eliminate opaque predicates, in this case the probability of precisely detecting an opaque predicate can be increased by augmenting the number of tries [14]. However randomized algorithms do not give an always trustful solution, but an answer that has an high probability of being precise. On the other hand, experimental evaluations on hybrid static/dynamic attacks show that breaking a single opaque predicate is rather time consuming, and may become unfeasible. We then introduce a novel methodology, based on formal program semantics and semantics approximation by abstract interpretation, to detect and then eliminate opaque predicates. Experimental evaluations show the efficiency of this new method of attack.

Attackers are malicious users that observe the behavior of the obfuscated program at different levels of abstraction with respect to the real program execution. The basic idea is to model attackers as abstract interpretations of the concrete program behaviour, i.e., the concrete program semantics. In this framework, an attacker is able to break an opaque predicate when the abstract detection of the opaque predicate is equivalent to its concrete detection. For opaque predicates as $\forall x \in \mathbb{Z} : n|f(x)$ and $\forall x \in \mathbb{Z} : f(x) \subseteq P$, this can be formalized as a completeness property of the underlying abstraction with respect to the function f. Completeness for an abstraction A with respect to some semantic function f means that no loss of precision is accumulated in the abstract computation of f on A with respect to its concrete computation. Abstract interpretation provides a systematic methodology for minimally refining an abstraction in order to make it complete for a given function. Thus, it turns out that completeness domain refinements provide here a systematic de-obfuscation technique that drives the design of abstractions, i.e., attackers, for disclosing opaque predicates.

2 Background

Notation. If $f : X^n \to Y$ is any n-ary function then its pointwise extension $f^p : \wp(X)^n \to \wp(Y)$ to the powerset is defined as $f^p(S_1, ..., S_n) \stackrel{\text{def}}{=} \{f(x_1, ..., x_n) \mid 1 \leq i \leq n,\ x_i \in S_i\}$. $\langle L, \leq, \vee, \wedge, \top, \bot \rangle$ denotes a complete lattice with ordering $\leq$, least upper bound (*lub*) $\vee$, greatest lower bound (*glb*) $\wedge$, greatest element $\top$ and least element $\bot$. Given an ordered set L the downward closure of $S \subseteq L$ is $\downarrow S \stackrel{\text{def}}{=} \{x \in L | \exists y \in S.x \leq y\}$, while the upward closure $\uparrow$ is dually defined. For $x \in L$, $\downarrow x$ is a shorthand for $\downarrow \{x\}$. Given $S \subseteq L$, $max(S) \stackrel{\text{def}}{=} \{x \in S \mid \forall y \in S.x \leq y \Rightarrow x = y\}$ is the set of maximal elements of S. Given any two functions $f, g : X \to L$, $f \sqsubseteq g$ denotes pointwise ordering, namely for any $x \in X$, $f(x) \leq g(x)$.

Abstract Interpretation. The basic idea of abstract interpretation is that the program behaviour at different levels of abstraction is an approximation of its formal semantics. The (concrete) semantics of a program is computed on the (concrete) domain $\langle C, \leq_C \rangle$, i.e., a complete lattice which models the values computed by programs. The partial ordering $\leq_C$ models relative precision between concrete values. An abstract domain $\langle A, \leq_A \rangle$ is a complete lattice which encodes an approximation of concrete program values. Abstract domains can be related to each other w.r.t. their relative degree of precision. Abstract domains are specified either by Galois connections (GCs), i.e., adjunctions, or by (upper) closures operators [7, 9]. Two complete lattices C and A form a Galois connection (C, α, γ, A), when $\alpha : C \to A$ and $\gamma : A \to C$ form an adjunction, namely $\forall a \in A, \forall c \in C : \ \alpha(c) \leq_A a \Leftrightarrow c \leq_C \gamma(a)$. α and γ are called, respectively, abstraction and concretization maps. An (upper) closure operator on C, or simply a closure, is an operator $\rho : C \to C$ which is monotone, idempotent, and extensive. We denote by $uco(C)$ the set of closures on C. When C is a complete lattice then $\langle uco(C), \sqsubseteq, \sqcap, \sqcup, \lambda x.\top, \lambda x.x \rangle$ is a complete lattice as well, where $\rho_1 \sqsubseteq \rho_2$ if and only if $\rho_2(C) \subseteq \rho_1(C)$, meaning that the abstract domain specified by ρ_1 is more precise than the abstract domain specified by ρ_2. Let us recall that each closure ρ is uniquely determined by the set of its fixpoints, given by its image $\rho(C)$. A set $X \subseteq C$ is the set of fixpoints of a closure operator if and only if X is a Moore family of C, i.e., $X = \mathcal{M}(X) \stackrel{\text{def}}{=} \{\wedge S | S \subseteq X\}$, where $\wedge \varnothing = \top \in \mathcal{M}(X)$. Given a GC (C, α, γ, A), $\rho = \gamma \circ \alpha$ is the closure corresponding to the abstract domain A.

Let (C, α, γ, A) be a GC, $f : C \rightarrow C$ a concrete function and $f^\sharp : A \rightarrow A$ an abstract function. $f^\sharp$ is a sound, i.e., correct, approximation of f if $\alpha \circ f \leq_A f^\sharp \circ \alpha$. When the soundness condition is strengthened to equality, i.e., when $\alpha \circ f = f^\sharp \circ \alpha$, the abstract function $f^\sharp$ is a complete approximation of f in A. This means that no loss of precision is accumulated in the abstract computation through $f^\sharp$. Given $A \in uco(C)$ and a semantic function $f : C \rightarrow C$, the notation $f^A \stackrel{\text{def}}{=} \alpha \circ f \circ \gamma$ denotes the best correct approximation of f in A [9]. It has been proved [12] that, given an abstraction A, there exists a complete approximation of $f : C \rightarrow C$ in A if and only if the best correct approximation f^A is complete. This means that completeness is an abstract domain property, namely that it depends on the structure of the abstract domain only. In particular, when an abstract domain is specified by a closure $\rho \in uco(C)$, we have that ρ is complete for f iff $\rho \circ f \circ \rho = \rho \circ f$ (soundness is instead encoded by $\rho \circ f \sqsubseteq \rho \circ f \circ \rho$). It turns out that an abstract domain $\rho \in uco(C)$ is complete for f if $\forall x \in \rho(C)$: $max(f^{-1}(\downarrow x)) \subseteq \rho(C)$, i.e., if ρ is closed under maximal inverse image of f. This leads to a systematic way for minimally refining an abstract domain in order to make it complete for a given semantic function [12]. The complete refinement of a domain ρ with respect to a function f is given by $\mathcal{R}_f(\rho) \stackrel{\text{def}}{=} gfp(\lambda X.\ \rho \sqcap \mathcal{M}(\cup_{y \in X} max(f^{-1}(\downarrow y))))$. It turns out that $\mathcal{R}_f(\rho)$ returns exactly the most abstract domain extending ρ and which is complete for f [12]. Thus, the completeness refinement adds the minimal amount of information needed to make the abstract domain complete. When f has more then one argument, for example when $f : C \times C \rightarrow C$, the maximal inverse image, i.e., $f^{-1}(x, y)$ is obtained by the union of the maximal inverse images of f for each fixed value of x and y [12]. For a set F of semantic functions, $\mathcal{R}_F(\rho)$ denotes the complete refinement of ρ for any function $f \in F$.

Opaque Predicates. A predicate is opaque if its outcome is known at embedding time, but it is hard for an attacker to deduce it [2,3]. The basic idea is that the insertion of opaque predicates in a program makes the program control flow difficult for an attacker to analyze. Opaque predicates find interesting applications not only in code obfuscation techniques [15], but also in software watermarking [23] and tamper-proofing [24]. There exist two major kinds of opaque predicates: true opaque predicates, denoted by P^T, that always evaluate *true*, and false opaque predicates, denoted by P^F, that always evaluate *false*. Opaque predicates can be derived from number theory [3], alias analysis [2], concurrency [6], etc. We focus here on opaque predicates based on number theory of the form $\forall x \in \mathbb{Z} : n | f(x)$. These predicates are applied in some major software protection techniques as code obfuscation [3], software watermarking [23], tamper-proofing [24] and secure mobile agents [19]. Moreover, this class of opaque predicates is used in recent implementations such as PLTO [25] — a binary rewriting system that transforms a binary program preserving the functionality — LOCO [17] — a tool for binary obfuscating and de-obfuscating transformations — and

SANDMARK [5] — a tool for software watermarking, tamper proofing and code obfuscation of Java programs.

3 Dynamic Attack

Dynamic attackers execute programs with several (but of course not all) different inputs and observe the paths followed after each conditional jump. Thus, a dynamic attacker classifies a conditional jump as controlled by a false/true opaque predicate if, during these executions, the false/true path is always taken. Therefore, a dynamic attacker detects all the executed opaque predicates, but, due to the limited set of inputs considered, it may classify a predicate as opaque while it is not, called a *false negative*. Let us measure the false negative rate of a dynamic attacker. We execute the SPECint2000 benchmarks (without adding opaque predicates) with the reference inputs, and then we observe the conditional jumps. We use DIOTA[1] [18] to identify conditional jumps that always follow the true path, the false path or take both of them.

Table 1. Execution after conditional jumps

	% only flth	*% only jump*	*% both ways*
bzip2	42	23	35
crafty	24	19	57
gap	39	21	40
gcc	36	18	46
gzip	36	24	40
mcf	43	32	25
parser	29	14	57
perlbmk	45	23	33
twolf	39	21	40
vortex	59	26	15
vpr	42	21	38
average	*39*	*22*	*39*

The benchmarks are listed in Table 1. For each benchmark, the percentage of regular conditional jumps that look like false/true opaque predicates are annotated in the first/second column, while the percentage of regular conditional jumps are reported in the third column. Benchmarks do not contain opaque predicates, so that the opaque predicates detected by dynamic attack are false negatives. This experimental evaluations show that a dynamic attacker has an average of false negative rate of 39% and 22%, respectively for false and true opaque predicates. An attacker can improve these results using its knowledge of the program functionality in order to generate different inputs that are likely to execute different program paths. This will be very time consuming. Another way is to generate dynamic test data to improve the condition/decision coverage (CDC)[2]. For complex programs, the CDC is at most 58% [20], so 42% of all conditions will be seen as opaque predicates or dead code by the attacker which is of course incorrect. This leads us to conclude that dynamic attacks are too imprecise.

[1] DIOTA: a dynamic instrumentation tool which keeps a running program unaltered at its original location and generate instrumented code on the fly somewhere else.

[2] Condition/decision coverage measures the percentage of conditional jumps that are executed true at least once and false at least once.

4 Brute Force Attack

In this section we study an hybrid static/dynamic brute force attack acting on assembly basic blocks[3], where the instructions of the opaque predicate are statically identified (static phase) and are then executed on all possible inputs (dynamic phase). Let us consider the following opaque predicate $\forall x \in \mathbb{Z} : 2|(x^2 + x)$. Let us remark that the implementation of this opaque predicate decomposes the function $x^2 + x$ into elementary functions such as square x^2 and addition $x + y$. We make the assumption that the instructions (that is, elementary functions) corresponding to an opaque predicate are always grouped together, i.e., there are no program instructions between them. The static phase aims at identifying the instructions corresponding to an opaque predicate. Thus, for each conditional jump j the attack considers the instruction i immediately preceding j. The dynamic phase then checks whether i and j give rise to an opaque predicate. If this is the case the predicate is classified as opaque. Otherwise, the analysis proceeds upward by considering the next instruction preceding i, until an opaque predicate is found or the instructions in the basic block terminate. In this latter case, the predicate is not opaque. The computational effort, measured as number of steps, of the attack is $n^2 * (2^w)^r$, where n is the number of instructions of the opaque predicate, r is the number of registers and w is the width of the registers used by the opaque predicate. Consider for example the above true opaque predicate compiled for a 32-bit architecture. The predicate is executed with all possible 2^{32} inputs. This compiled code is then executed under the control of GDB, a well known open-source debugger[4], with all 2^{32} inputs. In particular $2|(x^2 + x)$ can be written in five x86 instructions, so that for this architecture the computational effort to break this opaque predicate will be $5^2 * 2^{64}$. During the hybrid attack, two variables are needed as input for the addition, so that there are at most 2 registers taken as input during the attack, i.e. r=2, and the width of these registers is 32 bits, i.e. $w = 32$.

It would be interesting to measure the time needed by this attack to detect an opaque predicate. Let us consider the opaque predicate $\forall x \in \mathbb{Z} : 2|(x + x)$ and measure the time needed to detect it. In assembly, this opaque predicate in a 16-bit environment consists of three instructions. The execution under control of GDB of these three assembly instructions with all 2^{16} inputs takes 8.83 seconds on a 1.6 GHz Pentium M processor with 1 GB of main memory running RedHat Fedora Core 3. In this experimental evaluation, the static phase has been performed by hand, meaning that the starting instruction of the opaque predicate was given. This leads us to conclude that the hybrid static/dynamic approach is precise although it is noticeably time consuming.

5 Breaking Opaqueness by Abstract Interpretation

We introduce an approach based on abstract interpretation for detecting opaque predicates. This novel technique leads to a formal characterization of a class of attackers that are able to break a specific type of commonly used opaque predicates, i.e.,

[3] A basic block is a sequence of instructions with a single entry point, single exit point, and no internal branches.

[4] http://www.gnu.org/software/gdb/

$\forall x \in \mathbb{Z} : n|f(x)$. This result can then be generalized to a wider class of opaque predicates, i.e., $\forall x \in \mathbb{Z} : f(x) \subseteq P$ where P is a generic property of integer numbers. In this case, we provide a methodology for designing efficient attackers. Experimental evaluations show how this abstract interpretation-based approach significantly reduces the computational effort of the attacker.

5.1 Modeling Attackers

Attackers have different precision degrees, according to the accuracy they have in observing program behaviours. We show that abstract interpretation turns out to be a suitable framework for modeling attackers and for classifying them according to their level of precision [10, 11]. Let $\langle \wp(\mathbb{Z}), \subseteq \rangle$ be the concrete domain for an integer program variable. An attacker can be modeled by an abstract domain $A \in uco(\wp(\mathbb{Z}))$, which may precisely represent the level of abstraction of an attacker. In the following, A denotes an abstract domain with partial ordering relation $\leq_A$, abstraction/concretization maps $\alpha_A : \wp(\mathbb{Z}) \rightarrow A$ and $\gamma_A : A \rightarrow \wp(\mathbb{Z})$. For example, the following well-known abstract domains $Sign = \{\mathbb{Z}, \mathbb{Z}_{\geq 0}, \mathbb{Z}_{\leq 0}, 0, \varnothing\}$ and $Parity = \{\mathbb{Z}, even, odd, \varnothing\}$ can model different attackers. Modeling attackers by abstract domains allows us to compare them with respect to their level of abstraction. Consider two attackers $A_1, A_2 \in uco(\wp(\mathbb{Z}))$. If A_2 is an abstraction of A_1, i.e., $A_1 \sqsubseteq A_2$, then the attacker A_1 is more precise (i.e., concrete) than the attacker A_2 in observing the obfuscated program. In our model, an attacker A breaks an opaque predicate when the abstract detection of the opaque predicate is equivalent to its concrete detection. Abstract domains can encode a significant approximation of the concrete domain. Accordingly, we will show that abstract detection of opaque predicates may result significantly simpler.

Attackers for Predicates $n|f(x)$. Let us consider numerical true opaque predicates of the form: $\forall x \in \mathbb{Z} : n|f(x)$, namely the function $f : \mathbb{Z} \rightarrow \mathbb{Z}$ always returns a value that is a multiple of $n \in \mathbb{Z}$. This class of opaque predicates is used in major obfuscating tools such as SANDMARK [5] and LOCO [17], and in the software watermarking algorithm by Arboit [1], recently implemented by Collberg and Myles [23].

In order to detect that the predicate $n|f(x)$ is opaque one needs to check the concrete test $\mathrm{CT}^f \stackrel{\text{def}}{=} \forall x \in \mathbb{Z} : f(x) \in n\mathbb{Z}$, where $n\mathbb{Z}$ denotes the set of integers that are multiples of $n \in \mathbb{Z}$. Our goal is to devise an abstract interpretation-based method which allows to perform the test of opaqueness for f on a suitable abstract domain. We are therefore interested in abstract domains which are able to represent precisely the property of being a multiple of n, i.e., abstract domains $A \in uco(\wp(\mathbb{Z}))$ such that there exists some $a_n \in A$ such that $\gamma_A(a_n) = n\mathbb{Z}$. Let $f^\sharp : A \rightarrow A$ be an abstract function that approximates f on A. Then, the *abstract test* on A is defined as follows:

$$AT_A^{f^\sharp} \stackrel{\text{def}}{=} \forall x \in \mathbb{Z} : f^\sharp(\alpha_A(\{x\})) \leq_A a_n$$

Definition 1. $AT_A^{f^\sharp}$ is *sound* (*complete*) when $AT_A^{f^\sharp} \Rightarrow CT^f$ ($AT_A^{f^\sharp} \Leftrightarrow CT^f$).

When $AT_A^{f^\sharp}$ is complete we also say that the attack $\langle A, f^\sharp \rangle$ (or simply A when $f^\sharp$ is clear from the context) *breaks* the opaque predicate $\forall x \in \mathbb{Z} : n|f(x)$.

Theorem 1. *Consider A such that there exists $a_n \in A$: $\gamma_A(a_n) = n\mathbb{Z}$, then:*

(1) *If $f^\sharp$ is sound approximation of f on the singletons, that is $\forall x \in \mathbb{Z}$, $\alpha_A(\{f(x)\}) \leq_A f^\sharp(\alpha_A(\{x\}))$, then $AT_A^{f^\sharp}$ is sound.*
(2) *If $f^\sharp$ is complete approximation of f on the singletons, that is $\forall x \in \mathbb{Z}$, $\alpha_A(\{f(x)\}) = f^\sharp(\alpha_A(\{x\}))$, then $AT_A^{f^\sharp}$ is complete.*

Thus, the key point is to design a suitable abstract domain A together with a complete approximation $f^\sharp$ of f.

Abstract Functions. We already observed in Section 4 that a function $f : \mathbb{Z} \to \mathbb{Z}$ is decomposed into elementary functions, i.e. assembly instructions within some basic block. Following the same approach, let us assume that the function f can be expressed as a composition of elementary functions, namely $f = \lambda x.h(g_1(x, ..., x), ..., g_k(x, ..., x))$ where $h : \mathbb{Z}^k \to \mathbb{Z}$ and $g_i : \mathbb{Z}^{n_i} \to \mathbb{Z}$. More in general, each g_i can be further decomposed into elementary functions. For example, $f(x) = x^2 + x$ is decomposed as $h(g_1(x), g_2(x))$ where $h(x, y) = x + y$, $g_1(x) = x^2$ and $g_2(x) = x$. Let us consider the pointwise extensions of the elementary functions, which are still denoted, with a slight abuse of notation, by $h : \wp(\mathbb{Z})^k \to \wp(\mathbb{Z})$ and $g_i : \wp(\mathbb{Z})^{n_i} \to \wp(\mathbb{Z})$, and let us denote their composition by $F \stackrel{\text{def}}{=} \lambda X.h(g_1(X, ..., X), ..., g_k(X, ..., X)) : \wp(\mathbb{Z}) \to \wp(\mathbb{Z})$. For example, for the above decomposition $f(x) = x^2 + x = h(g_1(x), g_2(x))$, we have that $F : \wp(\mathbb{Z}) \to \wp(\mathbb{Z})$ is as follows: $F(X) = \{y^2 + z \mid y, z \in X\}$. Observe that F does not coincide with the pointwise extension f^p of f, e.g., $F(\{1, 2\}) = \{2, 3, 5, 6\}$ while $f^p(\{1, 2\}) = \{2, 6\}$. Let us also notice that F on singletons coincides with f, namely for any $x \in \mathbb{Z}$, $F(\{x\}) = f(x)$. Thus, the concrete test CT^f can be equivalently formulated as $\forall x \in \mathbb{Z} : F(\{x\}) \subseteq n\mathbb{Z}$.

Let $A \in uco(\wp(\mathbb{Z}))$ be an abstract domain such that there exists some $a_n \in A$ with $\gamma_A(a_n) = n\mathbb{Z}$. The attacker A approximates the computation of the function $F : \wp(\mathbb{Z}) \to \wp(\mathbb{Z})$ in a step by step fashion, meaning that A approximates every elementary function composing F. Thus, the abstract function $F^\sharp : A \to A$ is defined as the composition of the best correct approximations h^A and g_i^A on A of the elementary functions, namely:

$$F^\sharp(a) \stackrel{\text{def}}{=} \alpha_A(h(\gamma_A(\alpha_A(g_1(\gamma_A(a), ..., \gamma_A(a)))), ..., \gamma_A(\alpha_A(g_k(\gamma_A(a), ..., \gamma_A(a))))))$$

When the abstract test $AT_A^{F^\sharp}$ for $F^\sharp$ on A holds, the attacker modeled by the abstract domain A classifies the predicate $n|f(x)$ as opaque. It turns out that $F^\sharp$ is a correct approximation of F on A, namely $\alpha_A \circ F \sqsubseteq_A F^\sharp \circ \alpha_A$, and this guarantees the soundness of the abstract test $AT_A^{F^\sharp}$.

Corollary 1. *$AT_A^{F^\sharp}$ is sound.*

Consider for example the opaque predicate $\forall x \in \mathbb{Z} : 3|(x^3 - x)$, and the abstract domain A_{3+} in the figure below. A_{3+} precisely represents the property of being a multiple of 3, i.e. $3\mathbb{Z}$, and its negation, i.e. $\mathbb{Z} \smallsetminus 3\mathbb{Z}$.

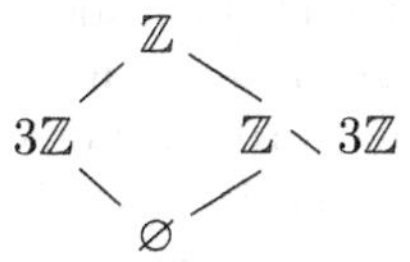

In this case, $f(x) = x^3 - x = h(g_1(x), g_2(x))$ where $h(x, y) = x - y$, $g_1(x) = x^3$ and $g_2(x) = x$, so that $F : \wp(\mathbb{Z}) \to \wp(\mathbb{Z})$ is given by $F(X) = \{y^3 - z \mid y, z \in X\}$. Hence, it turns out that $F^\sharp(3\mathbb{Z}) = 3\mathbb{Z}$ while $F^\sharp(\mathbb{Z} \smallsetminus 3\mathbb{Z}) = \mathbb{Z}$. Here, the abstract test $AT_{A_{3+}}^{F^\sharp}$ is sound but not complete, because $F^\sharp : A_{3+} \to A_{3+}$ is a sound but not complete approximation of f on the singletons. In fact, for $\{2\} \in \wp(\mathbb{Z})$, it turns out that $\alpha_{A_{3+}}(\{f(2)\}) = \alpha_{A_{3+}}(\{6\}) = 3\mathbb{Z}$ while $F^\sharp(\alpha_{A_{3+}}(\{2\})) = F^\sharp(\mathbb{Z} \smallsetminus 3\mathbb{Z}) = \mathbb{Z}$. Thus the abstract test $AT_{A_{3+}}^{F^\sharp}$, i.e., $\forall x \in \mathbb{Z} : F^\sharp(\alpha_{A_{3+}}(\{x\})) \leq 3\mathbb{Z}$ does not hold even if CT^f does. Thus, in general $AT_A^{F^\sharp}$ is sound but not complete, meaning that the attacker $\langle A, F^\sharp \rangle$ is not able to break the opaque predicate $\forall x \in \mathbb{Z} : n|f(x)$.

Recall that abstract domain completeness is preserved by function composition [12], i.e. if an abstract domain is complete for f and g then A is complete for $f \circ g$ as well. As a consequence, if an abstract domain A is complete for the elementary functions h and g_i that decompose F then A is complete also for their composition F. It turns out that completeness of an abstract domain A w.r.t. the elementary functions composing F guarantees that the attacker A is able to break the opaque predicate $\forall x \in \mathbb{Z} : n|f(x)$.

Corollary 2. *If A is complete for the elementary functions h and g_i composing F then $\langle A, F^\sharp \rangle$ breaks the opaque predicate $\forall x \in \mathbb{Z} : n|f(x)$.*

Let us consider the opaque predicate $\forall x \in \mathbb{Z} : 3|(x^3 - x)$ and the abstract domain 3-*arity* represented in the following figure.

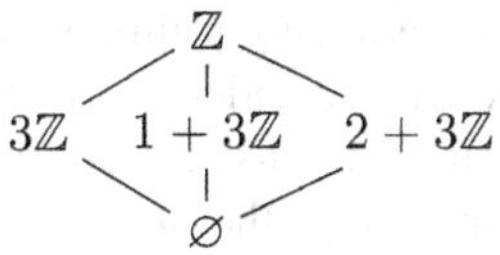

The function $f(x) = x^3 - x$ is decomposed as $h(g_1(x), g_2(x))$ where $h(x, y) = x - y$, $g_1(x) = x^3$ and $g_2(x) = x$. It turns out that the abstract domain 3-*arity* is complete for the pointwise extensions of h, g_1 and g_2, i.e. $\lambda\langle X, Y\rangle.X - Y$, $\lambda X.X^3$ and $\lambda X.X$, and therefore, by Corollary 2, the attacker 3-*arity* is able to break the opaque predicate $\forall x \in \mathbb{Z} : 3|(x^3 + x)$.

Lemma 1. *3-arity is complete for $\lambda X.X^3$, $\lambda X.X$ and $\lambda\langle X, Y\rangle.X - Y$.*

Experimental Results. A prototype of the above described attack based on the abstract domain *Parity* has been implemented using LOCO [17], a x86 tool for obfuscation/de-obfuscation transformations which is able to insert opaque predicates. This experimental evaluation has been conducted on the aforementioned 1.6 GHz Pentium M-based system. Each program of the SPECint2000 benchmark suite is obfuscated by inserting the following true opaque predicates: $\forall x \in \mathbb{Z} : 2|(x^2 + x)$ and $\forall x \in \mathbb{Z} : 2|(x + x)$.

It turns out that *Parity* is complete for addition, square and identity function, thus by Corollary 2, the abstract domain *Parity* models an attacker that is able to break these opaque predicates. In the obfuscating transformation each basic block of the input assembly program is split into two basic blocks. Then, LOCO checks whether the opaque predicate can be inserted between these two basic blocks: a liveness analysis is used here to ensure that no dependency is broken and that the obfuscated program is functionally equivalent to the original one. In particular, liveness analysis checks that the registers and the conditional flags affected by the opaque predicate are not live in the program point where the opaque predicate will be inserted. Moreover, our tool also checks by a standard constant propagation whether the registers associated to the opaque predicate are constant or not. If constant propagation detects that these are constant then the opaque predicate can be trivially broken and therefore is not inserted. Although liveness analysis and constant propagation are noticeably time-consuming, they are nevertheless necessary both to certificate functional equivalence between original and obfuscated program and to guarantee that the opaque predicate cannot be trivially broken by constant propagation. The algorithm used to detect opaque predicates is analogous to the brute force attack algorithm described in Section 4. Fig. 1 describes the basic block, by pseudo-code, which implements the opaque predicate $\forall x \in \mathbb{Z} : 2|(x^2 + x)$.

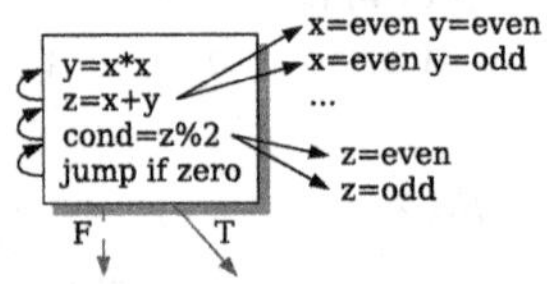

Fig. 1. Breaking $\forall x \in \mathbb{Z},\ 2|(x^2 + x)$

Let us describe how our de-obfuscation algorithm works. For each conditional jump j, `jump if zero` in the figure, we consider the instruction i which immediately precedes j, `cond=z%2` in the figure. The instructions j and i are abstractly executed on each value of the abstract domain (i.e. the attack). In the considered case of the attack modeled by *Parity*, both non-trivial values *even* and *odd* are given as input to `cond=z%2`. When `z` evaluates to *even*, `cond` evaluates to 0 and therefore the true path is followed. On the other hand, when `z` is evaluated to *odd*, `cond` evaluates to 1 and the false path is taken. Thus, i does not give rise to an opaque predicate, so that we need to consider the instruction `z=x+y` which immediately precedes i. The instruction `z=x+y` is binary and therefore we need to consider all the values in *Parity* × *Parity*. This process is iterated until an opaque predicate is detected or the end of the basic block is reached. In our case, the opaque predicate is detected when the algorithm analyses the instruction `y=x*x` because whether `x` is evaluated to *even* or *odd* the true path is taken. The number of computational steps needed for breaking one single opaque predicate by an attack based on an abstract domain A is $n^2 * d^r$, where n is the number of instructions composing the opaque predicate, r is the number of registers used by the opaque predicate and d is the number of abstract values in A. The reduction of the computational effort of the abstract interpretation-based attack with respect to the brute force attack can therefore be huge since the abstract domain can encode a very coarse approximation. In the considered example, the number of steps for detecting $\forall x \in \mathbb{Z} : 2|x + x$ through the abstract domain *Parity* results to be $3^2 * 2^2$. In fact, the opaque predicate consists of 3 instructions, uses 2 registers and *Parity* has 2 non-trivial abstract values. In Table 2 we show the results of the obfuscation/de-obfuscation process on the SPECint2000 benchmark suite. The first and second columns report respectively the number of opaque predicates inserted in each benchmark and the time needed for such obfuscation, while the third column lists the time needed to de-obfuscate. It turns out that the *Parity*-based de-obfuscation process is able to detect all the inserted opaque predicates. Let us recall that the brute force attack took 8.83 seconds to detect only one occurrence of the opaque predicate $\forall x \in \mathbb{Z} : 2|x+x$ in a 16-bit environment, while the abstract interpretation-based de-obfuscation attack took 8.13 seconds to de-obfuscate 66176 opaque predicates in a 32-bit environment.

Table 2. Timings of obfuscation and de-obfuscation

	# opaque pred	time obf	time deobf
bzip2	442	0.53	0.13
crafty	3298	35.18	0.47
gap	6659	8.59	1.02
gcc	28006	476.33	3.11
gzip	734	0.44	0.11
mcf	189	0.29	0.06
parser	2543	2.48	0.31
perlbmk	10255	42.71	1.23
twolf	3575	1.88	0.48
vortex	8269	8.79	0.91
vpr	2206	1.44	0.3
Total	66176	578.66	8.13

6 Designing Domains for Breaking Opaque Predicates

This section shows how the completeness domain refinements can be used to derive models of attackers which are able to break a given opaque predicate. Let us consider the opaque predicate $\forall x \in \mathbb{Z} : 3|(x^3 - x)$ and the attacker $A_3 \stackrel{\text{def}}{=} \{\mathbb{Z}, 3\mathbb{Z}\}$, that is the minimal abstract domain which represents precisely the property of being a multiple of 3. Recall that the function $f(x) = x^3 - x$ is decomposed as $h(g_1(x), g_2(x))$ where $h(x,y) = x - y$, $g_1(x) = x^3$ and $g_2(x) = x$. It turns out that A_3 is not able to break the above opaque predicate, since $F^\sharp : A_3 \rightarrow A_3$ is not a complete approximation of f on singletons. In fact, consider $\{2\} \in \wp(\mathbb{Z})$, it turns out that $\alpha_{A_3}(\{f(2)\}) = \alpha_{A_3}(\{6\}) = 3\mathbb{Z}$ while $F^\sharp(\alpha_{A_3}(\{2\})) = F^\sharp(\mathbb{Z}) = \mathbb{Z}$. Corollary 2 does apply here because A_3 is complete for g_1 and g_2 but not for h. However, as recalled in Section 2, completeness can be obtained by a domain refinement. We thus systematically transform A_3 by the completeness domain refinement w.r.t. $h = \lambda\langle X, Y\rangle.X - Y$. We obtain the abstract domain $\mathcal{R}_h(A_3)$ that models an attacker which is able to break $\forall x \in \mathbb{Z} : 3|(x^3 - x)$. As recalled in Section 2, the application of the completeness domain refinement adds to $A_{3\mathbb{Z}}$ the maximal inverse images under h of all its elements until a fixpoint is reached, that is for any fixed $X \subseteq \mathbb{Z}$ and a belonging to the current abstract domain, we iteratively add the following sets of integers: $max\{Z \subseteq \mathbb{Z} \mid Z - X \subseteq a\}$. It is not hard to verify that the following elements provide exactly the minimal amount information to add to A_3 in order to make it complete for h.

- if $X = \{0\}$ then: $max\{Z \subseteq \mathbb{Z} \mid Z - X \subseteq 3\mathbb{Z}\} = 3\mathbb{Z}$
- if $X = \{1\}$ then: $max\{Z \subseteq \mathbb{Z} \mid Z - X \subseteq 3\mathbb{Z}\} = 1 + 3\mathbb{Z}$
- if $X = \{2\}$ then: $max\{Z \subseteq \mathbb{Z} \mid Z - X \subseteq 3\mathbb{Z}\} = 2 + 3\mathbb{Z}$

Therefore, $\mathcal{R}_h(A_3) = \{\mathbb{Z}, 3\mathbb{Z}, 1+3\mathbb{Z}, 2+3\mathbb{Z}, \varnothing\} = 3\text{-}arity$. Let us notice that we were able to systematically obtain the attacker $3\text{-}arity$, which is able to break the opaque predicate, through a completeness refinement of the minimal abstract domain A_3.

It turns out that given $n \in \mathbb{N}$, the abstract domain $n\text{-}arity$, in the figure below, is complete for addition, difference and, for $k \in \mathbb{N}$, k-power (i.e., $\lambda X.X^k$). Therefore, by Corollary 2, the attacker $n\text{-}arity$ breaks the opaque predicates $\forall x \in \mathbb{Z}, n|f(x)$, where f is a polynomial function. Observe that the abstract domain $n\text{-}arity$ is an instance of Granger's domain of congruences [13].

Theorem 2. *The attacker $n\text{-}arity$ breaks all the opaque predicates of the following form: $\forall x \in \mathbb{Z}:\ n|f(x)$, where $f(x)$ is a polynomial function.*

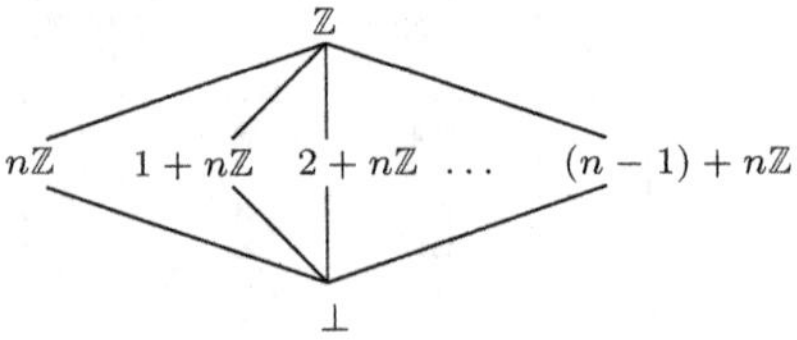

6.1 Breaking Opaque Predicates $P(f(x))$

Let us now consider the wider class $P(f(x))$ of opaque predicates where each predicate has the following form: $f(x) \subseteq P$, with $P \subseteq \mathbb{Z}$ and $f : \mathbb{Z} \to \mathbb{Z}$. It is possible to generalize the results of the previous sections, in particular Theorem 1, Corollary 1 and Corollary 2, to opaque predicates in $P(f(x))$. This is simply done by replacing the property $n\mathbb{Z}$ of being a multiple of n, with a general property P over integers. This allows us to provide a formal methodology for designing abstract domains that model attackers able to break opaque predicates in $P(f(x))$. Let $\forall x \in \mathbb{Z} : f(x) \subseteq P$ be an opaque predicate and let us consider the minimal abstract domain A_P that represents precisely the property P, i.e., $A_P \stackrel{\text{def}}{=} \{\mathbb{Z}, P\}$. As above, we assume that the function f can be expressed as a composition of elementary functions, namely $f = \lambda x.h(g_1(x, ..., x), ..., g_k(x, ..., x))$ where $h : \mathbb{Z}^k \to \mathbb{Z}$ and $g_i : \mathbb{Z}^{n_i} \to \mathbb{Z}$. Then, we compute the completeness domain refinement of A_P w.r.t. the set of elementary functions composing f, namely $\mathcal{R}_{h,g_1,\dots,g_k}(A_P)$. It turns out that the refined domain is able to break the opaque predicate $\forall x \in \mathbb{Z} : f(x) \subseteq P$.

Theorem 3. *The attacker modeled by the abstract domain $\mathcal{R}_{h,g_1,\dots,g_k}(A_P)$ breaks the opaque predicate $\forall x \in \mathbb{Z} : f(x) \subseteq P$.*

Thus, completeness domain refinements provide here a systematic methodology for designing attackers that are able to break opaque predicates of the form: $\forall x \in \mathbb{Z} : f(x) \subseteq P$.

The previous result is independent from the choice of the concrete domain $\mathbb{Z}$ and can be extended to a general domain of computation *Dom*.

Corollary 3. *Consider an opaque predicate* $\forall x \in Dom$: $f(x) \subseteq P$, *where* $f : Dom \to Dom$, $f = h(g_1(x,...,x),...,g_k(x,...,x))$, *and* $P \subseteq Dom$. *The abstract domain* $\mathcal{R}_{h,g_1,...,g_k}(\{Dom, P\})$ *is able to break opaque predicate.*

7 Conclusion and Future Work

In this work we propose an abstract interpretation-based approach to detect opaque predicates. It turns out that, considering opaque predicates of the form: $\forall x \in \mathbb{Z} : n|f(x)$, the ability of an attacker, i.e., an abstract domain, to break opaque predicates can be formalized as a completeness problem w.r.t. function f. Consequently completeness domain refinement can be used to derive efficient attackers. In particular it turns out that the abstract domain n-*arity* breaks the opaque predicate $\forall x \in \mathbb{Z} : n|f(x)$, where n ranges over $\mathbb{N}$ and f is a polynomial function. This result is then generalized to a wider class of opaque predicates of the form $\forall x \in \mathbb{Z} : f(x) \subseteq P$, where the attacker able to break the opaque predicate is obtained by completeness refinement of the abstract domain $A_P = \{\mathbb{Z}, P\}$.

The insertion of an opaque predicate code creates a path that is never taken. Notice that when the false path of a true opaque predicate contains another opaque predicate the degree of obfuscation of the transformation increases. The two opaque predicates interact with each other, and this dependence adds more confusion in the understanding of the original control flow of the program. Thus the insertion of dependent opaque predicates can be seen as a novel obfuscation technique.

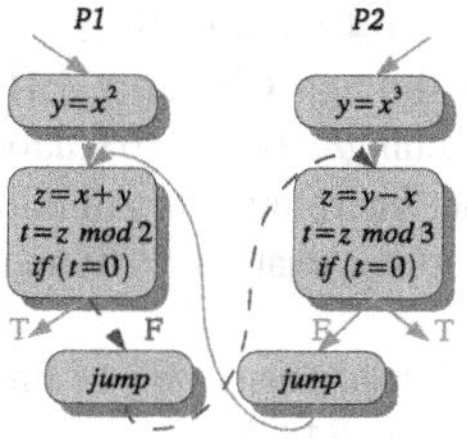

Fig. 2. Dependent opaque predicate

Consider for example the true opaque predicates $P1 : \forall x \in \mathbb{Z} : 2|(x^2 + x)$ and $P2 : \forall x \in \mathbb{Z} : 3|(x^3 - x)$ that interact with each other as depicted in the above figure. On the left-hand side we have the opaque predicate $P1$, while on the right-hand side we have $P2$, expressed in terms of elementary functions, i.e., assembly instructions. Observe that the false branch of predicate $P1$ enters the second basic block of predicate $P2$ and vice versa. The attacker modeled by the abstract domain *Parity* should be able to break opaque predicate $P1$. The problem is that *Parity* cannot break $P2$ and therefore we have an incoming edge on the second basic block of opaque predicate $P1$ coming from $P2$. This gives the idea of why we are no longer able to break opaque predicate $P1$ with the *Parity* domain. Therefore when there are opaque predicates that interact with each other the attacker needs to take into account these dependencies. Our guess is that a suitable attacker to handle this situation could probably be obtained by combining the abstract domains breaking the individual opaque predicates. The main problem is that one opaque predicate which is not breakable by our technique could protect breakable opaque predicates by interacting with these opaque predicates.

It would be also interesting to consider abstract domains that are more complex than the ones considered so far. Program properties that can be studied only on more complex domains could lead to the design of novel opaque predicates. Since these properties derive from a more complex analysis the corresponding opaque predicates should be more resilient to attacks. Consider for example the polyhedral abstract domain [8] and the abstract domain of octagons [21] for discovering properties of numerical variables.

Acknowledgments. The authors would like to thank the Fund for Scientific Research - Belgium - Flanders (FWO) and Ghent University for their financial support.

References

1. G. Arboit. A Method for Watermarking Java Programs via Opaque Predicates. In *Proc. Int. Conf. Electronic Commerce Research (ICECR-5)*, 2002.
2. C. Collberg, C. Thomborson and D. Low. Manufacturing Cheap, Resilient, and Stealthy Opaque Constructs. In *Proc. ACM POPL'98*, pp. 184-196, 1998.
3. C. Collberg, C. Thomborson and D. Low. A Taxonomy of Obfuscating Transformations. Technical Report 148, The University of Auckland, New Zealand, 1997.
4. C. Collberg, E. Carter, S. Debray, A. Huntwork, C. Linn and M. Stepp. Dynamic Path-Based Software Watermarking. In *Proc. ACM PLDI'04*, pp. 107-118, 2004.
5. C. Collberg, G. Myles and A. Huntwork. SandMark - A Tool for Software Protection Research. *IEEE Security and Privacy*, 1(4):40-49, 2003.
6. C. Collberg. CSc620: Security through Obscurity. Handouts of a course available at: www.cs.arizona.edu/ collberg/Teaching/620/ 2002/Handouts/Handout-13.ps, 2002.
7. P. Cousot and R. Cousot. Abstract interpretation: A unified lattice model for static analysis of programs by construction or approximation of fixpoints. In *Proc. ACM POPL'77*, pp. 238–252, 1977.
8. P. Cousot and N. Halbwachs. Automatic discovery of linear restraints among variables of a program. In *Proc. ACM POPL'78*, pp. 84–97, 1978.
9. P. Cousot and R. Cousot. Systematic design of program analysis frameworks. In *Proc. ACM POPL'79*, pp. 269–282, 1979.
10. M. Dalla Preda and R. Giacobazzi. Semantic-based code obfuscation by abstract interpretation. In *Proc. 32nd ICALP*, LNCS 3580, pp. 1325-1336, 2005.
11. M. Dalla Preda and R. Giacobazzi. Control Code Obfuscation by Abstract Interpretation. In *Proc. 3rd IEEE International Conference on Software Engineering and Formal Methods (SEFM'05)*, pp. 301-310, 2005.
12. R. Giacobazzi, F. Ranzato, and F. Scozzari. Making abstract interpretations complete. *J. ACM.*, 47(2):361-416, 2000.
13. P. Granger. Static analysis of linear congruence equality among variables of a program. In *Proc. Joint Conf. on Theory and Practice of Software Development (TAPSOFT'91)*, pp. 169-192, 1991.
14. J. Hormkovic. Algorithmics for Hard Problems. Springer-Verlag, 2002.
15. C. Linn and S. Debray. Obfuscation of Executable Code to Improve Resistance to Static Disassembly. In *Proc. 10th ACM Conference on Computer and Communications Security (CCS'03)*, 2003.

16. M. Madou, B. Anckaert, B. De Sutter and K. De Bosschere. Hybrid Static-Dynamic Attacks Against Software Protection Mechanisms. In *Proc. 5th ACM Workshop on Digital Rights Management (DRM'05)*, 2005.
17. M. Madou, L. Van Put and K. De Bosschere. Loco: An Interactive Code (De)Obfuscation tool. In *Proc. ACM SIGPLAN Workshop on Partial Evaluation and Program Manipulation (PEPM'06)*, 2006.
18. J. Maebe, M. Ronsse and K. De Bosschere. DIOTA: Dynamic Instrumentation, Optimization and Transformation of Applications. In *Proc. 4th Workshop on Binary Translation (WBT'02)*, 2002.
19. A. Majumdar and C. Thomborson. Securing Mobile Agents Control Flow Using Opaque Predicates. In *Proc. 9th Int. Conf. Knowledge-Based Intelligent Information and Engineering Systems (KES'05)*, 2005.
20. C. Michael, G. McGraw, M. Schatz and C. Walton. Genetic Algorithms for Dynamic Test Data Generation. In *Proc. ASE'97*, pp. 307-308, 1997.
21. A. Minè. The octagon abstract domain. In *Proc. Analysis, Slicing and Transformation (AST'01)*, pp. 310-319, 2001.
22. A. Monden, H. Iida, K. Matsumoto, K. Inoue and K. Torii. A Practical Method for Watermarking Java Programs. In *Proc. 24th Computer Software and Applications Conference*, pp. 191-197, 2000.
23. G. Myles and C. Collberg. Software Watermarking via Opaque Predicates: Implementation, Analysis, and Attacks. In *Proc. Int. Conf. Electronic Commerce Research (ICECR-7)*, 2004.
24. J. Palsberg, S. Krishnaswamy, M. Kwon, D. Ma, Q. Shao and Y. Zhang. Experience with Software Watermarking. In *Proc. 16th Annual Computer Security Applications Conference (ACSAC'00)*, pp. 308-316, 2000.
25. B. Schwarz, S. Debray and G. Andrews PLTO: A Link-Time Optimizer for the Intel IA-32 Architecture. In *Proc. Workshop on Binary Translation (WBT'01)*, 2001.

DO-Casl: An Observer-Based Casl Extension for Dynamic Specifications

Matteo Dell'Amico and Maura Cerioli

DISI–Dipartimento di Informatica e Scienze dell'Informazione
Università di Genova, Via Dodecaneso,
35, 16146 Genova, Italy
{dellamico, cerioli}@disi.unige.it

Abstract. We present DO-CASL, a new member of the CASL family of specification languages. It is an extension of CASL-LTL and it supports a methodology for conveniently writing loose specifications of observers on dynamic sorts. The need for such constructs arose during the development of a CASL library for distributed systems. Indeed, we have frequently used the same pattern of specification, in order to solve a generalization of the frame problem while using observers. The constructs we propose make the resulting specifications more readable, concise and maintainable. The semantics of our extension is given by reduction to standard CASL-LTL, which is, in turn, reducible to standard CASL whenever temporal logic is not used. A small prototype of the pre-processor from DO-CASL to CASL-LTL has been implemented as well.

1 Introduction

Middleware is widely and successfully used to support programmer productivity, by providing solutions to the most common tasks and abstractions of low-level concepts. Many software projects are developed by writing mainly "glue code", connecting the functionalities made available from middleware. Thus, the usefulness of a programming language or environment is greatly influenced by the quality and quantity of available libraries.

We argue that this is also true for specification languages. That is, besides libraries for standard datatypes (such as lists, sets or integers), software specifications also need libraries for middleware primitives. This way, the developer is lifted from the burden of axiomatization of components that can sometimes be very complex. Moreover, they are part of the context, not to be implemented anew for each system development.

Our work stems from the definition of a CASL-LTL library for decentralized distributed systems.

We decided to use one of the CASL family of languages, because of its large acceptance in the scientific community. Among them, we selected CASL-LTL, as it provides an intuitive representation of label transition systems and hence of dynamic systems.

M. Johnson and V. Vene (Eds.): AMAST 2006, LNCS 4019, pp. 96–110, 2006.

Our library supports a model driven development[1]: our specifications abstract away from concrete middleware to propose coherent sets of functionalities which are implemented, in most cases, by several infrastructures. Technically, this means that our specifications are loose. In order to allow different implementations and to get flexibility for further extensions of each given specifications by other primitives, we adopt an observer oriented style of specification. Observers are used to hide details of the internal states, which are only partially known in this layer of abstraction. Since we assume our knowledge of the needed observers to be incomplete, we do not deduce equality of two states from all the current observers yielding the same result on both.

Though our specifications are loose, we want to state the observable properties of the primitives, for the user to build upon them. Therefore, in this context loose does not mean underspecified.

For instance, in a specification of the primitives for connection to, and disconnection from the network, we want to axiomatize not only that immediately after a connection (disconnection) the system is connected (disconnected), but also that each connection is standing till the next (explicit) disconnection.

This is a very simple example of a common problem. It is, indeed, often the case that only a small subset of the transitions may affect the result of an observer. Therefore, the specifier should explicitly axiomatize the *frame assumption* (which, roughly speaking, is "nothing changes unless explicitly required") for the observers, in order to be able to deduce that their results are unaffected by most transitions. This problem is further complicated by the need for flexibility. Indeed, the definition of an observer may be in a specification, while that of some specific transitions affecting its result may be in another specification, extending or using the first one.

In this paper, we present an extension of CASL-LTL providing constructs to solve these problems, by automatically adding the axioms to capture the frame assumption. Thus, it enables a readable and compact style for the development of observer-oriented dynamic specifications. Our proposal is highly modular, because the information needed to state that some transition possibly affects the result of an observer are colocated with the definition of the transition itself. Thus, adding new transitions do not require to change the specification of the observer.

The semantics of the language we propose, DO-CASL (for Dynamic Observer-based CASL), is given by reduction to CASL-LTL. Among the different possible way to translate the language, we selected one that yields standard CASL if the input DO-CASL specification does not contain temporal logic axioms. We think this choice to be more readable for the average user and it surely allows to reuse the existing tools for CASL in most cases, while no tools for the temporal logic extension exist so far.

---

[1] Actually, the library includes specifications in different layers of abstraction, in order to support more detailed design, even committed to a specific technology. But, here we focus on the more abstract layer.

Paper Structure. Section 2 introduces the preliminaries about CASL, Sec. 3 describes the style of specifications adopted, Sec. 4 presents DO-CASL, and finally Sec. 5 shows how the language may be used in an extended example.

2 Casl and Casl-Ltl

The CASL algebraic specification language has been developed as central part of the CoFI initiative[2]. It provides constructs for declaring basic and structured specifications, whose semantics is a class of *partial* first-order models, and architectural specifications, whose semantics is, roughly speaking, a (higher-order) function over the semantics of basic specifications. Thus, the natural semantics of CASL specifications is the *loose* one: all the partial first-order structures satisfying its axioms are models of a basic specification. However, the models may be restricted to the initial (free) ones, by means of a structuring construct, so that methods based on initial semantics may be accommodated as well.

The building blocks of basic specifications are declarations of (sub)sorts, operations and predicates, giving a signature, and axioms on that signature. Operations may be total or *partial*, denoted by a question mark in the arity. CASL also accommodates subsorting; here, anyway, we do not explicitly use it.

The structuring operators are the usual in algebraic specification languages, like sum, (free) extension, renaming and hiding. We will use mostly union, extension and generic specifications. The latter being less standard, let us discuss a bit its semantics and usage. A generic specification is named and consists of

- a list of *formal parameters*, which are place-holder specifications to be replaced, in the instantiation phase, by more detailed specifications, the actual parameters, possibly using a *fitting morphism* to connect the symbols used in the formal parameters to those in the actual parameters;
- a list of *imports*, which are specifications to be used as they are, for instance that of integer numbers;
- a *body* specification, describing the features to be added to the parameters and the imports by the specification.

The result of an instantiation is, roughly speaking, the enrichment of the union of actual parameters and imports with the body, where body and parameters are translated using the fitting morphisms, if they exist.

For a complete description of CASL, we refer to [1].

Casl-Ltl and Generalized Labeled Transition Systems. It is important to note that CASL is one of a *family* of languages, sharing common constructs and their semantics. For instance, there are restrictions of CASL without partial functions, and/or subsorting, and/or predicates, so that the resulting languages may be translated in other less rich languages in order to reuse specialized tools. On the converse, there are extensions of CASL by constructs and corresponding

[2] See `http://www.brics.dk/Projects/CoFI` and `http://www.cofi.info/`

semantics to deal with specific problem. For instance, there is higher-order CASL (see e.g. [2]) and state-based CASL (see e.g. [3]).

In the sequel we will use CASL-LTL (see [4]), which is designed to describe *generalized labeled transition systems* (glts from now on).

A glts may be used to represent the evolution of a dynamic system. It consists of a set of *states* of the system, one of *labels*, one of *information* and finally the *transition relation*, representing the evolution capabilities of the system. Any element of the transition relation is a tuple consisting of the starting and the final states, a label, capturing what is relevant to the external world, and an information, for what is relevant only to the system itself. For instance, if a system is keeping track of the number of sent messages, the transition corresponding to sending all the messages in a queue will have the message list coded in the label and the number of sent messages in the info part, to be used to update the internal counter. Any state of the system corresponds to the process having an evolution tree determined by the transition system itself, where each branch is given by a transition of the system and represents a *capability* of moving of the parent state.

A glts may be specified by using CASL-LTL. Indeed, CASL-LTL allows to declare dynamic sorts, using the construct **dsort** *ds* **label** *l_ds* **info** *i_ds*. This CASL-LTL construct semantically corresponds to the declaration of the sorts *ds*, *l_ds*, and *i_ds* for the states, the labels and the information of the glts, and of the transition predicate **preds** $\_\_ :: \_\_ \xrightarrow{\_\_} \_\_ : i\_ds \times ds \times l\_ds \times ds$, as well.

Thus, each element s of sort *ds* in a model M (an algebra or first-order structure) of the above specification corresponds to a process modelled by a transition tree with initial state s determined by the glts $(i\_ds^M, ds^M, l\_ds^M, \_\_ :: \_\_ \xrightarrow{\_\_} \_\_^M)$[3].

The most important extension of CASL-LTL w.r.t. CASL is the enrichment of the logic by constructs from a branching-time CTL-style temporal logic, which effectively increase the expressive power of the language (see [5] and [4]).

In the sequel we will use an obvious shortcut when either the label part or the information one are irrelevant, dropping any reference to the immaterial aspect using transition predicates such as $\_\_ \xrightarrow{\_\_} \_\_ : ds \times l\_ds \times ds$ and $\_\_ :: \_\_ \rightarrow \_\_ : i\_ds \times ds \times ds$. The general case is computed from the shorter version, by adding a sort with just one element for the missing component and decorating all the transitions with that element.

3 Lessons Learned from Developing a Library for Distribution

Our CASL-library for distribution, as motivated in the Introduction, is hierarchically organized in a directed acyclic graph of refinements[4], where all the nodes are loose specifications, having as models all those middleware implementing some

[3] Given a Σ algebra A, and a sort s of Σ, s^A denotes the interpretation of s in A; similarly for the operation and predicates of Σ.

[4] Here, *refinements* has the traditional meaning of model-class inclusion. This property is guaranteed in our specifications by the use of the extension construct.

set of functionalities. Thus, the refinement in this context corresponds mostly to adding functionalities, not to making implementation decisions (though we also provide a few detailed specifications representing an individual middleware).

In order to achieve this result, we adopt an observational style, in the sense that we introduce functions and predicates to extract from the elements (representing internal states of the subsystems) the values of some of their aspects, which we regard as relevant for the applications to be built upon our infrastructure. However, we are not relying on the observers to distinguish elements, as in most observational approaches.

Indeed, by the nature of our library, the set of observers is continuously extended, as new aspects of the nodes and networks are introduced by the library specifiers and end-users. Thus, the fact that the current set of observers cannot distinguish between two elements is not a clue of their equality; it could as well be an indication of some aspect still to be taken into account. Therefore, our approach has in common with more traditional observational approaches (e.g., the pioneering [6]; we defer to [7] for further references) only the intuition of the black-box approach and the use of the word observer.

For instance let us consider the case of the most basic specification in our library, the one of *peer*, as a paradigmatic example of the difficulties we encountered in the development of the library and the solution we propose.

A *peer* is the abstraction of any node in a net. It has a persistent identity, the capability to connect to a net using a given address and to disconnect from the net. Thus, we leave underspecified how elements of the peer sort are made and introduce functions extracting the identity, address (possibly undefined if the peer is not connected), and online status from such elements, as in the following signature[5]:

sort *PeerId*, *Address*
dsort *Peer* **label** *PeerL*
ops $online : Address \rightarrow PeerL$
$offline :\rightarrow PeerL$
$id : Peer \rightarrow PeerId$
$addr : Peer \rightarrow ?\ Address$
preds $isOnline : Peer$

where we have (static) sorts, describing data types, like for instance the (totally unspecified) sort for peer identifiers, *PeerId*, or that for the labels of their transitions, *PeerL*. Moreover, we also have the dynamic sort *Peer*, representing the states of the nodes. Analogously, we have operations building some sort, like for instance *online* and *offline*, which denote particular labels, and we have observers, both operations like *id* and *addr*, and predicates like *isOnline*, used to extract, or observe, aspects of the peer states.

Now, we need to state two different kinds of axioms. First of all, we have the standard axioms, describing the effects of operations and transitions, such as asserting that after going online with an address a, the peer is actually online and its address is a. But, we also have to state that no transition is affecting the value of *id*, as the identity is persistent, that the only transitions affecting

[5] Notice the obvious adaptation to the case with silent information.

isOnline are those actually taking the peer on and off line, and that the address is persistent for each connection, so that it can change only if some connection or disconnection has taken place. These are quite different from the previous ones, from a logical point of view, because they express a property that the users usually implicitly assume: *each aspect of the status of the system changes only if forced to, by a transition explicitly modifying it.* But, there is no such a thing as an *implicit* assumption in specifications. Unless some axiom is imposed to guarantee it, there are models which do not satisfy it. In other words, we have to state a sort of *frame assumption* (see e.g. [8]) for some observers.

However, there are two important differences from standard frame assumption in our approach. First of all, we want to explicitly state that some properties of the system change and some do not for a given transition, but leave most of them underspecified, changing or not depending on the individual models. Thus, the frame assumption would hold only for a subset of the functions and predicates on the dynamic sorts, those we call *observers*.

Moreover, we need a flexible way to state the frame assumption to accommodate further refinements of the specification. Let us for instance consider the problem of stating the persistency of the address in each continuous connection. If we simply add the axiom[6]

$$\forall\, l : \mathit{PeerL};\ \forall\, p, p' : \mathit{Peer};\ \bullet\ \mathit{addr}(p) = \mathit{addr}(p')\ \mathit{if}$$
$$(p \xrightarrow{l} p' \wedge (\forall a : \mathit{Address}. \neg l = \mathit{online}(a)) \wedge \neg l = \mathit{offline})$$

then we forfeit the capability to add in an extension of this specification a label representing another way of connecting, for instance the connection without explicit address, for those cases where the address is dynamically provided by a server. Indeed, even if we add such labels, they cannot change the address of the peer, being different from *online* and *offline*.

This would be clearly unacceptable in our approach, where new refinements of the specifications can be added to represent richer middleware or to support more demanding applications. Following our observational approach, instead of stating that only transitions using some individual labels may affect an observer, we describe abstract properties on the labels and require that only the labels satisfying them can affect the predicate. For instance, in our example, the property of being online may be influenced by all the labels representing a connection or a disconnection, but by no others. Thus, we use again predicates on the labels and info to describe the category of actions they are representing and use these predicates in turn to state our weak form of frame assumption. In this way we achieve a separation of concerns and a higher level of modularity: at the moment when observers are introduced, the specifier has to decide with categories of transitions may change the observation. But, it is only at the moment of the definition of each individual transition, that is, of its label and info, that the decision about which are its categories has to be made.

It is worth noting that the actual labels become superfluous and can be dropped from the specifications of the abstract behavior of the middleware

---

[6] This approach is similar to the expansion of "not changed by other events" in [9].

classes. The actual labels appear, together with the axioms categorizing them w.r.t. the observers, in the lower level specifications, those representing individual middlewares and in the end-user specifications, where they are used to describe the moves of the system at the applicative level.

Thus, our toy example should be changed as follows

sort *PeerId*, *Address*
dsort *Peer* **label** *PeerL*
ops $id : Peer \to PeerId$
$addr : Peer \to? Address$
preds $goesOnline : PeerL$
$goesOffline : PeerL$
$isOnline : Peer$
$\forall\ l : PeerL;\ p, p' : Peer$

- $id(p) = id(p')\ if\ p \xrightarrow{l} p'$
- $addr(p) = addr(p')\ if\ (p \xrightarrow{l} p' \wedge \neg goesOnline(l) \wedge \neg goesOffline(l))$
- $def(addr(p))\ if\ p \xrightarrow{l} p' \wedge goesOnline(l)$

Though, technically, this is satisfactory, from a methodological point of view it is not. Indeed, the end user is required to add lots of trivial axioms of the form

- $obs(d) = obs(d')\ if\ (i : d \xrightarrow{l} d' \wedge \neg cat_1(l, i) \wedge \ldots \wedge \neg cat_n(l, i))$ to state

that the transition (s)he is introducing does not affect the result of most observers. But, the user should be more encouraged to focus on the pairs "transition + observer" where the transition is relevant to that observer, than on those where, being no relationship between the two components, things are not going to change and hence the corresponding axiom has to be issued.

We need a mechanism to clearly separate the axioms stating which category of transitions affects which observer, from those describing the effects and, moreover, to automatically add the axiomatization of the default behavior, where the observer values are not changing unless some transition affecting the corresponding aspect takes place.

4 DO-Casl

Let us introduce a syntactic short-cut, which does not require any change in the semantics, because the terms introduced by this new construct reduces to terms in standard CASL-LTL, and then CASL in the case no temporal logic is used in the specification. In the choice of the restrictions for such a construct, we have been guided by pragmatic considerations, choosing a generality sufficient to deal with all the cases in our library and, at the same time, not so extreme to make the translation in standard CASL difficult.

We start by adding a new production for the SPEC non-terminal of the grammar of the abstract syntax. The terms generated by this production will correspond to the dynamic specifications using observers, that is a special case of CASL-LTL basic specifications including one or more *observer blocks*.

Any observer block refers to a dynamic sort and encapsulates the definition of observers on that sort, together with categories of transitions which may affect them and axioms to express this dependency.

Definition 1. *The context free grammar of* DO-CASL *adds to that of* CASL-LTL *the following productions (terminal and non-terminal symbols):*

SPEC	::=	... | DSPEC
DSPEC	::=	dyn-spec DBASIC-ITEMS*
DBASIC-ITEMS	::=	BASIC-ITEMS | OBS-BLOCK
OBS-BLOCK	::=	obs-block DSORT-DECL OBS-ITEMS*
OBS-ITEMS	::=	OP-ITEMS | PRED-ITEMS | CATEGORY-ITEMS | VAR-ITEMS | CAT-AXIOM-ITEMS
CATEGORY-ITEMS	::=	category PRED-DECL+
CAT-AXIOM-ITEMS	::=	axiom-items CAT-FORMULA+
CAT-FORMULA	::=	CAT-QUANTIFICATION | CAT-IMPLICATION
CAT-QUANTIFICATION	::=	quantification universal VAR-DECL CAT-IMPLICATION
CAT-IMPLICATION	::=	implication PREDICATION AFFECTS | implication AFFECTS AFFECTS
AFFECTS	::=	affects VAR VAR PREDNAME | affects VAR VAR OPNAME

The concrete syntax is as close to CASL-LTL as possible.

Definition 2. *The concrete syntax for* DO-CASL *is the same as that of* CASL-LTL *for the terms common to both languages. For the newly added terms, the concrete syntax is as follows:*

- *a DSPEC starts with the keyword* ***dspec*** *and ends with* ***end_dspec*** [7]*;*
- *an OBS-BLOCK starts with the keyword* ***observe*** *and ends with* ***end_obs****;*
- *a CATEGORY-ITEMS starts with the keyword* ***cats****;*
- *the concrete syntax for CAT-AXIOM-ITEMS, CAT-QUANTIFICATION, CAT-FORMULA and CAT-IMPLICATION is the same as for the corresponding languages in* CASL*;*
- *(l,i)* affects *o is the concrete syntax for a term of the form* affects *l i o.*

Besides the static correctness of standard CASL, we also impose a few requirements.

Definition 3. *For an observer block referring to a dynamic sort* ***dsort*** *ds* ***label*** *l_ds* ***info*** *i_ds to be statically correct, we require that:*

- *all the enclosed operations and or predicates must have ds as (unique) source; in the following we will call them* observers on *ds;*
- *all declarations of predicates in a CATEGORY-ITEMS section must have* $l\_ds \times i\_ds$ *as source; in the following we will call them* categories on $l\_ds \times i\_ds$*;*
- *in each axiom,*

[7] Mandatory, to simplify parsing.

- *the predicate symbols in a* PREDICATION *must be declared as categories within the same block and their arguments must be local variables of appropriate sort;*
- *the operation and predicate names in an* AFFECTS *must be declared as observers within the same block;*
- *the variables in an* AFFECTS *must be of sort* l_ds *and* i_ds*, respectively.*

Since the source for observers and categories must agree with the static requirements, they could be deduced from the observer block head. Though requiring them to be explicitly stated is then redundant, and a possible source of error, we prefer to use this verbose syntax, because it is closer to the standard CASL declarations of functions and predicates and hence, supposedly, easier for the end-users. Specialized editors could be easily devised, which automatically supply the deducible sorts in the declaration of observers and categories, to save the users the pain of writing them.

Let us consider as an example the peer specification, using the syntactic sugar introduced so far to represent the observers. Notice that the operations *online* and *offline* have been dropped, because their role is filled by the corresponding predicates. Moreover, we give here the full specification, with also the axioms external to the block. Finally, note that the specification is parametric over the definition of the addresses (e.g., IPv4, IPv6, JXTA or Chord identifiers, etc.)

spec PEER [**sort** *Address*]=

dspec
sort *PeerId*
observe *Peer* **label** *PeerL*
  ops $id : Peer \to PeerId$
    $addr : Peer \to ? \ Address$
  preds $isOnline : Peer$
  cats $goesOnline : PeerL$
    $goesOffline : PeerL$
  $\forall\ l : PeerL$
  - $goesOnline(l) \Rightarrow l \text{ affects } isOnline$
  - $goesOffline(l) \Rightarrow l \text{ affects } isOnline$
  - $l \text{ affects } isOnline \Rightarrow l \text{ affects } addr$

end_obs
$\forall\ l : PeerL;\ p, p' : Peer$
- $isOnline(p') \text{ if } p \xrightarrow{l} p' \wedge goesOnline(l)$
- $\neg isOnline(p') \text{ if } p \xrightarrow{l} p' \wedge goesOffline(l)$
- $def(addr(p)) \text{ if } isOnline(p)$

end_dspec

Now, let us define the semantics of our constructs, by reduction to CASL-LTL. The basic intuition is to translate categories to standard predicates and to add a predicate *ad hoc* to represent the capability of affecting a given observer. Moreover, we want to keep only those models where such *ad hoc* predicates are minimal with respect to the axioms given in the observer block and the interpretation of the categories on the individual model. To get this result, we will intersect the models of the translated specification with those of a free, and hence minimal, specification for the *ad hoc* predicates.

Definition 4. *For a correct observer block obs_blk*[8]

observe ds **label** l_ds **info** i_ds

  ops $f_1 : ds \to? s_1; \ldots f_n : ds \to? s_n;$

  preds $p_1, \ldots, p_m : ds;$

  cats $pt_1, \ldots, pt_k : l\_ds \times i\_ds$

  vars $x_1^1, \ldots, x_1^{n_1} : s_1; \ldots; x_k^1, \ldots, x_k^{n_k} : s_k;$

  $\varphi_1 \ldots \varphi_h$

end_obs

its expansion, denoted DOCASL2CASL(obs_blk), is the following basic specification

dsort ds **label** l_ds **info** i_ds

ops $f_1 : ds \to? s_1; \ldots f_n : ds \to? s_n;$

preds $p_1, \ldots, p_m : ds;\ pt_1, \ldots, pt_k : l\_ds \times i\_ds$

  $\_aff\_f_1, \ldots, \_aff\_f_n, \_aff\_p_1, \ldots, \_aff\_p_m : l\_ds \times i\_ds$

$\forall\ l : l\_ds;\ i : i\_ds;\ d, d' : ds;$

  %% transitions not affecting $f_1 \ldots f_n, p_1 \ldots p_n$ leave the observer result unchanged

  $(\neg\_aff\_f_1(l, i)) \wedge i : d \xrightarrow{l} d' \Rightarrow f_1(d) = f_1(d')$

  ...

  $(\neg\_aff\_p_m(l, i)) \wedge i : d \xrightarrow{l} d' \Rightarrow (p_m(d) \Leftrightarrow p_m(d'))$

Moreover we will call free_aff(obs_blk) the 4-tuple

$$(SSp(obs\_blk), CSp(obs\_blk), OSp(obs\_blk), ASp(obs\_blk))$$

where

- *SSp(obs_blk) is* **sorts** *l_ds,i_ds*
- *CSp(obs_blk) is* **preds** $pt_1, \ldots, pt_k : l\_ds \times i\_ds$
- *OSp(obs_blk) is* **preds** $\_\mathrm{aff}\_f_1, \ldots, \_\mathrm{aff}\_f_n, \_\mathrm{aff}\_p_1, \ldots, \_\mathrm{aff}\_p_m : l\_ds \times i\_ds$
- *and ASp(obs_blk) is* $\forall\ x\_l\_ds : l\_ds; x\_i\_ds : i\_ds;$ $\mathrm{trans}(\varphi_1) \ldots \mathrm{trans}(\varphi_h)$, *where* trans *drops any variable quantification and transforms each occurrence of* (l, i) affects o *into* $\_\mathrm{aff}\_o(x\_l\_ds, x\_i\_ds)$, *where* x_l_ds *and* x_i_ds *are fixed variable names.*

A correct dynamic specification including observer blocks $obs\_blk_1 \ldots obs\_blk_n$ *translates to the specification given by leaving all the BASIC-ITEMS unaffected, by replacing each* obs_blk_i *by DOCASL2CASL*(obs_blk_i) *and adding at its end, the following fragment:*

and {

  SSp(obs_blk_1)... SSp(obs_blk_n)

  CSp(obs_blk_1)... CSp(obs_blk_n)

then free {OSp(obs_blk_1)... OSp(obs_blk_n)

  ASp(obs_blk_1)... ASp(obs_blk_n)}}

The models of the final fragment are first-order structures on a signature with as set of sorts all the sorts for labels and info from some observer block, no

[8] We are using only partial functions for simplicity, but total functions are allowed as well, of course.

operations, and as predicates both those representing categories and those introduced to translate the AFFECTS atoms. The models of this fragment may have any interpretation for the sorts and the predicates representing categories because these sorts and predicates are declared within the block with no axioms or operations. But, by definition of freenes, it has the minimal interpretation of the predicates representing the AFFECTS atoms compatible with the explicit axioms stated in the observer blocks. Since the axioms used in the free construct are all positive conditional (see e.g. [10]), this block is guaranteed to be consistent. By making the intersection of this model class with that of the expansion of the specification (which could be empty if the axioms imposed by the users are inconsistent), we get only those models where the common algebraic structure is interpreted in the same way. Thus, the label and info sorts and the category predicates satisfy the explicit axioms in the specification.

Let us see what is the expansion of our running example.

spec PEER [**sort** *Address*]=
sort *PeerId*
dsort *Peer* **label** *PeerL*
ops $id : Peer \to PeerId$
$\quad addr : Peer \to ?\ Address$
preds $isOnline : Peer$
$\quad goesOnline, goesOffline : PeerL$
$\quad \_aff\_id, \_aff\_addr, \_aff\_isOnline : PeerL$
$\forall\ l : PeerL; \forall\ p, p' : Peer$
- $\neg\_aff\_id(l) \wedge p \xrightarrow{l} p' \Rightarrow id(p) = id(p')$
- $\neg\_aff\_addr(l) \wedge p \xrightarrow{l} p' \Rightarrow addr(p) = addr(p')$
- $\neg\_aff\_isOnline(l) \wedge p \xrightarrow{l} p' \Rightarrow (isOnline(p) \Leftrightarrow isOnline(p'))$
- $isOnline(p')\ if\ p \xrightarrow{l} p' \wedge goesOnline(l)$
- $\neg isOnline(p')\ if\ p \xrightarrow{l} p' \wedge goesOffline(l)$
- $def(addr(p))\ if\ isOnline(p)$

end
and {
sorts *PeerL*
preds $goesOnline, goesOffline : PeerL$
then free {
preds $\_aff\_id, \_aff\_addr, \_aff\_isOnline : PeerL$
$\forall\ l : PeerL;$
- $goesOnline(l) \Rightarrow \_aff\_isOnline(l)$
- $goesOffline(l) \Rightarrow \_aff\_isOnline(l)$
- $\_aff\_isOnline(l) \Rightarrow \_aff\_addr(l)$ } }

Note that atoms of the form (l, i) *affects* o are well-formed only inside the observer block(s) where o is introduced. Thus, the information about which category of actions may change the value of an observer must be collected in the same block where the observer is declared. If the same observer obs is declared in two or more blocks (which must refer, then, to the same dynamic sort) within the same dynamic specification, then the resulting semantics is equivalent to having the two blocks merged in one. Indeed, the minimality of the predicate *_aff_*obs is described by the collection of all the *free_aff*(obs_blk_i) and hence it is immaterial in which block a category, or an axiom is given. On the contrary, if the same observer `obs` is declared in two or more blocks in different dynamic specifications, then the first occurrence completely defines the semantics of *_aff_*`obs` and the further definitions are either useless or harmful, if introducing inconsistences.

What does not need to be co-located with the observer first declaration, is the definition of the validity of the category predicates. Thus, labels and info introduced further on in the specification can affect an observer already defined.

A simple tool implementing the semantics given in Def. 4 is available on the web[9]. Such tool does not check syntax requirements; thus, it is only guaranteed to work on statically correct DO-CASL input. If the input specification does not contain any temporal logic formula, the result is a standard CASL specification,[10] which can be handled by CASL tools such as HETS[11].

5 Writing DO-CASL Specifications

In this section we will extend the already seen PEER specification, in order to give an example on how specifications in DO-CASL can be written.

We will show a simple specification of a P2P application where nodes can send messages each other. Received messages will be accessible in an "inbox" until they are deleted. We are making use of the CASL standard library (see e.g. [1]) for the *Set* and *Map* structured data types.

spec BASEPEER = PEER [**sort** *Address*]

The BASEPEER specification is just a shortcut. It is used in parametric specifications, which could otherwise get quite unwieldy when using a parametric specification as a parameter. We will systematically use this scheme in the following.

spec NET [BASEPEER]=
  SET [**sort** *PeerId* **fit** *Elem* $\mapsto$ *PeerId*] **and**
  MAP [**sort** *PeerId* **fit** S $\mapsto$ *PeerId*][**sort** *Peer* **fit** T $\mapsto$ *Peer*] **and**
  MAP [**sort** *PeerId* **fit** S $\mapsto$ *PeerId*][**sort** *PeerL* **fit** T $\mapsto$ *PeerL*]

then dspec
  sort *Net* = *Map* [*PeerId*, *Peer*]; *NetI* = *Map* [*PeerId*, *PeerL*];
  observe *Net* **info** *NetI*
    ops $dom : Net \to Set[PeerId]$
  end_obs
  $\forall id : PeerId; n, n' : Net; i : NetI$
    • $dom(i) \subseteq dom(n)$ *if* $i :: n \longrightarrow n'$ %% shortened notation with no label

[9] `http://www.disi.unige.it/person/DellamicoM/do-casl`

[10] The tool also expands the dynamic sort declarations from observer blocks. Thus, using the equivalent form **observe** *ds* **label** *l_ds* **info** *i_ds* **end_obs** to represent *ds* **label** *l_ds* **info** *i_ds*, it may be used also as a preprocessor for CASL-LTL, which is currently not supported at all, for specifications without temporal axioms.

[11] `http://www.informatik.uni-bremen.de/agbkb/forschung/formal_methods/CoFI/hets/`

- $lookup(id, n) \xrightarrow{lookup(id,i)} lookup(id, n')$ *if* $i :: n \longrightarrow n' \wedge id \in dom(i)$
- $lookup(id, n) = lookup(id, n')$ *if* $i :: n \longrightarrow n' \wedge \neg id \in dom(i)$

end_dspec

spec BASENET = NET [BASEPEER]

The NET specification models a *network* of peers which are part of it even if transiently offline.

Since a network is a closed system with no interactions with the outside world, its transitions are decorated with info only.

A *Net* is a mapping from peer identifiers to peers; in this way, it is guaranteed that no two nodes can have the same identifier within the same network. Analogously, *Net* transitions are mappings from node identifiers to peer transition labels, associating each moving node to the label decorating its local transition. Idle nodes do not belong to the domain of the *Net* transition.

We describe the *dom* operation on *Net* as an observer that is not influenced by any category of labels. This succinctly ensures that the identifiers represented in a network never change during the evolution of a network.

Our library handles a more general case, with nodes dynamically entering and leaving the network, by defining categories that influence the peers in the network.

spec MESSAGES [BASENET]=
SET [**sort** *Msg* **fit** $Elem \mapsto Msg$]

then dspec
observe *Peer* **label** *PeerL*
op $inbox : Peer \rightarrow Set[Msg]$
cats *receives, deletes* : *PeerL*
$\forall\, l : PeerL$

- $receives(l) \Rightarrow l$ *affects inbox*
- $deletes(l) \Rightarrow l$ *affects inbox*

end_obs
ops $orig, dest : Msg \rightarrow Address;$
$sent, received : PeerL \rightarrow Set[Msg]$
$\forall m : Msg; p, p' : Peer; l : PeerL; id : PeerId; n, n' : Net, i : NetI$

- $received(l) = \{\}$ *if* $\neg receives(l)$
- $deleted(l) = \{\}$ *if* $\neg deletes(l)$
- $inbox(p') = (inbox(p) - deleted(l)) \cup received(l)$ *if* $p \xrightarrow{l} p'$
- $isOnline(p)$ *if* $p \xrightarrow{l} p' \wedge isNonEmpty(received(l) \cup sent(l))$
- $dest(m) = addr(p)$ *if* $p \xrightarrow{l} p' \wedge m \in received(l)$
- $orig(m) = addr(p)$ *if* $p \xrightarrow{l} p' \wedge m \in sent(l)$
- $m \in received(lookup(id, i))$
  if $i :: n \rightarrow n' \wedge l \in range(i) \wedge m \in sent(l) \wedge id \in dom(n)$
  $\wedge\, addr(lookup(id, n)) = dest(m)$

end_dspec

spec BASEMESSAGES = MESSAGES [BASENET]

This specification enriches the peers with the capability of sending and receiving messages. Messages carry information about addresses of the sender and the recipient, respectively via the *orig* and *dest* operations. Whenever a peer sends a message, the destination immediately receives it (the recipient node has to be online, otherwise the sending operation cannot be executed). The *sent* and *received* operations are meant to extract the sent and received messages in a transition.

Moreover, an *inbox* observer has been defined. Received messages remain in it until they are erased, possibly after a command from a user. While a node can only send and receive messages while it is online, it can erase messages from its own inbox at any time.

In the MESSAGES specification we have adopted a different style in order to define the value of observers w.r.t. PEER. Indeed, in PEER the observers were defined with axioms of the form $obs(p') = \ldots\ if\ p \xrightarrow{l} p' \wedge cat_i(l)$, for each cat_i affecting *obs*. This style is safe, in the sense that no inconsistency with the minimality of the *_aff_obs* predicates having some $cat_i(l)$ in the premises may be introduced, but it is quite verbose.

In MESSAGES, the *inbox* observer is specified differently. Indeed, the axiom giving its value is not guarded by a $receives(l)$, nor by a $deletes(l)$. Thus, it could, potentially, introduce an inconsistency. However, since *received* and *deleted* yield the empty set on label not satisfying *receives* nor *deletes*, for such labels the axiom is equivalent to stating that the value of the observer in the source and target state is unchanged, as required by the frame assumption. This style of specification is more readable and concise, though more error prone.

Both styles are convenient in different settings and DO-CASL allows to use both.

spec FINALMESSAGES = BASEMESSAGES
then
  free generated type *PeerL* ::=
    online(Address) | *offline* | *send(Msg)* | *recv(Msg)* | *del(Msg)*
$\forall l : PeerL; a : Address; p, p' : Peer; m : Msg$
- $goesOnline(l) \Leftrightarrow (\exists a : Address \bullet l = online(a))$
- $addr(p') = a\ if\ p \xrightarrow{online(a)} p'$
- $goesOffline(l) \Leftrightarrow l = offline$
- $sent(send(m)) = \{m\}$
- $receives(l) \Leftrightarrow (\exists m : Msg \bullet m = recv(m))$
- $received(recv(m)) = \{m\}$
- $deletes(l) \Leftrightarrow (\exists m : Msg \bullet m = del(m))$
- $deleted(del(m)) = \{m\}$

end

We have concluded this section with an oversimplified example of a final definition of the transition labels. In this case, a transition can only be a simple operation like going online, offline, sending, receiving or deleting a message.

Notice that all the tedious axioms, such as $received(offline) = \{\}$ or $addr(p) = addr(p')\ if\ p \xrightarrow{del(m)} p'$, are inherited from the interplay between the, so to speak, *global* statements of BASEMESSAGES and the *local* definition of the category validity.

6 Conclusions and Further Work

We have introduced a specification style and some syntactic sugar supporting it in the CASL language, motivated by our experience with the definition of loose dynamic specifications.

The use of the resulting language, DO-CASL, is supported by a small prototype, compiling it in CASL, so that standard tools can be used on the resulting specifications.

Though we have fully developed the semantics of DO-CASL and, we hope, given convincing examples of DO-CASL usefulness in interesting realistic examples, the language itself has not been submitted formally as a CASL extension yet. We plan to do it if it is well received by the scientific community.

Acknowledgements

We thank Gianna Reggio for spending her valuable time in stimulating discussions on the subject of specification of dynamic system and for her kind support.

References

1. CoFI (The Common Framework Initiative): CASL Reference Manual. LNCS2960 (IFIP Series). Springer Verlag (2004)
2. Schröder, L., Mossakowski, T.: HASCASL: Towards integrated specification and development of Haskell programs. In Kirchner, H., Ringeissen, C., eds.: AMAST 2002 Proceedings. LNCS2422, Springer Verlag (2002) 99–116
3. Baumeister, H., Zamulin, A.V.: State-based extension of CASL. In Grieskamp, W., Santen, T., Stoddart, B., eds.: IFM 2000, Dagstuhl Castle, Germany, Proceedings. LNCS1945, Springer Verlag (2000) 3–24
4. Reggio, G., Astesiano, E., Choppy, C.: CASL-LTL: A CASL extension for dynamic reactive systems – Version 1.0 – Summary. Technical Report DISI-TR-03-36, Univ. of Genova (2003)
5. Costa, G., Reggio, G.: Specification of abstract dynamic-data types: A temporal logic approach. TCS **173** (1997) 513–554
6. Reichel, H.: Behavioural equivalence — a unifying concept for initial and final specification methods. In: Proc. 3rd. Hungarian CS Conference. (1981) 27–39
7. Cerioli, M.: Basic concepts. In: Algebraic System Specification and Development: Survey and Annotated Bibliography. 2nd edition, 1997. Number 3 in Monographs of the Bremen Institute of Safe Systems. Shaker (1998)
8. Borgida, A., Mylopoulos, J., Reiter, R.: On the frame problem in procedure specifications. IEEE Trans. Softw. Eng. **21** (1995) 785–798
9. Choppy, C., Reggio, G.: Using CASL to specify the requirements and the design: A problem specific approach. In Bert, D., Choppy, C., Mosses, P.D., eds.: Recent Trends in Algebraic Development Techniques, 14th International Workshop, WADT'99, Château de Bonas, France, 1999, Selected Papers. LNCS 1827, Springer-Verlag (2000) 104–123
10. Cerioli, M., Mossakowski, T., Reichel, H.: From total equational to partial first-order logic. In Astesiano, E., Kreowski, H.J., Krieg-Brückner, B., eds.: Algebraic Foundations of Systems Specification. IFIP State-of-the-Art Reports. Springer-Verlag (1999)

Model Transformations Incorporating Multiple Views

John Derrick[1] and Heike Wehrheim[2]

[1] Department of Computing, University of Sheffield, Sheffield, S1 4DP, UK
J.Derrick@dcs.shef.ac.uk
[2] Universität Paderborn, Institut für Informatik, 33098 Paderborn, Germany
wehrheim@uni-paderborn.de

Abstract. *Model transformations* are an integral part of OMG's standard for Model Driven Architecture. Model transformations are advocated to be *behaviour preserving*: platform specific models should adhere to platform independent descriptions developed in earlier design stages.

In this paper, we deal with models consisting of several *views* of a system. Often, in such a scenario, model transformations change just one view, and, although the overall transformation of all views is behaviour preserving, it is not behaviour preserving in isolation. To tackle this problem we develop a proof technique (and show its soundness) that allows one to consider just the view that has changed, and not the entire system. We focus specifically on one particular class of view-crossing transformations, namely on transformations conjunctively adding new constraints to a model.

1 Introduction

The Object Management Group's standard for model-driven architecture (MDA) defines *models* to be the core concept in software development. *Model transformations* are intended to provide the means for getting from high-level platform independent (PIM) to lower level platform specific models (PSM). Model transformations are expected to be *behaviour preserving*: lower-level models should preserve the behaviour of higher-level models. For complex systems, modelling usually has to take many different aspects into account [FKN+92, ZJ96]. A complex system has to fulfill several orthogonal requirements: on the static behaviour (data and operations), on its dynamic behaviour (adherence to protocols, scenarios), its timing behaviour, etc. Thus a model of a complex system will usually consist of descriptions of several *views*. Consequently, different kinds of model transformations operate on different views.

In this paper, we are interested in *proving correctness* of model transformations with respect to the intended property of behaviour preservation. To this end, we will look at models written in a *formal* specification language and give a formal definition of behaviour preservation. There is much recent activity on using formal notations to guarantee behaviour preservation, especially in the area of *refactorings* [MS04, MT04, PR03, SPTJ01, KHK+03]. These approaches

M. Johnson and V. Vene (Eds.): AMAST 2006, LNCS 4019, pp. 111–126, 2006.

either directly start with a formal model or operate on UML models, using a formal semantics for some of the UML diagrams. While most of these approaches *define* specific model transformations, we will take the contrary approach here and develop a technique for a posteriori *verifying* the correctness of a given transformation.

An ideal starting point for verifying model transformations in a formal context is the notion of *refinement*, used as the development method in formal methods [DB01, dE98]. Designed with the intention of guaranteeing *substitutability*, refinement is a suitable correctness criterion for model transformations. Most formal approaches to model transformations only apply to a single specification formalism, i.e., a single view in the model. The same is true for refinement. For models or specifications consisting of multiple views specified in different formalisms, different refinement concepts have to be applied in combination. This has led to extensive work on understanding how different notions of refinement fit together [He89, Jos88, BD02, DB03]. For example, one can show that a joint usage of data refinement (on a static view) and process refinement (on a dynamic view) guarantees behaviour preservation for models incorporating both views.

However, not all model transformations can be neatly divided into smaller transformations operating on a single view. Some transformations go across views, e.g., aspects are being moved between views or changes on one view are based on properties of other views. Such transformations will be the focus of this paper.

We encountered this problem in a project where UML was used to generate first a formal model, and from this Java code [MORW04] was developed. The translation to Java usually first necessitated alterations to the formal model, particularly the state-based view which made up most of the Java program. These transformations often turned out not to be refinements anymore (and thus not, individually, behaviour preserving), but – in connection with the other views – still were behaviour preserving. There was, however, no suitable proof technique for showing this at hand (besides those of computing the complete semantics of both models and directly comparing them). The purpose of this paper is to develop such proof techniques.

The particular formal notations we use here are Object-Z [Smi00] for data-oriented modelling (the static view) and the process algebra CSP [Hoa85, Ros98] for the dynamic view and for architecture specification. Our platform independent models are thus written in languages which allow abstract specification (nondeterminism, underspecification etc.) and leave out implementation details. For the platform specific models we use the same formalisms but then impose restrictions on the specification (no nondeterminism, bounded data types only, etc.) which mimic platform specific features found, e.g., in Java. We also use UML diagrams to describe the architecture of the overall system. Using the translation proposed in [MORW04], the proof technique for transformations developed here is thus also transferable to UML models when written in the profile employed in [MORW04].

The paper is structured as follows. Section 2 starts with a motivating example of transformations incorporating more than one view. The transformations that we deal with are changes on the static view which conjunctively add new constraints to the model. Section 3 presents our notion of correctness for model transformations. Section 4 explains the technique used for showing behaviour preservation for transformations incorporating multiple views, which consists of two aspects: a check for i/o independence of the new constraint and a trace check on projected components. Section 5 describes some tool-support for the technique, and in particular for the trace inclusion. Finally, Section 6 concludes and discusses related and future work.

2 Case Study

Our running example is the holonic manufacturing system [Weh00]. This system processes workpieces (initially stored in some in-store) by a number of machines and finally stores them in an out-store. The transportation of workpieces in the factory is carried out by autonomous (also called *holonic*) transportation units (abbreviated HTS = holonic transportation systems). Transportation is not regulated by a central control but achieved via negotiation between stores, machines and HTS'. Once a store or machine has a workpiece to be transported it asks the transportation units for offers about the costs of such a transportation. Having obtained all offers it chooses the one with the smallest cost and orders it. Our model shows just one small part of the case study concerning this negotiation (which we further simplify to ease readability).

2.1 Model 1 – PIM

The initial model consists of three views: A static view describing (just one) class with attributes and methods, a dynamic view describing a protocol of interactions for this class and a composite structure diagram describing the architecture of the system.

Static view. The static view is given in Object-Z, an object-oriented state-based specification formalism. The role of Object-Z in our model is similar to that played by class diagrams with OCL constraints in UML models: it describes variables and methods of classes with class invariants and pre- and postconditions. A class specification starts with a definition of the *interface* of the class, followed by a declaration of its *state* (its variables) together with an *init schema* giving restrictions on initial values of variables. After that a number of *method schemas* define the methods of the class. Here, primed and unprimed variables refer to the after and before states respectively; input and output parameters are decorated with ? and ! respectively, and are undecorated if they are used for addressing objects. The predicates appearing in the schemas of methods (`com_...`) state pre- and postconditions of methods (in UML, this would be specified in OCL). The precondition acts as a *guard* for method execution; outside the precondition method execution is refused (i.e., it is *blocked*).

Machine

$\texttt{method}\ \mathit{offer} : [\, h : \mathit{Hts};\ \mathit{cost?} : \mathit{Cost} \,]$
$\texttt{method}\ \mathit{order} : [\, h : \mathit{Hts} \,]$
$\texttt{local\_chan}\ \mathit{choose}$
...

$$\begin{array}{|l} \mathit{offers} : \mathrm{seq}(\mathit{Hts} \times \mathit{Cost}) \\ \mathit{orderTo} : \mathit{Hts} \\ \ldots \end{array}$$

Init
$$\begin{array}{|l} \mathit{offers} = \langle\,\rangle \end{array}$$

com_offer
$$\begin{array}{|l} \Delta(\mathit{offers}) \\ h : \mathit{Hts}, \mathit{cost?} : \mathit{Cost} \\ \hline \mathit{offers}' = \mathit{offers} \frown \langle (h, \mathit{cost?}) \rangle \end{array}$$

com_order
$$\begin{array}{|l} \Delta(\mathit{offers}) \\ h : \mathit{Hts} \\ \hline h = \mathit{orderTo} \land \mathit{offers}' = \langle\,\rangle \end{array}$$

com_choose
$$\begin{array}{|l} \Delta(\mathit{orderTo}) \\ \hline \exists\, i : 1..\#\mathit{offers}, n : \mathit{Cost} \bullet \\ \quad \mathit{offers}\ i = (\mathit{orderTo}', n) \land \forall\, j \neq i : n \leq \mathit{second}(\mathit{offers}\ j) \end{array}$$

...

This specification defines a class *Machine* which includes two public methods *offer* and *order* and a private method *choose*. It has a variable *offers* storing the name of the offering HTS' together with the cost, and a variable *orderTo* describing the destination HTS. The type *Hts* is the names of transportation units and *Cost* a fixed range of natural numbers. Method *offer* changes the variable *offers* (denoted by having *offers* in its delta-list), and appends the next offer to the end of the sequence. Method *order* orders the transportation unit which is currently assigned to attribute *orderTo* and empties sequence *offers*. Method *choose* chooses the HTS with the cheapest offer in the sequence and assigns it to *orderTo*. The static view also contains a number of classes for stores and transportation system, these are elided here.

Dynamic view. In the dynamic view every class in the system may, in addition, have a protocol regulating the allowed ordering of method invocation. For class *Machine* it is the following, given as a CSP process description[1].

$$\begin{array}{ll} \texttt{main} & = \mathit{FindHts};\ \mathit{main} \\ \mathit{FindHts} & = |||_{h \in \mathit{Hts}}\ \mathit{offer}.h;\ \mathit{choose} \rightarrow \mathit{order} \rightarrow \mathit{SKIP} \end{array}$$

[1] Alternatively, we could also have used a simple state machine for describing this protocol (and translate it to CSP in order to verify the model transformation later). Such a translation is employed in [RW03].

This process description specifies the protocol for class *Machine* to be a repeated execution of getting offers, followed by choosing an offer and ordering it. In the CSP, `main` defines the dynamic behaviour of *Machine*, $|||$ is the parallel interleaving operator, offers are obtained from the HTS' in parallel; $\rightarrow$ and ; are sequencing operators; *SKIP* denotes termination. Again we elide methods of *Machine* that were not included in the above class specification (e.g., those concerned with loading, unloading and processing workpieces).

Architecture. As a third view on our system we incorporate a UML 2.0 composite structure diagram (Figure 1) in our model. It describes the architecture of the system, again just giving the interconnections for *Machine*, which communicates with objects of class *Hts* via two ports corresponding to methods *offer* and *order*. The structure diagram also fixes the number of components in the system: here two machines and three transportation units.

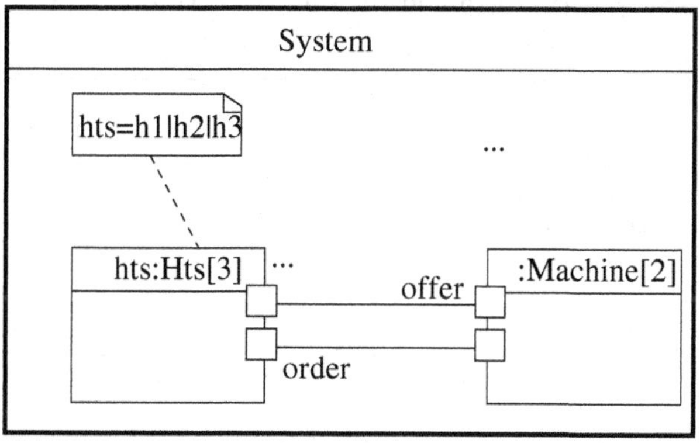

Fig. 1. Partial architecture of manufacturing system

These three specifications make up our first model.

2.2 Model 2 – PSM

Next, we modify our first model in order to improve its implementability in a programming language, i.e., we make a transformation towards a platform specific model. The target language in the project that motivated this work was Java, however, here we just assume that our target language requires

- methods to be *deterministic*,
- methods to be *totally defined*, and
- data types to be *bounded*.

This necessitates three changes in our model. The private method *choose* is non-deterministic since it chooses an arbitrary HTS out of those with the smallest number, and it is only partially defined since its predicate just fixes an outcome for nonempty sequences. The sequence *offers* used to store offers coming from transportation units is furthermore unbounded. We thus need the following transformations:

1. *Determinism.* The change necessary to make *choose* deterministic is the addition of $n = second(offers\ j) \Rightarrow i < j$ to its predicate. If more than one HTS has given a minimal offer, the first one in the sequence *offers* is chosen.
2. *Total methods.* For making *choose* total, we have to state what should happen in case that *choose* is called when the variable *offers* is empty. Here, we simply specify non-emptiness of *offers* to be a precondition for *choose* and thus add $offers \neq \langle\rangle$ to the schema.
3. *Bounded data types.* Finally, we have to fix an upper bound for the size of *offers*. We do so by adding a *class invariant* to the specification restricting the size of *offers* to three (we will soon see why three is sufficient): $\#offers \leq 3$. Note that we thus implicitly specify method *offer* to be blocked once the sequence has reached size three.

The dynamic view and the architecture remain unchanged. The resulting platform specific model thus consists of the following new class specification together with the same CSP process `main` and the same structure diagram.

Machine2

...

$$offers : \mathrm{seq}(Hts \times Cost)$$
$$orderTo : Hts$$

...

$$\#offers \leq 3$$

...

$\mathtt{com\_}choose$

$$\Delta(orderTo)$$

$$offers \neq \langle\rangle$$
$$\exists\, i : 1..\#offers, n : Cost \bullet$$
$$\quad offers\ i = (orderTo', n) \wedge \forall\, j \neq i : n \leq second(offers\ j)$$
$$\qquad \wedge\ n = second(offers\ j) \Rightarrow i < j$$

...

This completes the specification of our platform independent and platform specific model. Next, we are interested in showing correctness of the employed model transformation.

3 Behaviour Preservation and Refinement

For a *formal* proof of correctness of model transformations (which we aim at) we first of all need two things: a formal semantics for our models and a notion of correctness, based on this semantics.

Semantics. Each individual (Object-Z, CSP, etc) formal modelling notation has its own semantics. However, these notations are combined, and thus we need to determine how the semantics of the individual views make up the semantics of the entire model. In fact, this is constructed in two steps (see [Fis97, OW05] for a more detailed description) as follows. First, every Object-Z class is translated to CSP and combined with the CSP term of its protocol via parallel composition. The parallel composition is acting as a kind of conjunction here, both restrictions on the behaviour from the Object-Z and from the CSP are conjoined. The Object-Z part provides the data-dependent restrictions on the behaviour (and is thereby implicitly specifying possible orderings of method executions), whereas the CSP part focuses on data-independent orderings. Then, as the next step, the thus constructed CSP processes are combined according to the structure laid down in the structure diagram. For example, here we get

$$Ma = CSP(Machine) \parallel_{\{order, offer\}} main$$

as semantics of class *Machine*. The synchronisation set $\{order, offer\}$ attached to the parallel operator fixes the joint communication between the CSP processes. Assuming a definition of class *Hts* we then get the following out of the structure diagram as the semantics for the complete platform independent model (following a translation proposed in [FOW01]):

$$System = (Hts(h1) \mid\mid\mid Hts(h2) \mid\mid\mid Hts(h3)) \parallel_{\{offer, order\}} (Ma(1) \mid\mid\mid Ma(2))$$

In the same way we can obtain the semantics of the platform specific model which we refer to as *System*2.

Correctness criterion. Given the semantics of the combined model, we need to choose a suitable correctness criterion to determine which are acceptable model transformations. Specifically, model transformations are expected to preserve the overall behaviour of upper level models in platform specific models. While certain details are added, the changes made should still be unobservable to users of the system. This can either be achieved if the semantics of the two models are *equivalent* or by requiring the platform specific model to be a *refinement* of the platform independent model. Unless the added details result in equivalent models, it is likely that refinement will be the most suitable notion.

Furthermore, refinement is the notion of correctness employed in a formal development of programs. Refinement (provably) achieves the above required *substitutability*: if two models are in a refinement relationship, users cannot detect whether the original model or its refinement has been used. Refinement is therefore a suitable correctness criterion for evaluating model transformations in this context.

The modelling notations used for the static and for the dynamic view are both equipped with a well-defined theory of refinement. The notion of refinement associated with Object-Z (and Z, B, VDM etc) is *data refinement* [DB01], which allows changes of data types and operations as long as the change remains unobservable to a user of the system. We write $A \sqsubseteq_d C$ if the Object-Z class C

is a data refinement of class A. Note that the more abstract specification is on the left hand and its refinement on the right hand side.

The associated refinement concept for the dynamic view (CSP) is that of *process refinement* (failures-divergences refinement [Ros98]), denoted $P \sqsubseteq_p Q$ if Q is a process refinement of P. It basically allows to reduce nondeterminism in a process specification while keeping the required functionality. Like data refinement it guarantees substitutability and thus is suitable for model transformations. The semantic basis for the architectural view is CSP as well, thus process refinement is also applicable for changes on the architectural view. Both notions of refinement are moreover *transitive* which allows for a safe composition of model transformations.

So far this gives us refinement concepts for every individual view. Since the semantic domain for the entire system is CSP (or, to be precise, its semantic domain), behaviour preservation from one model to another is defined as a process refinement relationship between the models. The compositional way of defining the semantics allows for a smooth integration of the different refinement concepts, namely the following holds:

$$A \sqsubseteq_d C \Rightarrow CSP(A) \sqsubseteq_p CSP(C), \qquad \text{and}$$

$$P_1 \sqsubseteq_p P_2 \text{ and } Q_1 \sqsubseteq_p Q_2 \Rightarrow P_1 \mid\mid P_2 \sqsubseteq_p Q_1 \mid\mid Q_2$$

This allows us to separately use data refinement as the correctness criterion for model transformations on the static view, and process refinement for the dynamic and structural view while still achieving a process refinement relationship for the complete model.

This compositionality is sound, but not complete. That is there are valid transformations which cannot be studied in isolation in the individual views. This is the case for our example. We again look at the three changes made to our static view.

1. *Determinism.* The first change was introduced to make a method deterministic. This is a classical change allowed by data refinement: the nondeterminism in a specification is reduced and one possibility out of a number of possible outcomes of a method is chosen. The transformation can safely be qualified as being correct.
2. *Total methods.* The second change introduced a precondition for a method. This change modifies the behaviour of the class: while the method was always executable before (though with an undefined outcome in some cases), it is now sometimes refused. If the class is considered in isolation this change is observable by a user and not covered by data refinement.
3. *Bounded data types.* The third change is observable as well: the length of the sequence *offers* has been restricted and thus executability of method *offer* has implicitly been altered.

Of these three sub-transformations on individual views, only the first one (in the static view) is covered by the standard refinement concept. Hence $Machine \not\sqsubseteq_d Machine2$. In the following we will specifically treat the two problematic changes.

Both changes are of the same type: We have *conjunctively* added a new constraint (in our modelling language, a predicate) to the static view. Moreover, the changes *are* behaviour preserving when considered not in isolation but in connection with the two other views. The architectural view fixes the number of HTS' in the system (to be three) and the dynamic view for class *Machine* limits the number of *offer*s without an intervening *order* (which empties the sequence) to the number of HTS'. It furthermore allows *choose* only after some preceding *offer*, thus it will never be called on an empty sequence anyway. Thus when combining all views, model 2 does preserve the behaviour of model 1, that is,

$$System \sqsubseteq_p System2$$

indeed holds.

In the next section we develop a technique to tackle such cases. In particular, the technique avoids one having to compute the semantics of the two *complete* models (on which a comparison could of course show behaviour preservation), and instead just compares certain parts of the models. We thus aim to regain a certain degree of compositionality.

4 A Technique for Single View Transformations

In the sequel, our setting is always the following. We are given two models, both consisting of one or more[2] classes specified in Object-Z (static view) together with protocol descriptions in CSP (dynamic view) plus a structure diagram fixing the components in the system and their interconnections. We will consider changes to the static view which are *conjunctive* (as those above), i.e., which conjunctively add new restrictions to the static view. Denoting the class in the higher level model A and its corresponding class in the platform-specific model C we thus have transformations of the type

$$\begin{array}{c} A \\ \Downarrow \\ C : A \wedge Con \end{array}$$

where *Con* is an additional constraint (e.g., we generated *Machine*2 from *Machine* by conjunctively adding a constraint to its methods). Constraints can appear as pre- and postconditions in method specifications, as class invariants or as constraints on the initial state. The dynamic view and architecture stay as before. We let *Prot* denote the CSP process description for the protocol of class A. The role of the architectural view here is the instantiation of components: it fixes the number and names of components that the CSP part may refer to. We denote this instantiation by writing $Prot(arch)$. (In our example, the CSP process is parameterised in the set *Hts* which is instantiated to $\{h1, h2, h3\}$ by the structure diagram.) The problem to be tackled is thus the following:

[2] In case of more than one class, the technique simply has to be applied to every class.

Find conditions on A, Con and $Prot(arch)$ such that the following holds

$$CSP(A) \parallel Prot(arch) \sqsubseteq_p CSP(A \wedge Con) \parallel Prot(arch)$$

Our main theorem below isolates two conditions, C1 and C2 which are sufficient for this to hold.

In our case study the overall models were behaviour preserving since, although the static view changed, the CSP forbids the "new behaviour" anyway. To apply this type of argument in general, the conjunctive change has to be *i/o-independent.* A formal definition of i/o-independence based on a transition system semantics for Object-Z is given below, however, intuitively, a change is i/o-independent if the new behaviour does not depend on values of parameters of method invocations. This is sufficient to show that the overall models are behaviour preserving since the CSP part is not making restrictions on inputs and outputs of methods and can therefore not rule out i/o-dependent new behaviour.

The formal definition is the following, where for a trace $tr : Events^*$ over a set of method names M we first define $tr \upharpoonright M$ to be the sequence obtained by stripping off the values for parameters. For $tr = \langle offer.h1.5, offer.h2.8, order.h1\rangle$ we for instance have $tr \upharpoonright M = \langle offer, offer, order\rangle$.

Definition 1. *Let A and C be two Object-Z class specifications such that $C = A \wedge Con$. Let $T(A) = (Q_A, \longrightarrow_A, In_A)$ and $T(C) = (Q_C, \longrightarrow_C, In_C)$ be their transition systems. (By construction of the semantics and C being $A \wedge Con$ we have $Q_C \subseteq Q_A$).*

Con is i/o-independent *iff the following holds for all traces $tr : Events^*$, methods m of the class and possible parameters i and o:*

$$\begin{array}{l}\forall\, q_0 \in In_C, q_n \in Q_C : \\ q_0 \xrightarrow{tr}_C q_n, q_n \xrightarrow{m.i.o}\!\!\!\!\!\not\;\;_C \wedge q_n \xrightarrow{m.i.o}_A \\ \quad \Rightarrow \forall\, q_0' \in In_C, q_n' \in Q_C, i', o', tr' : Events^* : \\ \qquad\qquad q_0' \xrightarrow{tr'}_C q_n', tr \upharpoonright M = tr' \upharpoonright M \Rightarrow q_n' \xrightarrow{m.i'.o'}\!\!\!\!\!\not\;\;_C\end{array}$$

This definition relies on the on a transition system semantics for Object-Z and CSP specifications. For Object-Z specifications, this can either be derived from their corresponding CSP processes or given directly. For a CSP process P the transition system is derived via an operational semantics (see [Ros98]), i.e., $T(P) = (L_{CSP}, \longrightarrow, \{P\})$ with the states L_{CSP} being the set of all CSP terms, $\longrightarrow$ derived by the rules of the structured operational semantics and P itself being the initial state. The semantics for the combination of both parts is obtained by combining the semantics of the two views via CSP parallel composition, synchronising on all events in the intersection of the alphabets of the individual views. This has the effect that both the Object-Z and the CSP restrictions on the behaviour of the class are obeyed in the combination.

There is a proof technique which can be used to show i/o-independence of a conjunctive change of an Object-Z classes *without* computing the transition systems, namely by defining a simulation like relation between the two classes. States are equivalent in this relation if they are reached by the same method calls (regardless of i/o-parameters). One then has to check that on related states *Con* only blocks method calls due to their name, and not their i/o-parameters. For example, in our case study states are related if *offers* has the same size: the size of the sequence in a particular state does not depend on parameters of previous *offers* (and also not of *order* and *choose*), it just depends on the *number* of method executions. The addition of $Con = \#offers \leq 3$ then means that the ability to do an *offer*, *order* or *choose* method does not depend on the parameters of the methods, but just on which method is attempted. That is, if one $offer.h.c$ is blocked, they all are. The added precondition ($offers \neq \langle\rangle$) is similar.

In fact for our case study, we can syntacticly check this condition: with the methods given every predicate that only refers to the size of a sequence is i/o-independent (note that this does for instance not apply to sets[3]). Thus both new constraints $\#offers \leq 3$ and $offers \neq \langle\rangle$ are i/o-independent.

Once this has been established the check to be carried out is whether the CSP part sufficiently restricts the behaviour of the class so that the constraint does not introduce new behaviour. This is the case if

$$traces(Prot(arch)) \upharpoonright M \subseteq traces(CSP(C)) \upharpoonright M$$

holds. Here, M is the set of all method names, $traces(P)$ the set of all traces, i.e. sequences of events, that a CSP process P can perform and $\upharpoonright$ a projection operator on traces which removes all parameters from the events (e.g. $\langle offer.h1.5, offer.h2.8, order.h1\rangle \upharpoonright M = \langle offer, offer, order\rangle$). If the protocol only allows a subset of the traces of the constrained class C anyway, then the new behaviour of C (its new refusals) is not new when conjoining it with the protocol. Note that we only have to compare the projections of the traces here (omitting parameters) since we required i/o-independence.

These two conditions together are sufficient for checking correctness of such model transformations.

Theorem 1. *Let A, C be Object-Z classes such that $C = A \wedge Con$, and let $Prot(arch)$ be the CSP protocol description for the classes, instantiated according to some architectural specification. If*
(C1) Con is i/o-independent and
(C2) $traces(Prot(arch)) \upharpoonright M \subseteq traces(CSP(C)) \upharpoonright M$

then $CSP(A) \parallel Prot(arch) \sqsubseteq_p CSP(C) \parallel Prot(arch)$.

This gives us a technique for proving model transformations on the static view of the type $A \Rrightarrow A \wedge Con$ to be behaviour preserving: check condition C1 on A, C and Con and condition C2 on $Prot$ and C. Whereas the first condition can be checked on a single view, the second condition is a check across views. The next section explains a tool-supported way of checking this condition and exemplary shows this for our case study.

[3] Whereas the size of a sequence essentially depends on the number of elements appended to or removed from the sequence the size of a set depends on the actual elements that have been put in.

5 Checking Condition C2

A tractable way of checking condition C2 is by actually carrying out the translation of the Object-Z class to CSP, and afterwards using the CSP model checker FDR [FDR97] for checking trace inclusion. The translation of Object-Z to CSP follows an approach outlined in [FW99]. The class is translated into a CSP process (e.g., *Machine*2 becomes: process `OZ(offers,orderTo)` given below) parameterised with the variables of the class. The behaviour of this process is a choice (in CSP `[]`) over all possible method executions followed by a recursive call to the process possibly with modified instantiation of variables. The precondition of a method acts as a guard to the method execution (denoted `&`). The definition of data types (`HTS`) is derived from the structure diagram. This generic process is then instantiated with initial values as given by the Object-Z specification. The check for trace inclusion can be performed by FDR provided the state spaces of both processes (the protocol and the translated Object-Z class) are finite.

Below, the CSP specification of both the protocol and the Object-Z class can be found in the syntax of the FDR model checker.

```
-- declaration of data types
datatype HTS = h1 | h2 | h3
Hts = {h1,h2,h3}
Costs = {1..5}

-- declaration of channels
channel offer : Hts.Costs
channel order : Hts
channel choose
channel offerProj    -- channel offer without parameters
channel orderProj    -- channel order without parameters

-- CSP process of protocol
main = FindHts; main
FindHts =   (||| h : Hts @ offer.h?x -> SKIP); (choose -> order?x -> SKIP)

-- CSP process of class Machine2
Machine2 = OZ(<>,h1)
OZ(offers,orderTo) =
          (offers != <> & let
                              ot = first(min(offers))
                            within choose -> OZ(offers,ot))
       [] (length(offers) < 3 & offer?h?c -> OZ(offers^<(h,c)>,orderTo))
       [] (order!orderTo -> OZ(<>,orderTo))

-- projections of main and Machine2 to method names (by renaming)
mainProj =
   main[[offer.h.c<-offerProj,order.h<-orderProj|h<-Hts,c<-Costs]]
Machine2Proj =
   Machine2[[offer.h.c<-offerProj,order.h<-orderProj|h<-Hts,c<-Costs]]
```

Projection is defined by renaming all channels to ones without parameters. The check for trace inclusion can be carried out by asking FDR to verify the following assertion:

```
assert Machine2Proj [T= mainProj
```

This returns a positive answer. Hence we have now succeeded in showing both condition C1 and C2 for the two "non-refinement" changes on the static view. Thus the model transformations are valid according to our chosen notion of behaviour preservation and can - due to transitivity of refinement - safely be combined with any data refinement change, as for instance our first change to the model (determinism of method).

6 Conclusion

The aim of this paper has been to derive techniques whereby model transformations can be verified even if the sub-transformations on the individual views are not behaviour preserving. We have placed this work in the context of combinations of Object-Z and CSP, however, it should be clear that the techniques we have discussed could be transferable to other integrations. In particular, they are applicable to (parts of) UML models, e.g. when written in the profile proposed in [MORW04]. The work presented here covered the class of conjunctive changes, and this can be combined with changes of type data refinement and process refinement. Thus, more complex model transformations can be verified as well if they can be shown to be composed out of smaller, correct transformations.

Related work. Research on refinement and on model transformations is a very broad field. In relation to our work it might best be classified using the following two criteria. The first one is the *number of views* treated. There are (a large number of) approaches to refinement/refactoring/model transformations treating a single view only (or multiple views, but separately), e.g. [LB98, MS04, MT04, PR03], [SPTJ01, KHK$^+$03, BHTV04, KHE03, Whi02] and all classical definitions of refinement, but there is only a small number of approaches treating *multiple views* together [BPPT03, DS03, TS02]. A second criterion for classification are the techniques employed. While a lot of a work is carried out on *defining* model transformations, most with the ultimate goal of automating them, others are working on a posteriori *verifying* model transformations. The first category is sometimes also called the *operational* and the second one the *relational* approach. According to this classification our work falls into the category *verifying multiple view transformations.* We shortly comment on some related approaches.

Our approach uses a standard notion of refinement as a correctness criterion for model transformations. The use of refinement concepts can for instance also be found in [MS04] (Object-Z class refinement), [BHTV04] (refinement between graph transformation systems), [PR03] (refinement of data-flow architectures) and [TS02] (CSP process refinement). Transformations involving more than one view are treated in [BPPT03] where refactorings for class diagrams and (simple) consecutive modifications on state machines and sequence diagrams are defined. Relational approaches, i.e. validations of transformations, involving more than one view can be found in [DS03] where Object-Z/CSP classes are split, and in the work of Schneider and Treharne (e.g. [TS02]). The latter have considered the refinement of integration of CSP and B, and have discussed sufficient conditions such that the structure of a system changes. They for example isolate conditions by which refinements can be checked which are not compositional. This is similar to what we wish to achieve here. But the work looks at the general scenario, whereas our aim was to exploit the consequences of a particular situation.

Future work. A further paper will explore the extension to situations where we have simultaneous changes in the CSP part whereby new events are added and when new operations are added in the Object-Z part. We are furthermore interested in developing transformation *patterns* which can then by construction guarantee behaviour preservation. This is already partly the case since we had a syntactic check of one condition.

Acknowledgements

This work was supported by the Royal Society via an International Joint Project.

References

[BD02] Ch. Bolton and J. Davies. Refinement in Object-Z and CSP. In M. Butler, L. Petre, and K. Sere, editors, *IFM 2002: Integrated Formal Methods*, number 2335 in LNCS, pages 225–244, 2002.

[BHTV04] L. Baresi, R. Heckel, S. Thöne, and D. Varro. Style-Based Refinement of Dynamic Software Architectures. In *4th Working IEEE/IFIP Conference on Software Architecture (WICSA4)*, pages 155–164. IEEE Computer Society, 2004.

[BPPT03] P. Bottoni, F. Parisi-Presicce, and G. Taentzer. Coordinated distributed diagram transformation for software evolution. In Reiko Heckel, Tom Mens, and Michel Wermelinger, editors, *Electronic Notes in Theoretical Computer Science*, volume 72. Elsevier, 2003.

[DB01] J. Derrick and E. A. Boiten. *Refinement in Z and Object-Z*. Springer-Verlag, 2001.

[DB03] J. Derrick and E.A. Boiten. Relational concurrent refinement. *Formal Aspects of Computing*, 15(2-3):182–214, November 2003.

[dE98] W.-P. de Roever and K. Engelhardt. *Data Refinement: Model-Oriented Proof Methods and their Comparison*. CUP, 1998.

[DS03] J. Derrick and G. Smith. Structural Refinement of Systems Specified in Object-Z and CSP. *Formal Aspects of Computing*, 15(1):1 – 27, July 2003.

[FDR97] Formal Systems (Europe) Ltd. *Failures-Divergence Refinement: FDR2 User Manual*, Oct 1997.

[Fis97] C. Fischer. CSP-OZ: A combination of Object-Z and CSP. In H. Bowman and J. Derrick, editors, *Formal Methods for Open Object-Based Distributed Systems (FMOODS '97)*, volume 2, pages 423–438. Chapman & Hall, 1997.

[FKN+92] A. Finkelstein, J. Kramer, B. Nuseibeh, L. Finkelstein, and M. Goedicke. Viewpoints: A framework for integrating multiples perspectives in system development. *International Journal of Software Engineering and Knowledge Engineering*, 2(1):31–58, 1992.

[FOW01] C. Fischer, E.-R. Olderog, and H. Wehrheim. A CSP view on UML-RT structure diagrams. In H. Hussmann, editor, *FASE'01*, number 2029 in LNCS, pages 91–108. Springer, 2001.

[FW99] C. Fischer and H. Wehrheim. Model-checking CSP-OZ specifications with FDR. In K. Araki, A. Galloway, and K. Taguchi, editors, *Proceedings of the 1st International Conference on Integrated Formal Methods (IFM)*, pages 315–334. Springer, 1999.

[He89] Jifeng He. Process simulation and refinement. *Formal Aspects of Computing*, 1:229–241, 1989.

[Hoa85] C. A. R. Hoare. *Communicating Sequential Processes.* Prentice Hall, 1985.

[Jos88] M. B. Josephs. A state-based approach to communicating processes. *Distributed Computing*, 3:9–18, 1988.

[KHE03] J. Küster, R. Heckel, and G. Engels. Defining and and Validating Transformations of UML Models. In *IEEE Symposium on Visual Languages and Formal Methods*, pages 145–152. IEEE Computer Society, 2003.

[KHK+03] J. Koehler, R. Hauser, S. Kapoor, F. Wu, and S. Kumaran. A Model-Driven Transformation Method. In *EDOC 2003*, pages 186–197. IEEE Computer Society, 2003.

[LB98] K. Lano and J. Bicarregui. Semantics and Transformations for UML Models. In Jean Bézivin and Pierre-Alain Muller, editors, *UML*, volume 1618 of *Lecture Notes in Computer Science*, pages 107–119. Springer, 1998.

[MORW04] M. Möller, E.-R. Olderog, H. Rasch, and H. Wehrheim. Linking CSP-OZ with UML and Java: A Case Study. In *Integrated Formal Methods*, number 2999 in Lecture Notes in Computer Science, pages 267–286, March 2004.

[MS04] T. McComb and G. Smith. Architectural Design in Object-Z. In *Australian Software Engineering Conference (ASWEC 2004)*. IEEE Computer Society, 2004.

[MT04] T. Mens and T. Tourwé. A Survey of Software Refactoring. *IEEE Transactions on Software Engineering*, 30(2), 2004.

[OW05] E.-R. Olderog and H. Wehrheim. Specification and (property) inheritance in CSP-OZ. *Science of Computer Programming*, (55):227–257, 2005.

[PR03] J. Philipps and B. Rumpe. *Refactoring of Programs and Specifications*, pages 281–297. Kluwer Academic Publishers, 2003.

[Ros98] A. W. Roscoe. *The Theory and Practice of Concurrency.* Prentice Hall, 1998.

[RW03] H. Rasch and H. Wehrheim. Checking Consistency in UML Diagrams: Classes and State Machines. In *FMOODS 2003: Formal Methods for Open Object-based Distributed Systems*, number 2884 in LNCS, pages 229–243. Springer, 2003.

[Smi00] G. Smith. *The Object-Z Specification Language.* Kluwer Academic Publisher, 2000.

[SPTJ01] G. Sunyé, D. Pollet, Y. Le Traon, and J.-M. Jézéquel. Refactoring UML models. In Martin Gogolla and Cris Kobryn, editors, *UML 2001 - The Unified Modeling Language. Modeling Languages, Concepts, and Tools.*, volume 2185 of *LNCS*, pages 134–148. Springer, 2001.

[TS02] H. Treharne and S.A. Schneider. Communicating B machines. In *ZB2002: International Conference of Z and B Users*, volume 2272 of *LNCS*. Springer, 2002.

[Weh00] H. Wehrheim. Specification of an automatic manufacturing system – a case study in using integrated formal methods. In T. Maibaum, editor, *FASE 2000: Fundamental Aspects of Software Engineering*, number 1783 in LNCS, pages 334–348. Springer, 2000.

[Whi02] Jon Whittle. Transformations and software modeling languages: Automating transformations in uml. In Jean-Marc Jézéquel, Heinrich Hußmann, and Stephen Cook, editors, *UML*, volume 2460 of *Lecture Notes in Computer Science*, pages 227–242. Springer, 2002.

[ZJ96] P. Zave and M. Jackson. Where do operations come from? A multiparadigm specification technique. *IEEE Transactions on Software Engineering*, XXII(7):508–528, 1996.

Hyperfinite Approximations to Labeled Markov Transition Systems

Ernst-Erich Doberkat*

Chair for Software Technology
University of Dortmund
doberkat@acm.org

Abstract. The problem of finding an approximation to a labeled Markov transition system through hyperfinite transition systems is addressed. It is shown that we can find for each countable family of stochastic relations on Polish spaces a family of relations defined on a hyperfinite set that is infinitely close. This is applied to Kripke models for a simple modal logic in the tradition of Larsen and Skou. It follows that we can find for each Kripke model a hyperfinite one which is infinitely close.

1 Introduction and Motivation

The methods of non-standard analysis, originating with Robinson and Hewitt, and going back as far as Newton and Leibniz, permit to combine the simplicity of finite models with the elegance of general approaches. The method of relating these seemingly incompatible ways of looking at things may roughly be described through the following pattern (see e.g. [14]): first, the original problem is translated into a non-standard, preferably hyperfinite setting, then it is solved in this context, preferably with finite methods, and it is finally cast back into the original setting through techniques like taking standard parts. An instance of this approach to probabilistic problems may be studied with Anderson's beautiful representation of Brownian motion [16, § II.3]. In fact, probability theory and its applications are a field in which this setting in the past has turned out to be most fruitful. Not surprisingly, mathematical economics with its strong orientation towards measure theoretic methods is another field in which these methods bear fruit. For example, consider the case that there is a universe of customers with each individual having infinitely small influence, but that, as a whole, the customers' influence on the market is not negligible. It seems to be difficult to model this situation within the finite/infinite dichotomy; using methods from non-standard analysis, however, yields satisfactory models, see [20].

When looking at stochastic models in concurrency, a similar problem occurs: general models give general insight into the problems at hand, but finite models are easier to handle, although they sometimes tend to oversimplify the mathematical structure, thereby concealing its proper nature. It gives usually more insight into modeling a problem using a general, continuous model, but these models tend on the other hand in all

* Research funded in part by Deutsche Forschungsgemeinschaft, grant DO 263/8-3, *Algebraische Eigenschaften stochastischer Relationen*.

M. Johnson and V. Vene (Eds.): AMAST 2006, LNCS 4019, pp. 127–141, 2006.

their generality to be difficult to handle. The proverbial best of two worlds would be a model which argues from a finite point of view, but which also permits casting its results into a general framework.

This paper proposes a way of combining the infinite and general approach with a finite one. We show that each labeled Markov transition system is approximated by a hyperfinite one, hence by a system which is based on a finite scenario. This bridges the gap — appearing to be insurmountable — between seemingly inaccessible infinite probabilistic transition systems on the one hand and their finite, less complicated cousins on the other hand. We show that this idea may be carried over to probabilistic Kripke models for modal logic (which have at their very heart labeled Markov transition systems): given a model for a modal logic, we can find an infinitely close hyperfinite model.

Technically, we need to make a topological assumption for this to work: the state spaces of the transition systems should be Polish. These spaces underlie most investigations in this area anyway (take [6, 11, 7, 21] as samples). We cast the problem into a slightly more general form: given a countable family of stochastic relations between Polish spaces X and Y, we investigate finding stochastic relations on hyperfinite spaces that are infinitely close. This scenario is then utilized for the case of labeled Markov transition systems, and for Kripke models of a simple modal logic with a given set of actions.

The paper is organized as follows: section 2 discusses some preliminaries, it chiefly defines stochastic relations and transition systems. Section 3 shows that we can find a hyperfinite approximation for the general scenario and specializes this to transition systems; this section makes essential use of Loeb's and Anderson's work on the relation between standard and non-standard measures. Section 4 defines two sorts of Kripke models, and an approximation result is established as well. Related and further work are indicated in sections 5 and 6. Appendix A gives a very brief overview of the notions of non-standard analysis that are used in this paper.

2 Stochastic Relations

This section collects some basic facts from topology and measure theory for the reader's convenience and for later reference. It defines stochastic relations.

A *Polish space* $(X, \mathcal{T})$ is a topological space which is second countable, i.e., which has a countable dense subset, and which is metrizable through a complete metric. A *measurable space* $(X, \mathcal{A})$ is a set X with a σ-algebra $\mathcal{A}$. The *Borel sets* $\mathcal{B}(X, \mathcal{T})$ for the topology $\mathcal{T}$ are the smallest σ-algebra on X which contains $\mathcal{T}$. Given two measurable spaces $(X, \mathcal{A})$ and $(Y, \mathcal{B})$, a map $f : X \to Y$ is $\mathcal{A}$ - $\mathcal{B}$*-measurable* whenever $f^{-1}[B] \in \mathcal{A}$ for all $B \in \mathcal{B}$.

If the σ-algebras are the respective Borel sets of some topologies on X and Y, resp., then a measurable map is called *Borel measurable* or simply a *Borel* map. The real numbers $\mathbb{R}$ carry always the Borel structure induced by the usual topology which will usually not be mentioned explicitly when talking about Borel maps.

When the context is clear, we will write down topological or measurable spaces without their topologies or σ-algebras, resp., and the Borel sets are always understood with respect to a topology under consideration.

Denote for a measurable space $(X, \mathcal{A})$ by $\mathfrak{S}(X, \mathcal{A})$ the set of all subprobability measures on $(X, \mathcal{A})$ which is equipped with the *weak*-σ-algebra* for a measurable structure. The latter σ-algebra is the smallest σ-algebra on $\mathfrak{S}(X, \mathcal{A})$ which renders all maps $\mu \mapsto \mu(D)$ measurable, where $D \in \mathcal{A}$. If X is a Polish space, then $\mathfrak{S}(X)$ is Polish in the weak topology. This topology is the smallest topology on $\mathfrak{S}(X)$ which renders all evaluation maps $\mu \mapsto \int_X f\, d\mu$ continuous, where $f : X \to \mathbb{R}$ is continuous and bounded. It is well known that the weak*-σ-algebra constitutes the Borel sets of the weak topology.

Definition 1. *Given two Polish spaces X and Y, a* stochastic relation $K : X \rightsquigarrow Y$ between X and Y *is a Borel map from X to $\mathfrak{S}(Y)$.*

It can be shown that a stochastic relation is just a morphism in the Kleisli category for the monad that has the subprobability functor as its functorial part [13]. It is this analogy that makes a stochastic relation a *relation*: set theoretic relations are just the morphisms in the Kleisli category for a monad coming from the powerset functor. Note that we talk here about monads in the sense of categories; the later use of monads will refer to their use in non-standard topology[1]. In probability theory, a stochastic relation is known as a *sub-Markov kernel* or a *transition subprobability*. Hence $K : X \rightsquigarrow Y$ is a stochastic relation from X to Y iff

1. $K(x)$ is a subprobability measure on Y for all $x \in X$,
2. $x \mapsto K(x)(D)$ is a measurable map for each measurable set $D \subseteq Y$.

An $\mathcal{A}$ - $\mathcal{B}$- measurable map $f : X \to Y$ between the measurable spaces $(X, \mathcal{A})$ and $(Y, \mathcal{B})$ induces a map $\mathfrak{S}(f) : \mathfrak{S}(X, \mathcal{A}) \to \mathfrak{S}(Y, \mathcal{B})$ upon setting for $\mu \in \mathfrak{S}(Y, \mathcal{B})$ and $D \in \mathcal{B}$

$$\mathfrak{S}(f)(\mu)(D) := \mu(f^{-1}[D])$$

It is easy to see that $\mathfrak{S}(f)$ is measurable. This observation makes $\mathfrak{S}$ an endofunctor on the category of measurable spaces with measurable maps as morphisms [13, 10].

Hyperfinite stochastic relations are defined as the non-standard counterparts to finite stochastic relations. Clearly, if X and Y are finite sets, then a map $p : X \times Y \to [0, 1]$ is a stochastic relation iff $\forall x \in X : \sum_{y \in Y} p(x, y) \leq 1$ holds. Now a translation from finite to hyperfinite sets will require the same constraint translated, hence this time not in the unit interval $[0, 1]$ but in its non-standard counterpart $^*[0, 1]$. In addition we require such a relation to be an internal map; if domain and range of a relation can be represented in terms of standard notions, the relation should be as well. This leads to the following definition.

Definition 2. *Let X and Y be hyperfinite sets, then an internal map $k : X \times Y \to {}^*[0, 1]$ is called a* hyperfinite stochastic relation $k : X \rightsquigarrow Y$ *between X and Y iff*

$$\sum_{y \in Y} k(x, y) \leq 1$$

for each $x \in X$.

[1] The reader is referred to Appendix A for a very brief discussion of non-standard analysis.

We usually write $k(x)(y)$ rather than $k(x, y)$ and extend k to a map $X \times \mathcal{P}(Y) \to {}^*[0, 1]$ upon setting

$$k(x)(B) := \sum_{y \in B} k(x)(y).$$

Note that arithmetic and comparisons are in this case performed in ${}^*\mathbb{R}$.

Finally, we need the notion of invariant sets w.r.t. an equivalence relation.

Definition 3. *Let $\sim$ be an equivalence relation on a set M.*

1. *$E \subseteq M$ is called* $\sim$-invariant *iff $m \in E$ and $m \sim m'$ together imply $m' \in E$.*
2. *$\mathcal{INV}(\mathcal{C}, \sim) := \{E \in \mathcal{C} \mid E$ is $\sim$-invariant$\}$ are the $\sim$-invariant members of a family $\mathcal{C} \subseteq \mathcal{P}(M)$ of sets.*

Consequently, E is $\sim$-invariant iff $E = \bigcup_{m \in E} [m]_{\sim}$, thus iff E is a union of equivalence classes. The Borel sets that are invariant with respect to a countably generated equivalence relation are rather important in the theory of stochastic relations, see e.g. [6, 11], and will be shown to play an important rôle here for constructing approximations as well.

3 Approximating Labeled Markov Transition Systems

This section will demonstrate that a labeled Markov transition system can be approximated through a hyperfinite one.

Definition 4. *Let A be a countable set of actions, and assume that we have for each action $a \in \mathsf{A}$ a stochastic relation $K_a : S \rightsquigarrow S$. Then $(S, (K_a)_{a \in \mathsf{A}})$ is called a* labeled Markov transition system. *It is called* standard, *if S is Polish, and* hyperfinite, *if S is hyperfinite, and if in addition all K_a are internal maps.*

Thus a labeled Markov transition system models probabilistic state transitions: if the current state of the system is $s \in S$ and the action $a \in \mathsf{A}$ is taken, then $K_a(s)(B)$ is the probability for the new state to be a member of set B. Since $K_a(s)(S) < 1$ is not excluded, it is possible that the system does not enter a next state at all, for example when a computation is modeled that does not terminate. If we are working with a hyperfinite system, the transition laws K_a are internal maps, so we can assign to each state s' the probability $K_a(s)(s')$ with which it will be the next state after action a. Since S is internal as well, we may write $S = \langle (S_n) \rangle$ for sets S_n that are finite almost certainly (abbreviated as *a. c.*; see the explanation of terminology and notation for non-standard sets in Appendix A). Similarly, K_a can be written as $(K_{n,a})$ with $K_{n,a} : S_n \rightsquigarrow S_n$ a. c. These relations are then standard relations between finite sets, so that $K_a(s)(s')$ is really (the equivalence class of) a sequence $(r_n)_{n \in \mathbb{N}}$ of real numbers r_n with $K_{n,a}(s_n)(s'_n) = r_n$ a. c. We will demonstrate that we can approximate a standard labeled Markov transition system by a hyperfinite one. Consequently, we are able to express the general transition probabilities in a continuous state space through the probabilities for appropriately chosen discrete events.

When considering $K_a : S \rightsquigarrow S$ for Polish S, it will turn out that the respective spaces for the domain and for the range will be dealt with in different ways. Thus we take K_a

as an instance of a general stochastic relation $X \rightsquigarrow Y$ with Polish spaces X, Y that may happen to coincide. Neither are we bound formally too close to the actions; they form a countable set, so we technically treat for the time being the standard labeled Markov transition system $(S, (K_a)_{a \in \mathsf{A}})$ as an instance of a sequence $(K_n)_{n \in \mathbb{N}}$ of stochastic relations K_n with Polish domain and range spaces. Consequently, we fix in this section Polish spaces X and Y as the respective domains and ranges for a countable family $(K_n)_{n \in \mathbb{N}}$ of stochastic relations $K_n : X \rightsquigarrow Y$.

The idea of an approximation is then captured in the following definition. It is based on the following consideration. Suppose we have a stochastic relation $K : X \rightsquigarrow Y$. We want to find for each element x in the domain of a given stochastic relation an element from a hyperfinite set X_f that is infinitely close to x, and we want to find for the sub-probability $K(x)$ a reasonably structured sub-probability $k(x)$ on a hyperfinite set Y_f that is infinitely close to $K(x)$. Hence we can argue that the whole relation $k : X_f \rightsquigarrow Y_f$ is infinitely close to $K : X \rightsquigarrow Y$, and the former is based on finite components, as hyperfinite sets are. In order to formulate the notion of infinite closeness between two measures we refer to the Loeb construction $\mathsf{L}(\cdot)$. A brief summary of this construction is provided in Appendix A.

A formal translation of these ideas leads to this definition.

Definition 5. *Let* $\mathsf{K} = (X, Y, K)$ *be a stochastic relation with Polish* X, Y. *A* hyperfinite approximation to K *is a hyperfinite stochastic relation* (X_f, Y_f, κ) *with these properties:*

1. $X_f \subseteq {}^*X$ *and* $Y_f \subseteq {}^*Y$,
2. *for each* $x \in X$ *there exists* $\zeta \in X_f$ *with* $x = \mathsf{st}_X(\zeta)$ *and* $K(x) = \mathfrak{S}(\mathsf{st}_Y) \circ \mathsf{L}(\kappa(\zeta))$.

Topological properties will be crucial in what follows. To be specific, because Y is Polish, the measure spaces $(Y, \mathcal{B}(Y), K_n(x))$ that are induced by the measures $K_n(x) \in \mathfrak{S}(Y)$ have rather special and very convenient properties that are summarized below. The reader is referred to [18, Theorem 3.4.19] for a proof.

Proposition 1. *Let T be a Polish space, $\mu \in \mathfrak{S}(T)$ be a sub-probability measure. Then the measure space $(T, \mathcal{B}(T), \mu)$ has this property: For each Borel set $B \subseteq S$ and each $\epsilon > 0$ there exists a compact set $C \subseteq B$ with $\mu(B \setminus C) < \epsilon$.* $\dashv$

This property means that $(T, \mathcal{B}(T), \mu)$ is a *Radon space* (sometimes the property that the measure of each Borel set can be approximated by the measure of a compact subset is called *tightness*). We will need this property because we will refer to work done for approximating measures on Radon spaces [1, 2].

Given $x \in X, n \in \mathbb{N}$, each $K_n(x)$ is a subprobability measure on the Borel sets $\mathcal{B}(Y)$ of Y. By the usual standard procedure, $K_n(x)$ is extended to the *universally measurable sets* $\mathcal{U}(\mathcal{B}(Y))$ of Y; denote this extension again by $K_n(x)$. This family $\{K_n(x) \mid x \in X, n \in \mathbb{N}\}$ of measures may be approximated on a hyperfinite measure space.

Lemma 1. *There exists a hyperfinite set Y_f with a hyperfinite set algebra $\mathcal{A}_f \subseteq \mathcal{P}(Y_f)$, for each $x \in X$ a finitely additive measure $k_{n,f}(x)$ on $\mathcal{A}_f$ and a $\mathcal{A}_f - \mathcal{B}(Y)$-measurable map $S : Y_f \to Y$ with $K_n(x)(B) = \mathsf{L}(k_{n,f}(x))(S^{-1}[B])$ for all $n \in \mathbb{N}, x \in X$ and $B \in \mathcal{B}(Y)$.*

The proof of Lemma 1 follows Anderson's proof [1, Corollary 3.4]. Some minor modifications take the fact into account that we do want to approximate a whole family of measures rather than a single one. The present proof is broken into a series of auxiliary statements.

We show first that a suitable algebra exists on which we can define our measures. The argumentation is very similar to that used by Cutland in the proof of [4, Theorem 1.14] or by Anderson in [1, Corollary 3.4]. It is noted that the algebra constructed here is independent of any measure; this is different from Anderson's argument. We provide the argument explicitly for making the paper self-contained.

Lemma 2. *There exists a hyperfinite algebra $\mathcal{G}_f$ on $^*Y$ with*

$$\{^*C \mid C \in \mathcal{U}(\mathcal{B}(Y))\} \subseteq \mathcal{G}_f \subseteq {}^*\mathcal{U}(\mathcal{B}(Y)).$$

Proof. Given a finite subset $\mathcal{F}$ of $\mathcal{U}(\mathcal{B}(Y))$, there exists a finite algebra $\mathcal{A}_\mathcal{F}$ with $\mathcal{F} \subseteq \mathcal{A}_\mathcal{F}$. Consequently, the family of algebras $\{\mathcal{A}_{\{C\}} \mid C \in \mathcal{U}(\mathcal{B}(Y))\}$ has the finite intersection property, so the set

$$\bigcap\{^*\mathcal{A}_{\{C\}} \mid C \in \mathcal{U}(\mathcal{B}(Y))\}$$

is not empty by the Enlargement property (see Appendix).

The algebra $\mathcal{G}_f$ induces an equivalence relation $\sim_{\mathcal{G}_f}$ on $^*Y$ upon setting

$$y_1 \sim_{\mathcal{G}_f} y_2 \text{ iff } \forall A \in \mathcal{G}_f : y_1 \in A \Leftrightarrow y_2 \in A.$$

It is plain that each element of $\mathcal{G}_f$ is an $\sim_{\mathcal{G}_f}$-invariant set. The $\sim_{\mathcal{G}_f}$-equivalence class of y is denoted by $[y]_{\mathcal{G}_f}$. Since

$$[y]_{\mathcal{G}_f} = \bigcap\{A \mid A \in \mathcal{G}_f, y \in A\} \cap \bigcap\{^*Y \setminus A \mid A \in \mathcal{G}_f, y \notin A\},$$

and since the power set of a hyperfinite set is hyperfinite, the factor space is hyperfinite again.

The classes for near standard elements of $^*Y$ are related to the monads of the respective standard parts. This property is called *S-separation* in [1].

Lemma 3. *Let $y \in \mathsf{ns}(^*Y)$ be a near standard element of $^*Y$, then its class $[y]_{\mathcal{G}_f}$ is contained in the monad $\mathsf{monad}(\mathsf{st}_Y(y))$. Moreover, $\sim_{\mathcal{G}_f}$ is contained in the kernel of st_Y.*

Proof. Let T be an open neighborhood of $\mathsf{st}_Y(y)$, then by construction $\mathsf{monad}(\mathsf{st}_Y(y)) \subseteq {}^*T$, and since plainly $y \in \mathsf{monad}(\mathsf{st}_Y(y))$, we find that $y \in {}^*T$. The set T is open, hence it is universally measurable, consequently we know that $^*T \in \mathcal{G}_f$, thus $^*T$ is $\sim_{\mathcal{G}_f}$-invariant. But this entails $[y]_{\mathcal{G}_f} \subseteq {}^*T$. Since T was arbitrary, we conclude that

$$[y]_{\mathcal{G}_f} \subseteq \bigcap\{^*T \mid T \text{ open}, y \in T\} = \mathsf{monad}(\mathsf{st}_Y(y)).$$

The second part [1, Remark 3.2] is rather immediate: if $\mathsf{st}_Y(y_1) \in G$, and $y_1 \sim_{\mathcal{G}_f} y_2$, then $y_2 \in [y_1]_{\mathcal{G}_f} \subseteq \mathsf{monad}(\mathsf{st}_Y(y_1))$, thus $\mathsf{st}_Y(y_2) = \mathsf{st}_Y(y_1) \in G$.

Corollary 1. *The inverse image* $\mathsf{st}_Y^{-1}[G]$ *of an arbitrary* $G \subseteq Y$ *is* $\sim_{\mathcal{G}_f}$*-invariant.* $\dashv$

Proof. (of Lemma 1)

0. It can be shown that $\mathsf{ns}({}^*Y)$ is a universally measurable subset of ${}^*Y$ with

$$\forall x \in X : \mathsf{L}({}^*K(x))({}^*Y \setminus \mathsf{ns}({}^*Y)) = 0,$$

thus we may and do assume that ${}^*Y = \mathsf{ns}({}^*Y)$.

1. Construct $\mathcal{G}_f$ according to Lemma 2, define the hyperfinite set $Y_f := {}^*Y/\mathcal{G}_f$ and let $\mathcal{A}_f$ be the terminal algebra on Y_f with respect to the factor map $\eta_{\mathcal{G}_f} : {}^*Y \to Y_f$ and $\mathcal{INV}\left(\mathsf{L}(\mathcal{G}_f), \sim_{\mathcal{G}_f}\right)$, so that

$$\mathcal{A}_f = \{A \subseteq Y_f \mid \eta_{\mathcal{G}_f}^{-1}[A] \in \mathcal{INV}\left(\mathsf{L}(\mathcal{G}_f), \sim_{\mathcal{G}_f}\right)\}.$$

Define $S := \mathsf{st}_Y \circ \eta_{\mathcal{G}_f}^{-1}$, then S is well-defined by the second part of Lemma 3. Moreover, S is $\mathcal{A}_f$-$\mathcal{B}(Y)$-measurable. First, one notes that $\eta_{\mathcal{G}_f}^{-1}\left[\eta_{\mathcal{G}_f}[G]\right] = G$ for each $\sim_{\mathcal{G}_f}$-invariant G, and then one notes from [1, Theorem 3.3] that $\mathsf{st}_Y^{-1}[B] \in \mathcal{INV}\left(\mathsf{L}(\mathcal{G}_f), \sim_{\mathcal{G}_f}\right)$ for each Borel set $B \in \mathcal{B}(Y)$. This settles measurability.

2. Put for $n \in \mathbb{N}, x \in X$ and $A \in \mathcal{A}_f$ $k_{n,f}(x)(A) := {}^*K_n(x)(\eta_{\mathcal{G}_f}^{-1}[A])$, then $k_{n,f} : \mathcal{A}_f \to {}^*[0,1]$ is an internal map. We have for each $x \in X$ and each Borel set $B \in \mathcal{B}(Y)$

$$\begin{aligned} K_n(x)(B) &= \mathsf{L}({}^*K_n(x))(\mathsf{st}_Y^{-1}[B]) \\ &= \mathsf{L}(k_{n,f}(x))(\eta_{\mathcal{G}_f}\left[\mathsf{st}_Y^{-1}[B]\right]) \\ &= \mathsf{L}(k_{n,f}(x))(S^{-1}[B]). \end{aligned}$$

This caters for the range of the stochastic relations in question. Summarily, it says that we can find for $K_n(x)$ a measure defined on a hyperfinite algebra which comes infinitely close to $K_n(x)$ but has the characteristics of a finite measure, i.e., is concentrated on points. This construction can be performed in a way which makes the domain for the approximating measure independent of $K_n(x)$.

We will turn to the domain of $x \mapsto K_n(x)$ now and show that there can be found a hyperfinite set the members of which will be infinitely close to the members of X.

Lemma 4. *There exists a hyperfinite set* $F \subseteq {}^*X$ *so that*

1. *given* $x \in X$ *there exists* $\zeta \in F$ *such that* $\mathsf{st}_X(\zeta) = x$
2. $\mathfrak{S}(\mathsf{st}_X) \circ \mathsf{L}({}^*K_n)(\zeta) = K_n(\mathsf{st}_X(\zeta))$ *for all* $\zeta \in F, n \in \mathbb{N}$.

Proof. 0. It is no loss of generality to assume that K_n is a weakly continuous map $X \to \mathfrak{S}(Y)$ for each $n \in \mathbb{N}$, where the latter space has the weak topology. In fact, let $\mathcal{T}$ be the given topology on X. Since the map $K_n : X \to \mathfrak{S}(Y)$ is measurable, and since $\mathfrak{S}(Y)$ is a Polish space in the topology of weak convergence, we can find by [18, Corollary 3.2.6, Observation 2] a topology $\mathcal{T}'$ with $\mathcal{T} \subseteq \mathcal{T}'$ such that the Borel sets of the given topology coincide with the Borel sets of $\mathcal{T}'$ and such that each K_n is continuous with respect to $\mathcal{T}'$ and the weak topology on $\mathfrak{S}(Y)$.

1. We can find a hyperfinite set F_0 with $X \subseteq F_0 \subseteq {}^*X$ ([4, Theorem 1.14]). Since X is Polish, it has a countable dense subset $(x_n)_{n\in\mathbb{N}}$, hence by Comprehension (see Appendix) we can find an internal set $F := \{x_N \mid N \in {}^*\mathbb{N}\} \subseteq F_0$ that extends this sequence. Because F is an internal set with $F \subseteq F_0$, and F_0 is hyperfinite, F is hyperfinite as well.
2. Fix $n \in \mathbb{N}$. Since $K_n : X \to \mathfrak{S}(Y)$ is continuous when $\mathfrak{S}(Y)$ has the weak topology, we know that ${}^*K_n[\mathsf{monad}(x)] \subseteq \mathsf{monad}(K_n(x))$ holds for every $x \in X$ by [16, Proposition III.2.3] (the monad on the right hand side is taken with respect to the weak topology on $\mathfrak{S}(Y)$, the one on the left hand side with respect to X).
3. Fix $x \in X$. Since $(x_n)_{n\in\mathbb{N}}$ is dense, we can find for $x \in X$ an index $N \in {}^*\mathbb{N}$ (which is possibly infinite) such that $\mathsf{st}_X(x_N) = x$. Thus $x_N \in \mathsf{monad}(x)$, hence ${}^*K_n(x_N) \in \mathsf{monad}(K_n(x))$. Thus we have $K_n(x) = \mathsf{st}_{\mathfrak{S}(Y)}({}^*K_n(x_N))$. Because ${}^*K_n(x)({}^*Y \setminus \mathsf{ns}({}^*Y)) = 0$, we infer from [2, Lemma 2] that $\mathsf{st}_{\mathfrak{S}(Y)}({}^*K_n(x_N)) = \mathsf{L}({}^*K_n(x_N))$. Thus F is the desired hyperfinite set.

Summarizing, we have established that a stochastic relation on Polish spaces has a hyperfinite approximation:

Proposition 2. *For each countable set $(K_n)_{n\in\mathbb{N}}$ of stochastic relations $\mathsf{K}_n = (X, Y, K_n)$ over Polish spaces there exists a family $(\kappa_n)_{n\in\mathbb{N}}$ of hyperfinite stochastic relations $\kappa_n : X_f \rightsquigarrow Y_f$ so that (X_f, Y_f, κ_n) approximates (X, Y, K_n) for each $n \in \mathbb{N}$.*

Proof. 1. The remarks made at the beginning of the proof of Lemma 4 indicate that it is no loss of generality to assume that $K_n : X \to \mathfrak{S}(Y)$ is continuous. Construct the hyperfinite set X_f as in Lemma 4 and $(Y_f', \mathcal{A}_f', k_{n,f})$ as in Lemma 1. Pick from each class $[y]_{\mathcal{G}_f}$ an element, and collect all these elements in the set Y_f. This results in an internal set. Note that the singleton $\{[y]_{\mathcal{G}_f}\}$ is in $\mathcal{A}_f'$ for each $y \in Y_f$, since $\mathcal{A}_f'$ is an algebra, and since $[y]_{\mathcal{G}_f} \in \mathcal{INV}\left(\mathsf{L}(\mathcal{G}_f), \sim_{\mathcal{G}_f}\right)$.

2. Define for $\zeta \in X_f, y \in Y_f$ $\kappa_n(\zeta)(y) := k_{n,f}(\zeta)([y]_{\mathcal{G}_f})$, then $\kappa_n : X_f \times Y_f \to {}^*[0,1]$ is internal, since $k_{n,f}$ is. Consequently, the assertion follows from Lemma 4 and Lemma 1.

In particular we have

Corollary 2. *Each stochastic relation over Polish spaces has a hyperfinite approximation.* ⊣

Looking at relations $K_n : S \rightsquigarrow S$ that model state transitions, the construction leading to Proposition 2 does not warrant that the approximating hyperfinite relation $\kappa_n : X_f \rightsquigarrow Y_f$ may be used to model state transitions as well. This is so since the domain and the range of the approximating relation are different due to being constructed with different goals in mind: the approximation for the domain identifies a near standard element for each element in the domain, whereas the approximation for the range has a near standard element for the associated measure as a target.

We return to standard labeled Markov transition systems and show that we are able to approximate them through their hyperfinite cousins.

Corollary 3. *Let $(S, (K_a)_{a \in \mathsf{A}})$ be a standard labeled Markov transition system for a Polish state space S, then there exists a hyperfinite labeled Markov transition system $(S_f, (\kappa_a)_{a \in \mathsf{A}})$ so that κ_a approximates K_a for each $a \in \mathsf{A}$.*

Proof. Construct $S_{d,f}, S_{r,f}$ and $\kappa_a^0 : S_{d,f} \rightsquigarrow S_{r,f}$ as in the proof of Proposition 2. It is no loss of generality to assume that $S_{d,f}$ and $S_{r,f}$ are disjoint. Define $S_f := S_{d,f} \cup S_{r,f}$, then $S_f \subseteq {}^*S$ is hyperfinite, and define $\kappa_a : S_f \rightsquigarrow S_f$ through

$$\kappa_a(\sigma)(s) := \begin{cases} \kappa_a^0(\sigma)(s), & \sigma \in S_{d,f}, s \in S_{r,f}, \\ \delta_\sigma(s), & \text{otherwise.} \end{cases}$$

(δ_σ is the Dirac measure on σ). Then is it immediate that this constitutes a hyperfinite approximation to $K_a : S \rightsquigarrow S$.

4 Approximating Models

Given a probabilistic model $\mathcal{S}$ for a simple modal logic, we will approximate this model through a hyperfinite one. This means that for each formula φ and each state s of $\mathcal{S}$ which satisfies φ we can find a state in the hyperfinite model which is both infinitely close to s and which also satisfies φ. This property will be shown to hold conversely as well: if a state σ in the hyperfinite model satisfies a formula, then its standard part ${}^o\sigma = \mathsf{st}(\sigma)$ satisfies φ.

We work with a simple modal logic, which will be defined now. Fix a countable set P of atomic propositions, A is again the countable set of actions. The formulas of logic $\mathfrak{L}$ are defined through

$$\varphi ::= \top \mid p \mid \varphi \wedge \varphi \mid \neg\varphi \mid \langle a \rangle_q \varphi$$

Here $p \in \mathsf{P}$ is an atomic proposition, $a \in \mathsf{A}$ is an action, and $q \in \mathbb{Q} \cap [0, 1]$ is a rational number. The informal interpretation for state s satisfying formula $\langle a \rangle_q \varphi$ reads that we can make an a-move in a state s to a state that satisfies φ with probability greater than q.

Accordingly, a *standard model* $\mathcal{S} = (S, (K_a)_{a \in \mathsf{A}}, V)$ for logic $\mathfrak{L}$ is defined through a state space S, which is assumed to be a Polish space, for each action $a \in \mathsf{A}$ a stochastic relation $K_a : S \rightsquigarrow S$ and for each atomic proposition $p \in \mathsf{P}$ a Borel set $V(p) \in \mathcal{B}(S)$ of states in which p is assumed to hold. Define inductively the satisfaction relation $\models$ with respect to $\mathcal{S}$ together with the sets $[\![\varphi]\!]_{\mathcal{S}} := \{s \in S \mid \mathcal{S}, s \models \varphi\}$ of states in which formula φ holds:

1. $\mathcal{S}, s \models \top$ is true for all $s \in S$, thus $[\![\top]\!]_{\mathcal{S}} = S$.
2. $\mathcal{S}, s \models p \Leftrightarrow s \in V(p)$ for all $p \in \mathsf{P}$, thus $[\![p]\!]_{\mathcal{S}} = V(p)$.
3. $\mathcal{S}, s \models \varphi_1 \wedge \varphi_2$ iff $\mathcal{S}, s \models \varphi_1$ and $\mathcal{S}, s \models \varphi_2$, thus $[\![\varphi_1 \wedge \varphi_2]\!]_{\mathcal{S}} = [\![\varphi_1]\!]_{\mathcal{S}} \cap [\![\varphi_2]\!]_{\mathcal{S}}$.
4. $\mathcal{S}, s \models \neg\varphi$ iff $\mathcal{S}, s \models \varphi$ is false, thus $[\![\neg\varphi]\!]_{\mathcal{S}} = S \setminus [\![\varphi]\!]_{\mathcal{S}}$
5. $\mathcal{S}, s \models \langle a \rangle_q \varphi$ iff $K_a(s)([\![\varphi]\!]_{\mathcal{S}}) \geq q$, thus $[\![\langle a \rangle_q \varphi]\!]_{\mathcal{S}} = \{s \in S \mid K_a(s)([\![\varphi]\!]_{\mathcal{S}}) \geq q\}$.

The following is well known (e. g. [11, Lemma 5.1]):

Lemma 5. *Let $\mathcal{S}$ be a standard model for logic $\mathfrak{L}$. The set $[\![\varphi]\!]_{\mathcal{S}}$ is Borel for each formula φ.* ⊣

Standard models were introduced and discussed as probabilistic variants of Kripke models in [15, 6, 8] for a simple variant of the Hennessy-Milner logic and studied in connection with bisimulations. This notion and the study of bisimulations was then extended in [11] to general modal logics with operators of arbitrary arity [3].

A *hyperfinite model* $\mathcal{F} = (T, (k_a)_{a \in \mathsf{A}}, W)$ for logic $\mathfrak{L}$ is defined through a hyperfinite state space T, for each action $a \in \mathsf{A}$ a stochastic relation $k_a : T \rightsquigarrow T$ and for each atomic proposition $p \in \mathsf{P}$ a hyperfinite set $W(p) \subseteq T$ of states in which p is assumed to hold. The satisfaction relation $\models$ and the sets $[\![\cdot]\!]_{\mathcal{H}}$ are defined in exactly the same way for a hyperfinite model $\mathcal{H}$ as for a standard model, e.g. $\mathcal{H}, t \models \langle a \rangle_q \varphi$ iff $k_a(t)([\![\varphi]\!]_{\mathcal{H}}) \geq q$ holds (note that the comparison is done in ${}^*[0,1]$).

The following statement shows that we do not leave the realm of hyperfinite sets with $[\![\cdot]\!]_{\mathcal{H}}$; it is a companion to Lemma 5.

Lemma 6. *Let $\mathcal{H}$ be a hyperfinite model for logic $\mathfrak{L}$. The set $[\![\varphi]\!]_{\mathcal{H}}$ is a hyperfinite set for each formula φ.*

Proof. We proceed by induction on the structure of φ. There is nothing to show for $\top$, and for atomic propositions, and since the intersection of two hyperfinite sets is again hyperfinite, conjunction is covered, too. If $[\![\varphi]\!]_{\mathcal{H}}$ is hyperfinite, $[\![\neg\varphi]\!]_{\mathcal{H}} = S \setminus [\![\varphi]\!]_{\mathcal{H}}$ is hyperfinite as well (this is so because: if a set is not an element of an ultrafilter, its complement is).

So we need to demonstrate that $[\![\langle a \rangle_q \varphi]\!]_{\mathcal{H}} = \{t \in S \mid k_a(t)([\![\varphi]\!]_{\mathcal{H}}) \geq q\}$ is a hyperfinite set, provided $[\![\varphi]\!]_{\mathcal{H}}$ is one. The assumption that k_a is an internal map implies that for each hyperfinite set F the map $t \mapsto k_a(t)(F) = \sum_{t' \in F} k_a(t)(t')$ constitutes an internal function. From this observation the assertion follows.

The notion of an approximating model is quite straightforward. Roughly, a standard model $\mathcal{S}$ is approximated by a hyperfinite model $\mathcal{H}$ iff each state s with $\mathcal{S}, s \models \varphi$ is the standard part of a state σ of $\mathcal{H}$ with $\mathcal{H}, \sigma \models \varphi$, and vice versa, where φ is an arbitrary formula in $\mathfrak{L}$. To be specific:

Definition 6. *Let $\mathcal{S}$ be a standard model with state space S. Then the hyperfinite model $\mathcal{H}$ with state space S_f is a* hyperfinite approximation *to $\mathcal{S}$ iff $\mathsf{st}_{S_f}^{-1}[[\![\varphi]\!]_{\mathcal{S}}] = [\![\varphi]\!]_{\mathcal{H}}$ holds for all formulas φ of logic $\mathfrak{L}$, where st_{S_f} is the restriction of the standard map st_S to S_f.*

We show that for being an approximation to a given standard model it is sufficient to approximate the underlying transition system, and to test the validity sets for the atomic propositions (in addition to the requirement of approximation for the transition systems).

Proposition 3. *Let $\mathcal{H} = (S_f, (k_a)_{a \in \mathsf{A}}, W)$ be a hyperfinite, and $\mathcal{S} = (S, (K_a)_{a \in \mathsf{A}}, V)$ be a standard model. Assume that $(S_f, (k_a)_{a \in \mathsf{A}})$ approximates $(S, (K_a)_{a \in \mathsf{A}})$. If $W(p) = \mathsf{st}_{S_f}^{-1}[V(p)]$ for each atomic proposition $p \in \mathsf{P}$, then $\mathcal{H}$ is a hyperfinite approximation to $\mathcal{S}$.*

Proof. 1. The proof proceeds by induction on the formula φ. The cases $\top$ and p for $p \in \mathsf{P}$ are trivial, so are Boolean combinations of formulas.

2. Assume for the induction step that the assertion is true for formula φ, and let $a \in \mathsf{A}$, $0 \leq q \leq 1$ a rational number. If $s \in S, \sigma \in S_f$ with $\mathsf{st}_{S_f}(\sigma) = \mathsf{st}_S(\sigma) = s$, then

$$\begin{aligned} K_a(s)([\![\varphi]\!]_S) \geq q &\Leftrightarrow K_a(\mathsf{st}_S(\sigma))([\![\varphi]\!]_S) \geq q && (1)\\ &\Leftrightarrow \mathsf{L}(k_a(\sigma))(\mathsf{st}_{S_f}^{-1}\,[[\![\varphi]\!]_S]) \geq q && (2)\\ &\Leftrightarrow \mathsf{L}(k_a(\sigma))([\![\varphi]\!]_{\mathcal{H}}) \geq q && (3)\\ &\Leftrightarrow k_a(\sigma)([\![\varphi]\!]_{\mathcal{H}}) \geq q && (4) \end{aligned}$$

Equivalence (2) holds because k_a is an approximation to K_a, (3) is just the induction hypothesis, (4) refers to the construction of the Loeb measure (and the observation that $t \geq r$ iff $\mathsf{st}_{\mathbb{R}}(t) \geq r$ holds for $r \in \mathbb{R}, t \in {}^*\mathbb{R}$). But this means $\sigma \in [\![\langle a\rangle_q\varphi]\!]_{\mathcal{H}}$ iff $\mathsf{st}_{S_f}(\sigma) \in [\![\langle a\rangle_q\varphi]\!]_S$, establishing the assertion.

On account of being able to approximate labeled Markov transition systems, we are able to approximate standard models as well.

Proposition 4. *Let $\mathcal{S} = (S, (K_a)_{a\in\mathsf{A}}, V)$ be a standard model, then there exists a hyperfinite model $\mathcal{H}$ which approximates $\mathcal{S}$.*

Proof. Let $(S_f, (\kappa_a)_{a\in\mathsf{A}})$ be a hyperfinite approximation to $(S, (K_a)_{a\in\mathsf{A}})$ according to Corollary 3. Define the model $\mathcal{H} := (S_f, (\kappa_a)_{a\in\mathsf{A}}, W)$ with $W(p) := \mathsf{st}_{S_f}^{-1}\,[V(p)]$ for each $p \in \mathsf{P}$. Then Proposition 3 entails that this is a hyperfinite approximation.

Consider the theory $Th_{\mathcal{S}}(s) := \{\varphi \mid \mathcal{S}, s \models \varphi\}$ of a state s of $\mathcal{S}$. We see that we can find a hyperfinite model $\mathcal{H}$ with the property that $Th_{\mathcal{H}}(\sigma) = Th_{\mathcal{S}}(\mathsf{st}(\sigma))$ for each state σ of $\mathcal{H}$. In addition each state s is infinitely close to a state σ in $\mathcal{H}$ which has the same theory.

It should be noted that the logic $\mathfrak{L}$ does not enjoy properties that make it unique for the approximation discussed here. At the core of the discussion lies the approximability of a labeled Markov transition system by a hyperfinite one, around which the property the approximation of models may be grouped. Thus we could easily develop the same approximation result for a negation free logic, and we could even omit the atomic propositions, so that we would discuss the same logic that is taken e.g. in [6] as a starting point. Interestingly, we could replace conjunction by disjunction without having to be afraid of measure theoretic complications.

(EXCURSION: Conjunction $\varphi_1 \wedge \varphi_2$ of formulas implies $[\![\varphi_1 \wedge \varphi_2]\!]_{\mathcal{M}} = [\![\varphi_1]\!]_{\mathcal{M}} \cap [\![\varphi_2]\!]_{\mathcal{M}}$. This observation renders the set $\{[\![\varphi]\!]_{\mathcal{M}} \mid \varphi \text{ is a formula in the logic}\}$ a $\cap$-stable generator of a σ-algebra that has important model-theoretic properties; $\cap$-stability is important since it helps to apply the π-λ-Theorem that is so helpful in measure theory. END EXCURSION)

This discussion indicates that the approximation result given in Proposition 4 is a sample for a whole family of similar results, parameterized by the logic under consideration. We selected logic $\mathfrak{L}$ as a paragon; the proofs help illustrating the techniques which are used to cast the result from transition systems into the logical framework.

5 Related Work

An approximation of labeled Markov transition systems is given in [5] that approximate a systems using simpler ones; the approximants are given through suitable quotients in terms of a simple logic that is approximated by the system. This work improves upon a first metric approximation proposed in [7], that was oriented towards a logic as well. The approximation has been investigated with respect to convergence for various topologies that are important in domain theory in [21]. Since there is a natural interplay between convergence and hyperfinite approximation (non-standard methods were developed for modeling processes related to convergence, after all), these approximation results relate to the same spirit as the present proposal. It might be worth noting that the present approximation for transition systems has been developed independently of a particular logic; all that matters is given in terms of the measures and the underlying Polish topologies.

Another line of development is reported in the book [12] by Fajardo and Keisler. It discusses what is called *adapted spaces*. These spaces are given through a family of σ-algebras indexed by $[0, 1]$ and are of use in adapted probability logic as a logic for studying continuous time stochastic processes. Fajardo and Keisler investigate the relationship between the standard and the hyperfinite versions. This does not involve approximation properties of labeled Markov transition systems, and Kripke models for modal logics do not enter the discussion either.

6 Conclusion and Further Work

This paper proposes the use of hyperfinite models for approximating labeled Markov transition systems and, more general, for stochastic relations on Polish spaces. It is established that

- For each transition system a hyperfinite system can be found that is infinitely close to the given one; this also holds for each countable family of stochastic relations. Here closeness is not measured in terms of a metric but rather through infinitesimals.
- Each probabilistic Kripke model for a modal logic with a countable number of diamonds is approximated infinitely well by a hyperfinite one.

It is shown that the sets of theories for the standard model and for the hyperfinite approximation coincide. This brings about a slight *déjà-vu* to the equivalence relation that plays a rôle in the famous Hennessy-Milner Theorem on the equivalence of bisimulations and having the same sets of theories: we say that two models $\mathcal{E}$ and $\mathcal{F}$ are HM-equivalent iff

$$\{Th_{\mathcal{E}}(e) \mid e \text{ is a world in } \mathcal{E}\} = \{Th_{\mathcal{F}}(F) \mid f \text{ is a world in } \mathcal{F}\}$$

Then $\mathcal{E}$ and $\mathcal{F}$ are bisimilar iff they are HM-equivalent. This holds for modal logics under a light assumption [3], and can be shown to hold for their stochastic counterparts as well [11].

This discussion leads straight to the problem of characterizing the approximation behavior of morphisms. Suppose $\mathsf{F} : \mathsf{R}_1 \to \mathsf{R}_2$ is a morphism between stochastic relations, and suppose furthermore that the hyperfinite stochastic relations r_i are infinitely close to R_i $(i = 1, 2)$, can we find internal maps so that we have a morphism $\mathsf{f} : \mathsf{r}_1 \to \mathsf{r}_2$? This would permit translating constructions that are easily carried out for finite stochastic relations, and that can be translated to hyperfinite ones to the general Polish, and (via some standard constructions, see [10]) probably even to the analytic case. For example, it is not difficult to establish the existence of semi-pullbacks for hyperfinite relations, but it requires quite an effort doing so for the general Polish or analytic case [9]. A non-standard approximation would be of tremendous help here.

References

1. R. A. Anderson. Star-finite representations of measure spaces. *Trans. Amer. Math. Soc.*, 271(2):667 – 687, June 1982.
2. R. M. Anderson and S. Rashid. A nonstandard characterization of weak convergence. *Proc. Amer. Math. Soc.*, 69(2):327 – 332, May 1978.
3. P. Blackburn, M. de Rjike, and Y. Venema. *Modal Logic.* Number 53 in Cambridge Tracts in Theoretical Computer Science. Cambridge University Press, Cambridge, UK, 2001.
4. N. Cutland. Nonstandard measure theory and its applications. *Bull. London Math. Soc.*, 15:529 – 589, 1983.
5. V. Danos, J. Desharnais, and P. Panangaden. Labelled markov processes: Stronger and faster approximations. *Electr. Notes Theor. Comp. Sci.*, 87, 2004. 44 pages.
6. J. Desharnais, A. Edalat, and P. Panangaden. Bisimulation of labelled Markov-processes. *Information and Computation*, 179(2):163 – 193, 2002.
7. J. Desharnais, R. Jagadeesan, V. Gupta, and P. Panangaden. Approximating labeled Markov processes. In *Proc. 15th Ann. IEEE Symp. on Logic in Computer Science*, pages 95 – 106, June, 2000. IEEE Computer Society.
8. E.-E. Doberkat. Semi-pullbacks and bisimulations in categories of stochastic relations. In *Proc. ICALP'03*, volume 2719 of *Lect. Notes Comp. Sci.*, pages 996 – 1007, Berlin, 2003. Springer-Verlag.
9. E.-E. Doberkat. Semi-pullbacks for stochastic relations over analytic spaces. *Math. Struct. Comp. Sci.*, 15:647 – 670, 2005.
10. E.-E. Doberkat. Notes on stochastic relations. Research report, Department of Computer Science, University of Dortmund, January 2006.
11. E.-E. Doberkat. Stochastic relations: congruences, bisimulations and the Hennessy-Milner theorem. *SIAM J. Computing*, 35(3):590 – 626, 2006.
12. S. Fajardo and H. J. Keisler. *Model Theory of Stochastic Processes.* Number 14 in Lecture Notes in Logic. Association for Symbolic Logic, Natick, Mass., 2002.
13. M. Giry. A categorical approach to probability theory. In *Categorical Aspects of Topology and Analysis*, number 915 in Lect. Notes Math., pages 68 – 85, Berlin, 1981. Springer-Verlag.
14. H. J. Keisler. Infinitesimals in probability theory. In N. Cutland, editor, *Nonstandard analysis and its applications*, number 10 in London Mathematical Society Student Texts, pages 106 – 139. Cambridge University Press, Cambridge, UK, 1988.
15. K. G. Larsen and A. Skou. Bisimulation through probabilistic testing. *Information and Computation*, 94:1 – 28, 1991.
16. T. Lindstrøm. An invitation to nonstandard analysis. In N. Cutland, editor, *Nonstandard analysis and its applications*, number 10 in London Mathematical Society Student Texts, pages 1 – 105. Cambridge University Press, Cambridge, UK, 1988.

17. P. A. Loeb. Conversion to nonstandard measure spaces and applications to probability theory. *Trans. Amer. Math. Soc.*, 211:113 – 122, 1975.
18. S. M. Srivastava. *A Course on Borel Sets*. Graduate Texts in Mathematics. Springer-Verlag, Berlin, 1998.
19. K. D. Stroyan and W. A. J. Luxemburg. *Introduction to the theory of Infinitesimals*. Pure and Applied Mathematics. Academic Press, New York, 1976.
20. Y. Sun. Economics and nonstandard analysis. In P. A. Loeb and M. Wolff, editors, *Non-standard analysis for the Working Mathematician*, Mathematics and its Applications, pages 259 – 305. Kluwer Academic Publishers, Dordrecht, 2000.
21. F. van Breugel, M. Mislove, J. Ouaknine, and J. Worrell. Domain theory, testing and simulation for labelled Markov processes. *Theoret. Comp. Sci.*, 333:171 – 197, 2005.

A Appendix: Some Notions from Non-standard Analysis

We give a very brief summary of the constructions needed from non-standard analysis. For a comprehensive treatment, the reader is referred to e.g. [19], for a tutorial with an emphasis on measure and probability to [4, 16].

Internal Sets. Let S be a set and $\mathcal{U}$ be a free ultrafilter on $\mathbb{N}$. We say that a predicate P on $\mathbb{N}$ holds *almost certainly* (abbreviated by *a.c.*) iff $\{n \in \mathbb{N} \mid P(n)\} \in \mathcal{U}$. Define on $S^{\mathbb{N}}$ the equivalence relation $\varrho_{\mathcal{U}}$ through $(x_n)_{n\in\mathbb{N}}\ \varrho_{\mathcal{U}}\ (y_n)_{n\in\mathbb{N}}$ iff $x_n = y_n$ *a.c.* Denote by $[(x_n)]_{\mathcal{U}}$ the associated equivalence class, and by $^*S$ the factor space $S^{\mathbb{N}}/\varrho_{\mathcal{U}}$. Similarly, we define for a sequence $(A_n)_{n\in\mathbb{N}}$ of sets $\langle(A_n)\rangle := \{[(x_n)]_{\mathcal{U}} \mid x_n \in A_n\ a.c.\}$, and $^*A = \langle(A, A, A, \ldots)\rangle$. A is embedded to $^*A$ through $a \mapsto [(a)]_{\mathcal{U}}$.

Now construct for a set M a sequence $V_n(M)$ of sets inductively through

$$
\begin{aligned}
V_0(M) &:= M,\\
V_{n+1}(M) &:= V_n(M) \cup \mathcal{P}\left(V_n(M)\right),\\
V(M) &:= \bigcup_{n\in\mathbb{N}} V_n(M).
\end{aligned}
$$

$V(M)$ is called the *superstructure* associated with M. If $A \in V(^*S)$, then $^*A$ is called an *internal set* (over S); if A is of the form $\langle(B, B, \ldots)\rangle$ for some $B \in V(^*S)$, then A is called a *standard set*. It is known that a set is internal iff it is an element of some standard set [16, Lemma II.1.3]. Thus an internal set A has a representation as $\langle(A_n)\rangle$ for a suitable sequence $(A_n)_{n\in\mathbb{N}}$ of sets. All other sets are called *external*. Note that e. g. $\mathbb{N}$ is external (over $\mathbb{R}$). If $A = \langle(A_n)\rangle$ is an internal set such that the cardinalities $|A_n|$ are finite a.s. then A is called *hyperfinite*.

A map $f : A \to B$ is characterized as an internal map between the internal sets $A = \langle(A_n)\rangle$ and $B = \langle(B_n)\rangle$ through a sequence $(f_n)_{n\in\mathbb{N}}$ of maps so that $f_n(a_n) \in B_n$ a.c. whenever $a_n \in A_n$ a.c. A map $f : A \to B$ is extended in the obvious way to an internal map $^*f : {^*A} \to {^*B}$. When the context is clear, we occasionally omit the star from the notation.

We need two important properties:

Comprehension: For every internal set A and every function $f : \mathbb{N} \to A$ there is an internal function $g : {^*\mathbb{N}} \to A$ extending f.

Enlargement: If $(A_i)_{i \in I}$ is a collection of sets having the finite intersection property, then $\bigcap_{i \in I} {}^*A_i \neq \emptyset$.

Monads. If $(X, \mathcal{T})$ is an Hausdorff topological space, define for $x \in X$ the *monad* $\mathsf{monad}(x)$ *of* x as

$$\mathsf{monad}(x) := \bigcap \{{}^*G \mid x \in G \in \mathcal{T}\}.$$

By the Enlargement property above, $\mathsf{monad}(x) \neq \emptyset$. Since X is Hausdorff, distinct elements have disjoint monads; note that the monad depends on the topology. An element $y \in \bigcup_{x \in X} \mathsf{monad}(x)$ is called *near standard*; the set of all near standard elements of X is denoted by $\mathsf{ns}({}^*X)$. The standard part $\mathsf{st}_X(x)$ of $x \in \mathsf{ns}({}^*X)$ is the unique $y \in X$ with $x \in \mathsf{monad}(y)$; sometimes, ${}^o x$ is written for $\mathsf{st}_X(x)$.

The Loeb Construction. An *internal finitely additive measure* space $\mathbf{M} = (X, \mathcal{A}, \mu)$ consists of the internal set X, an internal algebra $\mathcal{A} \subseteq {}^*\mathcal{P}(X)$ on X and an internal function $\mu : \mathcal{A} \to {}^*\mathbb{R}_+$. The function μ is additive, thus $\mu(A \cup B) = \mu(A) + \mu(B)$, whenever $A, B \in \mathcal{A}$ are disjoint. The associated *Loeb space* $\mathsf{L}(\mathbf{M}) = (X, \mathsf{L}(\mathcal{A}), \mathsf{L}(\mu))$ is a measure space (in the usual sense) with these components:

1. $\mathsf{L}(\mathcal{A})$ is the universal completion of the σ-algebra $\sigma(\mathcal{A})$,
2. $\mathsf{L}(\mu)$ is the extension of the finitely additive measure ${}^o\mu : \mathcal{A} \to \mathbb{R}_+$ to the σ-algebra $\mathsf{L}(\mathcal{A})$ (where ${}^o\mu : A \mapsto \mathsf{st}_X(\mu(A))$ maps A to the standard part of $\mu(A)$).

The construction yielding the Loeb space $\mathsf{L}(\mathbf{M})$ from an internal finitely additive measure space $\mathbf{M}$ originates from [17] and is discussed at length e.g. in [4, § 3] or in [16, II.1]. For convenience, we use here the universal completion of a measure space, this is a little more restrictive than the commonly used completion induced by a particular measure.

State Space Reduction of Rewrite Theories Using Invisible Transitions

Azadeh Farzan and José Meseguer

Department of Computer Science
University of Illinois at Urbana-Champaign
{afarzan, meseguer}@cs.uiuc.edu

Abstract. State space explosion is the hardest challenge to the effective application of model checking methods. We present a new technique for achieving drastic state space reductions that can be applied to a very wide range of concurrent systems, namely any system specified as a rewrite theory. Given a rewrite theory $\mathcal{R} = (\Sigma, E, R)$ whose equational part (Σ, E) specifies some state predicates P, we identify a subset $S \subseteq R$ of rewrite rules that are P-invisible, so that rewriting with S does not change the truth value of the predicates P. We then use S to construct a reduced rewrite theory $\mathcal{R}/S$ in which all states reachable by S-transitions become identified. We show that if $\mathcal{R}/S$ satisfies reasonable executability assumptions, then it is in fact stuttering bisimilar to $\mathcal{R}$ and therefore both satisfy the same CTL^*_{-X} formulas. We can then use the typically much smaller $\mathcal{R}/S$ to verify such formulas. We show through several case studies that the reductions achievable this way can be huge in practice. Furthermore, we also present a generalization of our construction that instead uses a stuttering simulation and can be applied to an even broader class of systems.

1 Introduction

Although model checking is one of the most successful automated verification techniques, there are real limitations to its applicability in practice. These limitations are mostly related to the *state space explosion problem*. For example, as the number of processes considered in a distributed system grows, the associated state space may easily grow exponentially, particularly due to the system's concurrency. This can make it unfeasible to model check a system except for very small initial states, sometimes not even for those.

For this reason, a host of techniques to tame the state space explosion problem, which could be collectively described as *state space reduction techniques*, have been investigated: bisimulation techniques, partial order reduction (POR) techniques, abstraction techniques, and so on (see for example [20, 33, 22, 4, 9, 11, 19, 34, 32, 1, 21, 16]). The general idea is to transform the original system into a simpler one (typically bisimilar or at least similar to the original one) whose state space is small enough to model check properties. Transfer results then ensure that the same property holds in the original system.

M. Johnson and V. Vene (Eds.): AMAST 2006, LNCS 4019, pp. 142–157, 2006.

This paper proposes a new such state space reduction technique within the rewriting logic semantic framework, in which concurrent systems are formally specified as rewrite theories [27]. In such specifications, the set of states is specified as an algebraic data type by an equational theory (Σ, E), and the system's transitions are specified by rewrite rules R that are applied *modulo* the equations E. The rewrite theory specifying the system is then the triple $\mathcal{R} = (\Sigma, E, R)$. The fact that rewriting logic has been shown to be a very general and expressive semantic framework to specify concurrent systems [28, 25] makes our proposed state space reduction technique applicable to a very wide range of concurrent systems. Achieving a state space reduction typically requires discharging *proof obligations* to verify that the reduction is correct. In this regard, the fact that the state space is itself axiomatized by an equational theory (Σ, E) makes the tool-assisted discharging of such proof obligations using equational theorem proving techniques and tools much easier than if a non-logical specification formalism had been used instead.

Our technique is based on the idea of *invisible transitions*, that generalize a similar notion in POR techniques (see for example [5]). The basic setting is that we assume a rewrite theory $\mathcal{R} = (\Sigma, E, R)$ in which a certain set P of *state predicates* has been equationally axiomatized by some of the equations in E. $\mathcal{R}$ then has an associated Kripke structure, whose labeling function associates to each state (represented as an E-equivalence class of terms $[t]$ in the initial algebra $T_{\Sigma/E}$) all those predicates in P that hold in $[t]$ according to the equations E. We then call a rewrite rule r in R *P-invisible* if in any rewrite step $[t] \longrightarrow [t']$ using r the states $[t]$ and $[t']$ satisfy the *same* state predicates, i.e., they are labeled in the same way. Our state space reduction technique is then very simple: we identify a subset $S \subseteq R$ of rules such that all rules in S are invisible. We then define the *S-reduction* of $\mathcal{R} = (\Sigma, E, R)$ as the rewrite theory $\mathcal{R}/S = (\Sigma, E \cup S, R \backslash S)$, that is, we *turn all rules in S into equations*, thus collapsing the set of states from $T_{\Sigma/E}$ to the quotient $T_{\Sigma/E \cup S}$. The intuitive idea, therefore, is that all states that can be reached from a given state by repeated S-transitions can be *collapsed* into a single one. In practice, as we show by means of several case studies in Section 4, the reductions obtained this way can be huge.

However, the above technique must meet an important *executability requirement*. The point is that, for E an arbitrary set of equations, rewriting modulo E, which is the way transitions take place in the Kripke structure associated to $\mathcal{R} = (\Sigma, E, R)$, is in general undecidable. Therefore, to be able to execute and model check a rewrite theory in a rewriting logic language implementation such as Maude [6, 7] we must require that the equations E are confluent and terminating (perhaps modulo some axioms A) and that the rules R are strongly *coherent* with respect to the equations E [35]. Intuitively, the coherence requirement means that we can identify a state $[t]$ with the canonical form $can_E(t)$ of t by the equations E, and that rewriting with equations E and with rules R *commutes* in an appropriate sense, so that we can safely restrict our computations with R to only rewrite E-canonical forms. Therefore, the executability requirement for our technique is that $\mathcal{R}/S = (\Sigma, E \cup S, R \backslash S)$ should be executable, that is, that $E \cup S$ should be confluent and terminating, and that the rules $R \backslash S$ should be strongly coherent with respect to $E \cup S$ (perhaps modulo axioms A).

We show in Section 3 that the above-mentioned executability requirements on $\mathcal{R}/S$, besides being absolutely essential to model check $\mathcal{R}/S$ in practice, ensure a further very important property, namely that $\mathcal{R}$ and $\mathcal{R}/S$ are *stuttering bisimilar*, and therefore they satisfy exactly the same CTL^*_{-X} formulas. Furthermore, to make our technique applicable to cases where a suitable set S may not be available, we generalize it to allow enlarging a set of invisible rules S by adding new invisible rules not in R to get a superset $\widehat{S} \supseteq S$. This gives rise to a state space reduction $\widehat{\mathcal{R}}/\widehat{S}$ that is no longer stuttering bisimilar to $\mathcal{R}$ but is nevertheless similar to it. This still allows us to verify $ACTL^*_{-X}$ formulas for $\mathcal{R}$ if we can model check them for $\widehat{\mathcal{R}}/\widehat{S}$, but such model checking can now give rise to spurious counterexamples. We illustrate how this more general technique is also quite useful in practice by means of a client-server protocol in Section 4.3.

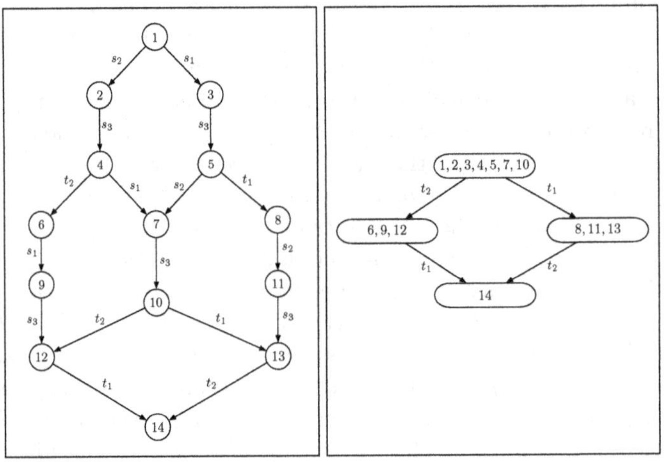

(a) (b)

Fig. 1. Restaurant State Space

We can make all these ideas concrete by means of an example which models the workflow in a simplistic restaurant with one waiter and two customers. Customers have a flag indicating their status (waiting, ordered, or eating), so a customer is represented as a pair $C(id, f)$ with id an identifier and f the flag. The waiter has also a status flag (free or order-taken). Therefore, the waiter is represented by a term of the form $W(f)$. The restaurant state is a set with a waiter and two customers, with set union represented by a binary associative and commutative juxtaposition operator "__". We have the following rewrite rules R in our theory $\mathcal{R} = (\Sigma, A, R)$, where A consists of the associativity and commutativity axioms for "__":

$$
\begin{array}{lll}
s_1 : W(\text{free})C(1,\text{waiting}) & \longrightarrow & W(\text{order-taken})C(1,\text{ordered}) \\
s_2 : W(\text{free})C(2,\text{waiting}) & \longrightarrow & W(\text{order-taken})C(2,\text{ordered}) \\
s_3 : W(\text{order-taken}) & \longrightarrow & W(\text{free}) \\
t_1 : W(\text{free})C(1,\text{ordered}) & \longrightarrow & W(\text{free})C(1,\text{eating}) \\
t_2 : W(\text{free})C(2,\text{ordered}) & \longrightarrow & W(\text{free})C(2,\text{eating})
\end{array}
$$

Figure 1 (a) shows the state space induced by the above rewrite rules from an initial state with the waiter free and the two customers waiting.

Let us assume that the property ϕ that we are interested in is: "eventually both customers eat". This property can be expressed as formula $\diamond(e_1 \wedge e_2)$ where e_i is true if the ith customer's status is "eating" and false otherwise. Rewrite rules s_1, s_2, and s_3 do not change the truth value of the predicates e_1 and e_2. One can observe that the rules in $S = \{s_1, s_2, s_3\}$ are confluent and terminating and $R \backslash S$ is strongly locally coherent [35] with respect to S modulo axioms. The reduced theory $\mathcal{R}/S$ (see the state space in Figure 1 (b), where each state represents an S-equivalence class) is then stuttering bisimilar to the theory $\mathcal{R}$.

Besides the small example used above to illustrate the main ideas, in Section 4 we show that our technique yields very drastic state space reductions in three more substantial case studies involving well-known algorithms and applications. Furthermore, in Section 5 we discuss in detail the discharging of the necessary proof obligations ensuring that a proposed S-reduction $\mathcal{R}/S$ is both correct and executable, and the kind of tool support necessary to facilitate such discharging activities. We end with a discussion of related work and some concluding remarks in Section 6.

2 Preliminaries

2.1 Termination, Confluence and Coherence in Rewrite Theories

A rewrite theory [27] is a triple $\mathcal{R} = (\Sigma, E, R)$ where (Σ, E) is an equational theory with signature Σ and equations E, and where R is a set of conditional rewrite rules of the form $l \longrightarrow r$ if C. In this paper we assume that C is always an equational condition. Intuitively, if a concurrent system is modeled as a rewrite theory $\mathcal{R} = (\Sigma, E, R)$, then the equational theory (Σ, E) defines the system *states* (terms in $T_{\Sigma/E}$) and the set of rewrite rules R specify the system's concurrent transitions.

Given two terms $u, v \in T_\Sigma$, a one-step rewrite $u \xrightarrow{\tau} v$ means that there is a rule $\tau : l \rightarrow r$ if C in R that can be applied to a subterm of u with a ground substitution θ such that $E \models \theta C$ and u rewrites to v by replacing the subterm $\theta(l)$ by the subterm $\theta(r)$. We write $u \xrightarrow{R} v$ to mean that there is a rule $\tau \in R$ such that $u \xrightarrow{\tau} v$. The notation $\xrightarrow{R}_*$ denotes the reflexive and transitive closure of the relation $\xrightarrow{R}$. Set Can_S includes all elements $x \in T_\Sigma$ such that no rule in S is enabled at x. We define $\xrightarrow{S}_! = \{(x,y) | x \xrightarrow{S}_* y \ \wedge \ y \in \text{Can}_S\}$ and $x \downarrow_S y \Leftrightarrow \exists z : x \xrightarrow{S}_* z \wedge y \xrightarrow{S}_* z$. Rewriting over equivalence classes modulo equations E is defined as follows: $[t]_E \xrightarrow{r} [t']_E$ if and only if there are terms

u and v such that $u \in [t]_E$ and $v \in [t']_E$ and $u \xrightarrow{r} v$. We define $\xrightarrow{R/E}$ by the equivalence $[t]_E \xrightarrow{R} [t']_E \Leftrightarrow t \xrightarrow{R/E} t'$.

A set $S \subseteq R$ of rewrite rules is *confluent* modulo E in the theory (Σ, E, R) if and only if $\forall t, t', t'' \in T_\Sigma : ([t]_E \xrightarrow{S}_* [t']_E \wedge [t]_E \xrightarrow{S}_* [t'']_E) \Rightarrow (\exists w : [t']_E \xrightarrow{S}_* [w]_E \wedge [t'']_E \xrightarrow{S}_* [w]_E)$. S is *terminating* if for all t there exists no infinite chain of rewriting $[t]_E \xrightarrow{S} \ldots \xrightarrow{S} \ldots$.

Definition 1. *[35] In a rewrite theory $\mathcal{R} = (\Sigma, E, R)$, where $E = E_0 \cup A$ with E_0 a (terminating) set of equations and A a set of equational axioms, R is called* locally strongly coherent *with respect to E_0 modulo A if*

$$(t \xrightarrow{R/A} t_1 \ \wedge\ t \xrightarrow{E_0/A} t_2) \Rightarrow (\exists t_3, t_4 :$$
$$t_2 \xrightarrow{E_0/A}_! t_3 \ \wedge\ t_3 \xrightarrow{R/A} t_4 \ \wedge\ t_4 \downarrow_{E_0/A} t_1)$$

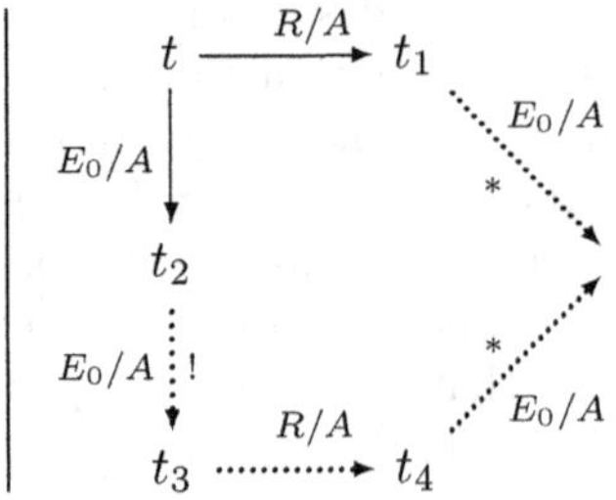

Strong local coherence is the main property to check to ensure executability of a rewrite theory $\mathcal{R} = (\Sigma, E_0 \cup A, R)$ when we have matching algorithms for the equational axioms A. Viry shows that if the equations E_0 are confluent and terminating modulo A, then strong local coherence implies a more general strong coherence property [35]. Strong coherence ensures that we can achieve the effect of rewriting with R in $E_0 \cup A$-equivalence classes by first computing the $E_0 \cup A$-canonical form modulo A, and then rewriting that canonical form with R modulo A.

2.2 Stuttering Simulations

Let us assume that the equational part (Σ, E) of a rewrite theory $\mathcal{R} = (\Sigma, E, R)$ defines, among other things a set P of *state predicates* on the initial algebra $T_{\Sigma/E}$[1]. We can then associate to $\mathcal{R}$ a *Kripke structure* [5] whose states are the set $T_{\Sigma/E,State}$ for some designated sort *State* of states, whose labeling function assigns to each state the predicates $p \in P$ that provably hold in it using E, and whose transition relation is the total closure $\xrightarrow{R}_\bullet$ of $\xrightarrow{R}$, that is, we make $\xrightarrow{R}$ into a total relation by adding identity transitions for each deadlock state. We can then interpret any temporal logic formula, say in CTL^* in $\mathcal{R}$, namely by interpreting it in its associated Kripke structure. For a more detailed presentation on the relations between rewrite theories, Kripke structures and temporal logic, with applications to model checking in Maude see [14].

We present some basic notions and results, used later on, about transition systems, Kripke structures, and stuttering (bi-)simulations between them that will apply in particular to the Kripke structures associated to rewrite theories.

[1] Note that all the equivalent states modulo E satisfy the same set of predicates.

Definition 2. *Let* $\mathcal{A} = (A, \xrightarrow{A})$ *and* $\mathcal{B} = (B, \xrightarrow{B})$ *be transition systems and let* $H \subseteq A \times B$ *be a relation. Given a path* π *in* $\mathcal{A}$ *and a path* ρ *in* $\mathcal{B}$*, we say that* ρ H*-matches* π *if there are strictly increasing functions* $\alpha, \beta : \mathbb{N} \to \mathbb{N}$ *with* $\alpha(0) = \beta(0) = 0$ *such that, for all* $i, j, k \in \mathbb{N}$*, if* $\alpha(i) \leq j < \alpha(i+1)$ *and* $\beta(i) \leq k < \beta(i+1)$*, it holds that* $\pi(j) H \rho(k)$*.*

Definition 3. *Given transition systems* $\mathcal{A}$ *and* $\mathcal{B}$*, a stuttering simulation of transition systems* $H : A \longrightarrow B$ *is a binary relation* $H \subseteq \mathcal{A} \times \mathcal{B}$ *such that if* aHb*, then for each path* π *in* $\mathcal{A}$ *starting at* a *there is a path* ρ *in* $\mathcal{B}$ *starting at* b *that* H*- matches* π*.*

Definition 4. *Given Kripke structures* $\mathcal{A} = (A, \xrightarrow{A}, L_A)$ *and* $\mathcal{B} = (B, \xrightarrow{B}, L_B)$ *over a set of predicates* P*, a stuttering* P*-simulation* $H : A \to B$ *is a stuttering simulation of transition systems* $H : (A, \xrightarrow{A}) \to (B, \xrightarrow{B})$ *such that if* aHb *then* $L_B(b) \subseteq L_A(a)$*. We call the stuttering* P*-simulation strict if* aHb *implies* $L_B(b) = L_A(a)$*.* H *is called a stuttering* P*-bisimulation if both* H *and* H^{-1} *are stuttering* P*-simulations.*

In [31], it is shown that (strict) stuttering simulations preserve the satisfaction of $ACTL^*_{-X}(P)$ formulas. Also, [24, 3] state that stuttering bisimulations preserve the satisfaction of $CTL^*_{-X}(P)$ formulas which can be derived by generalizing the results from [31].

3 Invisible Transitions and the $\mathcal{R}/S$ Reduction

Definition 5. *Given a rewrite theory* $\mathcal{R}=(\Sigma, E, R)$ *and having an equationally-defined set of atomic predicates* P*, a rewrite rule* $\tau : l \to r$ *if* C *in* R *is called* P***-invisible*** *if for any* $[t] \in T_{\Sigma/E}$ *and any* $u \in [t]$ *such that* $u \xrightarrow{\tau} v$*, then for each* $p \in P$ *we have* $[t] \models p \Leftrightarrow [v] \models p$*. We denote by* $Inv^P(R)$ *the set of all* P*-invisible rewrite rules of* R*.*

We call $\mathcal{R}/S = (\Sigma, S \cup E_0 \cup A, T = R \backslash S)$ the S-reduced theory of $\mathcal{R} = (\Sigma, E_0 \cup A, R)$. We are particularly interested in the S-reduced theory of $\mathcal{R}$ when $S \subseteq \mathrm{Inv}^P(R)$, $S \cup E_0$ is confluent and terminating modulo A, and T is coherent with respect to $S \cup E_0$ modulo A.

Theorem 1. *Let* $\mathcal{R} = (\Sigma, E_0 \cup A, R)$ *be a rewrite theory with* P *a set of equationally defined atomic predicates. Let* $S \subseteq R$ *be a set of* P*-invisible rules such that* $S \cup E_0$ *is confluent and terminating modulo* A*, and* $T = R \backslash S$ *is coherent with respect to* $S \cup E_0$ *modulo* A*. Then* $\mathcal{R}$ *and* $\mathcal{R}/S$ *are stuttering bisimilar.*

Proof. (sketch) The relation H on which the bisimilarity is based is defined by the quotient homomorphism $H : T_{\Sigma/E} \twoheadrightarrow T_{\Sigma/E \cup S}$. We need to prove that: (a) H is a stuttering simulation; and (b) that H^{-1} is so too. Since H maps deadlock states to deadlock states, and H^{-1} of a deadlock state always contains a deadlock state, we can disregard deadlocks.

(a) It suffices to show that for each path π in the underlying Kripke structure of the theory $\mathcal{R}$, $(T_{\Sigma/E}, \xrightarrow{R}_\bullet)$, there exists a stuttering equivalent path π' in the underlying Kripke structure of the S-reduced theory $\mathcal{R}/S$, $(T_{\Sigma/E\cup S}, \xrightarrow{S/\widehat{S}}_\bullet)$.

π must be of the following general form:

$$\pi : [s_0]_E \xrightarrow{S}_* [t_0]_E \xrightarrow{T} [s_1]_E \xrightarrow{S}_* [t_1]_E \xrightarrow{T} \dots \xrightarrow{T} [s_n]_E \xrightarrow{S}_* [t_n]_E \xrightarrow{T} \dots$$

Since the rules in S are P-invisible, we know that $L(s_i) = L(t_i)$ for all i. Also, observe that by collapsing the $\xrightarrow{S}_*$, we have $[s_i]_{E\cup S} = [t_i]_{E\cup S}$. Then the following path

$$\pi' : [t_0]_{S\cup E} \xrightarrow{T} [t_1]_{S\cup E} \xrightarrow{T} \dots \xrightarrow{T} [t_n]_{S\cup E} \xrightarrow{T} \dots$$

is stuttering equivalent to π and of course, by construction, it is a path in the underlying Kripke structure of $\mathcal{R}/S$.

(b) It suffices to show that for each path ρ in the underlying Kripke structure of the theory $\mathcal{R}/S$, $(T_{\Sigma/E\cup S}, \rightarrow_{R/S})$, there exists a stuttering equivalent path ρ' in the underlying Kripke structure of the reduced theory $\mathcal{R}$, $(T_{\Sigma/E}, \xrightarrow{R}_\bullet)$.

Assume that ρ is of the following general form:

$$\rho : [s_0]_{S\cup E} \xrightarrow{T} [s_1]_{S\cup E} \xrightarrow{T} \dots \xrightarrow{T} [s_n]_{S\cup E} \xrightarrow{T} \dots$$

We show by construction that there exists a stuttering equivalent path ρ' of the following form:

$$\rho' : [s'_0]_E \xrightarrow{S}_* [t_0]_E \xrightarrow{T} [s'_1]_E \xrightarrow{S}_* [t_1]_E \xrightarrow{T} \dots \xrightarrow{T} [s'_n]_E \xrightarrow{S}_* [t_n]_E \xrightarrow{T} \dots$$

where $s_0 = s'_0$ and for all i, $s_i \equiv_{S\cup E} s'_i$, and therefore $L(s_i) = L(s'_i)$. H^{-1} then relates the state $[s_i]_{S\cup E}$ to all the states on $[s'_i]_E \xrightarrow{S}_* [t_i]_E$ which by invisibility of S all satisfy the same set of predicates.

$[s_i]_{S\cup E} \xrightarrow{T} [s_{i+1}]_{S\cup E}$ implies that there are terms u_i and u_{i+1} such that $s_i \equiv_{S\cup E} u_i \xrightarrow{T} u_{i+1} \equiv_{S\cup E} s_{i+1}$. If $s'_i \equiv_{S\cup E} s_i$ (meaning $s'_i H s_i$), then there is a term t_i such that $s_i \xrightarrow{S\cup E}_! t_i$ and $s'_i \xrightarrow{S\cup E}_! t_i$. Since $u_i \equiv_{S\cup E} s_i$, by confluence of $S \cup E$, we have $u_i \xrightarrow{S\cup E}_! t_i$. Therefore, by T being coherent with respect to $S \cup E_0$ modulo A, there exists a term s'_{i+1} such that $t_i \xrightarrow{T} s'_{i+1}$ and $s'_{i+1} \downarrow_{S\cup E} u_{i+1}$. Since $u_{i+1} \equiv_{S\cup E} s_{i+1}$, we have $s'_{i+1} \equiv_{S\cup E} s_{i+1}$ (meaning that $s'_{i+1} H s_{i+1}$).

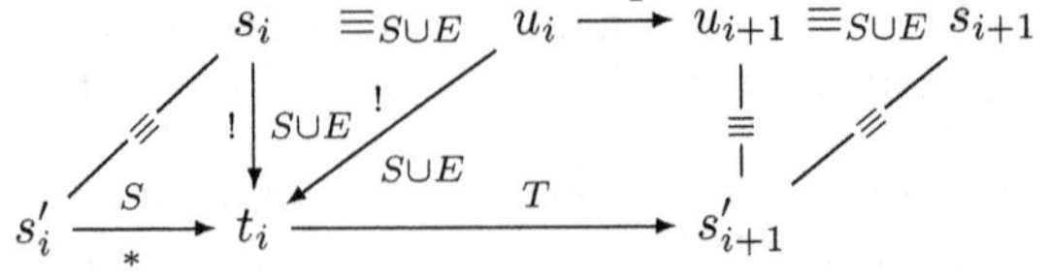

Start by letting $s'_0 = s_0$. Since $s'_0 = s_0$, it trivially holds that $s'_0 \equiv_{S\cup E} s_0$. Inductively construct the path according to the above diagram. Note that by viewing S steps as τ-transitions, the above argument also shows that $\equiv_{S \cup E}$ is a branching bisimulation relation [30]. □

We have shown that, under the theorem hypothesis, the reduced rewrite theory $\mathcal{R}/S$ is stuttering P-bisimilar with the original theory $\mathcal{R}$. Therefore, (see Section 2) for any $\phi \in CTL^*_{-X}(P)$[2], and any initial state $[t]_E$ we have

$$\mathcal{R}, [t]_E \models \phi \;\Leftrightarrow\; \mathcal{R}/S, [t]_{E \cup S} \models \phi$$

In practice, the reduced theory $\mathcal{R}/S$ can have a drastically smaller state space than $\mathcal{R}$, making model checking of $\mathcal{R}/S$ feasible when model checking of $\mathcal{R}$ is unfeasible.

In cases where the $\mathcal{R}/S$ construction cannot be carried out for lack of a suitable S satisfying the confluence condition in Theorem 1, we can nevertheless achieve a similar state space reduction with a relation H that is a stuttering *simulation*. For example the client-server reduction in Section 4.3 is achieved in this manner. The general method is as follows: we assume that we have a set of rules $S \subseteq R$ which are P-invisible ($S \subseteq \mathrm{Inv}^P(R)$), and $T = R \backslash S$ is coherent with respect to $S \cup E_0$ modulo A, and $S \cup E_0$ is terminating but not confluent modulo A. We then extend S to a set of rules $\widehat{S}$ with $S \subseteq \widehat{S}$, $\widehat{S} \not\subseteq R$, and where $\widehat{S}$ is still P-invisible, and $(R \backslash \widehat{S})$ is coherent with respect to $\widehat{S} \cup E_0$ modulo A, and furthermore, $E_0 \cup \widehat{S}$ is terminating and confluent modulo A. Consider now the rewrite theory $\widehat{\mathcal{R}} = (\Sigma, E_0 \cup A, R \cup \widehat{S})$. Since $\widehat{\mathcal{R}}$ has more rules than $\mathcal{R}$, if $\mathcal{R}$ is deadlock-free[3], that is, if any state $[t]_E$ can always be rewritten by R to a new state $[t']_E$, then the following proposition is easy to prove:

Proposition 1. *The identity homomorphism* $1_{T_{\Sigma/E_0 \cup A}} : T_{\Sigma/E_0 \cup A} \rightarrow T_{\Sigma/E_0 \cup A}$ *induces a* P*-simulation map from the underlying Kripke structure of* $\mathcal{R}$ *to that of* $\widehat{\mathcal{R}}$*.*

We can now apply Theorem 1 to $\widehat{\mathcal{R}}$ to obtain a stuttering P-bisimilar $\widehat{S}$-reduced theory $\widehat{\mathcal{R}}/\widehat{S}$. Since any simulation is a special case of a stuttering simulation, and stuttering simulations are closed under composition [31, 24], by composing the above simulation from $\mathcal{R}$ to $\widehat{\mathcal{R}}$ with the stuttering bisimulation from $\widehat{\mathcal{R}}$ to $\widehat{\mathcal{R}}/\widehat{S}$ generated by Theorem 1, we obtain a stuttering *simulation* from $\mathcal{R}$ to $\widehat{\mathcal{R}}/\widehat{S}$ and therefore we have

Theorem 2. *Under the above assumptions for any* $\phi \in ACTL^*_{-X}(P)$ *and any initial state* $[t]_{E_0 \cup A}$ *in* $\mathcal{R}$*, we have* $\widehat{\mathcal{R}}/\widehat{S}, [t]_{E_0 \cup \widehat{S} \cup A} \models \phi \;\Rightarrow\; \mathcal{R}, [t]_{E_0 \cup A} \models \phi$.

Therefore, if we can model check the property ϕ using the reduced theory $\widehat{\mathcal{R}}/\widehat{S}$, we are then guaranteed that ϕ holds in $\mathcal{R}$. See Section 4.3 for an example.

[2] Note that the simulation relations are *strict* in the sense that $aHb \Rightarrow L(a) = L(b)$ and therefore negation does not have to be excluded.

[3] Given a rewrite theory $\mathcal{R}$, we can always transform it into a bisimilar deadlock-free theory (see [29]). Therefore, there is no real loss of generality imposed by this requirement.

4 Case Studies

We present three case studies showing how the $\mathcal{R}/S$ and $\widehat{\mathcal{R}}/\widehat{S}$ constructions can be achieved in practice for real applications, leading to massive reductions in the state space. All the experiments have been performed with the Maude LTL model checker running on an Intel machine with a 2.6GHz processor and 4GB of memory running Linux.

4.1 Leader Election Protocol

We consider the simple case where the network is a ring consisting of n nodes, numbered from 1 to n in the clockwise direction. We want to investigate the LCR algorithm to select a leader. The informal description of this algorithm is as follows [23]:

> Each process sends its identifier around the ring. When a process receives an incoming identifier, it compares that identifier to its own. If the incoming identifier is greater than its own, it keeps passing the identifier; if it is less than its own, it discards the incoming identifier; if it is equal to its own the process declares itself the leader.

We can specify a rewrite theory modeling this algorithm by means of objects and messages, where the distributed state is a multiset of objects and messages built by an associative and commutative multiset union operator "__":

$$\begin{array}{lll}
s_0 : \langle I \rangle & \longrightarrow [I]\ (I \to I+1 \bmod N) & \\
s_1 : [I]\ (J \to I) & \longrightarrow [I] & \text{if } J < I \\
s_2 : [I]\ (J \to I) & \longrightarrow [I](J \to I+1 \bmod N) & \text{if } J > I \\
t : \ [I]\ (I \to I) & \longrightarrow \text{Leader}(I) &
\end{array}$$

where N is the number of processes on the ring and $\langle I \rangle$ is the initial state of process I. In the first phase (rewrite rule s_0), each process I sends its identifier to its neighbor and changes its format $[I]$ so that this is done only once. As soon as a process I receives its own identifier through the ring, the computation is over; it removes all the object and the message and outputs Leader(I). The messages are of the general form $(I \to J)$ where J is the identifier of the receiver and I is the integer content of the message.

The set $S = \{s_0, s_1, s_2\}$ can be shown to be confluent and terminating modulo associativity and commutativity. Let us assume that the property that we are interested in is that eventually some process will be elected as leader. This is expressed by means of a single atomic predicate, p, that is true in any state containing Leader(I). The rules in S are p-invisible, and t is coherent with respect to S modulo the associativity and commutativity axioms. Therefore, by Theorem 1, we can use the stuttering bisimilar reduction $\mathcal{R}/S$ to model check our property. Note that reducing $\mathcal{R}$ with the rewrite rule s_0 above (which can easily be shown to be confluent and terminating) collapses an N-dimensional cube (generated by rule s_0) into a path of length N, meaning that the number of states in $\mathcal{R}/\{s_0\}$ is reduced from 2^N to N, and the number of paths reduces from 2^N to 1. Table 1 shows the performance evaluation of model checking this problem before and after reduction using the Maude LTL model checker.

4.2 Distributed Spanning Tree

A *spanning tree* of an undirected graph $G = (V, E)$ is a tree (i.e., a connected acyclic graph) that consists entirely of undirected edges and contains every vertex of G. The distributed spanning tree problem tries to find a spanning tree for a given set of network nodes V that are connected by E. The asynchronous algorithm from [23] solves this as follows:

> There is a distinct node r that is initially marked and acts as the root. A marked node v asynchronously sends a message to each of its neighbors once and for all. An unmarked node v nondeterministically chooses one of the nodes who have sent it a message as its parent in the spanning tree, becomes marked, and discards all the other messages.

One possible way of specifying the above algorithm is by the following rewrite rules:

$$\begin{array}{lll} s_1 : [\, N \mid P,\ M\ NL] & \longrightarrow & [\, N \mid P,\ NL](M \leftarrow N) \\ t_1 : [N \mid \texttt{none},\ NL](N \leftarrow M) & \longrightarrow & [\, N \mid M,\ NL] \\ s_2 : [N \mid M,\ NL](N \leftarrow K) & \longrightarrow & [\, N \mid M,\ NL] \\ s_3 : [N \mid \texttt{root},\ NL](N \leftarrow K) & \longrightarrow & [\, N \mid \texttt{root}, NL] \end{array}$$

where the state is represented as a multiset (modulo associativity, commutativity, and identity) of nodes and messages. Each node is of the form $[N \mid P,\ L]$ where N is its unique identifier, P is its parent node (initially `none`), and L is the list of its neighbors (their identifiers to be exact). Variable M is of type integer which denotes a *known* parent and consequently cannot be `none` or `root`. The node with "root" as its parent is the root of the spanning tree. Let us assume that the property of interest is "to eventually reach a state in which every node has a parent". This property can be expressed using a single atomic predicate, p, that is false if there is a node with "none" as the parent. One can easily check that the set of rules $S = \{s_1, s_2, s_3\}$ is p-invisible, confluent, terminating modulo associativity, commutativity, and identity, and t_1 is coherent with respect to S modulo the same axioms. Since there are no equations (excluding the axioms) in the theory, one can turn these rules into equations and gain a huge reduction in the state space for model checking. Table 1 shows the performance evaluation of model checking this problem before and after this $\mathcal{R}/S$ reduction using the Maude LTL model checker.

4.3 A Distributed Client-Server System

Consider a system consisting of several clients and one server. The server has a log (a list) for incoming requests. The clients send a message to the server to request a service. When the server receives a request message, it sends the relevant client a message containing the requested material, and adds an entry

Table 1. Performance Results

Problem	Number of Nodes	Time	Space	Time (reduced)	Space (reduced)
Leader Election	10	3.6s	27633	0	2
	13	2.7m	506037	0	2
	14	19.3m	1329885	0	2
	15	–	–	0	2
Spanning Tree	3	0.02s	417	0	9
	4	10.2s	120183	0.01s	64
	5	–	–	0.17s	625
	6	–	–	0.5s	1296
	7	–	–	110.22s	117649
	8	–	–	99m	2097152
Client-Server	6	4.0s	125248	0.01	64
	7	81.4s	1753600	0.01s	128
	8	–	–	0.01s	256
	15	–	–	1.8s	32768
	20	–	–	5.3m	1048576

to its log (B) to keep track of this communication. The following set of rewrite rules model a simple version of this system:

$$\begin{aligned}
s_1 &: [N \mid M] && \longrightarrow \{N \mid M\}(\texttt{server} \leftarrow (N,M)) \\
s_2 &: (\texttt{server} \leftarrow (N,M))[\texttt{server} \mid B] && \longrightarrow [\,\texttt{server} \mid B\ (N,M)](N \leftarrow \texttt{serv}(M)) \\
t_1 &: (N \leftarrow \texttt{serv}(M))\{N \mid M\} && \longrightarrow \{N\}
\end{aligned}$$

where the state is a multiset (modulo associativity, commutativity, and identity of multiset union operator "__") of a server, clients, and messages. The server is indicated by identifier `server`. Clients each have an integer identifier N and another integer index M indicating the service they require from the server. Each client sends a message including its identifier and the index of the service to the server. The server replies back and logs the communication in its local list B. Assume that the property of interest is "a client that requires a service will eventually receive it". This property can be expressed by a set P of two atomic predicates, of which one indicates the requirement of the service and the other indicates the receipt. The set $\{s_1, s_2\}$ is P-invisible and a very good candidate for S, but because of the list nature of the buffer, these rules are not confluent. For the property of interest, it does not matter in what order the messages are buffered; but since the resulting buffer is different, confluence does not hold. If one assumes a lexicographical ordering on the buffer (pairwise comparison of the pairs (M, N)), then adding the following rule which always sorts the buffer

$$\begin{aligned}
s_3 : [\texttt{server} \mid B\ (N,M)\ (N',M')\ B'] \longrightarrow [\,\texttt{server} \mid B\ (N',M')\ (N,M)\ B'] \text{ if} \\
(N' > N) \vee ((N = N') \wedge (M' > M))
\end{aligned}$$

and makes the set $\widehat{S} = \{s_1, s_2, s_3\}$ confluent and terminating. It is also invisible, and t_1 is coherent with respect to $\widehat{S}$ modulo axioms. Therefore, one can reduce this theory to a theory of the form $\widehat{\mathcal{R}}/\widehat{S}$. Table 1 shows the performance evaluation of model checking this problem before and after reduction using the Maude LTL model Checker.

5 Discharging Proof Obligations

Typically, formal verification efforts using state space reduction techniques involve two separate tasks: (i) model checking the desired properties in the reduced model; and (ii) discharging *proof obligations* ensuring that the proposed reduction is indeed a correct reduction of the original system. We discuss here the proof obligations that must be verified to ensure the correctness of an S-reduction $\mathcal{R}/S$, and ways in which the discharging of such obligations can be assisted by formal tools. For $\mathcal{R}/S$ to be a correct reduction of $\mathcal{R}$ the following proof obligations must be discharged:

1. the rules S must be proved P-invisible;
2. $S \cup E_0$ must be shown confluent and terminating modulo A; and
3. the rules in $R \backslash S$ must be proved locally strongly coherent with respect to the equations $S \cup E_0$ modulo A.

Proving (1) is an inductive theorem proving task. Specifically, it amounts to proving that each state predicate $p \in P$ and also its negation $\neg p$ are both *invariants* for the rewrite theory $(\Sigma, E_0 \cup A, S)$. This can be reduced to proving a series of first-order formulas that must be shown to hold inductively in the equational specification $(\Sigma, E_0 \cup A)$; that is, to be satisfied in the initial model $T_{\Sigma/E_0 \cup A}$. Proofs can be assisted by any first-order inductive theorem prover. For Maude specifications Maude's ITP [8] can be used. The proof obligations for this task become considerably easier if the rules in S are *topmost*, that is, if all rewriting happens at the top of a term. Many rewrite theories whose state is a set or multiset of objects and messages, such as those in the case studies presented in this paper, can be transformed into bisimilar topmost rewrite theories.

Proving (2) can be done mechanically using standard termination and confluence checking tools that support reasoning modulo axioms A such as associativity and commutativity, and can in some cases handle conditional rules. Tools of this kind include, for example, CiME [10] (for both tasks) AProVE [17] (for termination), and for Maude specifications the Maude Termination Tool (MTT) [13] and the Maude Church-Rosser Checker [8].

There is a discussion on proving (3) in [35]. For most combinations of associativity, commutativity and identity axioms in A this task can be checked algorithmically when the rules are linear and unconditional. To the best of our knowledge the only tool available is Maude's Coherence Checker [12], which currently can only reason modulo commutativity axioms.

We now discuss briefly the proof obligations for the $\widehat{\mathcal{R}}/\widehat{S}$ reductions. To begin with, the same proof obligations (1)–(3) must be discharged, but now for $\widehat{\mathcal{R}}/\widehat{S}$ instead of $\mathcal{R}/S$. But that still leaves open the task of coming up with the rules $\widehat{S}$ in the first place. Two approaches are possible for this. On the one hand, as done in the case study of Section 4.3, one can use insight about the given specification to find a suitable $\widehat{S}$. On the other, it is also possible to automatically search for such a set $\widehat{S}$ by performing Knuth-Bendix (KB) completion modulo A on the equations $E_0 \cup S$ using any KB completion tool (modulo A) such as, for example, CiME [10].

6 Related Work and Conclusions

Broadly speaking, our work is related to all other state space reduction and abstraction techniques (see for example [20, 33, 22, 4, 9, 11, 19, 34, 32, 1, 21, 16]). We discuss below several approaches that are most closely related to our own.

Several *partial order reduction* (POR) techniques achieve a reduction to a representative subset of all states while preserving various types of bisimilarity. Some of these techniques [19, 34, 32, 1, 21, 16] exploit the notion of *invisibility*, an idea that is generalized here to arbitrary rewrite theories. A first main difference with the POR approach is that POR techniques are typically *dynamic* (all except [1, 21]), in the sense that the reduction is performed on-the-fly during the model checking and requires substantial changes to the underlying model checking algorithm (see [15, 18] for an exception to this); by contrast, our technique is a *static* method, since we generate the reduced rewrite theory and then model check it. Furthermore, it does not require any changes in the model checker. A second important difference is in the different levels of generality: POR techniques typically assume a conventional concurrent language with processes and consider invisible process transitions, whereas our approach is much more general: it does not rely on these assumptions, and applies to arbitrary rewrite theories.

Our method has also some similarities with a reduction technique presented in [2]. However, the settings are quite different, because [2] works in the framework of process algebras, whereas our technique works for arbitrary rewrite theories. Furthermore, the notion of invisibility used in [2] is not based on a certain set of predicates. Instead, in our case the invisibility depends on what state predicates are involved in the property that we want to model check. Also, the notion of confluence used in [2] is completely different from ours: we use the standard term-rewriting notion. The notion of coherence used in this work has some similarities with notion of *weak confluence* in [36] if one views the rules in S as τ-transitions. Moreover, their approach is dynamic, while ours is static. The symbolic prioritization in [2] is relevant to our work in two senses: (1) it is static, and (2) it is giving priority to some transitions over the rest, while we also in some sense give priority to some rules over the rest.

Our reduction technique is also closely related to other notions of abstraction and simulation used for reduction purposes in rewriting logic. In the case of *equational abstractions* [29] one begins with a rewrite theory $\mathcal{R} = (\Sigma, E_0 \cup A, R)$ and *adds extra equations* G to it to obtain an abstract theory $\mathcal{R}/G = (\Sigma, E_0 \cup G \cup A, R)$, so that we have a rewrite theory inclusion $\mathcal{R} \subseteq \mathcal{R}/G$. This technique is generalized in [26] to much more general *rewrite theory morphisms* $H : \mathcal{R} \longrightarrow \mathcal{R}'$ that need not be theory inclusions, give rise to simulations, and can be used for model checking purposes when $\mathcal{R}'$ is more abstract than $\mathcal{R}$. Our proposed technique is different from those in [29] and [26]. In our case the relationship between $\mathcal{R}$ and $\mathcal{R}/S$ cannot be understood as a theory *morphism*: it is only a theory *transformation*. This means that we now have a new state space reduction technique for rewrite theories that nicely complements those proposed in [29, 26].

Our technique makes essential use of Viry's notion of coherence [35] in rewrite theories. But we use the notion in precisely the *opposite way* than in Viry's work. The original purpose of coherence is to make a rewrite theory $\mathcal{R} = (\Sigma, E_0 \cup A, R)$ executable by *turning the equations* E_0 *into rules*. Strong coherence then guarantees that $\mathcal{R}$ and the resulting theory $(\Sigma, A, E_0 \cup R)$ are semantically equivalent. We do somehow the opposite: beginning with a rewrite theory $\mathcal{R} = (\Sigma, E_0 \cup A, R)$ we select a subset of rules $S \subseteq R$ and *turn those rules into equations* to obtain our reduced theory $\mathcal{R}/S = (\Sigma, E_0 \cup S \cup A, R \backslash S)$. We then check strong coherence of $\mathcal{R}/S$ for executability and stuttering bisimilarity purposes.

We can summarize our contributions as follows: we have presented a general method to reduce the state space of a concurrent system specified as a rewrite theory $\mathcal{R}$ by selecting a set S of P-invisible transition rewrite rules that, when turned into equations, yield a reduced theory $\mathcal{R}/S$. We have shown that if $\mathcal{R}/S$ satisfies reasonable executability assumptions it is stuttering bisimilar to $\mathcal{R}$ and therefore satisfies the same CTL^*_{-X} formulas under this bisimilarity. Several case studies presented show that $\mathcal{R}/S$ can have a drastically smaller state space in practice, making it feasible to model check properties for $\mathcal{R}$ by using $\mathcal{R}/S$ instead. We have also presented a method to obtain reductions of this kind using extra invisible rules not present in the original theory $\mathcal{R}$. The proof obligations that must be discharged to guarantee the correctness of our proposed reductions have also been discussed. Discharging them involves reasonable proof tasks that for the most part can be supported by existing formal tools.

This work is part of a broader effort to develop state space reduction techniques of wide applicability for concurrent systems specified as rewrite theories. In this sense, it complements earlier efforts to develop reduction techniques of this kind for rewrite theories [29, 26, 15]. It is however a new technique, different from earlier ones. In future work we plan to further develop the ideas presented here in two opposite directions. In a more general direction, we plan to investigate *weaker conditions* under which invisible transitions S can be used to reduce the state space. In a more specific direction, we plan to apply these techniques to *distributed object systems*, where we hope to exploit the more specific nature of those systems to obtain even more drastic reductions. Two other aspects that need to be further developed are: (i) building a stronger tool environment for checking proof obligations, particularly for checking coherence modulo more general axioms A; and (ii) developing a broader experimental base of case studies.

Acknowledgment. Research funded by ONR grant N00014-02-1-0715.

References

1. R. Alur, R. K. Brayton, T. A. Henzinger, S. Qadeer, and S. K. Rajamani. Partial-order reduction in symbolic state exploration. In *CAV*, volume 1254 of *LNCS*, pages 340 – 351, 1997.
2. Stefan Blom and Jaco van de Pol. State space reduction by proving confluence. In *CAV*, volume 2404 of *LNCS*, pages 596–609, 2002.
3. M.C. Browne, E. M. Clarke, and O. Grumberg. Characterizing finite kripke structures in propositional temporal logic. *Theoretical Computer Science*, 59:115 – 131, 1988.
4. E. M. Clarke, O. Grumberg, and D. E. Long. Model checking and abstraction. *ACM Transactions on Programming Languages and Systems*, 16(5):1512–1542, 1994.
5. Edmund M. Clarke, Orna Grumberg, and Doron A. Peled. *Model Checking.* MIT Press, 2001.
6. M. Clavel, F. Durán, S. Eker, P. Lincoln, N. Martí-Oliet, J. Meseguer, and J. Quesada. Maude: specification and programming in rewriting logic. *Theoretical Computer Science*, 285:187–243, 2002.
7. M. Clavel, F. Durán, S. Eker, P. Lincoln, N. Martí-Oliet, J. Meseguer, and C. Talcott. Maude Manual (Version 2.2). December 2005, `http://maude.cs.uiuc.edu`.
8. M. Clavel, F. Durán, S. Eker, and J. Meseguer. Building equational proving tools by reflection in rewriting logic. In *Proc. of the CafeOBJ Symposium*, April 1998.
9. M. A. Colón and T. E. Uribe. Generating finite-state abstractions of reactive systems using decision procedures. In *Computer Aided Verification*, volume 1427 of *Lecture Notes in Computer Science*, pages 293–304, 1998.
10. E. Contejean and C. Marché. CiME: Completion modulo E. In *RTA*, volume 1103 of *LNCS*, 1996.
11. D. Dams, R. Gerth, and O. Grumberg. Abstract interpretation of reactive systems. *ACM Transactions on Programming Languages and Systems*, 19:253–291, 1997.
12. F. Durán. Coherence checker and completion tools for Maude specifications. Manuscript, `http://maude.cs.uiuc.edu/papers`, 2000.
13. F. Durán, S. Lucas, J. Meseguer, C. Marché, and X. Urbain. Proving termination of membership equational programs. In *PEPM'04*, pages 147–158, 2004.
14. S. Eker, J. Meseguer, and A. Sridharanarayanan. The Maude LTL model checker and its implementation. In *SPIN'03*, volume 2648 of *LNCS*, pages 230 – 234, 2003.
15. A. Farzan and J. Meseguer. Partial order reduction for rewriting semantics of programming languages. In *WRLA06*, pages 56–75, 2006.
16. C. Flanagan and P. Godefroid. Dynamic partial order reduction for model checking software. In *Proceedings of POPL*, 2005.
17. J. Giesl, R. Thiemann, P. Schneider-Kamp, and S. Falke. Automated termination proofs with AProVE. In *RTA*, volume 3091 of *LNCS*, pages 210–220, 2004.
18. P. Godefroid. Model checking for programming languages using VeriSoft. In *POPL*, volume 174–186, 1997.
19. P. Godefroid and P. Wolper. A partial approach to model checking. In *Proceedings of Logic in Computer Science*, pages 406 – 415, 1991.
20. Y. Kesten and A. Pnueli. Control and data abstraction: The cornerstones of practical formal verification. *International Journal on Software Tools for Technology Transfer*, 4(2):328–342, 2000.
21. R. Kurshan, V. Levin, M. Minea, D. Peled, and H. Yenigun. Static partial order reduction. In *TACAS*, volume 1384, pages 345 – 357, 1998.

22. C. Loiseaux, S. Graf, J. Sifakis, A. Bouajjani, and S. Bensalem. Property preserving abstractions for the verification of concurrent systems. *Formal Methods in System Design*, 6:1–36, 1995.
23. N. A. Lynch. *Distributed Algorithms*. Morgan Kaufmann, 1996.
24. P. Manolios. *Mechanical Verification of Reactive Systems*. PhD thesis, University of Texas at Austin, August 2001.
25. N. Martí-Oliet and J. Meseguer. Rewriting logic: roadmap and bibliography. *Theoretical Computer Science*, 285:121–154, 2002.
26. N. Martí-Oliet, J. Meseguer, and M. Palomino. Theoroidal maps as algebraic simulations. In *WADT*, pages 126–143, 2004.
27. J. Meseguer. Conditional rewriting logic as a unified model of concurrency. *Theoretical Computer Science*, 96(1):73–155, 1992.
28. J. Meseguer. Research directions in rewriting logic. In *Computational Logic, NATO Advanced Study Institute, Marktoberdorf*. 1999.
29. J. Meseguer, M. Palomino, and N. Martí-Oliet. Equational abstractions. In *CADE*, volume 2741 of *LNCS*, pages 2–16, 2003.
30. R. De Nicola and F. Vaandrager. Three logics for branching bisimulation. *Journal of ACM*, 42(2), 1995.
31. M. Palomino, J. Meseguer, and N. Martí-Oliet. A categorical approach to simulations. In *CALCO*, pages 313–330, 2005.
32. D. Peled. Combining partial order reduction with on-the-fly model checking. In *CAV*, volume 818 of *LNCS*, pages 377 – 390, 1994.
33. H. Saïdi and N. Shankar. Abstract and model check while you prove. In *Computer Aided Verification*, volume 1633 of *LNCS*, pages 443–454, 1999.
34. A. Valmari. A stubborn attack on state explosion. In *CAV*, volume 531 of *LNCS*, pages 156 – 163, 1990.
35. P. Viry. Equational rules for rewriting logic. *Theoretical Computer Science*, 285:487–517, 2002.
36. M. Ying. Weak confluence and τ-inertness. *Theoretical Computer Science*, 238:465–475, 2000.

The Essence of Multitasking*

William L. Harrison

Department of Computer Science
University of Missouri
Columbia, Missouri, USA
harrison@cs.missouri.edu

Abstract. This article demonstrates how a powerful and expressive abstraction from concurrency theory—monads of resumptions—plays a dual rôle as a programming tool for concurrent applications. The article demonstrates how a wide variety of typical OS behaviors may be specified in terms of resumption monads known heretofore exclusively in the literature of programming language semantics. We illustrate the expressiveness of the resumption monad with the construction of an exemplary multitasking kernel in the pure functional language Haskell.

1 Introduction

Many techniques and structures have emigrated from programming language theory to programming practice (e.g., types, CPS, etc.), and this paper advocates that resumption monads make this journey as well. This work demonstrates how a natural (but, perhaps, under-appreciated) computational model of concurrency is used to construct multi-threaded concurrent applications suitable for formal verification. The expressiveness of resumption monads is illustrated by the construction of an exemplary multitasking operating system kernel with process forking, preemption, message passing, and synchronization constructs all requiring about fifty lines of Haskell 98 code[1]. And, because this machinery may be generalized as monad transformers, the functionality described here may be reused and refined easily.

The literature involving resumption monads [2, 3, 4, 5, 6, 7] focuses on their use in elegant and abstract mathematical semantics for programming languages. The current work advocates resumption monads as a useful abstraction for concurrent functional programming as well. **The contributions of this work are twofold:** (1) the formulation of typical concurrent operating system behaviors in terms of structures known heretofore in theoretical semantics literature and (2) a substantial case study illustrating this formulation within a higher-order functional programming language. The purpose of the case study, in part, is to provide an exposition so that the interested reader may grasp the theoretical literature more readily.

* This research supported in part by subcontract GPACS0016, System Information Assurance II, through OGI/Oregon Health & Sciences University.

[1] All the code presented in this paper is available online [1].

M. Johnson and V. Vene (Eds.): AMAST 2006, LNCS 4019, pp. 158–172, 2006.

A *resumption* [8] is stream-like construction similar to a continuation in that both tell what the "rest of the computation" is. However, resumptions are considerably less powerful than continuations—the *only* thing one may model with resumptions is multitasking computation. This conceptual economy makes concurrent applications structured with resumption monads easy to comprehend, modify, extend, and reason about. Specifically, we demonstrate how to construct a multitasking operating system kernel based on three monads and their operations (written here in categorical style):

$$\begin{array}{lll} St\ A &= Sto \rightarrow A \times Sto & \text{— state} \\ R\ A &= \mu X.\,(A + (St\ X)) & \text{— state+concurrency} \\ Re\ A &= \mu X.\,(A + (Req \times (Rsp \rightarrow St\ X))) & \text{— state+concurrency+interactive i/o} \end{array}$$

St is the familiar state monad, while *R* and *Re* are resumption monads providing what we call *basic* and *reactive* concurrency about which we will say much more below.

The structure of this article is as follows. After reviewing the related work below and the necessary background in Section 2, Section 3 describes in detail how resumption monads may be used to model multitasking concurrency. Section 4 presents a resumption-monadic semantics for a concurrent language extended with "signals"; a thread may signal the kernel to fork, suspend, preempt, print, send or receive a message, and acquire or release a semaphore and Section 5 describes the kernel on which these threads execute. Section 6 summarizes the work and outlines future directions.

Related Work. Functional languages are well-known for promoting mathematical reasoning about programs, and, perhaps because of this, there has been considerable research into their use for concurrent software such as OS kernels. The present work has this pedigree, yet fundamentally differs from it in at least one key respect: we explicitly encapsulate all effects necessary to the kernel with monads: input/output, shared state and preemptive multitasking concurrency.

The concurrency models underlying previous applications of functional languages to concurrent system software fall broadly into four camps. The first camp [9, 10, 11, 12] assumes the existence of a non-deterministic choice operator to accommodate "non-functional" situations where more than one action is possible, such as a scheduler choosing between two or more waiting threads. However, such a non-deterministic operator risks the loss of an important reasoning principle of pure languages—referential transparency—and considerable effort is made to minimize this danger. Non-determinism may be incorporated easily into the kernel presented here via the non-determinism monad, although such non-determinism is of a different, but closely related, form.

The second model uses "demand-driven concurrency" [13, 14] in which threads are mutually recursive bindings whose lazy evaluation simulates multitasking concurrency. Interleaving order is determined (in part) by the interdependency of these bindings. However, the demand-driven approach requires some alteration of the underlying language implementation to completely determine thread scheduling. Thread structure is entirely implicit—there are no atomic actions *per se.* Demand determines the extent to which a thread is evaluated—rather like the "threads" encoded by computations in the lazy state monad [15]. Thread structure in the resumption-monadic setting is explicit—one may even view a resumption monad as an abstract data type for threads. This exposed thread structure allows deterministic scheduling without changing the underlying language implementation as with demand-driven concurrency.

The third camp uses CPS to implement thread interleaving. Concurrent behavior may be modeled with first-class continuations [16, 17, 18, 19] because the explicit control over evaluation order in CPS allows multiple threads to be "interwoven" to produce any possible execution order. Claessen presents a formulation of this style using the CPS monad transformer [16], although without exploiting the full power of first-class continuations—i.e., he does not use *callcc* or *shift* and *reset*. While it is certainly possible to implement the full panoply of OS behaviors with CPS, it is also possible to implement much, much more—most known effects may be expressed via CPS [20]. This expressiveness can make programs in CPS difficult to reason about, rendering CPS less attractive as a foundation for software verification. Resumptions can be viewed as a disciplined use of continuations which allows for simpler reasoning.

The last camp uses a multi-threading paradigm called trampoline-style programming [21]. Programs in trampoline-style are organized around a single scheduling loop called a "trampoline." One attractive feature of trampolining is that it requires no appeal to first-class continuations. Of the four camps, trampolining is most closely related to the resumption-monadic approach described here. In [21], the authors motivate trampolining with a type constructor equivalent to the functor part of the basic resumption monad (described in Section 3.1 below), although the constructor is never identified as such.

The previous research relevant to this article involves those applications of functional languages where the concurrency model is explicitly constructed rather than inherited from a language implementation or run-time platform. There are many applications of functional languages to system software that rely on concurrency primitives from existing libraries or languages [22, 23]; as the modeling of concurrency is not their primary concern, no further comparison is made. Similarly, there are many concurrent functional languages—concurrent versions of ML, Haskell, and Erlang—but their concurrency models are built-in to their run-time systems and provide no basis of comparison to the current work. It may be the case, however, that the resumption-monadic framework developed here provides a semantic basis for these languages.

Resumptions are a denotational model of concurrency first introduced by Plotkin [8]; excellent introductions to this non-monadic form of resumptions are due to Schmidt [24] and Bakker [25]. Moggi was the first to observe that the categorical structure known as a monad supports modular semantic theories for programming languages and he showed how a sequential theory of concurrency could be expressed in the resumption monad [2]. The formulation of the basic resumption monad we use is due to Papaspyrou [3, 5], although other equivalent formulations exist [2, 26, 27].

2 Review: Monads

Monads are algebras just as groups or rings are algebras; that is, a monad is a type constructor (functor) with associated operators obeying certain equations—the well-known "monad laws" [28]. There are several formulations of monads, and we use one familiar to functional programmers called the Kleisli formulation: a monad M is given by an eponymous type constructor M and the *unit* operator, **return** : $a \rightarrow M\ a$, and the *bind* operator, $(\texttt{>>=}) : M\ a \rightarrow (a \rightarrow M\ b) \rightarrow M\ b$. We assume of necessity that the reader possesses familiarity with monads and their uses in modeling effects. Readers requiring further background should consult the references [2, 28]. We represent the monadic constructions here in the pure functional language Haskell 98 [29], although we would be equally justified using categorical notation or any other higher-order functional programming language.

A monad in Haskell typically consists of a data type declaration (defining the computational "raw materials" encapsulated by the monad) and definitions for the overloaded symbols (**return**) and (>>=) [29]. The state monad St, containing a single threaded state $Sto = Loc \rightarrow Int$, is declared:

$$\begin{array}{ll} \mathbf{data}\ St\ a = ST\ (Sto \rightarrow (a, Sto)) & \mathbf{return}\ v = ST\ (\lambda s.\ (v, s)) \\ deST\ (ST\ x) = x & (ST\ x) \texttt{>>=} f = ST(\lambda s.\ let\ (y, s') = (x\ s) \\ & \qquad\qquad\qquad\qquad in\ deST\ (f\ y)\ s') \end{array}$$

The state monad has operators for updating the state, u, getting the state, g, and reading a particular location, *getloc*:

$$\begin{array}{ll} u : (Sto \rightarrow Sto) \rightarrow St\ () & u\ \delta = ST\ (\lambda\ s.\ ((), \delta\ s)) \\ g : St\ Sto & g = ST\ (\lambda s.\ (s, s)) \\ getloc : Loc \rightarrow St\ Int & getloc\ x = g \texttt{>>=} \lambda\ \sigma.\ \mathbf{return}\ (\sigma\ x) \end{array}$$

Here, () is both the single element unit type and its single element. The "null" bind operator, $(\texttt{>>}) : M\ a \rightarrow M\ b \rightarrow M\ b$, is useful when the result of >>='s first argument is ignored: $x \texttt{>>}\ y = x \texttt{>>=} \lambda\ \_.\ y$.

Notational Convention. We suppress details of Haskell's concrete syntax when they are unnecessary to the presentation (in particular, instance declarations and class predicates in types). Haskell 98 reverses the standard use of (::) and (:) in that (::) stands for "has type" and (:) for list concatenation in Haskell 98. We will continue to use the standard interpretation of these symbols.

3 Concurrency Based on Resumptions

Two formulations of resumption monads are used here–what we call *basic* and *reactive* resumption monads. Both occur, in one form or another, in the literature [2, 3, 5, 26, 27]. The *basic* resumption monad (Section 3.1) encapsulates a notion of multitasking concurrency; that is, its computations are stream-like and may be woven together into single computations representing any arbitrary schedule.

The *reactive* resumption monad (Section 3.2) encapsulates multitasking concurrency as well, but, in addition, affords a request-and-response interactive notion of computation which, at a high-level, resembles the interactions of threads within a multitasking operating system.

To motivate resumptions, let's compare them with a natural model of concurrency known as the "trace model" [30]. The trace model views threads as (potentially infinite) streams of atomic operations and the meaning of concurrent thread execution as the set of all their possible thread interleavings. Imagine that we have two simple threads $a = [a_0, a_1]$ and $b = [b_0]$, where a_0, a_1, and b_0 are "atomic" operations, and, if it is helpful, think of such atoms as single machine instructions. According to the trace model, the concurrent execution $a \| b$ of threads a and b is denoted by the set[2] of all their possible interleavings:

$$\mathit{traces}\,(a \parallel b) = \{[a_0, a_1, b_0], [a_0, b_0, a_1], [b_0, a_0, a_1]\} \qquad (\ddagger)$$

This means that there are three distinct possible execution traces of $(a \parallel b)$, each of which corresponds to an interleaving of the atoms in a and b. Non-determinism in the trace model is reflected in the fact that $\mathit{traces}(a \parallel b)$ is a set consisting of multiple interleavings.

The trace model captures the structure of concurrent thread execution abstractly and is well-suited to formal characterizations of properties of concurrent systems (e.g., liveness). However, a gap exists between this formal model and an executable system: traces are streams of events, and each event is itself a place holder (i.e., what do the events a_0, a_1, and b_0 actually do?). Resumption monads bridge this gap because they are both a formal, trace-based concurrency model and may be directly realized and executed in a higher-order functional language.

The notion of computation provided by resumption monads is that of sequenced computation. A resumption computation has a stream-like structure in that it includes both a "head" (corresponding to the next action to perform) and a "tail" (corresponding to the rest of the computation)—very much like the execution traces in (‡). We now describe the two forms of resumption monads in detail.

3.1 Sequenced Computation and Basic Resumptions

This section introduces sequenced computation in monadic style, discussing the monad that combines resumptions with state. The monad combining resumptions with state is:

$$\begin{array}{ll}
\mathbf{data}\ R\ a & = \mathit{Done}\ a \mid \mathit{Pause}\ (\mathit{St}\ (R\ a)) \\
\mathbf{return} & = \mathit{Done} \\
(\mathit{Done}\ v)\ \texttt{>>=}\ f & = f\ v \\
(\mathit{Pause}\ r)\ \texttt{>>=}\ f & = \mathit{Pause}\,(r\ \texttt{>>=}_{St}\ \lambda\kappa.\,\mathbf{return}_{St}\,(\kappa\ \texttt{>>=}\ f))
\end{array} \qquad (*)$$

---
[2] This set is also *prefix-closed* in Roscoe's model, meaning that it includes all prefixes of any trace in the set. For the purposes of this exposition, we ignore this consideration.

Here, the bind operator for R is defined recursively using the bind and unit for the state monad (written above as $\gg=_{St}$ and $\mathbf{return}_{St}$, respectively). Some stateful computation—i.e., within "$r \gg=_{St} \ldots$"—takes place.

Returning to the trace model example from the beginning of this section, we can now see that R-computations are quite similar to the traces in (‡). The basic resumption monad has lazy constructors *Pause* and *Done* that play the rôle of the lazy list constructors cons (::) and nil ([ ]) in the traces example. If the atomic operations of a and b are computations of type $St\,()$, then the following computations of type $R\,()$ are the set of possible interleavings:

$$
\begin{array}{l}
Pause\,(a_0 \gg \mathbf{return}\,(Pause\,(a_1 \gg \mathbf{return}\,(Pause\,(b_0 \gg \mathbf{return}\,(Done\,()))))))\\
Pause\,(a_0 \gg \mathbf{return}\,(Pause\,(b_0 \gg \mathbf{return}\,(Pause\,(a_1 \gg \mathbf{return}\,(Done\,()))))))\\
Pause\,(b_0 \gg \mathbf{return}\,(Pause\,(a_0 \gg \mathbf{return}\,(Pause\,(a_1 \gg \mathbf{return}\,(Done\,()))))))
\end{array}
$$

where $\gg$ and **return** are the bind and unit operations of the St monad. While the stream version implicitly uses a lazy cons operation $(h :: t)$, the monadic version uses something similar: $Pause\,(h \gg \mathbf{return}\; t)$. The laziness of *Pause* allows infinite *computations* to be constructed in R just as the laziness of cons in $(h :: t)$ allows infinite *streams* to be constructed.

3.2 Reactive Concurrency

We now consider a refinement to the R monad allowing computations to signal requests and receive responses in a manner like the interaction between an operating system and processes. Processes executing in an operating system are interactive; processes are, in a sense, in a continual dialog with the operating system. Consider what happens when such a process makes a system call. (1.) The process sends a request signal q to the operating system for a particular action (e.g., a process fork). Making this request may involve blocking the process (e.g., making a request to an I/O device would typically fall into this category) or it may not (e.g., forking). (2.) The OS, in response to the request q, handles it by performing some action(s). These actions may be privileged (e.g., manipulating the process wait list), and a response code c will be generated to indicate the status of the system call (e.g., its success or failure). (3.) Using the information contained in c, the process continues execution.

How might we represent this dialog? Assume we have data types of requests and responses:

$$
\begin{array}{l}
\mathbf{data}\; Req = Cont \mid \langle \text{other requests} \rangle\\
\mathbf{data}\; Rsp = Ack \mid \langle \text{other responses} \rangle
\end{array}
$$

Both *Req* and *Rsp* are required to have certain minimal structure; the continue request, *Cont*, signifies merely that the computation wishes to continue, while the acknowledge response, *Ack*, is an information-free response. The following monad, *Re*, "adds" the raw material for interactivity to the monad R as follows:

$$\mathbf{data}\; Re\; a = D\; a \mid P\; (Req, Rsp \rightarrow (St(Re\; a)))$$

We coin the term *reactive* resumption to distinguish Re from R and use D and P instead of "*Done*" and "*Pause*", respectively. The notion of concurrency provided by Re formalizes the process dialog example described above. A paused Re-computation has the form $P(q, r)$, where q is a request signal in Req and r, if provided with a response code from Rsp, is the rest of the computation. The operations for Re are defined:

$$
\begin{array}{lcl}
\mathbf{return} & = & D \\
D\ v \ggg= f & = & f\ v \\
P\ (q, r) \ggg= f & = & P\ (q,\ \lambda\, rsp\, .\, (r\ rsp) \ggg=_{St} \lambda\, \kappa\, .\, \mathbf{return}_{St}\ (\kappa \ggg= f))
\end{array}
$$

In this article, we use a particular definition of the request and response data types Req and Rsp which correspond to the services provided by the operating system (more will be said about the use of these in Section 5):

$$
\begin{array}{lcl}
\mathbf{type}\ \mathit{Message} & = & \mathit{Int} \\
\mathbf{type}\ \mathit{PID} & = & \mathit{Int} \\
\mathbf{data}\ \mathit{Req} & = & \mathit{Cont} \mid \mathit{Sleep}_q \mid \mathit{Fork}_q\ \mathit{Process} \mid \mathit{Bcst}_q\ \mathit{Message} \\
 & & \mid \mathit{Rcv}_q \mid V_q \mid P_q \mid \mathit{Prnt}_q\ \mathit{String} \\
 & & \mid \mathit{PID}_q \mid \mathit{Kill}_q\ \mathit{PID} \\
\mathbf{data}\ \mathit{Rsp} & = & \mathit{Ack} \mid \mathit{Rcv}_r\ \mathit{Message} \mid \mathit{PID}_r\ \mathit{PID}
\end{array}
$$

Note that both Req and Rsp have *Cont* and *Ack*. The kernel in Section 5 will use the response *Ack* for several different requests. *Process* is defined in the next section.

Reactive resumption monads have two non-proper morphisms. The first, *step*, recasts a stateful computation as a resumption computation[3]:

$$
\begin{array}{l}
\mathit{step} \quad : \mathit{St}\ a \rightarrow \mathit{Re}\ a \\
\mathit{step}\ x = P\ (\mathit{Cont},\ \lambda \mathit{Ack}.\ x \ggg=_{St} (\mathbf{return}_{St} \circ D))
\end{array}
$$

The definition of *step* shows why we require that Req and Rsp have a particular shape including *Cont* and *Ack*, respectively; namely, there must be at least one request/response pair for the definition of *step*. Another non-proper morphism for Re allows a computation to raise a signal; its definition is:

$$
\begin{array}{ll}
\mathit{sig} \quad : \mathit{Req} \rightarrow \mathit{Re}\ \mathit{Rsp} & \mathit{sig}_i \quad : \mathit{Req} \rightarrow \mathit{Re}\ () \\
\mathit{sig}\ q = P(q,\ \mathbf{return}_{St} \circ \mathbf{return}_{Re}) & \mathit{sig}_i\ q = P\ (q, \lambda\_.\ \mathbf{return}_{St}(\mathbf{return}_{Re}\ ()))
\end{array}
$$

Furthermore, there are certain cases where the response to a signal is intentionally ignored, for which we use sig_i.

4 The Language of Threads

This section formulates an abstract syntax for kernel processes. Operating systems texts typically define threads as lightweight processes executed in the same

[3] For R, *step* is defined similarly: $\mathit{step}\ x = \mathit{Pause}(x \ggg=_{St} (\mathbf{return}_{St} \circ \mathit{Done}))$.

address space[4]. Events are abstract machine instructions—they read from and write to locations and signal requests to the operating system. Processes are infinite sequences of events, although it is straightforward to include finite (i.e., terminating) processes as well, but it suffices for our presentation to assume non-terminating, infinite processes.

$$\begin{aligned} \mathit{Process} &= \mathit{Event}\,;\,\mathit{Process} \\ \mathit{Event} &= \mathit{Loc}\texttt{:=}\mathit{Exp} \mid \texttt{bcast(}\mathit{Exp}\texttt{)} \mid \texttt{recv(}\mathit{Loc}\texttt{)} \mid \texttt{print(}\mathit{String}\texttt{,}\mathit{Exp}\texttt{)} \\ &\quad \mid \texttt{sleep} \mid \texttt{fork(}\mathit{Process}\texttt{)} \mid \texttt{P} \mid \texttt{V} \mid \texttt{kill(}\mathit{Exp}\texttt{)} \\ \mathit{Exp} &= \mathit{Int} \mid \mathit{Loc} \mid \texttt{pid} \end{aligned}$$

The *Exp* language is self-explanatory except for the `pid` expression that returns the process identifier of the calling process. The *Event* language has a simple assignment statement, l`:=`e, which evaluates its right-hand side, $e \in Exp$, and stores it in the location, $l \in Loc$, on the left-hand side. The language includes broadcast and receive primitives: `bcast(`e`)` and `recv(`l`)`. The event `bcast(`e`)` broadcasts the value of expression e, while `recv(`l`)` receives an available message in location l. There is also a process spawning primitive, `fork(`p`)`, producing a child process p executing in the same address space. The language has a single semaphore with test release operations, `P` and `V`. Finally, there is a process killing primitive, `kill(`*pid*`)`, that terminates the process with identifier *pid* (if such a process exists). Where the language and its semantics differ from previous work [5] is the inclusion of signals; that is, programs may request intervention from the kernel.

Figure 1 defines expressions, events, and processes with $\mathcal{E}[\![-]\!]$, $\mathcal{A}[\![-]\!]$, and $\mathcal{P}[\![-]\!]$, respectively. In most respects, this is a conventional store passing semantics in monadic form, the difference being that individual *St* actions (e.g., *getloc* x) are lifted to *Re* via the *step* function. *step* creates an "atomic" action out of a single *St* action, and $\mathcal{A}[\![-]\!]$ "chains together" one or two such actions. For example, $\mathcal{A}[\![\texttt{P}]\!]$ is the single kernel signal $(sig_i\, P_q)$, while $\mathcal{A}[\![x\texttt{:=}e]\!]$ chains together "$\mathcal{E}[\![e]\!]$" and "*store* x" with `>>=`. The meaning of a process, $\mathcal{P}[\![p]\!]$, is the infinite "chaining-together" of its event chains. These semantics are similar to published resumption-monadic language semantics [5] for CSP-like languages, differing only in the inclusion of signals (i.e., requests made with with *sig* and sig_i to be handled by the kernel).

5 The Kernel

This section describes the structure and implementation of a kernel providing a variety of services typical to an operating system. For the sake of comprehensibility, we have intentionally made this kernel simple; the goal of the present work is to demonstrate how typical operating system services may be represented using resumption monads in a straightforward and compelling manner. It should be clear, however, how more powerful or expressive operating system behaviors may be captured as refinements to this system.

[4] We use the terms "thread" and "process" interchangeably throughout.

$$[l{\mapsto}v] : (Loc{\to}Int){\to}Loc{\to}Int$$

$$[l{\mapsto}i]\ \sigma\ n = \begin{cases} i & l = n \\ \sigma\ n & l \neq n \end{cases}$$

$$\begin{aligned} &store \quad : Loc{\to}Int{\to}Re\ a \\ &store\ l\ i = (step{\circ}u)\ ([l{\mapsto}i]) \end{aligned}$$

$$\begin{aligned} &\mathcal{E}[\![-]\!] \quad : Exp \to Re\ Int \\ &\mathcal{E}[\![i]\!] \quad = \mathbf{return}\ i \\ &\mathcal{E}[\![x]\!] \quad = step\ (getloc\ x) \\ &\mathcal{E}[\![\mathtt{pid}]\!] = sig\ PID_q \mathbin{\texttt{>>=}} (\mathbf{return} \circ prj) \\ &\quad \mathbf{where}\ prj\ (PID_r\ pid) = pid \\ &\mathcal{P}[\![-]\!] \quad : Process \to Re\ () \\ &\mathcal{P}[\![e;p]\!] = \mathcal{A}[\![e]\!] \mathbin{\texttt{>>}} \mathcal{P}[\![p]\!] \end{aligned}$$

$$\begin{aligned} &\mathcal{A}[\![-]\!] \quad : Event \to Re\ () \\ &\mathcal{A}[\![x\mathtt{:=}e]\!] = \mathcal{E}[\![e]\!] \mathbin{\texttt{>>=}} store\ x \\ &\mathcal{A}[\![\mathtt{print}(m\mathtt{,}e)]\!] = \mathcal{E}[\![e]\!] \mathbin{\texttt{>>=}} print \\ &\quad \mathbf{where} \\ &\qquad out\ m\ v = m{+\!\!+}\text{``: ''}{+\!\!+}show\ v \\ &\qquad print \quad = sig_i \circ Prnt_q \circ (out\ m) \\ &\mathcal{A}[\![\mathtt{sleep}]\!] \quad = sig_i\ Sleep_q \\ &\mathcal{A}[\![\mathtt{fork}(p)]\!] \quad = sig_i\ (Fork_q\ p) \\ &\mathcal{A}[\![\mathtt{bcast}(x)]\!] = \mathcal{E}[\![x]\!] \mathbin{\texttt{>>=}} (sig_i \circ Bcst_q) \\ &\mathcal{A}[\![\mathtt{recv}(x)]\!] \quad = sig\ Rcv_q \mathbin{\texttt{>>=}} (store\ x) \circ prj \\ &\quad \mathbf{where}\ prj\ (Rcv_r\ m) = m \\ &\mathcal{A}[\![\mathtt{P}]\!] \quad = sig_i\ P_q \\ &\mathcal{A}[\![\mathtt{V}]\!] \quad = sig_i\ V_q \\ &\mathcal{A}[\![\mathtt{kill}(e)]\!] \quad = \mathcal{E}[\![e]\!] \mathbin{\texttt{>>=}} (sig_i \circ Kill_q) \end{aligned}$$

Fig. 1. Semantics of Expressions, Events, and Processes. All monad operations belong to the *Re* monad.

The structure of the kernel is given by the global system configuration and two mutually recursive functions representing the scheduler and service handler. The system configuration consists of a snapshot of the operating system resources; these resources are a thread waiting list, a message buffer, a single semaphore, an output channel, and a counter for generating new process identifiers. The system configuration is specified by:

$$\begin{aligned} \mathbf{type}\ System = (&[(PID, Re\ ())], && \text{— waiting list} \\ &[Message], && \text{— message buffer} \\ &Semaphore, && \text{— } Semaphore{=}Int\text{, 1 initially} \\ &String, && \text{— output channel} \\ &PID) && \text{— identifier counter} \end{aligned}$$

The first component is the waiting list consisting of a list of pairs: (pid, t). Here, pid is the unique process identifier of thread t. The second component is a message buffer where messages are assumed to be single integers and the buffer itself is a list of messages. Threads may broadcast messages, resulting in an addition to this buffer, or receive messages from this buffer if a message is available. There is a single semaphore, and individual threads may acquire or release this lock. The semaphore implementation here uses busy waiting, although one could readily refine this system configuration to include a list of processes blocked waiting on the semaphore. The fourth component is an output channel (merely a *String*) and the fifth is a counter for generating process identifiers.

The types of a scheduler and service handler are:

$$\begin{aligned} &sched \quad : System \to R\ () \\ &handler : System \to (PID, Re\ ()) \to R\ () \end{aligned}$$

A *sched* morphism takes the system configuration (which includes the waiting list), picks the next thread to be run, and calls the *handler* on that thread. The *sched* and *handler* morphisms translate reactive computations—i.e., those interacting threads typed in the *Re* monad present in the wait list—into a single, interwoven scheduling typed in the basic R monad. The range in the typings of *sched* and *handler* is $R\,()$ precisely because the requested thread interactions have been mediated by *handler*.

From the waiting list component of the system configuration, the scheduler chooses the next thread to be serviced and passes it, along with the system configuration, to the service handler. The service handler performs the requested action and throws the remainder of the thread and the system configuration (possibly updated reflecting the just-serviced request) back to `sched`. The scheduler/handler interaction converts reactive *Re* computations representing threads into a single basic R computation representing a particular schedule.

There are many possible choices for scheduling algorithms—and, hence, many possible instances of *sched*—but for our purposes, round robin suffices:

$$\begin{array}{ll} \mathit{rrobin} : \mathit{System} \rightarrow R\,() & \\ \mathit{rrobin}\ ([], \_, \_, \_, \_) & = \mathit{Done}\ () \quad \text{— stop when no threads} \\ \mathit{rrobin}\ (((i,t) :: w), mq, s, o, g) & = \mathit{handler}\ (w, mq, s, o, g)\ (i, t) \end{array}$$

The handler fits entirely in Figure 2. A call $(\mathit{handler}\ \mathit{sys}\ (i, P(q, r)))$ responds to query q based on the contents of *sys* and follows the same pattern:

$$P(q,r) \ \rightarrow\ \mathit{Pause}(\langle \mathit{St}\ \text{action}\rangle\ ;\ \mathbf{return}_{St}\ (\mathit{rrobin}\ \mathit{sys}'))$$

Here, the "$\langle St$ action$\rangle$" is a (possibly empty) *St* computation determined by r and sys' is the *System* configuration reflecting changes to kernel resources necessary to handling q. Each *handler* branch is discussed in detail below and the labels (a.)-(l.) refer to lines within Figure 2.

Basic Operation. When *handler* encounters a thread which is completed (i.e., the thread is a computation of the form $D\_$), it simply calls the scheduler with the unchanged system configuration (a.). If the thread wishes to continue (i.e., it is of the form $P(\mathit{Cont}, r)$), then *handler* acknowledges the request by passing *Ack* to r (b.). As a result, the first atom in r is scheduled, and the rest of the thread (written $\mathit{next}\ (i, \kappa)$ above) is passed to the scheduler.

Dynamic Scheduling. A thread may request suspension with the Sleep_q signal (c.); the handler changes the Sleep_q request to a *Cont* and reschedules the thread. The effect of this is to delay the thread by one scheduling cycle. An obvious refinement of this service would include a counter field within the Sleep_q request and use this field to delay the thread through multiple cycles.

A thread may request to spawn a new thread (d.). The child process is $(g, \mathcal{P}[\![p]\!])$ for new identifier g. Then, both parent and child thread are added back to the waiting list. We introduce the "continue" helper function, *cont*, that takes a partial thread, r, and a response code, *rsp*, and creates a thread which receives and continues on the response code *rsp*. Another useful service (á la Unix

```
        handler : System → (PID, Re ()) → R ()
        handler (w, mq, s, o, g) (i, t) =
        case t of
(a.)      (D v)            → rrobin (w, mq, s, o, g)
(b.)      (P(Cont, r))     → Pause (r Ack >>=_St λκ. return_St (next (i, κ))
            where
              next t = rrobin (w++[t], mq, s, o, g)
(c.)      (P(Sleep_q, r))  → Pause (return_St next)
            where
              next = rrobin (w++[(i, P(Cont, r))], mq, s, o, g)
(d.)      (P(Fork_q p, r)) → Pause (return_St next)
            where
              parent = (i, cont r Ack)
              child  = (g, P⟦p⟧)
              next   = rrobin (w++[parent, child], mq, s, o, g + 1)
(e.)      (P(Bcst_q m, r)) → Pause (return_St next)
            where
              next = rrobin (w++[(i, cont r Ack)], mq++[m], s, o, g)
(f.)      (P(Rcv_q, r)) | (mq == []) → Pause (return_St next)
            where
              next = rrobin (w++[(i, P(Rcv_q, r))], [], s, o, g)
(g.)      (P(Rcv_q, r)) | otherwise  → Pause (return_St next)
            where
              next = rrobin (w++[(i, cont r (Rcv_r m))], ms, s, o, g)
              m    = head mq
              ms   = tail mq
(h.)      (P(Prnt_q msg, r)) → Pause (return_St next)
            where
              next = rrobin (w++[(i, P(Cont, r))], mq, s, o++msg, g)
(i.)      (P(P_q, r))        → Pause (return_St next)
            where
              next     = if s > 0 then goahead else tryagain
              goahead  = rrobin (w++[(i, P(Cont, r))], mq, s − 1, o, g)
              tryagain = rrobin (w++[(i, P(P_q, r))], mq, s, o, g)
(j.)      (P(V_q, r))        → Pause (return_St next)
            where
              next = rrobin (w++[(i, cont r Ack)], mq, s + 1, o, g)
(k.)      (P(PID_q, r))      → Pause (return_St next)
            where
              next = rrobin (w++[(i, cont r (PID_r i))], mq, s, o, g)
(l.)      (P(Kill_q j, r))   → Pause (return_St next)
            where
              next             = rrobin (wl′, mq, s, o, g)
              wl′              = filter (exit j) (w++[(i, cont r Ack)])
              exit i (pid, t)  = pid /= i
        cont       : (Rsp → St (Re a)) → (Rsp → Re a)
        cont r rsp = P (Cont, λAck. r rsp)
```

Fig. 2. The Request Handler

`fork` system call) would include a response $\mathit{Fork}_r\ \mathit{Bool}$ in *Rsp* to distinguish child and parent processes.

Asynchronous Message Passing. For a thread to broadcast m (e.), the message is simply appended to the message queue. When a Rcv_q signal occurs and the message queue is empty, then the thread must wait (f.) and so is put back on the thread list. Note that, rather than busy-waiting for a message, the message queue could contain a "blocked waiting list" for threads waiting for the arrival of messages, and, in that scenario, the handler could wake a blocked process whenever a message arrives. If there is a message m in the message queue, then it is passed to the thread (g.).

Printing. When a print request ($\mathit{Prnt}_q\ \mathit{msg}$) is signaled (h.), then the string *msg* is appended to the output channel *out* and the rest of the thread, $P(\mathit{Cont}, r)$, is passed to the scheduler. An alternative could use the "interactive output" monad formulation for R: $R\ A\ =\ \mu X.\,(A\ +\ (\mathit{String}\times S\ X))$ instead of encoding the output channel as the string o.

Synchronization Primitives. Requesting the system semaphore (i.) will succeed if $s > 0$, in which case the requesting thread will continue with the semaphore decremented; if $\not> 0$, the requesting thread will suspend. These possible outcomes are bound to *goahead* and *tryagain* in the following *handler* clause, and *handler* chooses between them based on the current value of s: Note that this implementation uses busy waiting merely for simplicity's sake. One could easily implement more efficient strategies by including a queue of waiting threads with the semaphore. A thread may release the semaphore (j.) without blocking. Note this semaphore is *general* rather than *binary*, meaning that the counter s may have as its value any non-negative integer rather than just 0 or 1.

Process Id Request. A thread may request its identifier i (k.), which is simply passed to it in $\mathit{cont}\ r\ (\mathit{PID}_r\ i)$.

Preemption. One thread may preempt another by sending it a kill signal reminiscent of the Unix (`kill -9`) command; this is implemented by the *handler* declaration at line (l.). Upon receiving the signal $\mathit{Kill}_q\ j$, the thread with process identifier j (if one exists) is removed from the waiting list using the Haskell built-in function $\mathit{filter} : (a{\to}\mathit{Bool}){\to}[a]{\to}[a]$. In a call ($\mathit{filter}\ b\ l$), *filter* returns those elements of list l on which b is true (in order of their occurrence in l).

Time Behavior of $\texttt{>>=}_R$ and $\texttt{>>=}_{Re}$. Because the bind operations for R and *Re* are both $O(n)$ in the size of their first arguments, one can write programs that, through the careless use of the bind, end up with quadratic (or worse) time complexity. Note, however, the kernel avoids this entirely by relying on co-recursion in the definition of *handler*.

Executing the kernel. An R computation may be projected to *St* with:

$$
\begin{aligned}
&\mathit{run} : R\,a \to \mathit{St}\,a\\
&\mathit{run}\,(\mathit{Done}\,v) \ = \mathbf{return}_{\mathit{St}}\ \ v\\
&\mathit{run}\,(\mathit{Pause}\,\varphi) = \varphi\ \texttt{>>=}_{\mathit{St}}\ \mathit{run}
\end{aligned}
$$

Running the kernel on initial processes $p_1, \ldots, p_n$ is accomplished with

$$run\ (rrobin\ ([\mathcal{P}[\![p_1]\!], \ldots, \mathcal{P}[\![p_n]\!]], [], 1, "", 0))$$

Sample executions are provided in the code base [1].

6 Conclusions

As of this writing, resumptions as a model of concurrency have been known for thirty years and, in monadic form, for almost twenty. Yet, unlike other techniques and structures from language theory (e.g., continuations, type systems, etc.), resumptions have evidently never found wide-spread acceptance in programming practice. This is a shame, because resumptions—especially in monadic form—are a natural and beautiful organizing principle for concurrent applications: they capture exactly what one needs to write and think about multitasking programs—and no more! Resumptions capture precisely the intuition that threads are potentially infinite sequences of atoms interacting according to some discipline. The framework presented here has been applied in a number of diverse settings and expresses a broad sampling of concurrent behaviors. This is solid evidence that resumptions express the true essence and structure of multitasking computation.

Although the kernel is, of necessity, simple, it does demonstrate both the wide scope of concurrent behaviors expressible with resumption monads and the ease with which such behaviors may be expressed. To be sure, more efficient implementations and realistic features may be devised (e.g., by eliminating busy-waiting). As each of the three monads may be generalized as monad transformers, instances of this kernel inherit the software engineering benefits of monad transformers that one would expect—namely, modularity, extensibility, and reusability. Such kernel instances may be extended by either application of additional monad transformers or through refinements to the resumption monad transformers themselves. Such refinements are typically straightforward; to add a new service to the kernel of Section 5, for example, one merely extends the *Req* and *Rsp* types with a new request and response and adds a corresponding *handler* definition. The kernel in Section 5 may, in fact, be viewed as the result of multiple extensions to a core "basic operation" kernel (i.e., one having only a *Cont* request).

The framework developed here has been applied to such seemingly diverse purposes as language-based security [31] and systems biology [32]; each of these applications is an instance of this framework. The difference is evident in the request and response data types *Req* and *Rsp*. Recent work in systems biology applies process calculae as a descriptive mechanism for biological processes and structures [33]. As an alternative foundation, the resumption-monadic framework discussed here has been applied to the formal modeling of the life cycles of autonomous, intercommunicating cellular systems [32]. Each cell has some collection of possible actions describing its behavior with respect to itself and its environment. The actions of the photosynthetic bacterium *Rhodobacter Sphaeroides* are reflected in the request and response types:

data *Req* = *Cont* | *Divide* | *Die* | *Sleep* | *Grow* |*LightConcentration*
data *Rsp* = *Ack* | *LightConcRsp Float*

Each cell may undergo physiological change (e.g., cell division) or react to its immediate environment (e.g., to the concentration of light in its immediate vicinity). The kernel instance here also maintains the physical integrity of the model.

The kernel presented here confronts many "impurities" considered difficult to accommodate within a pure, functional setting—concurrency, state, and i/o—which are all members of the so-called "Awkward Squad" [34]. In Haskell, these real world impurities are swept, in the memorably colorful words of Simon Peyton Jones, into a "giant sin-bin" called the *IO* monad[5]. But is *IO* truly a monad (i.e., does it obey the monad laws)? All of these impurities have been handled individually via various monadic constructions (consider the manifestly incomplete list [2, 35]) and the current approach combines some of these constructions into a single monad. While it is not the intention of the current work to model the awkward squad as it occurs in Haskell, the techniques and structures presented here point the way towards such models.

References

1. Harrison, W.: (The Essence of Multithreading Codebase) Available from `www.cs.missouri.edu/~harrison/EssenceOfMultitasking`.
2. Moggi, E.: An Abstract View of Programming Languages. Technical Report ECS-LFCS-90-113, Department of Computer Science, Edinburgh University (1990)
3. Papaspyrou, N., Maćoš, D.: A Study of Evaluation Order Semantics in Expressions with Side Effects. Journal of Functional Programming **10**(3) (2000) 227–244
4. Hancock, P., Setzer, A.: Interactive programs in dependent type theory. In: Proceedings of the 14th Annual Conference of the EACSL on Computer Science Logic, London, UK, Springer-Verlag (2000) 317–331
5. Papaspyrou, N.: A Resumption Monad Transformer and its Applications in the Semantics of Concurrency. In: Proceedings of the 3rd Panhellenic Logic Symposium. (2001) An expanded technical report is available from the author by request.
6. Krstic, S., Launchbury, J., Pavlovic, D.: Categories of processes enriched in final coalgebras. In: Proceedings of FOSSACS'01. (2001) 303–317
7. Jacobs, B., Poll, E.: Coalgebras and Monads in the Semantics of Java. Theoretical Computer Science **291**(3) (2003) 329–349
8. Plotkin, G.: A Powerdomain Construction. SIAM Journal of Computation **5**(3) (1976) 452–487
9. Henderson, P.: Purely functional operating systems. In: Functional Programming and Its Applications: an Advanced Course. Cambridge Univ. Press (1982) 177–191
10. Stoye, W.: Message-based Functional Operating Systems. Science of Computer Programming **6**(3) (1986) 291–311
11. Turner, D.: An Approach to Functional Operating Systems. In: Research Topics in Functional Programming. (1990) 199–217

[5] See his slides "*Lazy functional programming for real: Tackling the Awkward Squad*" available at `research.microsoft.com/Users/simonpj/papers/marktoberdorf/Marktoberdorf.ppt`.

12. Cupitt, J.: The Design and Implementation of an Operating System in a Functional Language. PhD thesis, Computing Laboratory, Univ. Kent (1992)
13. Carter, D.: Deterministic Concurrency. PhD thesis, Department of Computer Science, University of Bristol (1994)
14. Spiliopoulou, E.: Concurrent and Distributed Functional Systems. Technical Report CS-EXT-1999-240, Univ. Bristol (1999)
15. Launchbury, J., Peyton Jones, S.: Lazy functional state threads. In: Proceedings of PLDI'94. (1994) 24–35
16. Claessen, K.: A poor man's concurrency monad. Journal of Functional Programming **9**(3) (1999) 313–323
17. Wand, M.: Continuation-based multiprocessing. In: Proceedings of the 1980 ACM Conference on LISP and Functional Programming, ACM Press (1980) 19–28
18. Flatt, M., Findler, R., Krishnamurthi, S., Felleisen, M.: Programming languages as operating systems (or revenge of the son of the lisp machine). In: Proceedings of ICFP'99. (1999) 138–147
19. van Weelden, A., Plasmeijer, R.: Towards a strongly typed functional operating system. In: Proceedings of IFL'02. Volume 2670 of LNCS. (2002)
20. Filinski, A.: Representing monads. In: Proceedings of POPL '94. (1994)
21. Ganz, S., Friedman, D., Wand, M.: Trampolined style. In: Proceedings of ICFP'99. (1999) 18–27
22. Harper, R., Lee, P., Pfenning, F.: The Fox project: Advanced language technology for extensible systems. Technical Report CMU-CS-98-107, CMU (1998)
23. Alexander, D., Arbaugh, W., Hicks, M., Kakkar, P., Keromytis, A., Moore, J., Gunder, C., Nettles, S., Smith, J.: The switchware active network architecture. IEEE Network (1998)
24. Schmidt, D.: Denotational Semantics: A Methodology for Language Development. Allyn and Bacon, Boston (1986)
25. Bakker, J.d., Vink, E.d.: Control Flow Semantics. Foundations of Computing Series. The MIT Press (1996)
26. Espinosa, D.: Semantic Lego. PhD thesis, Columbia University (1995)
27. Filinski, A.: Representing layered monads. In: Proceedings of POPL '99. (1999)
28. Liang, S.: Modular Monadic Semantics and Compilation. PhD thesis, Yale University (1998)
29. Peyton Jones, S., ed.: Haskell 98 Language and Libraries, the Revised Report. Cambridge University Press (2003)
30. Roscoe, W.: Theory and Practice of Concurrency. Prentice-Hall (1998)
31. Harrison, W., Hook, J.: Achieving information flow security through precise control of effects. In: 18th IEEE Computer Security Foundations Workshop (CSFW05). (2005)
32. Harrison, W., Harrison, R.: Domain specific languages for cellular interactions. In: Proceedings of the 26th Annual IEEE International Conference on Engineering in Medicine and Biology. (2004)
33. A. Regev, E. Panina, W.S.L.C.E.S.: Bioambients: An abstraction for biological compartments. Theoretical Computer Science **325**(1) (2004) 141–167
34. Peyton Jones, S.: Tackling the Awkward Squad: Monadic Input/Output, Concurrency, Exceptions, and Foreign-language Calls in Haskell. In: Engineering Theories of Software Construction. (2000) 47–96
35. Peyton Jones, S., Wadler, P.: Imperative functional programming. In: Proceedings of POPL'93. (1993) 71–84

The Substitution Vanishes

Armin Kühnemann and Andreas Maletti*

Institute for Theoretical Computer Science, Department of Computer Science
Dresden University of Technology,
D–01062 Dresden, Germany
{kuehne, maletti}@tcs.inf.tu-dresden.de

Abstract. Accumulation techniques were invented to transform functional programs, which intensively use append functions (like inefficient list reversal), into more efficient programs, which use accumulating parameters instead (like efficient list reversal). In this paper we present a generalized and automatic accumulation technique that also handles programs operating with unary functions on arbitrary tree structures and employing substitution functions on trees which may replace different designated symbols by different trees. We show that this transformation does not deteriorate the efficiency with respect to call-by-need reduction.

1 Introduction

The sequence of trees in Figure 1 illustrates the stepwise growth of a tree, where in every step in parallel every occurrence of a symbol A (and B, respectively) is substituted by a tree $(D\ A)$ (and $(T\ A\ B\ A)$, respectively).

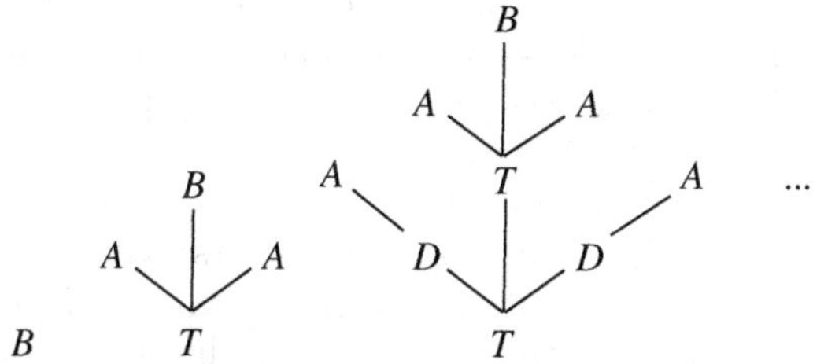

Fig. 1. Stepwise growth of a tree

In a functional program p_{non} (formulated e.g. in Haskell), this substitution can be defined by a ternary function g that takes the "previous tree" and the two kinds of "fresh branches" $(D\ A)$ and $(T\ A\ B\ A)$ as arguments. Additionally, a unary function f generates as many nested calls of g as the argument of f

* Research of this author supported by DFG under grants GK 334 and KU 1290/2-4.

M. Johnson and V. Vene (Eds.): AMAST 2006, LNCS 4019, pp. 173–188, 2006.

indicates, where natural numbers are represented by a nullary Z and a unary S, i.e. the *initial expression* $(f\ (S^n\ Z))$ generates n nested calls of g.[1]

$$
\begin{array}{ll}
f\ Z & = B \\
f\ (S\ x_1) & = g\ (f\ x_1)\ (D\ A)\ (T\ A\ B\ A) \\
g\ A\ y_1\ y_2 & = y_1 \\
g\ B\ y_1\ y_2 & = y_2 \\
g\ (D\ x_1)\ y_1\ y_2 & = D\ (g\ x_1\ y_1\ y_2) \\
g\ (T\ x_1\ x_2\ x_3)\ y_1\ y_2 & = T\ (g\ x_1\ y_1\ y_2)\ (g\ x_2\ y_1\ y_2)\ (g\ x_3\ y_1\ y_2)
\end{array}
$$

Unfortunately, p_{non} has cubic time-complexity in the input number n, since g has to process n *intermediate results* with sizes $1^2, 2^2, \ldots, n^2$, respectively (though they are not explicitly constructed in a call-by-need evaluation).

Therefore we would prefer the following program p_{acc} which has linear time-complexity in n to evaluate the (modified) initial expression $(f\ (S^n\ Z)\ A\ B)$:

$$
\begin{array}{ll}
f\ Z\ y_1\ y_2 & = y_2 \\
f\ (S\ x_1)\ y_1\ y_2 & = f\ x_1\ (D\ y_1)\ (T\ y_1\ y_2\ y_1)
\end{array}
$$

Since p_{acc} uses its second and third argument to accumulate step by step a "D-branch" and an output tree, respectively, we call p_{acc} an *accumulative* program, whereas we call p_{non} *non-accumulative.* Techniques which transform non-accumulative into accumulative programs are called *accumulation.*

In the case that substitutions on tree structures are restricted to append functions on list structures, there is a long history of research on accumulation: Already in [6] it is shown in the context of transforming programs into iterative (tail-recursive) form, how non-accumulative programs can be transformed (non-automatically) into their accumulative versions. In [3, 18] a similar technique for linear recursive functions is presented. Other non-automatic realizations of accumulation are given e.g. in [4, 14]. Finally, the transformation of [25] is fully automatic and is accompanied by an efficiency analysis. The crucial laws, on which the transformation of [25] is based, can be found in our paper in a similar form. All mentioned techniques essentially rely on the properties of the monoid of lists with append. This fact is detailed in [5].

Our automatic transformation technique is more general in two aspects: we consider (i) arbitrary tree structures (instead of lists) as input and output, and (ii) substitutions on trees (instead of append) which additionally may replace different designated symbols by different trees. On the other hand, our technique is restricted to unary functions (apart from substitutions), though also in [25] the only example program involving a non-unary function could not be optimized. Hence the scope of our technique includes the unary functions in the examples

[1] Since there is only one unary input symbol S for f, the actual parameters $(D\ A)$ and $(T\ A\ B\ A)$ of g are unique. Hence, an alternative version of g could avoid its formal parameters y_1 and y_2 and directly use $(D\ A)$ and $(T\ A\ B\ A)$ in its A- and B-rules, respectively. In a more elaborate example with different unary input symbols for f the actual parameters of g may be different and hence the formal parameters are essential. For convenience we avoided to blow up our example into this direction.

of [25] (in particular, inefficient list reversal). Moreover, restricting recursive calls of the unary functions to be primitive-recursive will guarantee that in contrast to [25] no substitutions appear in transformed programs anymore. Our efficiency result is based on exactly this fact.

For this purpose, we view functional programs like p_{non} as special 2-*modular tree transducers* [10]. Every function in module 1 (like f in p_{non}) is unary and is defined by a case analysis on the root symbol c of its argument t. The right-hand side of the equation for f and c may contain (primitive-recursive) calls of functions in module 1 on subtrees of t and arbitrary calls of the (only) function in module 2. The function in module 2 (like g in p_{non}) is a *substitution function*, i.e. designated nullary *substitution constructors* are replaced by parameters and other constructors are left unchanged. In [16] it was shown, how such programs can be transformed into *macro tree transducers* [8, 9], in which every function (like f in p_{acc}) may have arbitrary rank and is defined by a case analysis on the root symbol c of its first argument t. The right-hand side of the equation for f and c may contain (nested) recursive function calls on subtrees of t.

The accumulation technique of [16] is divided into three indirect transformation steps which are mainly based on a composition result [9, 15] for *top-down tree transducers* [19, 21] with macro tree transducers and on the associativity of substitution functions. Although the resulting programs avoid substitution functions, they are not always more efficient than the original programs. In [16] and in the present paper the efficiency is measured in the number of performed call-by-need reduction steps. This point of view, which neglects the actual complexity of rule applications, is also taken in, e.g., [20, 23, 24].

In [16] also the reverse transformation is presented. Both transformations together induce that the classes of macro tree transducers and of the special 2-modular tree transducers have the same computational power. Although the reverse transformation deteriorates in general the efficiency, it also has practical relevance: In [12] it is extended to a *deaccumulation* technique which is useful to improve the automatic verification of functional (and even imperative) programs.

In [17] the deficiencies of the accumulation technique in [16] were solved by presenting a direct transformation which additionally employs let-expressions to avoid causes of inefficiency. Moreover, it was shown in [17] that the transformation does not deteriorate the efficiency. To this end, a call-by-need reduction on term graphs was defined and compared for the original and resulting program. The efficiency result is based on the fact that the number of applications of functions in module 1 of the original program equals the number of function applications in the resulting program. Hence, the applications of the substitution function in module 2 of the original program are saved!

We simplify the presentation of [17] by avoiding an explicit call-by-need reduction and by adopting a technique of [20, 23, 24], where function applications (in [23, 24] for (compositions of) macro tree transducers) additionally produce special "ticking symbols" in order to make the number of performed (call-by-name) reduction steps visible in the output. Instead of a call-by-need reduction relation on term graphs which (implicitly) uses sharing to avoid that unevaluated function arguments are copied, we use a nondeterministic reduction relation on expressions with an explicit denotation for sharing (cf., e.g., [2, 1]). Unfortunately, this *explicit sharing* does not prevent that our nondeterministic reduction relation creates a shared subexpression e such that e contains function applications and e is not relevant in the overall expression (in the sense that call-by-need reduction would delete all references to e). To avoid that ticking symbols which are generated by e are counted (and thus reduction steps needed to evaluate e), we additionally use a *counting function* which takes care of such nonrelevant subexpressions. Hence the concepts of *explicit sharing* and the *counting function* provide a new technique to count call-by-need reduction steps.

2 Preliminaries

We denote the set of natural numbers including 0 by $\mathbb{N}$ and the set $\mathbb{N} - \{0\}$ by $\mathbb{N}_+$. For every $m \in \mathbb{N}$, the set $\{1, \ldots, m\}$ is denoted by $[m]$. The cardinality of a set K is denoted by $card(K)$. We use the sets $X = \{x_0, x_1, x_2, x_3, \ldots\}$, $Y = \{y_1, y_2, y_3, \ldots\}$, and $Z = \{z_1, z_2, z_3, \ldots\}$ of *variables*. For every $n \in \mathbb{N}$, let $X_n = \{x_1, \ldots, x_n\}$, $Y_n = \{y_1, \ldots, y_n\}$, and $Z_n = \{z_1, \ldots, z_n\}$. Let $\Rightarrow$ be a binary relation on a set K and $n \in \mathbb{N}$. Then, $\Rightarrow^n$ denotes the n-fold composition and $\Rightarrow^*$ the transitive, reflexive closure of $\Rightarrow$. If $k \Rightarrow^* k'$ for $k, k' \in K$ and there is no $k'' \in K$ such that $k' \Rightarrow k''$, then k' is called a *normal form of k with respect to* $\Rightarrow$, which is denoted by $nf(\Rightarrow, k)$, if it exists and if it is unique.

A *ranked alphabet* is a pair $(C, rank)$ where C is a finite set and *rank* is a mapping which associates with every symbol $c \in C$ a natural number called the *rank* of c. We simply write C instead of $(C, rank)$ and assume *rank* as implicitly given. The set of symbols of C with rank n is denoted by $C^{(n)}$ and if $c \in C^{(n)}$, we also use the notation $c^{(n)}$. The set of *trees* (or *terms*) *over C indexed by* (*a set of variables*) U, denoted by $T_C(U)$, is the smallest subset $T \subseteq (C \cup U \cup \{(,)\})^*$ such that $U \subseteq T$ and for every $c \in C^{(n)}$ with $n \in \mathbb{N}$ and $t_1, \ldots, t_n \in T$: $(c\ t_1 \ldots t_n) \in T$. If $c \in C^{(0)}$, we write just c instead of (c). The set $T_C(\emptyset)$ is abbreviated by T_C. If R is the set of rules of a term rewriting system, then $\Rightarrow_R$ denotes the (nondeterministic) reduction relation induced by R. If there is at most one occurrence of a variable v in a term t, then we call t *linear* in v.

For a term t, pairwise distinct variables $v_1, \ldots, v_n$, and terms $t_1, \ldots, t_n$, we denote by $t[v_1/t_1, \ldots, v_n/t_n]$ the term that is obtained from t by substituting for every $i \in [n]$ every occurrence of v_i in t by t_i. We abbreviate $[v_1/t_1, \ldots, v_n/t_n]$ by $[v_i/t_i]$, if the involved variables and terms are clear from the context. We use a linear, "substitution-like" notation for term graphs to express the sharing of subgraphs: $e[z_1 \rightsquigarrow e_1, \ldots, z_n \rightsquigarrow e_n]$ denotes a term graph, in which for every occurrence of z_i in the subgraph denoted by e there is a directed edge from the direct ancestor node of z_i to the root node of the subgraph denoted by e_i.

For the rest of the paper, let $n \in \mathbb{N}_+$.

Definition 1. Let C be a ranked alphabet and $U \in \{Z_n, \emptyset\}$. The set $E_{C,n}(U)$ of *C-expressions with sharing* (*and free variables of U*) is defined by:

- For every $z_j \in U$: $z_j \in E_{C,n}(U)$.
- For every $c \in C^{(k)}$ and $e_1, \ldots, e_k \in E_{C,n}(U)$: $(c\, e_1 \ldots e_k) \in E_{C,n}(U)$.
- For every $e_1, \ldots, e_n \in E_{C,n}(U)$ and $e \in E_{C,n}(Z_n)$: $e[z_1 \rightsquigarrow e_1, \ldots, z_n \rightsquigarrow e_n] \in E_{C,n}(U)$.

The set $E_{C,n}$ of *C-expressions with sharing* is defined as $E_{C,n}(\emptyset)$. □

Note that a C-expression with sharing $e \in E_{C,n}(U)$ can be considered as a tree $e \in T_{C'}(U)$ where $C' = C \cup \{(\cdot[z_1 \rightsquigarrow \cdot, \ldots, z_n \rightsquigarrow \cdot])^{(n+1)}\}$ is the ranked alphabet obtained from C by adding a new $(n+1)$-ary symbol. Thus we can employ the notions and notations, which we introduced for trees, also for C-expressions.

Example 2. Let $C = \{A^{(0)}, B^{(0)}, D^{(1)}, T^{(3)}\}$. Then,

$$
\begin{array}{ll}
(T\ z_1\ z_2\ z_1) & \in E_{C,2}(Z_2),\\
(T\ z_1\ z_2\ z_1)[z_1 \rightsquigarrow (D\ z_1), z_2 \rightsquigarrow (T\ z_1\ z_2\ z_1)] & \in E_{C,2}(Z_2),\\
(T\ z_1\ z_2\ z_1)[z_1 \rightsquigarrow (D\ z_1), z_2 \rightsquigarrow (T\ z_1\ z_2\ z_1)] & \\
\qquad [z_1 \rightsquigarrow (D\ A), z_2 \rightsquigarrow (T\ A\ B\ A)] & \in E_{C,2},
\end{array}
$$

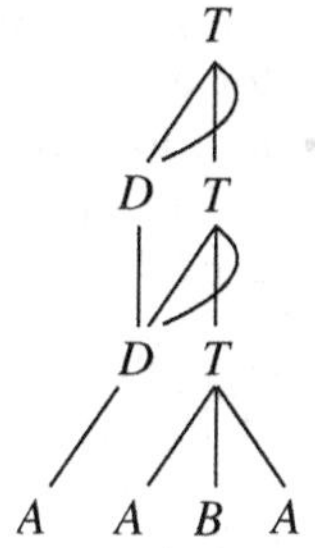

and the latter represents the depicted term graph. □

If $e_1, \ldots, e_n$ are clear from the context, then we abbreviate an expression of the form $e[z_1 \rightsquigarrow e_1, \ldots, z_n \rightsquigarrow e_n]$ by $e[z_i \rightsquigarrow e_i]$.

In the following we define a function on C-expressions with sharing, which constructs trees by unfolding all sharings.

Definition 3. Let C be a ranked alphabet and $U \in \{Z_n, \emptyset\}$. The function $\underline{tree} : E_{C,n}(U) \to T_C(U)$ is defined as follows:

- For every $z_j \in U$: $\underline{tree}(z_j) = z_j$.
- For every $c \in C^{(k)}$ and $e_1, \ldots, e_k \in E_{C,n}(U)$: $\underline{tree}(c\, e_1 \ldots e_k) = (c\ \underline{tree}(e_1) \ldots \underline{tree}(e_k))$.
- For every $e_1, \ldots, e_n \in E_{C,n}(U)$ and $e \in E_{C,n}(Z_n)$: $\underline{tree}(e[z_i \rightsquigarrow e_i]) = \underline{tree}(e)[z_i/\underline{tree}(e_i)]$. □

We call $z_i \in Z_n$ a *free occurrence in* $e \in E_{C,n}(Z_n)$, if z_i occurs in $\underline{tree}(e)$. Note that this clarifies the scope of a sharing. The scope of $z_1, \ldots, z_n$ in an expression $e[z_i \rightsquigarrow e_i]$ is limited to the free occurrences of $z_1, \ldots, z_n$ in e.

3 Nonaccumulating and Accumulating Tree Transducers

First we define nonaccumulating tree transducers as functional source language. Nonaccumulating tree transducers are special 2-modular tree transducers [10, 16].

Definition 4. An *n-nonaccumulating tree transducer* (for short *n-ntt*) is a tuple $M = (F, Sub, C, \Pi, R_1, R_2, r_{in})$, where

- F is a ranked alphabet (of *function symbols*) with $F = F^{(1)}$,
- $Sub = \{sub^{(n+1)}\}$ (*substitution function*),

– C is a ranked alphabet (of *constructors*),
– $\Pi = \{\Pi_1, \ldots, \Pi_n\} \subseteq C^{(0)}$ with $card(\Pi) = n$ (the *substitution constructors*)

such that F, Sub, and C are pairwise disjoint,

– R_1 is a set of *rules* such that for every $f \in F$ and $c \in C^{(k)}$ there is exactly one rule $f\ (c\, x_1 \ldots x_k) = rhs_{R_1,f,c}$ with $rhs_{R_1,f,c} \in RHS(F, Sub, C, X_k)$, where for every $k \in I\!N$, $RHS(F, Sub, C, X_k)$ is the smallest set RHS such that
  - for every $f \in F$ and $i \in [k]$: $(f\ x_i) \in RHS$,
  - for every $r_0, \ldots, r_n \in RHS$: $(sub\ r_0 \ldots r_n) \in RHS$, and
  - for every $c \in C^{(l)}$ and $r_1, \ldots, r_l \in RHS$: $(c\, r_1 \ldots r_l) \in RHS$,
– R_2 is a set of *rules* such that
  - for every $j \in [n]$ there is the rule $sub\ \Pi_j\ y_1 \ldots y_n = y_j$
  - and for every $c \in (C - \Pi)^{(k)}$ there is the rule $sub\ (c\, x_1 \ldots x_k)\ y_1 \ldots y_n = c\ (sub\ x_1\ y_1 \ldots y_n) \ldots (sub\ x_k\ y_1 \ldots y_n)$,
– $r_{in} \in RHS(F, Sub, C, X_1)$ is the *initial right-hand side*. □

Since every function is defined by recursion on its first argument (i.e., the only argument in case of F-functions), this argument is called *recursion argument.* The other arguments are called *context arguments.* The set RHS formalizes the description of right-hand sides found in the introduction. The initial right-hand side r_{in} serves as *call pattern* for the n-ntt, where x_1 acts as a placeholder for the actual input tree. Note that the concept of n-ntts (with one substitution function of rank $n+1$) can easily be generalized to a model with several substitution functions. This, however, does not increase the computational power.

In the following examples we will avoid rules, which are never used.

Example 5. $M_{non} = (F, Sub, C, \Pi, R_1, R_2, r_{in})$ is a 2-ntt where $F = \{f\}$, $Sub = \{g^{(3)}\}$, $C = \{S^{(1)}, Z^{(0)}, A^{(0)}, B^{(0)}, D^{(1)}, T^{(3)}\}$, $\Pi_1 = A$, $\Pi_2 = B$, R_1 and R_2 contain the f-rules and g-rules, respectively, of p_{non}, and $r_{in} = (f\ x_1)$. □

Now we present n-ntts with sharings as abstractions for our functional source programs. In contrast to functional programs, where in a call-by-need reduction the sharing of expressions which are bound to variables of rules is performed implicitly, in n-ntts with sharings the sharing is performed explicitly, whenever there is the risk to copy unevaluated expressions (cf., e.g., [2, 1]). This concerns only the context arguments of substitution functions (since other arguments are not copied or are constructor trees). To denote explicit sharing in a right-hand side of a rule or in a sentential form, we also use expressions with sharing. Thus, because of possibly nested substitution functions during an evaluation, also expressions with sharing may occur in the recursion argument of substitution functions. Hence they must be handled by a special rule. Actually, n-ntts with sharings could be considered as special "2-modular tree-to-graph transducers". See [10, 11] for the concepts of modular tree transducers and top-down tree-to-graph transducers, respectively. Note that the additional sharing mechanism does not change the computational power of n-ntts, but may improve efficiency.

Definition 6. An *n-nonaccumulating tree transducer with sharings* (for short *n-sntt*) is a tuple $M = (F, Sub, C, \Pi, R_1, R_2, r_{in})$, where

- F, Sub, C, Π, R_1, and r_{in} are defined as in Definition 4,
- R_2 is a set of *rules* such that
  - for every $j \in [n]$ there is the rule $\quad sub\ \Pi_j\ y_1 \dots y_n = y_j$,
  - for every $c \in (C - \Pi)^{(k)}$ there is the rule[2]
    $sub\ (c\ x_1 \dots x_k)\ y_1 \dots y_n$
    $= (c\ (sub\ x_1\ z_1 \dots z_n) \dots (sub\ x_k\ z_1 \dots z_n))[z_i \rightsquigarrow y_i]$,
  - and there is the rule[3]
    $sub\ x_0[z_i \rightsquigarrow x_i]\ y_1 \dots y_n = x_0[z_i \rightsquigarrow (sub\ x_i\ z_1 \dots z_n)][z_i \rightsquigarrow y_i]$. □

Note that in the last rule in the previous definition *sub* walks into $x_1, \dots, x_n$, but not into x_0. This is due to the fact that every instantiation of $x_0[z_i \rightsquigarrow x_i]$ was generated by an inner occurrence of *sub* which already handled the substitution constructors in x_0. Moreover, it is easily seen that the inner occurrence of *sub* does not introduce substitution constructors in x_0 because only calls of the form $(sub\ x_i\ z_1 \dots z_n)$ can occur in x_0. We further note that an application of the last rule represents a short cut, since a call-by-need reduction on term graphs would (i) walk stepwise through the expression bound to x_0 and would (ii) end up with different occurrences of *sub* at different occurrences of a z_i (thus performing several runs on the expression bound to x_i).

Example 7. $\tilde{M}_{non} = (F, Sub, C, \Pi, R_1, R_2, r_{in})$ is a 2-sntt, where $F = \{f\}$, $Sub = \{g^{(3)}\}$, $C = \{S^{(1)}, Z^{(0)}, A^{(0)}, B^{(0)}, D^{(1)}, T^{(3)}\}$, $\Pi_1 = A$, $\Pi_2 = B$, R_1 contains the f-rules of p_{non} and R_2 contains rules

$$\begin{aligned}
&g\ A\ y_1\ y_2 &&= y_1\\
&g\ B\ y_1\ y_2 &&= y_2\\
&g\ (D\ x_1)\ y_1\ y_2 &&= D\ (g\ x_1\ y_1\ y_2)\\
&g\ (T\ x_1\ x_2\ x_3)\ y_1\ y_2 &&= (T\ (g\ x_1\ z_1\ z_2)\ (g\ x_2\ z_1\ z_2)\ (g\ x_3\ z_1\ z_2))[z_i \rightsquigarrow y_i]\\
&g\ x_0[z_i \rightsquigarrow x_i]\ y_1\ y_2 &&= x_0[z_i \rightsquigarrow (g\ x_i\ z_1\ z_2)][z_i \rightsquigarrow y_i],
\end{aligned}$$

and $r_{in} = (f\ x_1)$. Let $R = R_1 \cup R_2$. Then,

$$\begin{aligned}
f\ (S^3\ Z) &\Rightarrow_R^5 g\ (g\ (T\ A\ B\ A)\ (D\ A)\ (T\ A\ B\ A))\ (D\ A)\ (T\ A\ B\ A)\\
&\Rightarrow_R^4 g\ (T\ z_1\ z_2\ z_1)[z_1 \rightsquigarrow (D\ A), z_2 \rightsquigarrow (T\ A\ B\ A)]\ (D\ A)\ (T\ A\ B\ A)\\
&\Rightarrow_R (T\ z_1\ z_2\ z_1)[z_1 \rightsquigarrow (g\ (D\ A)\ z_1\ z_2), z_2 \rightsquigarrow (g\ (T\ A\ B\ A)\ z_1\ z_2)]\\
&\qquad [z_1 \rightsquigarrow (D\ A), z_2 \rightsquigarrow (T\ A\ B\ A)]\\
&\Rightarrow_R^6 (T\ z_1\ z_2\ z_1)[z_1 \rightsquigarrow (D\ z_1), z_2 \rightsquigarrow (T\ z_1\ z_2\ z_1)[z_1 \rightsquigarrow z_1, z_2 \rightsquigarrow z_2]]\\
&\qquad [z_1 \rightsquigarrow (D\ A), z_2 \rightsquigarrow (T\ A\ B\ A)].
\end{aligned}$$
□

Our main transformation will deliver accumulating tree transducers with sharings, which could be considered as special "macro tree-to-graph transducers". See [9, 11] for the concepts of macro tree transducers and top-down tree-to-graph transducers, respectively.

[2] If c is nullary or unary, then the explicit sharing will be avoided in examples.

[3] If $n = 1$, then the explicit sharing $[z_i \rightsquigarrow y_i]$ could be avoided.

Definition 8. An *n-accumulating tree transducer with sharings* (for short *n-satt*) is a tuple $M = (F, C, R, r_{in})$, where

- F is a ranked alphabet (of *function symbols*) with $F = F^{(n+1)}$,
- C is a ranked alphabet (of *constructors*), such that F and C are disjoint,
- R is a set of *rules* such that for every $f \in F$ and $c \in C^{(k)}$ there is exactly one rule $f\ (c\, x_1 \dots x_k)\ y_1 \dots y_n = rhs_{R,f,c}$ with $rhs_{R,f,c} \in RHS'(F, C, X_k, Y_n)$, where for every $j \in [n]$, the right-hand side $rhs_{R,f,c}$ is linear in y_j, and for every $k \in I\!N$ and $U \in \{Y_n, Z_n, \emptyset\}$, the set $RHS'(F, C, X_k, U)$ is the smallest set such that
  - for every $f \in F$, $i \in [k]$, and $r_1, \dots, r_n \in RHS'(F, C, X_k, U)$: $(f\ x_i\ r_1 \dots r_n) \in RHS'(F, C, X_k, U)$,
  - for every $c \in C^{(l)}$ and $r_1, \dots, r_l \in RHS'(F, C, X_k, U)$: $(c\, r_1 \dots r_l) \in RHS'(F, C, X_k, U)$,
  - for every $r_1, \dots, r_n \in RHS'(F, C, X_k, U)$ and $r_0 \in RHS'(F, C, X_k, Z_n)$: $r_0[z_i \rightsquigarrow r_i] \in RHS'(F, C, X_k, U)$, and
  - for every $u \in U$: $u \in RHS'(F, C, X_k, U)$,
- $r_{in} \in RHS'(F, C, X_1, \emptyset)$ is the *initial right-hand side*. □

The linearity condition in the previous definition will be called *context-linearity*. Note that it guarantees that no unevaluated subexpressions are copied in a nondeterministic reduction relation. This fact will be needed in Subsections 4.3 and 5.3, where n-satts are realized by functional programs under call-by-need evaluation, such that the number of performed reduction steps is equal.

Example 9. $M_{acc} = (F', C, R, r'_{in})$ is a 2-satt, where $F' = \{f^{(3)}\}$, $C = \{S^{(1)}, Z^{(0)}, A^{(0)}, B^{(0)}, D^{(1)}, T^{(3)}\}$, R contains the rules

$$\begin{aligned} f\ Z\ y_1\ y_2 &= y_2 \\ f\ (S\ x_1)\ y_1\ y_2 &= (f\ x_1\ (D\ z_1)\ (T\ z_1\ z_2\ z_1))[z_i \rightsquigarrow y_i] \end{aligned}$$

and $r'_{in} = (f\ x_1\ A\ B)$. Then,

$$\begin{aligned} f\ (S^3\ Z)\ A\ B &\Rightarrow_R\ (f\ (S^2\ Z)\ (D\ z_1)\ (T\ z_1\ z_2\ z_1))[z_1 \rightsquigarrow A, z_2 \rightsquigarrow B] \\ &\Rightarrow_R\ (f\ (S\ Z)\ (D\ z_1)\ (T\ z_1\ z_2\ z_1)) \\ &\qquad [z_1 \rightsquigarrow (D\ z_1), z_2 \rightsquigarrow (T\ z_1\ z_2\ z_1)][z_1 \rightsquigarrow A, z_2 \rightsquigarrow B] \\ &\Rightarrow_R^2\ (T\ z_1\ z_2\ z_1)[z_1 \rightsquigarrow (D\ z_1), z_2 \rightsquigarrow (T\ z_1\ z_2\ z_1)] \\ &\qquad [z_1 \rightsquigarrow (D\ z_1), z_2 \rightsquigarrow (T\ z_1\ z_2\ z_1)][z_1 \rightsquigarrow A, z_2 \rightsquigarrow B]. \end{aligned}$$ □

For every n-sntt $M = (F, Sub, C, \Pi, R_1, R_2, r_{in})$ with $R = R_1 \cup R_2$ and for every n-satt $M = (F, C, R, r_{in})$, $\Rightarrow_R$ is locally confluent, because there are no critical pairs. Similarly to modular tree transducers [10] and macro tree transducers [9], $\Rightarrow_R$ is also terminating, since every rule application to a function symbol with its recursion argument t delivers only (i) new function symbols with subtrees of t as recursion arguments or (in the case of an n-sntt) (ii) occurrences of the substitution function which does not call any other function and also "strictly walks down" on its recursion argument. Thus, for every $t \in T_C$, $nf(\Rightarrow_R, r_{in}[x_1/t])$ exists. Moreover, there are no function symbols in this normal form, because all functions are exhaustively defined on their possible recursion arguments (in particular on all outputs of functions which are nested in their recursion arguments). Hence, the normal form is a C-expression with sharing.

Definition 10. Let $M = (F, Sub, C, \Pi, R_1, R_2, r_{in})$ be an n-sntt with $R = R_1 \cup R_2$ or let $M = (F, C, R, r_{in})$ be an n-satt. The *tree transformation computed by* M is the function $\tau_M : T_C \to T_C$, which is for every $t \in T_C$ defined by $\tau_M(t) = \underline{tree}(nf(\Rightarrow_R, r_{in}[x_1/t]))$. □

4 Accumulation Technique

Our transformation technique consists of three steps: (i) a preprocessing step which abstracts n-ntts into n-sntts, (ii) the main transformation on the level of tree transducers with sharings (transforming n-sntts into n-satts), and (iii) a postprocessing step which realizes n-satts as functional programs. Since the pre- and postprocessing steps are relatively simple compared to the main transformation, they will be presented only informally.

4.1 Preprocessing

Explicit sharings which were introduced in the previous section, are added. More exactly, the *sub*-rules of Definition 4 are replaced by those of Definition 6. Note that this preprocessing step simplifies our efficiency considerations; the main transformation could also take n-ntts as inputs.

4.2 Main Transformation

The main transformation processes an n-sntt M and yields an n-satt M'. The construction introduces a new $(n+1)$-ary function symbol f for every function symbol f of M. The context arguments of the new f shall store replacements for the substitution constructors. Intuitively speaking, a call like $(f\ t\ e_1 \dots e_n)$ should evaluate (using M') to the result of $(sub\ (f\ t)\ e_1 \dots e_n)$ (evaluated using M). Thereby the intermediate result that is produced by the call $(f\ t)$ is avoided. The formalization of this intuitive relation can be found in Lemma 13. The construction uses an auxiliary function $\underline{sub}$ to transform right-hand sides of rules thereby evaluating substitutions at compile time.

Definition 11. Let $M = (F, Sub, C, \Pi, R_1, R_2, r_{in})$ be an n-sntt. First, we define the set $\overline{R_2}$ of *transformation rules* which contains

- for every $j \in [n]$ a rule $\quad \underline{sub}\ \Pi_j\ y_1 \dots y_n = y_j$,
- for every $c \in (C - \Pi)^{(k)}$ a rule[4]
  $$\begin{aligned} &\underline{sub}\ (c\ x_1 \dots x_k)\ y_1 \dots y_n \\ = &\ (c\ (\underline{sub}\ x_1\ z_1 \dots z_n) \dots (\underline{sub}\ x_k\ z_1 \dots z_n))[z_i \leadsto y_i], \end{aligned}$$
- for every $f \in F$ a rule $\quad \underline{sub}\ (f\ x_0)\ y_1 \dots y_n = f\ x_0\ y_1 \dots y_n$,

[4] If c is nullary or unary, then the explicit sharing will be avoided in examples.

– and the rule[5]
$\underline{sub}\ (sub\ x_0\ x_1 \dots x_n)\ y_1 \dots y_n$
$= (\underline{sub}\ x_0\ (\underline{sub}\ x_1\ z_1 \dots z_n) \dots (\underline{sub}\ x_n\ z_1 \dots z_n))[z_i \rightsquigarrow y_i]$.

Then, the *n-satt constructed from M by accumulation* is defined as $acc(M) = (acc(F), C, acc(R_1), acc(r_{in}))$, where

– $acc(F) = \{f^{(n+1)} \mid f \in F\}$,
– $acc(R_1)$ contains for every $f \in acc(F)$ and $c \in C^{(k)}$ the rule $f\ (c\ x_1 \dots x_k)\ y_1 \dots y_n = nf(\Rightarrow_{\overline{R_2}}, \underline{sub}\ rhs_{R_1,f,c}\ y_1 \dots y_n)$,
– $acc(r_{in}) = nf(\Rightarrow_{\overline{R_2}}, \underline{sub}\ r_{in}\ \Pi_1 \dots \Pi_n)$. □

It can be shown easily that $acc(M)$ is in fact a well-defined n-satt. In particular, the context-linearity is induced by the fact that the $\underline{sub}$-rules do not copy variables. Given the above intuition, the rules for $\underline{sub}$ should be straightforward: The first rule avoids the explicit construction of Π_j-symbols. The second rule is standard and the third rule encodes our intuition. Finally, the fourth rule represents the "associativity" of substitutions. Note the similarity of these rules to the laws (1), (2), (∗), and (3), respectively, of [25]. In the second and fourth rule we use explicit sharings in order to avoid that occurrences of F-functions are copied and thus are executed more than once in the transformed program.

Example 12. Let $\tilde{M}_{non} = (F, Sub, C, \Pi, R_1, R_2, r_{in})$ be the 2-sntt from Example 7. Then, the set $\overline{R_2}$ contains the rules

$$
\begin{aligned}
&\underline{g}\ A\ y_1\ y_2 &&= y_1\\
&\underline{g}\ B\ y_1\ y_2 &&= y_2\\
&\underline{g}\ (D\ x_1)\ y_1\ y_2 &&= D\ (\underline{g}\ x_1\ y_1\ y_2)\\
&\underline{g}\ (T\ x_1\ x_2\ x_3)\ y_1\ y_2 &&= (T\ (\underline{g}\ x_1\ z_1\ z_2)\ (\underline{g}\ x_2\ z_1\ z_2)\ (\underline{g}\ x_3\ z_1\ z_2))[z_i \rightsquigarrow y_i]\\
&\underline{g}\ (f\ x_0)\ y_1\ y_2 &&= f\ x_0\ y_1\ y_2\\
&\underline{g}\ (g\ x_0\ x_1\ x_2)\ y_1\ y_2 &&= (\underline{g}\ x_0\ (\underline{g}\ x_1\ z_1\ z_2)\ (\underline{g}\ x_2\ z_1\ z_2))[z_i \rightsquigarrow y_i]
\end{aligned}
$$

and the 2-satt constructed from $\tilde{M}_{non}$ by accumulation is defined as $\tilde{M}_{acc} = acc(\tilde{M}_{non}) = (acc(F), C, acc(R_1), acc(r_{in}))$, where $acc(F) = \{f^{(3)}\}$, $acc(R_1)$ contains the following rules with underlined left- and right-hand sides

$$
\begin{aligned}
&\underline{f\ Z\ y_1\ y_2} = nf(\Rightarrow_{\overline{R_2}}, \underline{g}\ B\ y_1\ y_2) = \underline{y_2},\\
&\underline{f\ (S\ x_1)\ y_1\ y_2} = nf(\Rightarrow_{\overline{R_2}}, \underline{g}\ (g\ (f\ x_1)\ (D\ A)\ (T\ A\ B\ A))\ y_1\ y_2)\\
&= nf(\Rightarrow_{\overline{R_2}}, (\underline{g}\ (f\ x_1)\ (\underline{g}\ (D\ A)\ z_1\ z_2)\ (\underline{g}\ (T\ A\ B\ A)\ z_1\ z_2))[z_i \rightsquigarrow y_i])\\
&= \underline{(f\ x_1\ (D\ z_1)\ (T\ z_1\ z_2\ z_1)[z_i \rightsquigarrow z_i])[z_i \rightsquigarrow y_i]}
\end{aligned}
$$

and $acc(r_{in}) = nf(\Rightarrow_{\overline{R_2}}, \underline{g}\ (f\ x_1)\ A\ B) = (f\ x_1\ A\ B)$. □

The correctness proof[6] of the main transformation is based on the following lemma which formalizes our intuition from the beginning of this subsection.

[5] If $n = 1$, then the explicit sharing could be avoided.
[6] Available at `www.orchid.inf.tu-dresden.de/gdp/conferences/amast06.shtml`

Lemma 13. Let $M = (F, Sub, C, \Pi, R_1, R_2, r_{in})$ be an n-sntt and $acc(M) = (acc(F), C, acc(R_1), acc(r_{in}))$. For every $f \in F$ and $t \in T_C$:

$$\underline{tree}(nf(\Rightarrow_{R_1 \cup R_2}, sub\ (f\ t)\ z_1 \ldots z_n)) = \underline{tree}(nf(\Rightarrow_{acc(R_1)}, f\ t\ z_1 \ldots z_n)). \quad \square$$

Theorem 14. Let M be an n-sntt. Then, $\tau_M = \tau_{acc(M)}$. $\square$

4.3 Postprocessing

Finally, an n-satt resulting from the main transformation is translated into a functional program by replacing in the right-hand sides of rules and in the initial right-hand side the explicit sharings with let-expressions. More exactly, an expression of the form

$$r[z_1 \rightsquigarrow r_1, \ldots, z_n \rightsquigarrow r_n] \qquad \text{is replaced by} \qquad let\ \{v_1 = \overline{r_1}; \ldots; v_n = \overline{r_n}\}\ in\ \overline{r},$$

where $v_1, \ldots, v_n$ are fresh variables (which can be obtained using tree-structured addresses in the translation process) and $\overline{r_1}, \ldots, \overline{r_n}, \overline{r}$ result from recursively replacing explicit sharings in $r_1, \ldots, r_n, r$, respectively, and additionally using $v_1, \ldots, v_n$ instead of the free occurrences of $z_1, \ldots, z_n$ in r.

Example 15. Let $\tilde{M}_{acc}$ be the 2-satt of Example 12. Postprocessing translates $f\ (S\ x_1)\ y_1\ y_2 = (f\ x_1\ (D\ z_1)\ (T\ z_1\ z_2\ z_1)[z_i \rightsquigarrow z_i])[z_i \rightsquigarrow y_i]$ into the rule

$$\begin{aligned} f\ (S\ x_1)\ y_1\ y_2 = {} & let\ \{v_1 = y_1; v_2 = y_2\} \\ & in\ f\ x_1\ (D\ v_1)\ (let\ \{v_{11} = v_1; v_{12} = v_2\}\ in\ (T\ v_{11}\ v_{12}\ v_{11})). \square \end{aligned}$$

A more elaborate translation could simplify (or even avoid) some let-expressions, e.g. if z_j does not occur or occurs only once freely in r or if $r_j = z_j$ or $r_j = y_j$ (i.e. we have $z_j \rightsquigarrow z_j$ or $z_j \rightsquigarrow y_j$). For the case $r_j = y_j$ note that the resulting program will be again treated call-by-need, and hence y_j is shared implicitly.

Example 16. Instead of constructing the rule as in Example 15, the following rule can be used (cf. also the introduction):

$$f\ (S\ x_1)\ y_1\ y_2 = f\ x_1\ (D\ y_1)\ (T\ y_1\ y_2\ y_1) \quad \square$$

5 Efficiency Non-deterioration by Accumulation

Our aim is to show the efficiency non-deterioration for call-by-need reduction. Unfortunately, it is technically difficult to formally compare the number of steps caused by deterministic reduction relations (cf. e.g. [17]). Hence we will base our comparison on the nondeterministic reduction relations for n-sntts and n-satts.

Therefore we first present a mechanism such that the number of call-by-need reduction steps caused by the R_1-rules of an n-ntt M equals the number of "relevant" nondeterministic reduction steps caused by the R_1-rules of the corresponding n-sntt $\tilde{M}$: In both reduction relations the copying of unevaluated applications of F-functions is avoided (by implicit and explicit sharing, respectively). But, whereas the deletion of a useless unevaluated application of an F-function is performed automatically in a call-by-need reduction, the nondeterministic reduction relation for $\tilde{M}$ either simply evaluates such an application and later moves the result into a subexpression e_i of an expression of the form $e[z_i \rightsquigarrow e_i]$ or, vice versa, the reduction relation for $\tilde{M}$ first moves it into an e_i of an expression $e[z_i \rightsquigarrow e_i]$, where it is evaluated later. In both situations the normal form of e will not contain a free occurrence of z_i, but nevertheless the useless evaluation is performed! In order to consider only the relevant R_1-reduction steps, in our mechanism (i) every application of an R_1-rule will additionally generate a special symbol $\circ$ and (ii) in the normal form only those $\circ$-symbols are counted by a function $\underline{step}$, which do not occur in a subexpression e_i of an expression $e[z_i \rightsquigarrow e_i]$, where z_i does not occur freely in e.

Then we use the same counting mechanism for the n-satt $acc(\tilde{M})$ in order to prove that the number of relevant R_1-reduction steps of $\tilde{M}$ equals the number of relevant reduction steps of $acc(\tilde{M})$. Together with a final argumentation that the postprocessing phase does not change the number of reduction steps, we obtain the desired efficiency result. Note that our comparison procedure does not consider the R_2-reduction steps of M or $\tilde{M}$, which do not occur in $acc(\tilde{M})$ and hence are saved by accumulation!

In the following we assume that $\circ$ is a new unary symbol and for every ranked alphabet C we define $C^\circ = C \cup \{\circ\}$.

Definition 17. Let $M = (F, Sub, C, \Pi, R_1, R_2, r_{in})$ be an n-sntt. The n-sntt $M^\circ = (F, Sub, C^\circ, \Pi, R_1^\circ, R_2^\circ, r_{in})$ is defined by

- if R_1 contains a rule $f\ (c\, x_1 \ldots x_k) = rhs_{R_1,f,c}$,
  then R_1° contains a rule $f\ (c\, x_1 \ldots x_k) = \circ\ rhs_{R_1,f,c}$, and
- R_1° contains for every $f \in F$ a (dummy; never applied) rule $f\ (\circ\, x_1) = \ldots$,
- R_2° contains the rules of R_2, and
- R_2° contains the rule $sub\ (\circ\, x_1)\ y_1 \ldots y_n = \circ\ (sub\ x_1\ y_1 \ldots y_n)$.

Let $M=(F, C, R, r_{in})$ be an n-satt. The n-satt $M^\circ=(F, C^\circ, R^\circ, r_{in})$ is defined by

- if R contains a rule $f\ (c\, x_1 \ldots x_k)\ y_1 \ldots y_n = rhs_{R,f,c}$,
  then R° contains a rule $f\ (c\, x_1 \ldots x_k)\ y_1 \ldots y_n = \circ\ rhs_{R,f,c}$, and
- R° contains for every $f \in F$ a (dummy) rule $f\ (\circ\, x_1)\ y_1 \ldots y_n = \ldots$. □

Note that by the additional sub-rule of R_2° in the previous definition the $\circ$-symbols produced by R_1°-rules are retained.

Definition 18. Let C be a ranked alphabet.
The function $\underline{step} : E_{C^\circ,n}(Z_n) \to I\!N$ is defined as follows:

- For every $e \in E_{C^\circ,n}(Z_n)$: $\underline{step}(\circ\ e) = 1 + \underline{step}(e)$.
- For every $c \in C^{(k)}$ and $e_1, \ldots, e_k \in E_{C^\circ,n}(Z_n)$:
  $\underline{step}(c\, e_1 \ldots e_k) = \sum_{i=1}^{k} \underline{step}(e_i)$.

– For every $e_1, \ldots, e_n, e \in E_{C^\circ,n}(Z_n)$:
$\underline{step}(e[z_i \rightsquigarrow e_i]) = \underline{step}(e) + \sum_{i=1}^{n}(\underline{step}(e_i) * \underline{rel}(z_i, e))$.
– For every $i \in [n]$: $\underline{step}(z_i) = 0$.

The function $\underline{rel} : Z_n \times E_{C^\circ,n}(Z_n) \longrightarrow \{0, 1\}$ is for every $i \in [n]$ and $e \in E_{C^\circ,n}(Z_n)$ defined by $\quad \underline{rel}(z_i, e) = 1 \quad$ iff $\quad z_i$ occurs in $\underline{tree}(e)$. □

Example 19. Since the phenomenon of non-relevant subexpressions does not occur in our running example, we choose an artificial example here:

$$\begin{aligned} & \underline{step}((\circ\, z_1)[z_1 \rightsquigarrow (\circ\, A)][z_1 \rightsquigarrow (\circ\, A)]) \\ = & \underline{step}((\circ\, z_1)[z_1 \rightsquigarrow (\circ\, A)]) + \underline{step}(\circ\, A) * \underline{rel}(z_1, (\circ\, z_1)[z_1 \rightsquigarrow (\circ\, A)]) \\ = & \underline{step}(\circ\, z_1) + \underline{step}(\circ\, A) * \underline{rel}(z_1, (\circ\, z_1)) + 1 * 0 = 1 + 1 * 1 + 1 * 0 = 2 \end{aligned}$$ □

Now we have to consider again our three transformation steps, where the second step involves a formal proof and the first and last step are argumentations concerning call-by-need reduction, which we avoided to define formally.

5.1 Preprocessing

For an n-ntt $M = (F, Sub, C, \Pi, R_1, R_2, r_{in})$ and a term $t \in T_C$ we will denote by $cbn_{R_1}(t)$ the number of R_1-reduction steps which are used to reduce $r_{in}[x_1/t]$ to a term graph corresponding to $nf(\Rightarrow_{R_1 \cup R_2}, r_{in}[x_1/t])$ with a call-by-need reduction. Let $\tilde{M} = (F, \{sub\}, C, \Pi, \tilde{R}_1, \tilde{R}_2, \tilde{r}_{in})$ result from M by preprocessing. Then we have to argue that

$$cbn_{R_1}(t) = \underline{step}(nf(\Rightarrow_{\tilde{R}_1^\circ \cup \tilde{R}_2^\circ}, \tilde{r}_{in}[x_1/t])).$$

First we only consider the F-functions: Since every application of a rule in $\tilde{R}_1^\circ$ delivers exactly one occurrence of $\circ$, the number of occurrences of $\circ$ in $nf(\Rightarrow_{\tilde{R}_1^\circ}, \tilde{r}_{in}[x_1/t])$ equals the number of applied $\tilde{R}_1$-steps to calculate $nf(\Rightarrow_{\tilde{R}_1}, \tilde{r}_{in}[x_1/t])$. This number is in turn equal to the number of applied R_1-steps to calculate a term graph corresponding to $nf(\Rightarrow_{R_1}, r_{in}[x_1/t])$ with call-by-need, because occurrences of F-functions are not nested and hence are neither copied nor deleted.

Now we additionally consider the substitution function sub: (i) Occurrences of F-functions in the recursion argument of sub are neither copied nor deleted in a reduction by $R_1 \cup R_2$. Correspondingly, in a reduction by $\tilde{R}_1^\circ \cup \tilde{R}_2^\circ$ occurrences of $\circ$ in the recursion argument of sub are exactly once reproduced and counted by $\underline{step}$. (ii) In a call-by-need reduction by $R_1 \cup R_2$ no application of an F-function inside a context argument of sub is copied and also in a reduction by $\tilde{R}_1^\circ \cup \tilde{R}_2^\circ$ (with explicit sharing) no corresponding occurrence of $\circ$ is copied. (iii) But, in a call-by-need reduction by $R_1 \cup R_2$, every R_1-step which constitutes a subgraph of the term graph corresponding to $nf(\Rightarrow_{R_1}, r_{in}[x_1/t])$, such that the subgraph occurs in a deleted context argument position j of an occurrence of sub will not be executed, whereas a reduction by $\tilde{R}_1^\circ \cup \tilde{R}_2^\circ$ may behave differently: either the occurrence of $\circ$ in $nf(\Rightarrow_{\tilde{R}_1^\circ}, \tilde{r}_{in}[x_1/t])$ that corresponds to the above R_1-step is also deleted (by a sub-rule on a Π_i with $i \neq j$) or it is shifted into a subexpression e_j of an expression of the form $e[z_i \rightsquigarrow e_i]$ in which z_j does not occur freely in e and hence it is not counted by $\underline{step}$.

5.2 Main Transformation

The proof[7] of efficiency non-deterioration of the main transformation is based on the following lemma. Note the similarity of this lemma to Lemma 13: Instead of the reduction relations of M and $acc(M)$, their "$\circ$-generating versions" are used here. Moreover, instead of calculating the output tree by $\underline{tree}$, the number of $\circ$-symbols is counted by $\underline{step}$.

Lemma 20. Let $M = (F, Sub, C, \Pi, R_1, R_2, r_{in})$ be an n-sntt and $acc(M) = (acc(F), C, acc(R_1), acc(r_{in}))$. For every $f \in F$ and $t \in T_C$:

$$\underline{step}(nf(\Rightarrow_{R_1^\circ \cup R_2^\circ}, sub\ (f\ t)\ z_1 \ldots z_n)) = \underline{step}(nf(\Rightarrow_{acc(R_1)^\circ}, f\ t\ z_1 \ldots z_n)). \quad \square$$

Theorem 21. Let $M = (F, Sub, C, \Pi, R_1, R_2, r_{in})$ be an n-sntt and $acc(M) = (acc(F), C, acc(R_1), acc(r_{in}))$. Then, for every $t \in T_C$:

$$\underline{step}(nf(\Rightarrow_{R_1^\circ \cup R_2^\circ}, r_{in}[x_1/t])) = \underline{step}(nf(\Rightarrow_{acc(R_1)^\circ}, acc(r_{in})[x_1/t])). \quad \square$$

5.3 Postprocessing

Let-expressions do not cause additional reduction steps, rather they denote in functional languages explicit sharings. Thus, if we denote for an n-satt $M = (F, C, R, r_{in})$, for $\tilde{R}$ and $\tilde{r}_{in}$ obtained from R and r_{in}, respectively, by introducing let-expressions, and for a term $t \in T_C$, by $cbn_{\tilde{R}}(t)$ the number of call-by-need reduction steps which are used to reduce $\tilde{r}_{in}[x_1/t]$ to a term graph corresponding to $nf(\Rightarrow_R, r_{in}[x_1/t])$ with $\tilde{R}$, then it suffices to argue that

$$cbn_{\tilde{R}}(t) = \underline{step}(nf(\Rightarrow_{R^\circ}, r_{in}[x_1/t])).$$

The introduction of let-expressions does not change the copying or deletion properties of the rules in R. Moreover, in a call-by-need reduction by $\tilde{R}$ no function application inside some context argument is copied, and because of the context-linearity of R and hence also of R°, no corresponding occurrence of $\circ$ is copied in a reduction by R°. Hence there is only one main difference in the calculations of $\tilde{R}$ and R°: On the one hand, in a call-by-need reduction by $\tilde{R}$ a function application is not evaluated, if it occurs in a subexpression e_j of an expression of the form $let\ \{v_1 = e_1; \ldots; v_n = e_n\}\ in\ e$, in which e is evaluated and v_j does not occur in e. On the other hand, the corresponding function application in a subexpression e'_j of an expression of the form $e'[z_i \rightsquigarrow e'_i]$ (where $v_1, \ldots, v_n$ correspond to $z_1, \ldots, z_n$, respectively) is evaluated by R° and a symbol $\circ$ is produced. But since z_j does not occur freely in e', the produced $\circ$ is not counted by $\underline{step}$.

6 Future Work

We have proved that our accumulation technique does not deteriorate the efficiency, where efficiency is measured in the number of performed call-by-need

[7] Available at `www.orchid.inf.tu-dresden.de/gdp/conferences/amast06.shtml`

reduction steps. This point of view neglects the actual complexity of reduction steps. In particular, we weigh applications of unary functions against applications of the corresponding functions with accumulating parameters. A more elaborate efficiency measure could be based on weighted reduction steps, e.g., by using more than one $\circ$-symbol for a rule with parameters. Furthermore, it would be interesting to develop a syntactic characterization of those programs, for which the time-complexity is changed by accumulation (like in our running example).

In [22] list manipulating operations, in particular append, are eliminated by employing shortcut deforestation [13, 7] instead of tree transducer composition as in [16]. To this end, the technique from [22] does not only abstract from list constructors, but also from the list manipulating operations. We believe that this transformation can be generalized to eliminate also tree manipulating operations as, e.g., substitutions. But, as already stated in the Conclusion of [22], "a general statement about the relation between the runtimes of original and transformed programs is hard to make". Nevertheless it would be interesting to compare such a transformation with our approach.

Acknowledgment

The authors would like to thank Janis Voigtländer for suggesting an improved linear notation for term graphs and for carefully reading a draft of this paper.

References

1. M. Abadi, L. Cardelli, P.-L. Curien, and J.-J. Levy. Explicit substitutions. *Journal of Functional Programming*, 1(4):375–416, 1991.
2. Z.M. Ariola, M. Felleisen, J. Maraist, M. Odersky, and P. Wadler. A call-by-need lambda calculus. In *POPL'95, Proceedings*, pages 233–246. ACM Press, 1995.
3. F.L. Bauer and H. Wössner. *Algorithmic Language and Program Development.* Springer–Verlag, 1982.
4. R.S. Bird. The promotion and accumulation strategies in transformational programming. *ACM TOPLAS*, 6(4):487–504, 1984.
5. E.A. Boiten. The many disguises of accumulation. Tech. Report 91-26, Dept. of Informatics, University of Nijmegen, 1991.
6. R.M. Burstall and J. Darlington. A transformation system for developing recursive programs. *J. Assoc. Comput. Mach.*, 24:44–67, 1977.
7. O. Chitil. Type inference builds a short cut to deforestation. In *ICFP'99, Paris, France, Proceedings*, volume 34, pages 249–260. ACM Sigplan Notices, 1999.
8. J. Engelfriet. Some open questions and recent results on tree transducers and tree languages. In R.V. Book, editor, *Formal language theory; perspectives and open problems*, pages 241–286. New York, Academic Press, 1980.
9. J. Engelfriet and H. Vogler. Macro tree transducers. *J. Comput. Syst. Sci.*, 31:71–145, 1985.
10. J. Engelfriet and H. Vogler. Modular tree transducers. *Theoret. Comput. Sci.*, 78:267–304, 1991.
11. J. Engelfriet and H. Vogler. The translation power of top-down tree-to-graph transducers. *J. Comput. Syst. Sci.*, 49:258–305, 1994.

12. J. Giesl, A. Kühnemann, and J. Voigtländer. Deaccumulation — Improving provability. In *ASIAN'03, Mumbai, India, Proceedings*, volume 2896 of *LNCS*, pages 146–160. Springer-Verlag, 2003.
13. A. Gill, J. Launchbury, and S.L. Peyton Jones. A short cut to deforestation. In *FPCA'93, Copenhagen, Denmark, Proceedings*, pages 223–231. ACM Press, 1993.
14. J. Hughes. A novel representation of lists and its application to the function "reverse". *Information Processing Letters*, 22:141–144, 1986.
15. A. Kühnemann. Comparison of deforestation techniques for functional programs and for tree transducers. In *FLOPS'99, Tsukuba, Japan, Proceedings*, volume 1722 of *LNCS*, pages 114–130. Springer-Verlag, 1999.
16. A. Kühnemann, R. Glück, and K. Kakehi. Relating accumulative and non-accumulative functional programs. In *RTA'01, Utrecht, The Netherlands, Proceedings*, volume 2051 of *LNCS*, pages 154–168. Springer-Verlag, 2001.
17. A. Maletti. Direct construction and efficiency analysis for the accumulation technique for 2-modular tree transducers. Master's thesis, TU Dresden, 2002.
18. H. Partsch. *Specification and Transformation of Programs – A Formal Approach to Software Development*. Springer–Verlag, 1990.
19. W.C. Rounds. Mappings and grammars on trees. *Math. Syst.Th.*, 4:257–287, 1970.
20. D. Sands. A naïve time analysis and its theory of cost equivalence. *Journal of Logic and Computation*, 5(4):495–541, 1995.
21. J.W. Thatcher. Generalized^2 sequential machine maps. *J. Comput. Syst. Sci.*, 4:339–367, 1970.
22. J. Voigtländer. Concatenate, reverse and map vanish for free. In *ICFP'02, Pittsburgh, USA, Proceedings*, pages 14–25. ACM Press, 2002.
23. J. Voigtländer. Formal efficiency analysis for tree transducer composition. Tech. Report TUD-FI04-08, TU Dresden, 2004. To appear in *Theory Comput. Syst.*
24. J. Voigtländer. *Tree Transducer Composition as Program Transformation*. PhD thesis, TU Dresden, 2004.
25. P. Wadler. The concatenate vanishes. Note, University of Glasgow, 1987 (Revised, 1989).

Decomposing Interactions

Juliana Küster Filipe Bowles

School of Computer Science, The University of Birmingham
Edgbaston, Birmingham B15 2TT, UK
J.Bowles@cs.bham.ac.uk

Abstract. In UML 2.0 sequence diagrams have been considerably extended and are now fundamentally better structured. Interactions in sequence diagrams can be structured using so-called interaction fragments, including **alt** (alternative behaviour), **par** (parallel behaviour), **neg** (forbidden behaviour), **assert** (mandatory behaviour) and **ref** (reference another diagram). The operator **ref** in particular greatly improves the way diagrams can be decomposed. In previous work we have given a semantics to a subset of sequence diagrams using labelled event structures, a true-concurrent model that naturally captures alternative and parallel behaviour. In this paper, we expand that work to address refinement and show how to obtain a refined model by means of a powerful categorical construction over two categories of labelled event structures. The underlying motivation for this work is reasoning and verification of complex scenario-based inter-object behavioural models. We conclude the paper with a discussion on future work.

1 Introduction

In UML 2.0 sequence diagrams have been considerably extended and are now fundamentally better structured. Interactions in sequence diagrams can be structured using interaction fragments which include operators for capturing alternative behaviour (**alt**), parallel behaviour (**par**), negative or forbidden behaviour (**neg**), the only valid or mandatory behaviour (**assert**), and reference or include the behaviour described in another diagram (**ref**).

One of the most significant improvements of sequence diagrams in UML 2.0 is the possibility to decompose a diagram into subparts. UML already allowed this for class diagrams and state diagrams, but there was no standard way to decompose a sequence diagram in UML 1.x. Needless to say, the ability to decompose diagrams is crucial for managing the complexity of interactions in large-scale systems.

Interactions become more complex because interacting elements may be added to interactions as designs evolve and are elaborated, or simply because more messages, more alternatives and more variants are added. The addition of operators such as **alt** and **par** help, but there is still a need to move all the details of an interaction and show it on a separate diagram, allowing this interaction to be reused in several diagrams. In UML 2.0 decomposition is facilitated in two ways:

M. Johnson and V. Vene (Eds.): AMAST 2006, LNCS 4019, pp. 189–203, 2006.

referencing interaction fragments and decomposing lifelines. These mechanisms are borrowed directly from Message Sequence Charts (MSCs) [1].

In previous work [2] we have given a semantics to a subset of sequence diagrams using labelled event structures, a true-concurrent model that naturally captures alternative and parallel behaviour. In this paper, we expand that work to address refinement and show how to obtain a refined model by means of a powerful categorical construction over two categories of labelled event structures. This enables reasoning and verification of complex scenario-based inter-object behavioural models. Moreover, the distributed logic used in [2] to express interactions is in fact hierarchical and can be used in addition to express properties over unrefined and refined models. This is, however, beyond the scope of the present paper. Our main contribution in this paper is the automatic construction of refined sequence diagram models, in other words a novel categorical semantics for the decomposition mechanisms available in UML 2.0 sequence diagrams. Despite the fact that these mechanisms have been borrowed from MSCs, and MSCs have a well known standard algebraic semantics (see [3]), this semantics is not complete as it does not cover the mentioned decomposition mechanisms.

The paper is structured as follows. In Section 2, we give an overview of sequence diagrams in UML 2.0. In Section 3 we introduce our underlying semantic model, namely labelled event structures, describe the categorical properties of two categories of event structures, and present a powerful categorical construction for composing event structures. Section 4 recalls how to use event structures for modelling sequence diagrams, and is followed by Section 5 where we show how to build a refined model for a given sequence diagram that uses decomposition mechanisms. It defines a reference refinement diagram that serves as an input for the categorical construction of Section 3.3. The paper finishes with some concluding remarks and ideas for future work.

2 Sequence Diagrams in UML 2.0

Sequence diagrams are the more commonly used diagram for capturing inter-object behaviour. Graphically, a sequence diagram has two dimensions: an horizontal dimension representing the instances participating in the scenario, and a vertical dimension representing time. An instance can correspond to a particular object or a role played in the interaction. A role may be a part of a collaboration and/or an internal part of a structured class, component or subsystem. An instance has a vertical dashed line called *lifeline* representing the existence of the instance at a particular time; the order of events along a lifeline is significant denoting, in general, the order in which these events will occur.

A *message* is a communication between two instances which can cause an operation to be invoked, a signal to be raised, an instance to be created or destroyed. Messages are shown as horizontal arrows from the lifeline of one instance to the lifeline of another. A message specifies the kind of communication between instances (synchronous or asynchronous), and the sender and receiver event occurrences associated to it. In this paper, we do not consider lost and found messages. We do, however, allow the message sender or receiver to be unspecified using what is called a *gate*.

A sequence diagram is enclosed in a frame and the five-sided box at the upper lefthand corner names the sequence diagram. Further, interactions can be structured using so-called interaction fragments. Each interaction fragment has at least one operator held in the five-sided box at the upper left corner of the fragment. There are several possible operators, for example, **alt** (alternative behaviour), **par** (parallel behaviour), **neg** (forbidden behaviour), **assert** (mandatory behaviour), **loop** (repeated behaviour), **ref** (reference another diagram), and so on. Depending on the operator used, an interaction fragment consists of one or more *operands*. In the case of **neg, assert**, **loop** and **ref** the fragment has exactly one operand, whilst for most other operators it has several. Fig. 1 shows an example of a sequence diagram using UML 2.0 constructs.

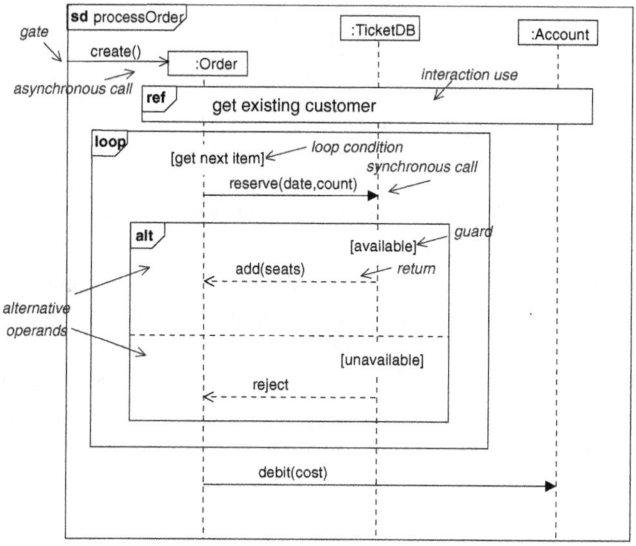

Fig. 1. A sequence diagram

The semantics of an interaction operator is described informally in the UML 2.0 superstructure specification [4]. Below we give the meaning of some operators used in this paper.

sd names a sequence diagram

ref references an interaction fragment which appears in a different diagram. This fragment is called an *interaction use*.

alt represents a choice of behaviour. At most one of the operands will execute. The operand that executes must have a guard expression that evaluates to true at this point in the interaction. If several guards are true, one of them is selected nondeterministically for execution.

par represents a parallel merge between the behaviours of the operands. The event occurrences of the different operands can be interleaved in any way as long as the ordering imposed by each operand as such is preserved.

In this paper, we are mainly interested in the formal semantics of the reference operator. Other operators have been dealt with in [2] by providing a true-concurrent semantics to sequence diagrams.

As mentioned earlier, decomposition in UML 2.0 sequence diagrams is possible by referencing interaction fragments (called interaction uses) and lifeline decomposition. In Fig. 1 we have seen one example of the former, and Fig. 2 shows both cases. In this example, a corresponds to a part or instance that is

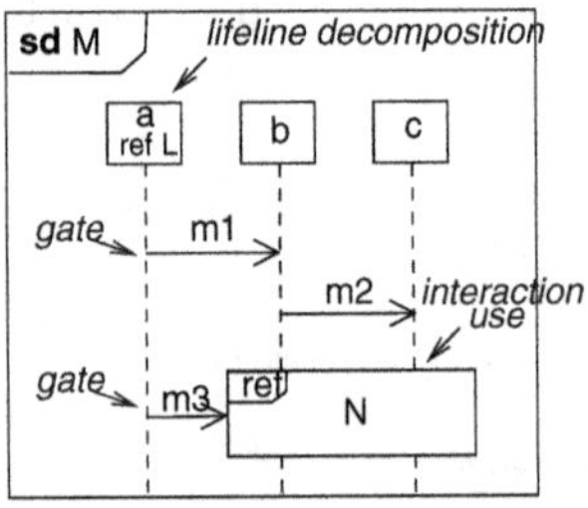

Fig. 2. Decomposition in sequence diagrams

itself decomposed in diagram L. This means that any point along the lifeline of a corresponds to a gate, and the instances sending messages m1 and m3 (not necessarily the same) are left unspecified. The sender of these messages is only known to diagram L. Similarly, the receiver of message m3 is only known to diagram N. In diagram M, we know that instances b and c are involved in the interaction N. Fig. 3 shows both referenced diagrams. Lifeline decomposition is particularly

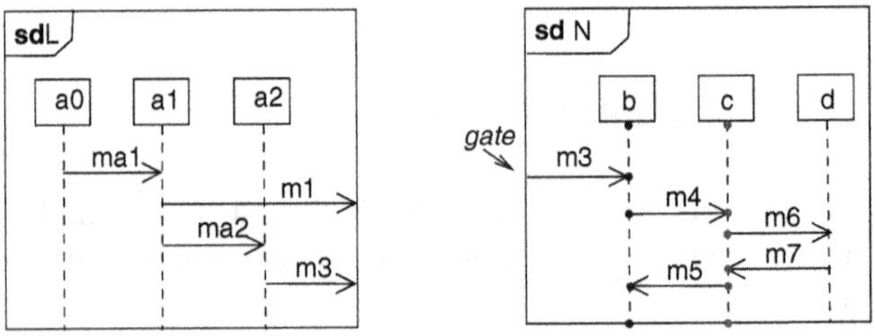

Fig. 3. Lifeline decomposition (left) and referenced interaction fragment (right)

useful when modelling component-based systems where the internals of a component (in this case a) are intentionally hidden. For example, we may want to replace a by a similar or updated component, and even if the interactions described in diagram L change, diagram M itself remains unchanged.

We borrow the concept *location* introduced in Live Sequence Charts (LSCs) [5] which is missing in sequence diagrams, but is useful semantically. *Locations* are the points in the lifeline of an instance which correspond to the occurrence of events (see the circles on the lifelines b and c of diagram N in Fig. 3). For sequence diagrams, we consider that all instances (except lifeline decomposition instances like instance a in Fig. 2) have at least two locations: an initial location (corresponding to the beginning of the diagram or instance creation) and a final location (corresponding to the end of the diagram or instance destruction). Further locations are associated with the sending and receiving of messages, the beginning and the end of interaction fragments, and conditions (state invariants or interaction constraints)[1]. The locations along a single lifeline and within an interaction operand are ordered top-down; therefore, a *partial order* is induced among these locations determining the order of execution. Notice that locations from different operands of an **alt** or **par** fragment are not ordered in any way. In the first case they are part of different execution traces whereas in the second case they are to be executed in parallel.

3 The Model

3.1 Event Structures: Basic Notions

We recall some basic notions on the model we use, namely *labelled prime event structures* [6].

Prime event structures, or event structures for short, allow the description of distributed computations as event occurrences together with relations for expressing causal dependency and nondeterminism. The first relation is called *causality*, and the second *conflict*. The causality relation implies a (partial) order among event occurrences, while the conflict relation expresses how the occurrence of certain events excludes the occurrence of others. Consider the following definition of event structures.

Event Structure. An *event structure* is a triple $E = (Ev, \rightarrow^*, \#)$ where Ev is a set of events and $\rightarrow^*, \# \subseteq Ev \times Ev$ are binary relations called *causality* and *conflict*, respectively. Causality $\rightarrow^*$ is a partial order. Conflict $\#$ is symmetric and irreflexive, and propagates over causality, i.e., $e\#e' \rightarrow^* e'' \Rightarrow e\#e''$ for all $e, e', e'' \in Ev$. Two events $e, e' \in Ev$ are *concurrent*, $e \; co \; e'$ iff $\neg(e \rightarrow^* e' \vee e' \rightarrow^* e \vee e\#e')$.

From the two relations defined on the set of events, a further relation is derived, namely the *concurrency* relation *co*. As stated, two events are concurrent if and only if they are completely unrelated, i.e., neither related by causality nor by conflict.

In our approach to inter-object behaviour specification, we will consider a restriction of event structures sometimes referred to as *discrete* event structures. An event structure is said to be *discrete* if the set of previous occurrences of an event in the structure is finite.

[1] Conditions have been considered in [2] and are omitted in this paper.

Discrete Event Structure. Let $E = (Ev, \rightarrow^*, \#)$ be an event structure. E is a *discrete event structure* iff for each event $e \in Ev$, the *local configuration* of e given by $\downarrow e = \{e^{'} \mid e^{'} \rightarrow^* e\}$ is finite.

The finiteness assumption of the so-called local configuration is motivated by the fact that system computations always have a starting point, which means that any event in a computation can only have finitely many previous occurrences.

Consequently, we are able to talk about immediate causality in such structures. Two events e and $e^{'}$ are related by *immediate* causality if there are no other event occurrences in between. Formally, if $\forall_{e^{''} \in Ev}(e \rightarrow^* e^{''} \rightarrow^* e^{'} \Rightarrow (e^{''} = e \vee e^{''} = e^{'}))$ holds. If $e \rightarrow^* e^{'}$ are related by immediate causality then e is said to be an *immediate predecessor* of $e^{'}$ and $e^{'}$ is said to be an *immediate successor* of e. We may write $e \rightarrow e^{'}$ instead of $e \rightarrow^* e^{'}$ to denote immediate causality. Furthermore, we also use the notation $e \rightarrow^+ e^{'}$ whenever $e \rightarrow^* e^{'}$ and $e \neq e^{'}$.

Hereafter, we only consider discrete event structures.

Configuration. Let $E = (Ev, \rightarrow^*, \#)$ be an event structure and $C \subseteq Ev$. C is a *configuration* in E iff it is both (1) conflict free: for all $e, e^{'} \in C$, $\neg(e \# e^{'})$, and (2) downwards closed: for any $e \in C$ and $e^{'} \in Ev$, if $e^{'} \rightarrow^* e$ then $e^{'} \in C$. A maximal configuration denotes a run. A run is sometimes called life cycle.

Finally, in order to use event structures to provide a denotational semantics to languages, it is necessary to link the event structures to the language they are supposed to describe. This is achieved by attaching a labelling function to the set of events. A generic labelling function is as defined next.

Labelling Function. Let $E = (Ev, \rightarrow^*, \#)$ be an event structure, and L be an arbitrary set. A *labelling function* for E is a total function $l : Ev \rightarrow L$ mapping each event into an element of the set L.

An event structure together with a labelling function defines a so-called labelled event structure.

Labelled Event Structure. Let $E = (Ev, \rightarrow^*, \#)$ be an event structure, L be a set of labels, and $l : Ev \rightarrow L$ be a labelling function for E. A *labelled event structure* is a pair $(E, l : Ev \rightarrow L)$.

We will see in Section 4 that when using event structures for modelling sequence diagrams in UML 2.0, the labelling function indicates whether an event represents sending or receiving a message, a condition, the beginning or end of an interaction fragment.

3.2 Categorical Properties of Event Structures

In this section we describe some categorical properties of event structures. We assume the reader is familiar with basic categorical constructions.

In order to define a category of event structures, we need a concept of morphism between event structures. Morphisms on event structures have been defined in, for example, [6], as follows.

Event Structure Morphism. Let $E_i = (Ev_i, \rightarrow_i^*, \#_i)$ for $i = 1, 2$ be event structures, and $C \subseteq Ev_1$ an arbitrary subset of events. A *morphism* from E_1 to E_2 consists of a partial function $h : Ev_1 \rightarrow Ev_2$ on events satisfying both (1) if C is a configuration in E_1 then $h(C)$ is a configuration in E_2, and (2) for all $e, e' \in C$, if $h(e), h(e')$ are defined and $h(e) = h(e')$ then $e = e'$.

The notion of event structure morphism as given before preserves the concurrency relation, as has been proved in [6]. The intuition is that the occurrence of an event is matched (synchronised) with the occurrence of its image. However, such a definition (condition (1)) is too strong for our purposes.

Event structures and event structure morphisms as defined above constitute a category **ev** presented in [6], among others. This category is known to have both products and coproducts whereby the first models parallel composition and the second nondeterministic choice. A coproduct construction is given for instance in [6]. A product in the category is more tricky and difficult to define in a direct way. A categorical construction for the product of event structures has been is given in [7] making use of a new notion of preconfigurations. Alternatively, the product in **ev** can be derived from the product of trace languages and the coreflection from event structures to trace languages [6]. Nevertheless, the category of event structures **ev** has both products and equalizers, and is therefore complete (a proof of the later can be found in [8]). Consequently, we know that it has pullbacks.

By contrast, **ev** is not cocomplete, because this category does not have coequalizers. Consequently, pushouts do not always exist. This is due to the fact that event structure morphisms may map events in conflict into the same event. Indeed, injectivity is only assumed on configurations. We require a more rigid notion of a morphism in order to have coequalizers and consequently pushouts.

Consider the following notion of a so-called *communication* event structure morphism.

Communication Event Structure Morphism. Let $E_i = (Ev_i, \rightarrow_i^*, \#_i)$ for $i = 1, 2$ be event structures. A *communication* event structure morphism from E_1 to E_2 consists of a total function $h : Ev_1 \rightarrow Ev_2$ on events preserving $\rightarrow_1^+$ and $\#_1$.

Notice that a communication morphism is *total* instead of partial. Moreover, injectivity is no longer required over configurations but guaranteed over sequential substructures as a consequence of the relations being preserved. This makes the communication morphism notion more rigid than the previous one. However, configurations do not have to be mapped into configurations. As a communication morphism preserves $\rightarrow^+$, a sequential configuration is mapped into a subset of events contained in a configuration. Recall that $\rightarrow^*$ is obtained from the reflexive closure of $\rightarrow^+$. Moreover, preserving $\rightarrow^+$ instead of $\rightarrow^*$ guarantees that distinct events related by causality are mapped into distinct events related by causality as well. Finally, a communication morphism preserves conflict but not necessarily concurrency.

Event structures and communication morphisms constitute a category **cev**. Furthermore, this category is complete, has coproducts and, under certain conditions, coequalizers. Indeed, **cev** has coequalizers for two morphisms $f, g : Ev_1 \rightarrow Ev_2$ provided these morphisms are injective, $f(Ev_1) \cap g(Ev_1) = \emptyset$ and for any $e_1 \in Ev_1$, $f(e_1)\ co_2\ g(e_1)$. Consequently, **cev** has pushouts under the same conditions. A detailed analysis on this category can be found in [8].

Finally, let the associated categories for labelled event structures be $\mathcal{L}(\mathbf{ev})$ and $\mathcal{L}(\mathbf{cev})$. The category $\mathcal{L}(\mathbf{ev})$ has the properties of its underlying category **ev**, and is therefore complete but not cocomplete. Similarly, $\mathcal{L}(\mathbf{cev})$ is complete, has coproducts, and coequalizers for a pair of communication morphisms under the same conditions as the underlying unlabelled category.

3.3 Categorical Construction

From the categories introduced earlier, a well defined categorical construction has been obtained in [8] to describe synchronous concurrent composition. To simplify the presentation of the construction we deal with the unlabelled categories **ev** and **cev** instead.

The main idea behind this construction is the following. A product in **ev** denotes parallel composition but is, however, far more than is needed because it consists of the events of both structures in isolation and all possible synchronisations. This product does not have much relevance for practical applications in the sense that we usually want to synchronise some events (according to their labels) but not all of them. Pullbacks (which exist for **ev** and $\mathcal{L}(\mathbf{ev})$) may be understood as constrained products. However, here indicated events are synchronised as desired but remaining events are combined in all possible ways (synchronised or left in isolation). A pullback in **ev** gives us the desired model only when both event structures are to be fully synchronised. By contrast, a coproduct in **cev** (and $\mathcal{L}(\mathbf{cev})$) denotes now full concurrent composition. A pushout in the same category would give us concurrent composition with synchronisation as intended. However, pushouts only exist under certain conditions and we have to make sure that we fulfill these conditions to make use of this construct.

Synchronisation Diagram. Let E_1 and E_2 be two event structures. A *synchronisation diagram* for E_1 and E_2 is given by a triple $S = (E_{synch}, f_1, f_2)$ where E_{synch} is a nonempty event structure, and f_i with $i \in \{1, 2\}$ are two surjective event structure morphisms such that $f_i : Ev_i \rightarrow Ev_{synch}$, and satisfying $f_1(Ev_1) = f_2(Ev_2)$. Moreover, E_{synch} is called the *synchronisation event structure* of E_1 and E_2.

The synchronisation diagram tells us how the two models relate. If a synchronisation diagram is not definable we say that the models are not composable. We only consider composable models herein.

Categorical Construction. Let E_1 and E_2 be two event structures with a synchronisation diagram given by $S = (E_{synch}, f_1, f_2)$ where $f_i : Ev_i \rightarrow Ev_{synch}$ for $i \in \{1, 2\}$. Let $E_i^{'}$ be the maximal event substructure of E_i such that $f_{i|E_i'}$ is a total morphism. Then doing the pullbacks in **ev** and the pushout in **cev** as depicted, we obtain the concurrent composition of E_1 and E_2, written $E_1 \times_{synch} E_2$, in accordance with the synchronisation diagram S.

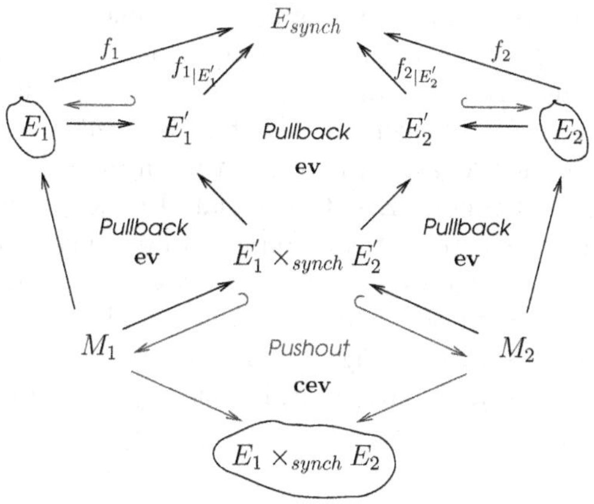

The interesting aspect of the above construction is that it combines pullbacks in **ev** and one final pushout in **cev** in such a way that the pullbacks are done over fully synchronised event structures and we always obtain morphisms in **cev** satisfying the necessary conditions for the existence of the the final pushout. We will see in Section 5 how it can be reused for the present purpose.

4 Event Structures for Sequence Diagrams

In [2] we have shown how labelled event structures can be used to provide a model for sequence diagrams. Here we only provide the general idea.

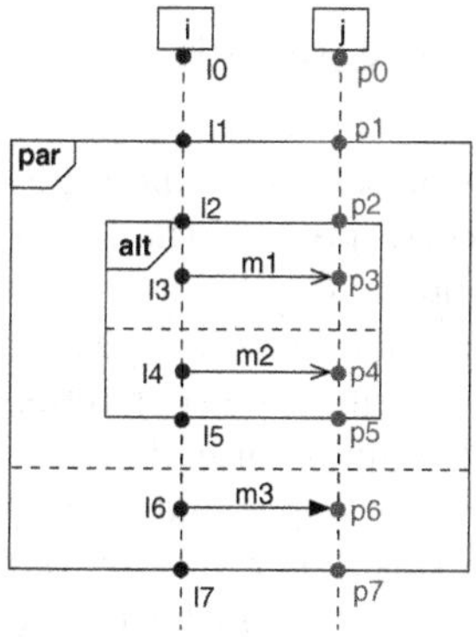

Fig. 4. A case of nested par-alt fragments

Consider the sequence diagram in Fig. 4 showing the interaction between two instances i and j within nested **par-alt** fragments. To obtain the corresponding event structure model, we want to associate events to the locations of the diagram and determine the relations between those events to reflect the meaning of the diagram. The end location of an **alt** fragment is problematic. If it corresponded to one event then this event would be in conflict with itself due to the fact that in a prime event structure conflict propagates over causality. This would, however, lead to an invalid model since conflict is irreflexive. We are therefore forced to copy events for locations marking the end of **alt** fragments, as well as for all locations that follow. Events associated to locations that fall within a **par** fragment are concurrent. Synchronous communication is denoted by a shared event whereas asynchronous communication is captured by immediate causality between the send event and receive event. The expected model for the diagram of Fig. 4 is as shown in Fig. 5, where events e_i or e_j denote events of instance i or j respectively. In particular, location l_x (p_x) in Fig. 4 is associated to event e_{i_x} (e_{j_x}) or several copies $e_{i_{x1}}, \ldots, e_{i_{xn}}$ ($e_{j_{x1}}, \ldots, e_{j_{xn}}$). Event e is a

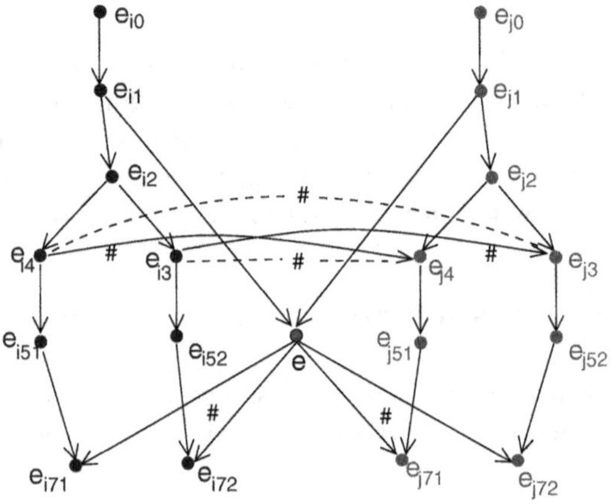

Fig. 5. A case of nested par-alt fragments - semantic model

shared event denoting synchronisation by message $m3$. The events e_{i_3} and e_{i_4} are in conflict because they correspond to the execution of alternative operands (similarly $e_{i_3} \# e_{j_4}$, $e_{j_3} \# e_{i_4}$ and $e_{j_3} \# e_{j_4}$).

We have mentioned before, that we will use the labelling function of labelled event structures to indicate whether an event represents sending or receiving a message, a condition, or the beginning or end of an interaction fragment. We are not considering conditions in this paper and therefore disregard that case. The only considered fragments in this paper are **alt**, **par** and **ref**.

Let D be a set of diagram names, I_d be the set of instances participating in the interaction described by $d \in D$, and g denote an unspecified instance or gate with $g \in I_d$ for all $d \in D$. Let $F_D = \{d, par, alt, ref(d)\}$ where $d \in D$, and $\overline{F_D} = \{\overline{d}, \overline{par}, \overline{alt}, \overline{ref(d)}\}$. We use par (or $\overline{par}$) as a label of an event associated to the location marking the beginning (or end) of a **par** fragment. In particular, events associated to initial (or end) locations of a diagram d have labels d (or $\overline{d}$). Let Mes be a set of message labels. The labelling function for diagram d is a total function defined over:

$$\mu_d : Ev \rightarrow I_d \times (Mes \times \{s,r\} \cup F_D \cup \overline{F_D} \cup ref(D) \times (Mes \times \{s,r\}))$$
$$\cup\, I_d \times (Mes \cup ref(D) \times Mes) \times I_d$$

The first part of the codomain is used to describe asynchronous messages (possibly at a referenced fragment) or beginning/end of fragments, whilst the second part of the codomain deals with synchronous messages (possibly at a referenced fragment).

For the example of Fig. 5, a few labels are as follows: $\mu(e_{i_2}) = (i, alt)$ (beginning of an **alt**), $\mu(e_{j_2}) = (j, alt)$, $\mu(e_{i_3}) = (i,(m_1,s))$ (asynchronous send), $\mu(e_{j_4}) = (j,(m_2,r))$ (asynchronous receive), $\mu(e) = (i, m_3, j)$ (synchronisation between i and j on m_3), $\mu(e_{i_{71}}) = \mu(e_{i_{72}}) = (i, \overline{par})$ (end of an **alt**). The label of an event associated to a gate location at a referenced fragment is an element in $I_d \times ref(D) \times (Mes \times \{s,r\})$. For example, consider the gate location of message `m3` in diagram `M` in Fig 6. The label of the associated event is given by $\mu(e) = (g, ref(n),(m_3,r))$ where g denotes a gate or unspecified instance.

Finally, for a diagram $m \in D$, a model is a labelled event structure $M = (E_m, \mu_m)$.

5 Modelling Refinement

In this section we show how to model refinement of event structures and consequently give a semantics to sequence diagram decomposition. This is done by defining an appropriate synchronisation diagram for the categorical construction introduced earlier. The expected refined model is obtained automatically from this construction.

Consider Fig. 6 where diagram `M` has a reference to diagram `N`. Notice that as before, each instance has locations along its lifeline, and additionally *gates* (here the target of message `m3` in `M` and the source of message `m3` in `N`) have locations as well.

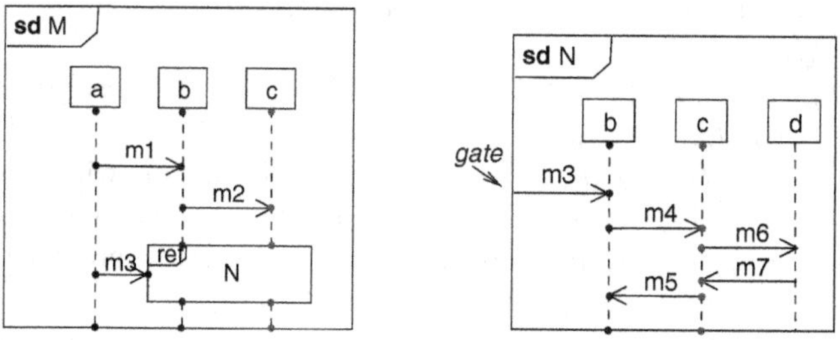

Fig. 6. Two sequence diagrams in UML 2.0 where `M` references `N`

Assume the associated models given in Fig. 7. Notice that in the model for `M`, event e (associated to the target location of message `m3`) is not associated in any way to the events of instances b and c and is concurrent to these events. The idea is that only in the refined model concrete relations are introduced. Only events

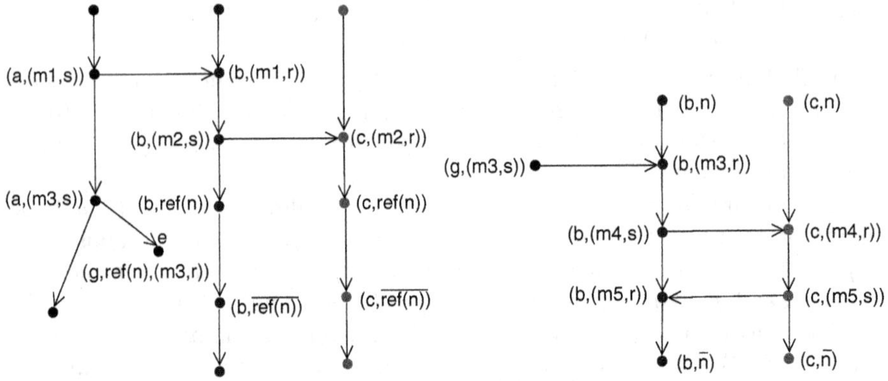

Fig. 7. Model for diagram M (left) and model for diagram N (right)

of instances *known* to M should be visible in a refined model for this diagram. This means that we hide all events relating to instance d from the model of N. It is the restricted model of diagram N that we show on Fig. 7 (right). Nonetheless, this is not an essential restriction for what follows.

Reference Refinement Diagram. Let $m, n \in D$, $M = (E_m, \mu_m)$ be a model for diagram m, and n be the referenced diagram in m to be refined within M. Let $N = (E_n, \mu_n)$ be the model for n. A *reference refinement diagram* for M and N is a synchronisation diagram $S = (E_{ref}, f_m, f_n)$ with $f_m : Ev_m \to Ev_{ref}$, $f_n : Ev_n \to Ev_{ref}$ both surjective and such that for $e \in Ev_m$ and $e' \in Ev_n$, $f_m(e)$ and $f_n(e')$ are defined and $f_m(e) = f_n(e')$ iff one of the following cases applies:

1. For $i \in I_m \cap I_n$, $\mu_m(e) = (i, \underline{ref(n)})$ and $\mu_n(e') = (i, n)$
2. For $i \in I_m \cap I_n$, $\mu_m(e) = (i, \overline{ref(n)})$ and $\mu_n(e') = (i, \overline{n})$
3. For $i \in I_n$ and $m_1 \in Mes$, $\mu_m(e) = (g, ref(n), (m_1, c))$ with $c \in \{s, r\}$ and $\mu_n(e') = (i, (m_1, c))$
4. For $i \in I_m$, $j \in I_n$ and $m_1 \in Mes$, if $\mu_m(e) = (i, ref(n), m_1, g)$ and $\mu_n(e') = (g, m_1, j)$ then $f_m(e) = f_n(e')$
5. For $i \in I_m$, $j \in I_n$ and $m_1 \in Mes$, $\mu_m(e) = (i, (m_1, c))$ with $c \in \{s, r\}$ and $\mu_n(e') = (g, (m_1, c))$ and such that there exist $e_1 \in Ev_m$ and $e_2 \in Ev_n$ with $\mu_m(e) = (g, ref(n), (m_1, \overline{c}))$ and $\mu_n(e') = (j, (m_1, \overline{c}))$ where $\overline{c}$ is the converse of c and $(e \to_m e_1 \wedge e' \to_n e_2) \vee (e_1 \to_m e \wedge e_2 \to_n e')$

The morphism f_m is only defined for events containing $ref(n)$ in their label (cases 1-4), or events that are in immediate causality (due to asynchronous communication) with events associated to gate locations on $ref(n)$ (case 5). The morphism f_n is only defined for events associated to initial and final locations in n or denoting communication with a gate.

The above reference refinement diagram does not take into account that diagram m may have a finite number of repeated references (say k) to diagram

n. In this case, we would consider a synchronisation diagram between M and $R = (E_r, \mu_r) = \coprod_k (E_n, \mu_n)$ where R is the coproduct in $\mathcal{L}(\mathbf{cev})$ of k identical structures.

We can now apply the categorical construction mentioned in Section 3.3 to the reference refinement diagram above and obtain automatically the expected refined model. Consider the reference refinement diagram as given in Fig. 8 for the models of Fig. 7, where f_m and f_n are defined over the following events

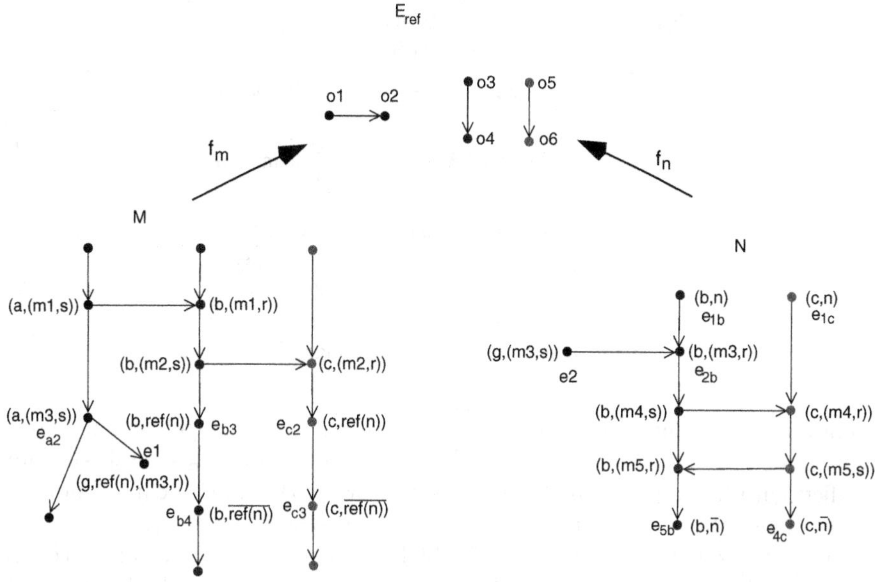

Fig. 8. Reference Refinement Diagram

only: $f_m(e_{a2}) = f_n(e_2) = o_1$, $f_m(e_1) = f_n(e_{2b}) = o_2$, $f_m(e_{b3}) = f_n(e_{1b}) = o_3$, $f_m(e_{c2}) = f_n(e_{1c}) = o_5$, $f_m(e_{b4}) = f_n(e_{5b}) = o_4$ and $f_m(e_{c3}) = f_n(e_{4c}) = o_6$.

It is not difficult to see that this is a valid reference refinement diagram: f_m and f_n are morphisms in **ev** and $f_m(Ev_m) = f_n(Ev_n)$.

Given this diagram we apply the categorical construction and obtain the diagram shown in Fig. 9 (for space reasons we cannot show the intermediate steps of the construction). The refined model contains events from Ev_m, events from Ev_n and pairs of events (e_1, e_2) where $e_1 \in Ev_m$ and $e_2 \in Ev_n$. The pairs of events correspond to the events synchronised through the reference refinement diagram. The relations are as expected preserved in the refined model. For the labelling function, individual events have the same label as before, and for synchronised events:

$$\mu(e_1, e_2) = \begin{cases} \mu_n(e_2) \Leftarrow \mu_n(e_2) \in \{(i, n), (i, \overline{n}), (i, (m_1, s)), (i, (m_1, r))\} \\ \mu_m(e_1) \Leftarrow \mu_m(e_1) \in \{(i, (m_1, s)), (i, (m_1, r))\} \\ (i, m_1, j) \Leftarrow \mu_m(e_1) = (i, ref(n), m_1, g) \wedge \mu_n(e_2) = (g, m_1, j) \end{cases}$$

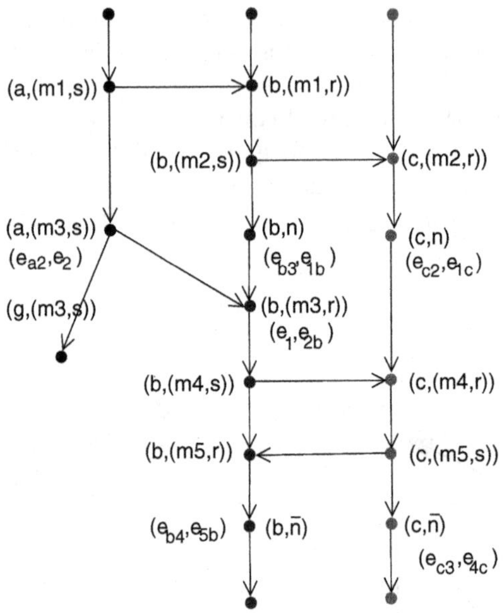

Fig. 9. Refined Model for Diagram M

Essentially synchronised events always keep the label of the refinement model event except in the case where the label of the event in Ev_n contains an unspecified gate instance. A further optimisation over this model can be done to remove (hide) those events labelled (i, n) or $(i, \overline{n})$ where $i \in \{b, c\}$.

Finally, the same approach works for lifeline decomposition (see instance a in Fig. 2). In this case we generally have a simpler unrefined model in the sense that there are only events for the gate locations along the lifeline (these events are concurrent) and at most related by immediate causality with another event of the unrefined model in case of asynchronous communication.

6 Conclusions

In this paper, we extend the sequence diagram semantics given in [2] to address refinement, and show how to obtain a refined model by means of a powerful categorical construction over two categories of labelled event structures. In UML terminology, we give a semantics to the new interaction decomposition mechanisms aiming at formal reasoning and the verification of complex scenario-based inter-object behavioural models. In general, the intention of this work is to offer automatic tools to analyse the properties of a UML model in such a way that software designers are not concerned with the details of the semantic model, and the feedback from these tools is reflected again at the UML model.

Existing work providing a trace-based semantics to UML 2.0 sequence diagrams includes [9, 10]. Whereas the former does not actually deal with decomposition, the latter only addresses one kind of decomposition (referencing).

Similarly, the standard algebraic semantics given to MSCs in [3] does not deal with decomposition. Our work is therefore, to the best of our knowledge, novel. A further difference lies in our choice of a true-concurrent semantic model.

We are currently exploring an extension of sequence diagrams and their semantics to allow *backtracking* of interactions. This is on the one side motivated by what is already possible in state diagrams, and on the other side by the idea of reversible interactions as found in biological systems and molecular interactions. Furthermore, it is a natural idea for describing interactions for web applications where the user may at any moment in time press the "back" and "forward" buttons of a web browser thus altering the expected interaction between the user and the application. Semantically, this requires an extension of our current distributed logic described in [2] to additionally express reversible behaviour. The main ideas are fundamentally close to the reversible extension of CCS called RCCS [11].

References

1. ITU: Recommendation Z.120: Message Sequence Chart (MSC). (1999)
2. Küster-Filipe, J.: Modelling concurrent interactions. Theoretical Computer Science **351**(2) (2006) Algebraic Methodology and Software Technology Special Issue.
3. Reniers, M.: Message Sequence Charts:Syntax and Semantics. PhD thesis, Eindhoven University of Technology (1998)
4. OMG: UML 2.0 Superstructure Specification. document ptc/04-10-02, available from www.uml.org. (2004)
5. Harel, D., Marelly, R.: Come, Let's Play: Scenario-based Programming Using LSCs and the Play-Engine. Springer (2003)
6. Winskel, G., Nielsen, M.: Models for Concurrency. In Abramsky, S., Gabbay, D., Maibaum, T., eds.: Handbook of Logic in Computer Science, Vol. 4, Semantic Modelling. Oxford Science Publications (1995) 1–148
7. Vaandrager, F.: A simple definition for parallel composition of prime event structures. Technical Report CS-R8903, Centre for Mathematics and Computer Science, P.O. Box 4079, 1009 AB Amsterdam, The Netherlands (1989)
8. Küster-Filipe, J.: Foundations of a Module Concept for Distributed Object Systems. PhD thesis, Technische Universität Braunschweig, Germany (2000)
9. Cengarle, M., Knapp, A.: Operational semantics of UML 2.0 interactions. Technical Report TUM-I0505, Institut für Informatik, TU München (2005)
10. Haugen, Ø., Husa, K., Runde, R., Stølen, K.: STAIRS towards formal design with sequence diagrams. Journal of Software and System Modeling **4**(4) (2005) 355–357
11. Danos, V., Krivine, J.: Reversible communicating systems. In Gardner, P., Yoshida, N., eds.: CONCUR 2004 - Concurrency Theory, 15th International Conference. Volume 3170 of LNCS., Springer (2004) 292–307

Verification of Communication Protocols Using Abstract Interpretation of FIFO Queues

Tristan Le Gall, Bertrand Jeannet, and Thierry Jéron

IRISA/INRIA Rennes, Campus de Beaulieu,
35042 Rennes cedex, France

Abstract. We address the verification of communication protocols or distributed systems that can be modeled by Communicating Finite State Machines (CFSMs), i.e. a set of sequential machines communicating via unbounded FIFO channels. Unlike recent related works based on acceleration techniques, we propose to apply the Abstract Interpretation approach to such systems, which consists in using approximated representations of sets of configurations. We show that the use of regular languages together with an extrapolation operator provides a simple and elegant method for the analysis of CFSMs, which is moreover often as accurate as acceleration techniques, and in some cases more expressive. Last, when the system has several queues, our method can be implemented either as an attribute-independent analysis or as a more precise (but also more costly) attribute-dependent analysis.

1 Introduction

Communicating Finite State Machines (CFSMs) [1, 2] is a simple model to describe distributed systems exchanging messages over an asynchronous network. This model consists of finite state processes that exchange messages via unbounded FIFO queues. Indeed, unbounded queues provide a useful abstraction that simplifies the semantics of specification languages, and frees the protocol designer from implementation details related to buffering policies and limitations. As a consequence, it is used to define the semantics of standardized protocol specification languages such as SDL and Estelle [3]. Despite its simplicity, the CFSM model cannot be easily verified: reachability is undecidable for CFSM [1], since unbounded queues can be used to simulate the tape of a Turing Machine.

Analysis of communicating systems. Two fundamental approaches have been followed for the analysis of communicating systems in general. One consists of eliminating the need for analyzing FIFO queues contents by adopting a partial order semantics or a so-called *true concurrency* model: when one process sends a message to another process, one just records the information that the emission precedes the reception. The seminal work about event structures [4] leads later to scenario-based models like (High-level) Message Sequence Charts [5, 6] incorporated in UML. The second approach, on which this paper focuses, consists in considering a model with explicit FIFO queues, namely the CFSM model described above, and in analyzing their possible contents during the execution of the system.

M. Johnson and V. Vene (Eds.): AMAST 2006, LNCS 4019, pp. 204–219, 2006.

The undecidability of the reachability of CFSM [1] does not prevent any verification attempt, but requires to give up with at least one of the following properties of an ideal method: an ideal method should indeed be (*i*) general (*i.e.* address any CFSM system), (*ii*) always terminate, and (*iii*) deliver exact results. Two main directions have mainly been explored so far: the first one abandon property (*i*) by simplifying the model or considering only a subclass of it, whereas the second one prefer to abandon property (*ii*) by looking only for efficient semi-algorithms that may not terminate but deliver exact results "often enough". *Lossy channels systems* illustrate both directions. They are CFSMs where the channels can lose messages at any time. Those systems are easier to verify than perfect channels systems [7]: the reachability problem is decidable, but there is no effective algorithm to compute the reachability set. However, an on-the-fly analysis semi-algorithm based on *simple regular expressions* is given in [8]. This algorithm can "accelerate" loops, that is, it is able to compute the effect of any *meta transition* (loops in the control transition systems). The termination problem remains because the number of loops is potentially infinite. This *acceleration* approach has been generalized to standard CFSMs systems (*cf.* section 3), leading to various semi-algorithms applying the acceleration principle on different representations for queues contents.

We propose here an alternative tradeoff to face the undecidability problem, which is to keep generality and termination (properties (*i*) and (*ii*)) and to give up with the exactness of the results (property (*iii*)). When analyzing CFSMs, this consists in replacing in dataflow equations, sets of FIFO channel configurations by abstract properties belonging to a lattice. Such a transformation results in conservative approximations: we will be able to prove a safety property, or the non-reachability of a state, but not to prove that a property is false or that a state is effectively reachable. The abstractions we propose in this paper are all based on regular languages, which exhibit among nice properties the closure under all Boolean operations, and a canonical representation with deterministic and minimized finite automata.

Contributions. We show in this paper that our abstract-interpretation based method presents several advantages: it is arguably technically less involving than acceleration-based techniques, it often returns exact results on cases where the acceleration techniques terminate, and relevant information in the other cases where the acceleration techniques do not terminate and do not provide any result, either because the control structure of the system is too intricate, or because the reachable set cannot be represented with the chosen representation. Our method can also be seen as complementary to acceleration techniques when they fail. Last, although acceleration techniques have been applied to other infinite datatypes (counters [9], etc), it is not clear whether they can be easily *combined*, whereas general methods are available for combining different abstract domains.

Outline. We introduce in section 2 the model of communicating finite state machines, and the analysis problem we address, namely reachability analysis. We discuss the related works in section 3. We then explain our approach for the reachability analysis of CFSMs in the case of one FIFO channel (section 4).

In section 5 we generalize it to the case of several FIFO channels. We implemented our method and we present in section 6 a few case studies on which we experimented it, and we compare it with other techniques.

2 Finite Automata and Communicating Finite State Machines

Finite automata. A finite automaton is a 5-tuple $\mathcal{M} = (Q, \Sigma, Q_0, Q_f, \rightarrow)$ where Q is a finite set of states, Σ a finite alphabet, $Q_0, Q_f \subseteq Q$ are the sets of initial and final states, and $\rightarrow \subseteq Q \times \Sigma \times Q$ is the transition relation. The relation $\rightarrow$ is extended on words as the smallest relation $\Rightarrow \subseteq Q \times \Sigma^* \times Q$ satisfying: (i) $\forall q \in Q : q \overset{\epsilon}{\Rightarrow} q$ and (ii) if $q \xrightarrow{a} q'$ and $q' \overset{w}{\Rightarrow} q''$, then $q \overset{a \cdot w}{\Rightarrow} q''$. $\mathcal{M}$ is deterministic if $Q_0 = \{q_0\}$ and if $\rightarrow$ defines a function $Q \times \Sigma \rightarrow Q$. A word $w \in \Sigma^*$ is *accepted by* $\mathcal{M}$ if $\exists q_0 \in Q_0, \exists q_f \in Q_f : q_0 \overset{w}{\Rightarrow} q_f$. The language $L(\mathcal{M})$ accepted by $\mathcal{M}$ is the set of accepted words. Conversely, given a regular language $L \in \wp(\Sigma^*)$, the unique (up to isomorphism) minimal deterministic automaton (MDA) accepting L is denoted by $\mathcal{M}(L)$. The set of regular languages on alphabet Σ is denoted by $\mathcal{R}(\Sigma)$. Given an automaton $\mathcal{M} = (Q, \Sigma, Q_0, Q_f, \rightarrow)$ and an equivalence relation $\simeq$ on its states, $\mathcal{M}/\simeq\; = (Q/\simeq, \Sigma, \widetilde{Q_0}, \widetilde{Q_f}, \widetilde{\rightarrow})$ denotes the *quotient automaton* defined in the usual way : the states of $\mathcal{M}/\simeq$ are the equivalence classes of $\simeq$, $\overline{q} \in Q/\simeq$ is an initial (resp. final) state if one state of this equivalence class is an initial (resp. final) state of $\mathcal{M}$, and $(\overline{q}, a, \overline{q'}) \in \widetilde{\rightarrow}$ if $\exists q \in \overline{q}, \exists q' \in \overline{q'}, (q, a, q') \in \rightarrow$. For any equivalence relation $\simeq$, we have $L(\mathcal{M}) \subseteq L(\mathcal{M}/\simeq)$.

Definition 1 (CFSM). *A Communicating Finite State Machine is given by a tuple* (C, Σ, c_0, Δ) *where:*

- C *is a finite set of locations (control states)*
- $\Sigma = \Sigma_1 \cup \Sigma_2 \cup \cdots \cup \Sigma_n$ *is a finite alphabet of messages, where* Σ_i *denotes the alphabet of messages that can be stored in queue* i*;*
- $c_0 \in C$ *is the initial location;*
- $\Delta \subseteq C \times A \times C$ *is a finite set of transitions, where* $A = \bigcup_i \{i\} \times \{!, ?\} \times \Sigma_i$ *is the set of actions. An action can be*
  - *either an output* $i!m$*: "the message* m *is sent through the queue* i*";*
  - *or an input* $i?m$*: "the message* m *is received from the queue* i*".*

In the examples, we define CFSMs in terms of an asynchronous product of finite state machines (FSMs) reading and writing on queues.

Example 1. The connexion/deconnexion protocol between two machines is the following (Fig. 1): the client can open a session by sending the message open to the server. Once a session is open, the client may close it on its own by sending the message close or on the demand of the server if it receives the message disconnect. The server can read the request messages open and close, and ask for a session closure.

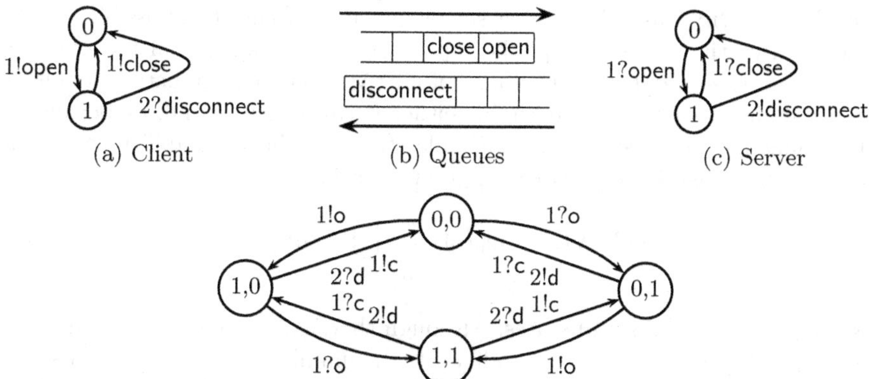

(a) Client (b) Queues (c) Server

(d) Global CFSM: product of client and server processes

Fig. 1. The connexion/deconnexion protocol

Semantics. The semantics of a CFSM (C, Σ, c_0, Δ) is given as a labelled transition system (LTS) $\langle Q, Q_0, A, \rightarrow \rangle$ where

- $Q = C \times \Sigma_1^* \times \cdots \times \Sigma_n^*$ is the set of states;
- $Q_0 = \{\langle c_0, \varepsilon, \ldots, \varepsilon \rangle\}$ is the set of the initial states;
- A is the alphabet of actions (*cf.* Def. 1).
- $\rightarrow$ is defined by the two rules:

$$\frac{(c_1, i!m, c_2) \in \Delta \qquad w_i' = w_i \cdot m}{\langle c_1, w_1, \ldots, w_i, \ldots, w_n \rangle \rightarrow \langle c_2, w_1, \ldots, w_i', \ldots, w_n \rangle}$$

$$\frac{(c_1, i?m, c_2) \in \Delta \qquad w_i = m.w_i'}{\langle c_1, w_1, \ldots, w_i, \ldots, w_n \rangle \rightarrow \langle c_2, w_1, \ldots, w_i', \ldots, w_n \rangle}$$

A global state of a CFSM is thus a tuple $\langle c, w_1, \ldots, w_n \rangle \in C \times \Sigma_1^* \times \cdots \times \Sigma_n^*$ where c is the current location and w_i is a finite word on Σ_i representing the content of queue i. At the beginning, all queues are empty, so the *initial state* is $\langle c_0, \varepsilon, \ldots, \varepsilon \rangle$. The reflexive transitive closure $\rightarrow^*$ is defined as usual. A state $\langle c, w_1, \ldots, w_n \rangle$ is *reachable* if $\langle c_0, \varepsilon, \ldots, \varepsilon \rangle \rightarrow^* \langle c, w_1, \ldots, w_n \rangle$. The *reachability set* is the set of all states that are reachable. Computing this set is the purpose of the reachability analysis. We can achieve this computation by solving a fix-point equation, as shown in the next paragrah.

Forward collecting semantics and reachability analysis of a CFSM. The forward collecting semantics defines the semantics of a system in terms of its reachable set. A set of states $X \in \wp(Q) = \wp(C \times \Sigma_1^* \times \cdots \times \Sigma_n^*)$ can be viewed as a map $X : C \rightarrow \wp(\Sigma_1^* \times \cdots \times \Sigma_n^*)$ associating a control state c with a language $X(c)$ representing all possible contents of the queues when being in the control state c. The forward semantics of actions $[\![a]\!] : \wp(\Sigma_1^* \times \cdots \times \Sigma_n^*) \rightarrow \wp(\Sigma_1^* \times \cdots \times \Sigma_n^*)$ is defined as:

$$[\![i!m]\!](L) = \{\langle w_1, \ldots, w_i \cdot m, \ldots, w_n \rangle | \langle w_1, \ldots, w_i, \ldots, w_n \rangle \in L\} \qquad (1)$$

$$[\![i?m]\!](L) = \{\langle w_1, \ldots, w_i, \ldots, w_n \rangle | \langle w_1, \ldots, m \cdot w_i, \ldots, w_n \rangle \in L\} \qquad (2)$$

$[\![i!m]\!]$ (resp. $[\![i?m]\!]$) associates to a set of queues contents the possible queues contents after the output (resp. the input) of the message m on the queue i, according to the operational semantics of CFSM. Using the inductive definition of reachability — a state is reachable either because it is initial, or because it is the immediate successor of a reachable state —, the reachability set RS is defined as the least solution of the fixpoint equation

$$\forall c \in C,\ X(c) = Q_0(c) \cup \bigcup_{(c',a,c)\in\Delta} [\![a]\!](X(c')) \tag{3}$$

where Q_0 is the initial set of states. Although there is no general algorithm that can compute exactly such a reachability set [1], a number of semi-algorithms that compute the reachability set in some cases have been designed and are described in the next section.

3 Related Works

Semi-algorithms based on the acceleration paradigm. The acceleration paradigm is a popular paradigm for infinite state systems, which we describe in the specific case of CFSM. Eq. (3) is difficult to solve in presence of cycles in the control graph, because iterative solving using Kleene's theorem will not converge. Now, assuming a canonical representation $\mathcal{L}$ for queue contents, given a loop $\theta \stackrel{\triangle}{=} c = c_0 \xrightarrow{a_1} c_1 \xrightarrow{a_2} \dots \xrightarrow{a_k} c_k = c$ and a language $L \in \mathcal{L}$, we may compute in a single step the effect of the loop θ, i.e. finding a language $[\![\theta^*]\!](L) \in \mathcal{L}$ representing the set of states that can be reached from a state in L following the loop θ an arbitrary number of times. Then, when exploring the state space, we can substitute the entire loop by the single *meta-transition* θ^*. However, even if each loop may be accelerated, we still have to explore an infinite transition system since there is an infinite number of loops. We may exploit some termination conditions [10] or use heuristics that lead to semi-algorithms: for example, we may "flatten" the transition system and find a proper exploration order [9]. In the cases of systems with FIFO channels, this technique has been applied with different kind of representations, depicted in Tab. 1. Usually only forward analysis has been studied. Observe that when several channels are involved in a loop, with some representations, the acceleration is not always possible. [11] provides a detailed comparison of the cited references.

Algorithms based on transducer iterations. Instead of extrapolating sequences of values, one may also extrapolate the full relations $L_{i+1} = R(L_i)$ linking two successive terms, represented as a regular transducer R (in this case, the full state is encoded as a regular word). The computation of the transducer R^* allows the computation of the reachability set. This *regular model-checking* paradigm [14] has mainly been applied to networks of finite state machines. A method to compute the transducer R^* is given in [15], but will not work for any CFSM. [16, 17] define extrapolation operators to compute an over-approximation of R^*, but has experimented them only on one lossy FIFO system [17].

Table 1. Acceleration techniques on CFSMs

queue	representation and typical example	attr.[a] dependent	acceleration with[b] single/several queue	ref.
lossy	SRE[1] : $\sum(a+\epsilon)+(a_1+\ldots+a_m)^*$	no	always / always	[8]
perfect	SLRE[2]: $\sum a_1a_2(b_1b_2)^*a_3(b_3)^*(b_4)^*\ldots$	no	always / sometimes	[11]
perfect	QDD[3] : n-dim regular expression	yes	always / sometimes	[12]
perfect	CQDD[4]: $\sum a_1^{p_1}a_2^{p_2}x_1^{q_1}x_2^{q_2} \mid p_1+2q_1 \leq p_2+q_2$	yes	always / always	[13]

[a] yes if one expression for all queues, no if one expression for each queue
[b] ability to exactly compute the effect of meta-transition
[1] Simple Regular Expressions [2] SemiLinear Regular Expressions
[3] Queue Decision Diagrams [4] Constrained QDD, using Presburger formulas

Decidable subclasses of CFSMs. Reachability has been shown decidable for *monogeneous* [18], *linear* [19] or half-duplex [20] CFSMs. Allowing the channels to be lossy makes also the problem decidable [21, 7]. A recent research direction focuses on probabilisitic lossy channels [22].

Approximated techniques. Besides techniques based on the generation of finite abstract models that are then model-checked,

abstractions have also been experimented on FIFO queues using the classical dataflow analysis framework, hence restricting to lattices of properties satisfying the ascending chain condition (*i.e.* there is no infinite ascending chain). For instance, [23] proposes an analysis that infers the emptiness property and the possible values of the first element to be read in queues. [24] proposes a "widening operator" for decreasing sequences of regular languages, in the same spirit as [16]. However it does not guarantee the convergence of the sequence.

4 Analyzing Systems with Only One Queue

In this section we consider the simple case of CFSMs with a single FIFO queue, on which we describe our method based on abstract interpretation [25].

With a single queue, the concrete state-space has the structure $C \rightarrow \wp(\Sigma^*)$, and it will be abstracted by the set $C \rightarrow A$, where A is an abstract lattice equipped with a meaning or concretization function $\gamma : A \rightarrow \wp(\Sigma^*)$ (*i.e.* γ is monotone and $\gamma(\bot) = \emptyset$). We will consider for A the set of regular languages $\mathcal{R}(\Sigma)$ over Σ, with $\gamma : \mathcal{R}(\Sigma) \rightarrow \wp(\Sigma^*)$ being the identity. This simple solution presents several interesting properties:

- $\mathcal{R}(\Sigma)$ is closed under union, intersection, negation and semantic transformers $[\![!m]\!]$ (corresp. to concatenation) and $[\![?m]\!]$ (corresp. to the derivative operator of [26]). Moreover, $Q_0 = \{\langle c_0, \epsilon\rangle\}$ is regular, so that all operators involved in Eq. (3) can be transposed to $\mathcal{R}(\Sigma)$ without loss of information.
- From a computational point of view, regular languages have as a standard canonical representation the minimal deterministic automaton (MDA) recognizing them.

As a consequence, we only have to define a suitable widening operator to ensure convergence of iterative resolution of Eq. (3). Indeed, the lattice $\mathcal{R}(\Sigma)$ does not satisfy the ascending chain condition and is even not complete[1]: it is well-known that the monotone sequence $L_n = \{a^k b^k \mid k \leq n\}$ converges towards a context-free language which is not regular.

Generally speaking, a widening operator is a binary operator $\nabla : A \times A \to A$ satisfying technical conditions (*c.f.* proposition 1) that ensure, in the context of the iterative resolution of a fixpoint equation $X = F(X)$, that the sequence $X_0 = F(\bot), X_{i+1} = X_i \nabla F(X_i)$ converges in a finite number of steps towards a post-fixpoint of F. In general, a widening operator tries to capture and to extrapolate the difference between its two arguments X_i and $F(X_i)$, by making the hypothesis that the difference will be repeated in the sequence X_i, $F(X_i)$, $F(F(X_i))$,.... The main difference with acceleration techniques is that the widening, at least in its basic definition, does not exploit the semantic function F (which is defined by the analyzed system), but is defined solely on abstract values. This is both a weakness — it is then more difficult to make a good or even an exact guess, and a strength — a highly complex function F is not a difficulty, whereas acceleration-based techniques may fail in such cases (non-flat systems, nested loops, ...).

4.1 Widening Operator

In our case, the choice of ∇ is all the more important as all approximations performed by the analysis will depend on its application. Because of the FIFO operations, the widening operator should remain precise for both the begining and the end of the queue. It also should induce intuitive approximations. In [27], a widening operator for regular languages was mentioned. We will adapt this operator to regular languages representing the content of a FIFO channel.

This widening operator will be based on an extensive and idempotent operator $\rho_k : \mathcal{R}(\Sigma) \to \mathcal{R}(\Sigma)$ (*i.e.* $\rho_k(X) \supseteq X$ and $\rho_k \circ \rho_k = \rho_k$), where $k \in \mathbb{N}$ is a parameter. ρ_k will induce a widening operator defined by $X_1 \nabla_k X_2 = \rho_k(X_1 \cup X_2)$. Thus, the proposed widening does not work by extrapolating a difference, but by simplifying the regular languages generated during the iterative resolution. The operator ρ_k is defined on a language L by considering the automaton $\mathcal{M}(L)$ quotiented by a bisimulation up to depth k.

Definition 2 (Bisimulation of depth k). *Let $(Q, \Sigma, Q_0, Q_f, \to)$ be a minimal deterministic automaton and* col $: Q \to [1..N]$ *a color function defining an equivalence relation $q_1 \approx_{\mathrm{col}} q_2 \Leftrightarrow \mathrm{col}(q_1) = \mathrm{col}(q_2)$. For $k \geq 0$, the smallest bisimulation of depth k finer than $\approx_{\mathrm{col}}$ is defined inductively by: $\forall q_1, q_2 \in Q$,*

$q_1 \approx_0^{\mathrm{col}} q_2$ *iff* $q_1 \approx_{\mathrm{col}} q_2$

$$q_1 \approx_{k+1}^{\mathrm{col}} q_2 \text{ iff } \begin{cases} q_1 \approx_k^{\mathrm{col}} q_2 \\ \forall a \in \Sigma, \forall q_1' \in Q,\ q_1 \xrightarrow{a} q_1' \implies \exists q_2' \in Q \ :\ q_2 \xrightarrow{a} q_2' \wedge q_1' \approx_k^{\mathrm{col}} q_2' \\ \forall a \in \Sigma, \forall q_2' \in Q,\ q_2 \xrightarrow{a} q_2' \implies \exists q_1' \in Q \ :\ q_1 \xrightarrow{a} q_1' \wedge q_1' \approx_k^{\mathrm{col}} q_2' \end{cases}$$

[1] It is precisely because $A = \mathcal{R}(\Sigma)$ is not complete that we cannot define an abstraction function $\alpha : \wp(\Sigma^*) \to \mathcal{R}(\Sigma)$ as it is usually done in abstract interpretation.

In this section, we consider the *standard color function*, which uses $N = 4$ colours for separating initial and final states from other states:

$$\mathrm{col}(q) = 1 \text{ if } q \in Q_0 \cap Q_f,\ 2 \text{ if } q \in Q_f \setminus Q_0,\ 3 \text{ if } q \in Q_0 \setminus Q_f,\ 4 \text{ otherwise} \quad (4)$$

Definition 3 (Operator ρ_k^{col}.). *Given a bisimulation relation $\approx_k^{\mathrm{col}}$ of depth k the operator $\rho_k^{\mathrm{col}} : \mathcal{R}(\Sigma) \to \mathcal{R}(\Sigma)$ is defined by quotienting the MDA of L:*

$$\rho_k^{\mathrm{col}}(L) = \mathcal{L}(\mathcal{M}(L)/ \approx_k^{\mathrm{col}})$$

ρ_k^{col} is extensive as being defined by a quotient automaton, and it is idempotent as a consequence of $\approx_k^{\mathrm{col}}$ being a bisimulation relation. As $\approx_{k+1}^{\mathrm{col}} \subseteq \approx_k^{\mathrm{col}}$, we also have $\forall L \in \mathcal{R}(\Sigma) : \rho_{k+1}(L) \subseteq \rho_k(L)$. However, ρ_k is not monotone, as shown by the following example: $a^4 \subseteq a^4 + a^2b$, but $\rho_1(a^4) = a^3a^*$ is not comparable to $\rho_1(a^4 + a^2b) = a^4 + a^2b$.

Definition 4 (Widening operator ∇_k^{col}). *Given an integer $k \geq 0$ and a color function* col, *we define a binary operator $\nabla_k^{\mathrm{col}} : \mathcal{R}(\Sigma) \times \mathcal{R}(\Sigma) \to \mathcal{R}(\Sigma)$:*

$$L_1 \nabla_k^{\mathrm{col}} L_2 \triangleq \rho_k^{\mathrm{col}}(L_1 \cup L_2)$$

Proposition 1. *∇_k^{col} is a widening operator for $\mathcal{R}(\Sigma)$ in the sense of [25]:*

1. *$L_1 \cup L_2 \subseteq L_1 \nabla_k^{\mathrm{col}} L_2$;*
2. *For any increasing chain ($L_0 \subseteq L_1 \subseteq \dots$), the increasing chain defined by $L'_0 = L_0$, $L'_{i+1} = L'_i \nabla_k^{\mathrm{col}} L_{i+1}$ is not strictly increasing (it stabilizes after a finite number of steps).*

This property ensures the global correctness of our analysis [25].

Proof. 1. The language recognized by a quotient automaton is a superset of the language of the initial automaton. 2. Given a deterministic automaton $(Q, \Sigma, Q_0, Q_f, \to)$ and a color function col : $Q \to [1..N]$, we have $|Q/ \approx_k^{\mathrm{col}} | \leq N^{|\Sigma|^{k+1}} \times 2^{|\Sigma|^k}$ (proved in [28]). Thus the set $\{\rho_k^{\mathrm{col}}(L) \,|\, L \in \mathcal{R}(\Sigma)\}$ is finite.

4.2 Effects of the Widening Operator

We analyze here in detail the effect of the extensive operator ρ_k on a language, using the color function of Eq. (4).

Sum of languages: If $L = L_1 \cup L_2$, the widening operator may merge some subwords of L_1 with subwords of L_2. For instance, $\rho_1(aax + bay) = (a + b)a(x + y)$; we thus lose the property " we have an 'a' at the beginning of the queue iff we have an 'x' at the end".

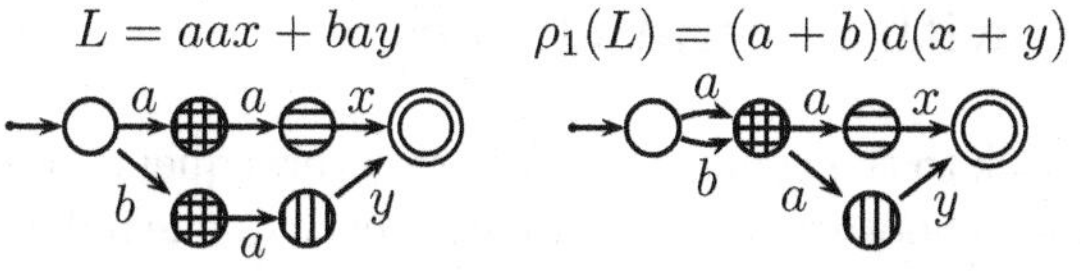

Repetition: an important effect of ρ_k is to introduce Kleene closures in regular expressions. We have $\rho_k(a^n) = \begin{vmatrix} a^{k+2}a^* & \text{if } k < n-2 \\ a^n & \text{otherwise} \end{vmatrix}$: the repetition of a letter beyond some number is thus abstracted by an unbounded repetition. The same happens for the repetition of bounded-length words: for $n \geq 3$, $\rho_k((a_1 \ldots a_k)^n) = (a_1 \ldots a_k)(a_1 \ldots a_k)^*$. If the system allows arbitrarily-long channel contents, this approximation can guess the limit of the fix-point computation. If a letter is repeated at different places, the two Kleene stars may be merged: for instance $\rho_1(ax^3bx^3c) = ax^+(bx^+)^*c$, instead of the (preferable) ax^+bx^+c:

$L = ax^3bx^3c$ $\qquad\qquad$ $\rho_1(L) = ax^+(bx^+)^*c$

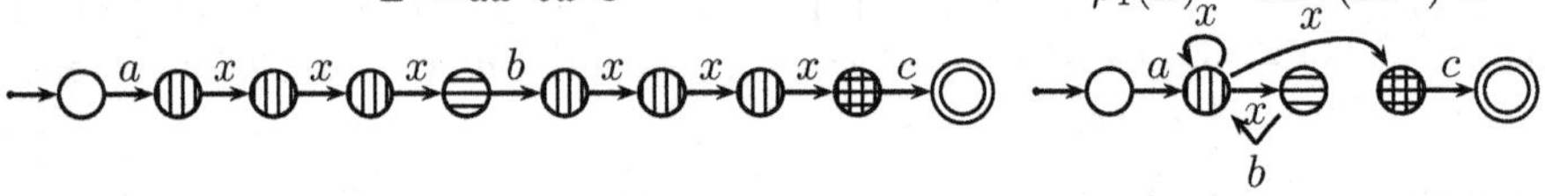

One can improve the widening for the two previous situations, by considering a color function col2 which also separates states according to the set of letters already encountered from the initial states. One has $\rho_1^{\text{col2}}(L) = ax^+bx^+c$. This allows to propagate non-local properties in the FIFO queue.

Suffixes and prefixes: we have the following properties:

Proposition 2. *[28] L and $\rho_k(L)$ have the same set of prefixes of length 1 and the same set of suffixes of length less or equal to k.*

Thus, the k last messages written in a queue are not abstracted. As a consequence, we wait enough before trying to capture some regularity with the operator ρ_k. Notice than one can improve the result for prefixes by combining forward with *backward* bisimulation relations.

Surprisingly, this simple widening has not yet been experimented for the analysis of CFSMs. Our contribution here is to adapt for FIFO queues the widening mentioned in [27], by choosing an appropriate color function, and to demonstrate its practical relevance in this context (*c.f.* section 6).

4.3 Complexity of the Analysis

The operations we perform on finite automata are polynomial and rather efficient in practice. The complexity of our analysis depends also on the number of steps of the fixpoint computation. This number is quite small on the examples of section 6 (≤ 12, with $\rho_{k\leq 2}$), but the theoretical bound is exponential in the size of the alphabet and double-exponential in k. We conjecture than even on larger examples, the practical complexity remains much below this bound.

5 Systems with Several Queues

We now come back to the general case where several queues are to be analyzed. In this case, we must choose whereas we analyse each queue independently, using the method of the previous section, or we analyse all the queues together. In the first case, according to the classification of [29], we obtain an *attribute-independent* analysis based on a *non-relational* abstraction, because properties on different queues are not inter-related. In the second case, we obtain an *attribute-dependent* analysis based on a *relational* abstraction, in which one can represent properties like "queue 1 contains 'a' messages iff queue 2 contains 'b' messages". We propose both solutions.

Concrete representation. In the previous section, a configuration was a word. Now a configuration is defined by n words $w_1, \dots, w_n$ which can be represented:

1. as a vector of words $\langle w_1, \dots, w_n \rangle$
2. as a single word $w_1 \sharp \dots \sharp w_n$ obtained by concatenation and the addition of a separation letter $\sharp$
3. or as an "interleaved" word $w_1^0 \dots w_n^0 w_1^1 \dots w_n^1 \dots$

The third representation is used for representing sets of unbounded integer vectors with NDDs [30], but it is not suited to the FIFO operations. We will consider the two first representations that naturally define two different analyses.

5.1 Non-relational Abstraction

Here we adopt the view of a configuration as a vector of words, and we abstract each component independently: we take

$$A^{nr} = \mathcal{R}(\Sigma_1) \times \dots \times \mathcal{R}(\Sigma_n)$$

as an abstract lattice, ordered component-wise. The meaning function γ^{nr} : $A^{nr} \rightarrow \wp(\Sigma_1^* \times \dots \times \Sigma_n^*)$ is defined by

$$\gamma^{nr}(\langle L_1, \dots, L_n \rangle) = \gamma(L_1) \times \dots \times \gamma(L_n)$$

The widening ∇_k of section 4 is extended to A^{nr} component-wise: $\langle L_1, \dots, L_n \rangle \nabla_k \langle L'_1, \dots, L'_n \rangle = \langle L_1 \nabla_k L'_1, \dots, L_n \nabla_k L'_n \rangle$, which defines a proper widening operator. Sending or receiving a message on the queue i consists in modifying the component i of the abstract value. In this lattice, the least upper bound ("the union") is no longer exact, because of the cartesian product. For example, the upper bound of the values $\langle a, x \rangle$ and $\langle b, y \rangle$ is the language $\langle a+b, x+y \rangle$. Hence, the loss of information is no longer only due to the widening operator.

5.2 Relational Abstraction

If we adopt instead the view of a configuration as a concatenated word, we obtain the QDD representation of [12], to which we apply the principles of section 4:

$$A^r = \mathcal{R}(\Sigma \cup \{\sharp\}) \tag{5}$$

$$\gamma^r(X) = \{\langle w_1, \dots, w_n \rangle \in \Sigma_1^* \times \dots \times \Sigma_n^* \mid w_1 \sharp \dots \sharp w_n \in X\} \tag{6}$$

We implicitly restrict A^r to sets of concatenated words of the form described above. The only difference with [12] is the use of widening instead of acceleration. This representation allows to represent relations or dependencies between queues. For instance the language L of Fig. 2 encodes the relation "the queue 1 starts with an 'a' iff the queue 2 contains an 'x' ".

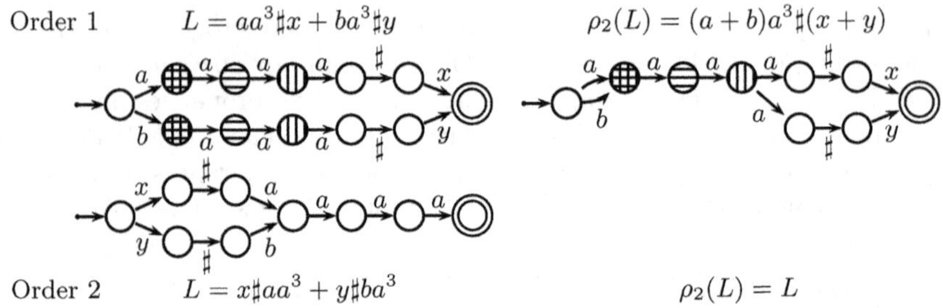

Fig. 2. Widening and ordering of queues in concatenated words

Operations. The union, intersection and inclusion test operations are the natural extensions of their counterpart for an automaton representing a single queue. However, we have to adapt the operations $[\![i!m]\!]$, $[\![i?m]\!]$ and ∇_k. As each word recognized by a MDA $M = \mathcal{M}(L)$ with $L \in \mathcal{R}(\Sigma \cup \{\sharp\})$ is a concatenated word separated by $\sharp$ letters, each state $q \in Q$ of M can be associated to one queue-content by a function $\eta : Q \to [1..n]$, and can be characterized as initial, and/or final for this queue [12, 28]. Given such a partition, the operations $[\![i!m]\!]$ and $[\![i?m]\!]$ are easily implemented. Concerning the widening operator, it should avoid to merge the different queue contents, and preserve the invariant that each word has $n-1$ $\sharp$ letters. We thus adapt the standard color function, which uses now $N = 4n$ colours:

$$\mathrm{col}(q) = \begin{cases} 4*\eta(q)-3 & \text{if } q \text{ is both an initial and a final state for the queue } \eta(q) \\ 4*\eta(q)-2 & \text{if } q \text{ is a final (but not initial) state for the queue } \eta(q) \\ 4*\eta(q)-1 & \text{if } q \text{ is an initial (but not final) state for the queue } \eta(q) \\ 4*\eta(q) & \text{otherwise} \end{cases} \quad (7)$$

Impact of the ordering. A natural question arises: to which extent is our relational analysis dependent on the chosen ordering for queues ? All the exact operations, which do not lose information, do not depend on it. However, the widening is dependent on the ordering of queues, as shown by the example of Fig. 2. Consequently, our analysis depends on the ordering. A widening operator which would be independent of the ordering would have been more satisfactory, but we did not find out yet such a widening operator, with good properties w.r.t. precision and efficiency (see the discussion in [28]).

6 Experiments and Comparisons

The approach we followed for the analysis was to sacrifice exactness for universality of the analyzed model and convergence guarantee. Of course such an approach is relevant only if the approximations introduced are not too strong, and if they still allow to obtain interesting results. In order to perform this experimental evaluation, we implemented both the non-relational and the relational abstractions, and we connected them to a generic fixpoint calculator, that takes

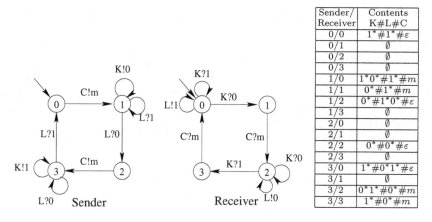

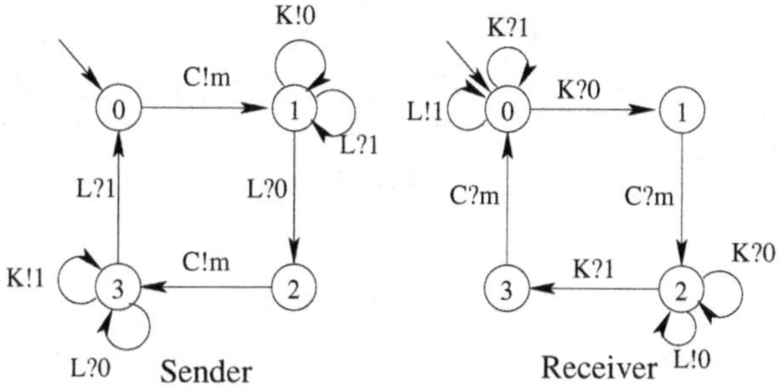

Sender/ Receiver	Contents K#L#C
0/0	$1^*\#1^*\#\varepsilon$
0/1	$\emptyset$
0/2	$\emptyset$
0/3	$\emptyset$
1/0	$1^*0^*\#1^*\#m$
1/1	$0^*\#1^*\#m$
1/2	$0^*\#1^*0^*\#\varepsilon$
1/3	$\emptyset$
2/0	$\emptyset$
2/1	$\emptyset$
2/2	$0^*\#0^*\#\varepsilon$
2/3	$\emptyset$
3/0	$1^*\#0^*1^*\#\varepsilon$
3/1	$\emptyset$
3/2	$0^*1^*\#0^*\#m$
3/3	$1^*\#0^*\#m$

Fig. 3. The alternating bit protocol

care of the iterative resolution of fixpoint equations and applies widening following the principles of [31]. All our experiments used the ∇_1 widening operator based on the standard color function, and returned their result in less than 1 sec. on a 2 GHz Intel™ Pentium computer. The fixpoint was obtained in 7 to 12 iteration steps, depending on the examples.

The Alternating Bit Protocol (ABP) is a data-transmission protocol, between a sender S and a receiver R. S transmits some data package m through a FIFO channel C and R and S exchange some information (one-bit messages) through two channels K and L (Fig. 3). We performed a relational analysis of the CFSM modeling this protocol (Fig. 3). It shows that some control states are not reachable and that there is at most one message in data channel C. As in [12, 32], we obtain the exact result. Notice that in this case, a simpler non-relational analysis delivers the same results.

The connexion/deconnexion protocol, defined in Example 1, demonstrates the usefulness of a relational analysis:

Relational Analysis

Client/ Server	Queue 1 # Queue 2
0/0	$(co)^*(oc)^*\#\varepsilon\ +\ c(oc)^*\#d$
1/0	$(co)^*(oc)^*o\#\varepsilon\ +\ (co)^*\#d$
0/1	$c(oc)^*\#\varepsilon$
1/1	$(co)^*\#\varepsilon$

Non-Relational Analysis

Client/ Server	Queue 1	Q.2
0/0	$o^* + (o^*c)^+(\varepsilon + o^+ + o^+c)$	d^*
1/0	$(o^*c)^*o^+$	d^*
0/1	$o^* + (o^*c)^+(\varepsilon + o^+ + o^+c)$	d^*
1/1	$o^+ + o^*(co^+)^+$	d^*

The result given by the relational analysis happens to be the exact reachability set, unlike the non-relational one. The non-relational analysis misses the fact that there is at most one d in the second queue, which induces many approximations.

A non-regular example. Our abstraction can deal with cases where the reachability set is not regular. Let us consider the CFSM depicted in Fig. 4. Each process can send a message a or c, and a synchronisation is guaranteed by the messages b and d.

In location (0/0), the content of the queues will be $a^n\sharp\varepsilon\sharp c^n\sharp\varepsilon$ with $n \geq 0$. This set is non-regular, and thus cannot be represented by a regular expression. Our method will find an over-approximation of the exact reachability set. In location (0/0) the queue-content we found is represented by the language :

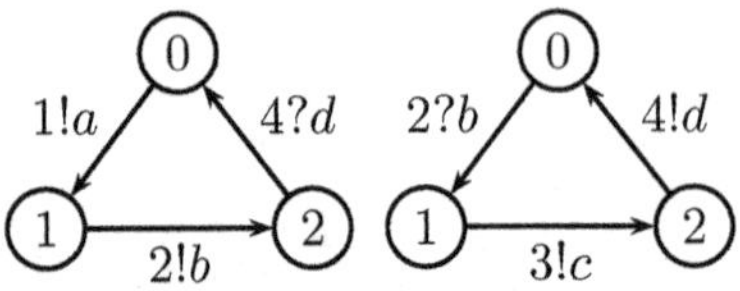

Fig. 4. A non-regular protocol

$$L_{(0/0)} = \varepsilon\sharp\varepsilon\sharp\varepsilon\sharp\varepsilon + a\sharp\varepsilon\sharp c\sharp\varepsilon + aaa^*\sharp\varepsilon\sharp ccc^*\sharp\varepsilon$$

This example shows that our method may give a good over-approximation of a non-regular reachability set.

A protocol with nested loops is depicted in Fig. 5, which is an abstraction of systems exchanging frames composed of several packets.

The sender first sends a start message, then sends any number of a messages and ends the frame with an end message. The receiver can read any message at any time.

Our analysis shows that, when the sender is in location 0, the content of the queue is :

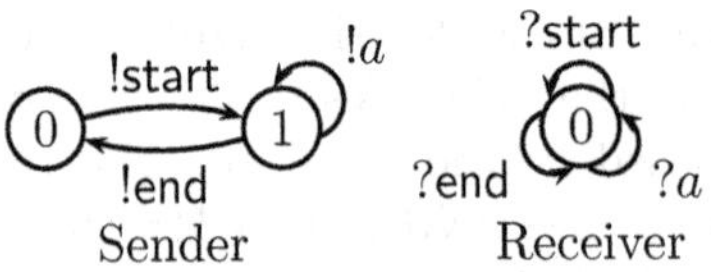

Fig. 5. Nested loop

$$L_0 = \varepsilon + (s + \varepsilon)a^*e(sa^*e)^*$$

Here the ability of representing regular expressions with nested Kleene closures is important; in this case we even obtain the exact reachability set.

Comparison with acceleration techniques. In Tab. 2 we compare the techniques mentioned in Tab. 1 with our non-relational and relational analysis, on the 4 previous examples. We did not consider the method of [8], which assumes lossy channels.

- *yes* means that the reachability analysis gives the exact result.
- *no* means that the reachability analysis does not terminate.
- *approx* means that the reachability analysis gives an over-approximation of the reachability set.

The only case where our relational method gives less satisfactory result than another method, which is also the only case where the result is not exact, is the Non-Regular protocol. On this protocol, the CQDD method can compute the exact reachability set $\bigcup_{n\geq 0} a^n\sharp\epsilon\sharp c^n\sharp\epsilon$, whereas we approximate it, using ∇_k, by $\bigcup_{0\leq n\leq k+2} a^n\sharp\epsilon\sharp c^n\sharp\epsilon \ \cup \ a^{k+2}a^*\sharp\epsilon\sharp c^{k+2}c^*\sharp\epsilon$, which is not so bad. On the other hand, none of the other methods delivers results for all the examples.

This comparison is experimental, and should be completed in the future with larger examples. However, it is very difficult to *prove* the superiority of an analysis that uses a widening operator, as pointed out by [33]. From a theoretical point of view, we can make two statements. First, we can partially order the *expressiveness of the representations* (which does not necessarily induce a corresponding

Table 2. Comparison of acceleration techniques with our 2 analysis

Example	Acceleration techniques			Regular languages with widening	
	SLRE [11]	QDD [12]	CQDD [13]	non-relational	relational
(1) ABP protocol	yes	yes	yes	yes	yes
(2) Conn./deconn.	approx[a]	yes	yes	approx[a]	yes
(3) Non-regular	no[a,b,c]	no[b,c]	yes	approx[a,c]	approx[c]
(4) Nested loops	no[c]	yes	no[c]	yes	yes

[a] non-relational representation [b] counting loops [12] that cannot be accelerated
[c] exact set not representable

ordering of the analyses in terms of accuracy). Following Tab. 1, we have that SLRE is the less expressive, QDD and our relational method are equivalent, and are uncomparable to CQDD. Second, proposition 2 implies a (modest) partial completeness result: if in a CFSM the length of the FIFO queues is bounded by l, then taking $k \geq n \cdot l$ for the widening ∇_k lead to exact results.

7 Conclusion

In this paper, we showed how to perform reachability analysis of CFSMs using an Abstract Interpretation approach and the notion of relational/non-relational analysis [29]. Our method can be applied to any CFSM and always terminates. It is technically simple, based on standard Abstract Interpretation technique and well-known concepts like regular languages and bisimulation of depth k. Despite of its simplicity, that we see as a strength, our method is often as accurate as acceleration techniques on standard examples, and it can deal with counting loops [12]. It is however unable to certify by itself whether the obtained result is exact or not (which is a limitation common to abstract interpretation techniques). Last but not least, we think that our approach is more amenable to the combination of FIFO channels with other unbounded datatypes, like counters, in the spirit of [34]. Indeed, it seems very difficult to accelerate loops where FIFO operations are guarded by numerical tests on counters and where counters are conversely updated depending on the FIFO queues contents.

For CFSMs, our method is a good alternative to acceleration based techniques. The two approaches may actually be seen as complementary. Typically, one can first try to get the exact reachability set using acceleration techniques and then apply our method in case of failure. A more interesting combination consists in using acceleration techniques to add meta-transitions in the original model, when possible, and to apply our method to the augmented system.

In the future we plan to explore two directions: the first one is to combine the abstraction for FIFO queues with abstractions for numerical variables, in order to attack the verification of more realistic models. The second one is to consider CFSM with infinite alphabets. This is required for the many protocols that use "tokens" to uniquely identify different frames. These tokens are typically assumed to belong to an infinite set.

Acknowledgments

We thank the anonymous AMAST referees for their careful comments and suggestions.

References

1. Brand, D., Zafiropulo, P.: On communicating finite-state machines. J. ACM **30** (1983) 323–342
2. Bochmann, G.V.: Finite state description of communication protocols. IEEE Computer Society Press, Los Alamitos, CA, USA (1995)
3. Turner, K.J.: Using Formal Description Techniques: An Introduction to Estelle, Lotos, and SDL. John Wiley & Sons, Inc., New York, NY, USA (1993)
4. Nielsen, M., Plotkin, G., Winskel, G.: Petri nets, event structures and domains, part 1. Theoretical Computer Science **13** (1981)
5. ITU-TS: ITU-TS Recommendation Z.120: Message Sequence Chart (MSC). (1999)
6. Reniers, M., Mauw, S.: High-level message sequence charts. In Cavalli, A., Sarma, A., eds.: Proc. of the 8th SDL Forum. (1997)
7. Cécé, G., Finkel, A., Iyer, S.P.: Unreliable channels are easier to verify than perfect channels. Information and Computation **124** (1996) 20–31
8. Abdulla, P., Bouajjani, A., Jonsson, B.: On-the-fly analysis of systems with unbounded, lossy FIFO channels. In: Computer Aided Verification (CAV '98). Volume 1427 of LNCS. (1998)
9. Bardin, S., Finkel, A., Leroux, J., Petrucci, L.: FAST: Fast Acceleration of Symbolic Transition systems. In: Computer Aided Verification (CAV'03). Volume 2725 of LNCS. (2003)
10. Boigelot, B., Godefroid, P., Willems, B., Wolper, P.: The power of QDDs. In: Static Analysis Symposium (SAS'97). Volume 1302 of LNCS. (1997)
11. Finkel, A., Iyer, S.P., Sutre, G.: Well-abstracted transition systems: application to FIFO automata. Information and Computation **181** (2003) 1–31
12. Boigelot, B., Godefroid, P.: Symbolic verification of communication protocols with infinite state spaces using QDDs. FMSD **14** (1997) 237–255
13. Bouajjani, A., Habermehl, P.: Symbolic reachability analysis of FIFO-channel systems with nonregular sets of configurations. Theor. Comp. Science **221** (1999)
14. Abdulla, P., Jonsson, B., Nilsson, M., Saksena, M.: A survey of regular model checking. In: CONCUR'04. Volume 3170 of LNCS. (2004)
15. Bouajjani, A., Jonsson, B., Nilsson, M., Touili, T.: Regular model checking. In: Computer Aided Verification (CAV'00). Volume 1855 of LNCS. (2000)
16. Boigelot, B., Legay, A., Wolper, P.: Iterating transducers in the large. In: Computer Aided Verification (CAV'03). Volume 2725 of LNCS. (2003)
17. Bouajjani, A., Habermehl, P., Vojnar, T.: Abstract regular model checking. In: Computer Aided Verification (CAV'04). Volume 3114 of LNCS. (2004)
18. Memmi, G., Finkel, A.: An introduction to FIFO nets-monogeneous nets: a subclass of FIFO nets. Theoretical Computer Science **31** (1985)
19. Finkel, A., Rosier, L.: A survey on the decidability questions for classes of FIFO nets. In: Eur. Workshop on Applications and Theory of Petri Nets. Volume 340 of LNCS. (1987)
20. Cécé, G., Finkel, A.: Verification of programs with half-duplex communication. Information and Computation **202** (2005)

21. Abdulla, P., Jonsson, B.: Verifying programs with unreliable channels. Information and Computation **127** (1996)
22. Abdulla, P., Bertrand, N., Rabinovich, A., Schnoebelen, P.: Verification of probabilistic systems with faulty communication. Inf. and Comp. **202** (2005)
23. Peng, W., Puroshothaman, S.: Data flow analysis of communicating finite state machines. ACM Trans. Program. Lang. Syst. **13** (1991) 399–442
24. Lesens, D., Halbwachs, N., Raymond, P.: Automatic verification of parameterized linear networks of processes. In: Principles of Programming Languages (POPL'97), ACM Press (1997)
25. Cousot, P., Cousot, R.: Abstract interpretation and application to logic programs. Journal of Logic Programming **13** (1992)
26. Brzozowski, J.A.: Derivatives of regular expressions. Journal of the ACM **1** (1964)
27. Feret, J.: Abstract interpretation-based static analysis of mobile ambients. In: Static Analysis Symposium (SAS'01). Volume 2126 of LNCS. (2001)
28. Jeannet, B., Jeron, T., Le Gall, T.: Abstracting interpretation of FIFO channels. Technical Report 5784, INRIA (2005)
29. Jones, N., Muchnick, S.: Complexity of flow analysis, inductive assertion synthesis, and a language due to Dijkstra. In Jones, N., Muchnick, S., eds.: Program Flow Analysis: Theory and Applications. Prentice-Hall (1981)
30. Wolper, P., Boigelot, B.: An automata-theoretic approach to Presburger arithmetic constraints. In: Static Analysis Symposium (SAS'95). Volume 983 of LNCS. (1995)
31. Bourdoncle, F.: Efficient chaotic iteration strategies with widenings. In: Int. Conf. on Formal Methods in Progr. and their Applications. Volume 735 of LNCS. (1993)
32. Abdulla, P.A., Annichini, A., Bouajjani, A.: Symbolic verification of lossy channel systems: Application to the bounded retransmission protocol. In: Tools and Algorithms for Construction and Analysis of Systems (TACAS'99). (1999)
33. Su, Z., Wagner, D.: A class of polynomially solvable range constraints for interval analysis without widenings and narrowings. In: Tools and Algorithms for the Construction and Analysis of Systems (TACAS'04). Volume 2988. (2004)
34. Jeannet, B., Halbwachs, N., Raymond, P.: Dynamic partitioning in analyses of numerical properties. In: Static Analysis Symposium, SAS'99. Volume 1694 of LNCS. (1999)

Assessing the Expressivity of Formal Specification Languages*

Natalia López, Manuel Núñez, and Ismael Rodríguez

Dept. Sistemas Informáticos y Programación
Universidad Complutense de Madrid, 28040 Madrid, Spain
{natalia, mn, isrodrig}@sip.ucm.es

Abstract. Formal modelling languages are powerful tools to systematically represent and analyze the properties of systems. A myriad of new modelling languages, as well as extensions of existing ones, are proposed every year. We may consider that a modelling language is useful if it allows to represent the *critical* aspects of systems in an expressive way. In particular, we require that the modelling language allows to accurately discriminate between *correct* and *incorrect* behaviors concerning critical aspects of the model. In this paper we present a method to assess the suitability of a modelling language to define systems belonging to a specific domain. Basically, given a system, we consider alternative correct/incorrect systems and we study whether the representations provided by the studied modelling language keep the distinction between correct and incorrect as each alternative system does.

1 Introduction

Hundreds of languages have been proposed to formally describe any kind of systems. A lot of them differ only in some aspect concerning the way some features are represented or interpreted (e.g. timed automata [AD94] versus temporal process algebras [NS94] versus timed Petri nets [Zub80], generative versus reactive probabilistic systems [GSS95], Markovian [Hil96, BG98] versus non-markovian stochastic models [LN01, BG02], and so on). Since the number of possible ways to deal with each feature is high, each lineal combination of these possibilities eventually leads to a new language. Thus, a clear and well-defined criterion to asses the utility of a language to model systems belonging to some domain would be very useful. In this line, we could ask ourselves which characteristics we want in a given formal method [AR00]. Informally speaking, a language is good to model a class of systems if it allows to create models where critical features are suitably represented. In terms of correction, a model should be able to perform what the original system does, and should not do what the system does not. For instance, if a system must perform the action a only if the variable x is equal to 10 then a model specifying only that *a may be performed* would not be accurate enough. Besides, if that action a must be performed only

* Research partially supported by the Spanish MCYT project TIC2003-07848-C02-01 and the Junta de Castilla-La Mancha project PAC-03-001.

M. Johnson and V. Vene (Eds.): AMAST 2006, LNCS 4019, pp. 220–234, 2006.

after 5 seconds then a model where that requirement is not included would not be suitable. Following these ideas, a modelling language is suitable to define a class of systems if it discriminates desirable and undesirable behavior (almost) as the corresponding modelled systems do. In order to check it, we will (semi-)automatically compare the behavior of systems and their models. In particular, we will compare the correct and incorrect behaviors each of them may expose.

There is a *testing technique* that can inspire the creation of a *new* methodology that actually fits into our purposes. *Mutation Testing* [Ham77, How82, BM99] allows to estimate the power of a test suite to assess the (in-)correctness of an implementation with respect to a specification. Basically, mutation testing consists in facing tests with several *mutants* of the specification, that is, alternative specifications where some aspect is modified. Mutants are created from the specification by introducing modifications that simulate typical programming errors. Then, the set of tests to be assessed is applied to each mutant and we observe the capacity of tests to *kill* mutants, that is, to detect erroneous behaviors in mutant specifications that are actually wrong. Let us note that a mutant could be *correct*, that is, equivalent to the original specification. Unfortunately, to check the correctness of a mutant is not decidable. So, this technique is, in general, semi-automatic. Our method, inspired in the previous idea, can be basically described as follows: Given a real system (defined in some language) we create some mutants (in the same language) that might behave incorrectly. Then, we apply the *modelling language under assessment* to create models of both the original system and their mutants. If the modelling language were not expressive enough, then several systems with different behaviors (taken from the mutant systems and/or the original system) could converge to a single model. If these systems were either all correct or all incorrect then it would not be a problem that the modelling language provides a single model for them. This is so because the conversion might not have lost relevant aspects. However, if some of the systems that converged to a single model were correct and others were not then the modelling language is losing relevant characteristics that delimit the difference between correct and incorrect. Let us note that, in general, no modelling language allows to express in a *natural* way exactly the same things than the language used for the system definition.

Our measure of the expressivity and suitability of a language to define a given system will be based on the relation between the correctness of the mutants and the correctness of their corresponding models. For example, let us consider 1000 mutants of a system. Let us suppose that 900 mutants are incorrect with respect to the system and let us consider the 900 models of these mutants. If 300 of them are also models of the original system then the modelling language is not very suitable for defining this class of systems because the modelling phase loses aspects that are critical to define the *border* between correct and incorrect. Other possible unsuitability criterion is the following: If 300 of these models of incorrect mutants are *equivalent* to the original system or its model (or *more/less* restrictive in a sense that can be considered valid) then the modelling language

```
    Program P
1:  x = 0;
2:  while (no message is received) {
3:     wait for a random delay
         of [0,2] minutes;
4:     if (x==1) then {
5:       send('A');
6:       x = 0;
7:       }
8:     else {
9:       x = 1;
10:      };
11:    };
```

```
Mutants of P

M1: 3: wait for a random delay
          of [0.5,1.5] minutes;
M2: 4: if (x==2) then {

M3: 6: x = 2;      M4: 6: x = 1;
```

Fig. 1. Program example

is unsuitable. Since our methodology can be applied to other systems of the same domain to measure the suitability of the target language to describe them, it may help to check whether a new language makes a relevant contribution to express the critical aspects of systems belonging to a specific domain.

The rest of the paper is structured as follows. In the next section we introduce a simple example to motivate the definition of our methodology. In Section 3 we formally present the main concepts of our methodology. Next, in Section 4 we study some formal properties relating the concepts previously introduced. Finally, in Section 5 we present our conclusions and some lines for future work.

2 Motivating Example

In this section we illustrate our method with a running example. A program P, written in a given language L, is depicted in Figure 1. Until a message is received, it iteratively performs two operations: First, it waits for a delay between 0 and 2 minutes (we assume that the random variable denoting the delay is *uniformly* distributed); next, it sends the message A one out of two times. Let us note that, from an external observer's point of view, the behavior of P actually consists in iteratively sending the message A after random delays between 0 and 4 minutes, until a message is received from the environment.

In Figure 1 we also give four *mutants* of P. Each of them differs from P in a single code line. We will apply the following *correctness criterion*: A mutant M_i is correct with respect to P if its external behavior coincides with the one from P. We suppose that *external behavior* concerns only sent messages and delays between them. Since the maximal delay between two consecutive A messages in M_1 is 3, M_1 cannot produce some behaviors included in P. So, M_1 is incorrect.

M_2 is incorrect as well: For instance, it can wait 10 minutes until a message is received without sending any message. M_3 is correct because its behavior is not affected by the change. Finally, M_4 is incorrect: *All* the times the loop is taken, except the first one, we have that `X` is 1, so an `A` message is sent. Hence, the delay between messages `A` is equal to, at most, 2 minutes, while delays can take 4 minutes in P.

Let us consider three modelling languages L_1, L_2, and L_3. Each of them misses some aspect that is actually considered in the language L: Only *fix* temporal delays can be represented in L_1, L_2 does not use any *variables* to govern the behavior of systems, and L_3 cannot represent any *temporal delay* at all. We will study and compare the suitability of these languages to model the program P. In particular, we will study whether each of the languages properly captures the (subtle) aspects delimiting the border between correct and incorrect behaviors. In order to do it, each language will face the definition of each mutant of P, and we will study whether the (in-)correctness of each alternative properly remains in the models domain provided by each language.

The modelling language L_1 represents most program aspects exactly in the same way as L does. The only difference is that L_1 does not allow to represent *random* temporal delays. Instead, all temporal delays must denote a *deterministic* amount of time. In particular, the translation of a program from L to L_1 is done as follows: For any random delay we consider its *mean* expected delay. For example, line 3 in P is translated into "`wait for 1 minute`." As a result, the models of P and M_1 in language L_1 actually *coincide*. Since P is correct but M_1 is not, this collision denotes that L_1 does not properly represent this behavior. The collision of two different systems, being one of them correct and the other incorrect, into a single *model*, is called *collision mistake* in our framework (the formal definition is given in Section 3). Basically, it denotes that some critical details are lost in translation.

Let us consider the correctness criterion we use in the domain of L and let us apply it to the domain of L_1 models. Despite M_1 is an incorrect mutant, its model in L_1 produces exactly the same external behavior as the model of P. In fact, let us note that the *fix* delay placed before the `if` statement is 1 minute long in *both* models. That is, the model of M_1 is *correct* with respect to the model of P, but the system it comes from is not correct with respect to P. We denote the situation where an incorrect system leads to a correct model, or viceversa, by *model mistake* (see Section 3). Basically, it shows that the correctness criterion is not properly preserved in the model domain.

Regarding the modifications induced by M_2, M_3, and M_4, they are properly represented in L_1. This is because L_1 represents all lines but line 3 exactly as L does. In particular, these mutants do not produce any of the mistakes considered before. On the one hand, their models differ from each other and from the model of P, so there do not lead to any collision mistake. On the other hand, models of incorrect mutants are also incorrect in the domain of models, so they do not produce model mistakes. In particular, the model of M_2 does not produce any message, and the (fix) delay between messages in the model of M_4 is 1 minute.

However, the fix delay in the model of P is 2 minutes. Finally, since both M_3 and its model are correct, M_3 does not yield a model mistake. Summarizing, in both approaches 3 out of 4 mutants do not produce mistakes and are correctly represented by L_1.

We will conclude the presentation of this example at the end of Section 3, once all the concepts underlying our methodology have been formally introduced. In particular, we will use L_2 and L_3 to create models of P and its mutants, and we will compare the suitability of each language to represent P.

3 Formal Framework

In this section we present the basic concepts of our methodology and show how they are applied to assess the suitability of a modelling language to describe a class of systems. First, we introduce the concept of *language*. In our framework a language is defined by a *syntax*, allowing to construct the appropriate words (i.e., programs), as well as a *semantics*, associating semantic values to syntactic expressions.

Definition 1. A *language* is a pair $\mathcal{L} = (\alpha, \beta)$, where α denotes the *syntax* of $\mathcal{L}$ and β denotes its *semantics*. Let $\texttt{Systems}(\mathcal{L})$ denote the set of all words conforming to α. Then, β is a total function $\beta : \texttt{Systems}(\mathcal{L}) \longrightarrow B$, where B is the *semantic domain* for $\mathcal{L}$. □

Next we present the concept of *correctness* of a system. Correctness is defined in the domain of semantic values and it is given in terms of a comparison between values. So, we may say that a semantic value b_1 is correct *with respect to* b_2. For example, this might mean that both semantic values represent *bisimilar* behaviors. The correctness relation is not necessarily symmetric. For example, it may define a *conformance* relation where the behavior of a system must be a subset of that exposed by another. Besides, it could be defined in terms of the set of semantic values where some required property holds. In general, if c is a *correctness criterion* then $b \in c(b')$ means that b is correct with respect to b'. Let us note that by using this criterion we implicitly define which aspects of a system will be considered *critical*. For example, if this criterion says that a system is correct regardless of its temporal performance, then temporal issues are not considered critical in our analysis. In the following definition, $\mathcal{P}(X)$ denotes the powerset of the set X.

Definition 2. Let $\mathcal{L} = (\alpha, \beta)$ be a language with $\beta : \texttt{Systems}(\mathcal{L}) \longrightarrow B$. A *correctness criterion* for the language $\mathcal{L}$ is a total function $c : B \longrightarrow \mathcal{P}(B)$.

If for all $b \in B$ we have $b \in c(b)$ then c is a *reflexive* criterion. If for all $b, b' \in B$, $b' \in c(b)$ implies $b \in c(b')$, then c is a *symmetric* criterion. If for all $b, b', b'' \in B$, $b' \in c(b)$ and $b'' \in c(b')$ imply $b'' \in c(b)$, then c is a *transitive* criterion. We say that a criterion c is an *equivalence criterion* if it is reflexive, symmetric, and transitive. □

According to our methodology, we will create *mutants* to check whether they behave as their corresponding models. Thus, we will be able to assess whether the modelling language properly discriminates correct and incorrect behaviors. This will provide us with a measure of its suitability. In order to create mutants we introduce modifications in the original systems. These modifications *substitute* some subexpressions of the system by others. Given a string of symbols and a substitution, this function returns the set of strings resulting from the application of the substitution to the string at any point. In the next definition we formally present this concept. For any set of symbols Σ, we consider that Σ^* denotes the set of all finite strings of symbols belonging to Σ.

Definition 3. Let Σ be a set of symbols. Let $e = e_1 \cdots e_n$ and $e' = e'_1 \cdots e'_m$ be sequences in Σ^*. The *term substitution function* for e and e', denoted by $[e/e']$, is a function $[e/e'] : \Sigma^* \longrightarrow \mathcal{P}(\Sigma^*)$ where for any $w = w_1 \cdots w_p$ we have

$$[e/e'](w) = \left\{ w_1 \cdots w_k e'_1 \cdots e'_m w_{k+n+1} \cdots w_p \,\middle|\, \begin{array}{c} 1 \leq k \leq p \,\wedge \\ w = w_1 \cdots w_k e_1 \cdots e_n w_{k+n+1} \cdots w_p \end{array} \right\} \square$$

The creation of mutants will be defined in terms of a function. Given a system and a set of term substitution functions, this function generates a set of mutants by applying substitutions of the set to the system.

Definition 4. Let $\mathcal{L} = (\alpha, \beta)$ be a language and $\mathtt{S}$ be a set of term substitution functions. A *mutation function* for $\mathcal{L}$ and $\mathtt{S}$ is a total function $M : \mathtt{Systems}(\mathcal{L}) \longrightarrow \mathcal{P}(\mathtt{Systems}(\mathcal{L}))$, where for any $a \in \mathtt{Systems}(\mathcal{L})$ and $b \in M(a)$ we have $b \in \sigma(a)$ for some $\sigma \in \mathtt{S}$. $\square$

From now on we will assume that for any mutation function M for $\mathcal{L}$ and any $a \in \mathtt{Systems}(\mathcal{L})$ we have $a \in M(a)$. This assumption will help to deal with a system and its mutants in a more compact way.

The *translation* of a system from the original system language into the modelling language under assessment will be also represented by means of a function. Let us note that this function, in general, will not be injective: Some systems with different behaviors could be represented by the same model. Hence, the modelling language may lose some details. If these details are not considered critical in the analysis, then losing them is not necessarily bad. On the contrary, eliminating irrelevant details in models may help to create more handleable models. However, if the translation into the modelling language loses critical details, then the modelling language may not be suitable. As we will see below, these situations will be detected within our framework.

Definition 5. Let $\mathcal{L}_1 = (\alpha_1, \beta_1), \mathcal{L}_2 = (\alpha_2, \beta_2)$ be two languages. A *translation function* from $\mathcal{L}_1$ to $\mathcal{L}_2$ is a total function $T : \mathtt{Systems}(\mathcal{L}_1) \longrightarrow \mathtt{Systems}(\mathcal{L}_2)$. $\square$

Now we are provided with all the needed machinery to present the concepts used in our methodology. We propose two alternative criteria to check whether a modelling language is suitable to represent relevant details of systems. Next we present the first one: If a correct system and an incorrect one (taken from the set of mutants and the original system) converge to a single model then

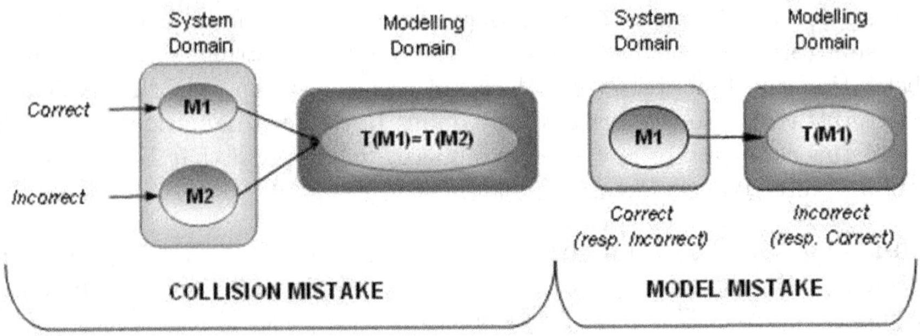

Fig. 2. Correctness Criteria

relevant details concerning the correctness are lost. In particular, the translation misses details that make the difference between correct and incorrect. Hence, we say that it is a *collision mistake* in the representation of the system by the modelling language. This situation is graphically presented in Figure 2 (left). In the following definition the cardinal of a set X is denoted by $|\,X\,|$.

Definition 6. Let $\mathcal{L}_1 = (\alpha_1, \beta_1), \mathcal{L}_2 = (\alpha_2, \beta_2)$ be two languages, $\mathtt{S}$ be a set of term substitution functions, and M be a mutation function for $\mathcal{L}_1$ and $\mathtt{S}$. Let T be a translation function from $\mathcal{L}_1$ to $\mathcal{L}_2$. Let c_1 be a correctness criterion for $\mathcal{L}_1$ and $s \in \mathtt{Systems}(\mathcal{L}_1)$. We say that $m \in M(s)$ is a *collision mistake* for s, M, T, and c_1, denoted by $\mathtt{CMistake}^{s,M,T,c_1}(m)$, if there exists $m' \in M(s)$, with $T(m) = T(m')$, such that either

(a) $\beta_1(m) \in c_1(\beta_1(s))$ and $\beta_1(m') \notin c_1(\beta_1(s))$ or
(b) $\beta_1(m) \notin c_1(\beta_1(s))$ and $\beta_1(m') \in c_1(\beta_1(s))$.

In this case, we say that m' is a *collision pair* of m for s, M, T, and c_1, and we denote it by $m' \updownarrow^{s,M,T,c_1} m$. The *number of collision mistakes* of s for M, T, and c_1, denoted by $\mathtt{NumCM}^{M,T,c_1}(s)$, is defined as $|\,\{m \mid m \in M(s) \wedge \mathtt{CMistake}^{s,M,T,c_1}(m)\}\,|$. □

For the sake of notation simplicity, we will omit some parameters when we assume that they are unique in the context. Hence, we will sometimes simply write $\mathtt{CMistake}(m)$, $m' \updownarrow m$, and $\mathtt{NumCM}(s)$. Next we define our second mechanism to check whether a modelling language fails to express a system. Now, a correctness criterion will also be applied to the modelling language. For instance, an equivalence criterion may be used, that is, a reflexive, symmetric, and transitive criterion. We say that the modelling language produces a *model mistake* in the translation of the original system if the correctness criteria of the original language and the modelling language map (in-)correct systems in a different way. In particular, if a *correct* (resp. *incorrect*) system is translated into a model that is considered *incorrect* (resp. *correct*), then we detect a failure in the representation of the system by the modelling language. This situation is graphically presented in Figure 2 (right).

Definition 7. Let $\mathcal{L}_1 = (\alpha_1, \beta_1), \mathcal{L}_2 = (\alpha_2, \beta_2)$ be two languages, $\mathtt{S}$ be a set of term substitution functions, M be a mutation function for $\mathcal{L}_1$ and $\mathtt{S}$, T be a translation function from $\mathcal{L}_1$ to $\mathcal{L}_2$, c_1 and c_2 be correctness criteria for $\mathcal{L}_1$ and $\mathcal{L}_2$, respectively, and $s \in \mathtt{Systems}(\mathcal{L}_1)$. We say that $m \in M(s)$ is a *model mistake* for s, M, t, c_1, and c_2, denoted by $\mathtt{MMistake}^{s,M,t,c_1,c_2}(m)$, if either

(a) $\beta_1(m) \in c_1(\beta_1(s))$ and $\beta_2(t(m)) \notin c_2(\beta_2(t(s)))$ or
(b) $\beta_1(m) \notin c_1(\beta_1(s))$ and $\beta_2(t(m)) \in c_2(\beta_2(t(s)))$.

The *number of model mistakes* of s for M, t, c_1, and c_2, denoted by $\mathtt{NumMM}^{M,t,c_1,c_2}(s)$, is given by $\mid \{m \mid m \in M(s) \wedge \mathtt{MMistake}^{s,M,t,c_1,c_2}(m)\} \mid$. □

Again, we will omit some parameters when they can be inferred from the context. Now we present mechanisms to find the mistakes in the representation of a system, regardless of the used criterion. In both cases, we define the *suitability* of a modelling language to represent a system as the ratio of mutants that are *correctly* translated.

Definition 8. Let $\mathcal{L}_1 = (\alpha_1, \beta_1), \mathcal{L}_2 = (\alpha_2, \beta_2)$ be languages, $\mathtt{S}$ be a set of term substitution functions, M be a mutation function for $\mathcal{L}_1$ and $\mathtt{S}$, T be a translation function from $\mathcal{L}_1$ to $\mathcal{L}_2$, c_1 and c_2 be correctness criteria for $\mathcal{L}_1$ and $\mathcal{L}_2$, respectively, and $s \in \mathtt{Systems}(\mathcal{L}_1)$. The *collision suitability* of $\mathcal{L}_2$ to define s under M, T, and c_1, denoted by $\mathtt{CSuit}(s, \mathcal{L}_2, M, T, c_1)$, is defined as

$$\mathtt{CSuit}(s, \mathcal{L}_2, M, T, c_1) = \frac{\mid M(s) \mid - \mathtt{NumCM}(s)}{\mid M(s) \mid}$$

The *model suitability* of $\mathcal{L}_2$ to define s under M, T, c_1, and c_2, denoted by $\mathtt{MSuit}(s, \mathcal{L}_2, M, T, c_1, c_2)$, is defined as

$$\mathtt{MSuit}(s, \mathcal{L}_2, M, T, c_1, c_2) = \frac{\mid M(s) \mid - \mathtt{NumMM}(s)}{\mid M(s) \mid}$$

□

Let us remark that in order to measure collision suitability no correctness criterion is applied to the modelling language $\mathcal{L}_2$, that is, c_2 is ignored. The previous concepts can be extended to deal with *sets* of systems belonging to some specific domain. Thus, our framework is not only useful to assess the suitability of a language with respect to a specific system but also with respect to a family of systems. For instance, we could assess the suitability of a modelling language to describe data link layer network protocols or distributed cryptographic protocols. A modelling language unable to represent temporal delays (resp. data encryption) would produce several collision and/or model mistakes and would be unsuitable. In each case, a representative set of systems in the domain has to be chosen and applied. Depending on the relevance of each system and its fitness in the domain, it should have a different weight in the overall assessment. In the following definition we *overload* the functions presented in the previous definition to deal with *sets*.

Definition 9. Let $\mathcal{L}_1 = (\alpha_1, \beta_1)$ and $\mathcal{L}_2 = (\alpha_2, \beta_2)$ be two languages, $\mathtt{S}$ be a set of term substitution functions, and M be a mutation function for $\mathcal{L}_1$ and $\mathtt{S}$. Let T be a translation function from $\mathcal{L}_1$ to $\mathcal{L}_2$ and c_1, c_2 be correctness criteria for $\mathcal{L}_1$ and $\mathcal{L}_2$, respectively. Let $\mathcal{S} = \{(s_1, w_1), \ldots, (s_n, w_n)\} \in \mathcal{P}(\mathtt{Systems}(\mathcal{L}_1) \times (0..1])$ be a set of pairs of systems and *weights* for these systems, where for all $1 \leq i \leq n$ we have $w_i > 0$ and $\sum_{1 \leq i \leq n} w_i = 1$. We define the *collision suitability* of $\mathcal{L}_2$ to define $\mathcal{S}$ under M, T, and c_1 as

$$\mathtt{CSuit}(\mathcal{S}, \mathcal{L}_2, M, T, c_1) = \sum_{1 \leq i \leq n} w_i \cdot \mathtt{CSuit}(s_i, \mathcal{L}_2, M, T, c_1)$$

We define the *model suitability* of $\mathcal{L}_2$ to define $\mathcal{S}$ under M, T, c_1, and c_2 as

$$\mathtt{MSuit}(\mathcal{S}, \mathcal{L}_2, M, T, c_1, c_2) = \sum_{1 \leq i \leq n} w_i \cdot \mathtt{MSuit}(s_i, \mathcal{L}_2, M, T, c_1, c_2)$$

□

3.1 Concluding Our Motivating Example

Let us consider the modelling language L_2. It represents random delays exactly as L does, but it considers neither program variables nor their effects. In particular, it converts any condition on variables into a *non-deterministic* choice. For instance, concerning the conditional sentence in line 4 of the program P presented in Figure 1, the model of P in L_2 just denotes that *either of the branches is non-deterministically chosen.* In fact, since variables are ignored in L_2, the choice just consists in choosing between sending `A` or not.

Let us note that the models of the programs P and M_1 in L_2 do not coincide because their different versions of line 3 stay in L_2. For the same reason, the model of M_1 is different to the model of the *correct* mutant M_3, so there is no collision mistake with M_1. Moreover, there is not a *model* mistake with M_1 either, but the reason is quite different. Let us note that the model of P in L_2 can produce *any* delay between `A` messages. Since the `if` choice is implemented as a non-deterministic choice, the `else` statement can be taken *any* number of times before the next message `A` is sent. Thus, any delay can be cumulated before sending each message. However, the model of M_1 cannot produce its *first* message until 0.5 units of time pass. That is, the model of M_1 is not equivalent to the model of P. Hence, in spite of the collateral effects of the absence of variables, the (wrong) temporal behavior of M_1 is properly detected.

Let us remark that the mutants of M_2, M_3, and M_4 only differ from P in the way they deal with variables. Hence, the models of M_2, M_3, and M_4 in L_2 coincide with the model of P. The collision with the original program is not a collision mistake for M_3, because M_3 is actually correct. However, it is so for the incorrect mutants M_2 and M_4. Moreover, since these models coincide with the model of the correct mutant M_3, M_3 is also involved in a collision mistake. Regarding the correctness in the model domain, we find the same problem for the incorrect mutants M_2 and M_4: The models of M_2 and M_4 are *correct* with respect to the *model* of P (they have the same external behavior), but the respective systems M_2 and M_4 are incorrect with respect to P. On the other hand, since the model of M_3 is correct, M_3 does not produce any model mistake. So, the number of properly represented mutants with L_2 is 1 and 2, for collision and model mistakes, respectively.

Finally, we consider the modelling language L_3. This language does not represent temporal delays (neither deterministically nor probabilistically), though variables are represented and used exactly as in L. Since the line 3 is ignored by L_3, P and the incorrect mutant M_1 collide by producing the same model in L_3, which in turn is a collision mistake. Besides, since both models coincide, we have that the model of M_1 is equivalent to the model P (though M_1 is not equivalent to P). Hence, we also have a model mistake (as we will study in Section 4, a collision mistake between two models induces a model mistake in *at least* one of the involved systems).

Regarding M_2 and M_4, they lead to models that are different to both the model of P and the model of the correct mutant M_3. That is, M_2 and M_4 do not induce any collision mistake. The correct mutant M_3 does not produce a collision mistake either. Concerning model mistakes, M_2 does not produce them because its model cannot produce any message at all, which is not equivalent to the behavior of the model of P. However, the model of the incorrect M_4 is actually *equivalent* to the model of P. Let us note that, since temporal delays are not concerned by L_3, the external behavior of L_3 only reveals the number of times the message `A` is sent. Both the model of P and the model of M_4 can send `A` any number of times. So, they present the same external behavior and we have a model mistake. Summarizing, if we consider collision mistakes then 3 out of 4 mutants are free of them. However, if model mistakes are concerned then only 2 out of 4 mutants are properly represented.

In the next tables we compare the performance of each language to define the program P and its mutants. The first table shows the mistakes of each mutant as well as the mistakes of P. Obviously, P cannot yield a model mistake (its model is correct by definition). However, if the model of P coincides with the model of an incorrect mutant, then P is actually involved in a collision mistake. The second table, for each kind of mistake, divides the number of mistake-free systems by the total number of systems (both P and the mutants are considered).

	L_1 (*det. time*)		L_2 (*no vars*)		L_3 (*no time*)	
	collision m?	model m?	collision m?	model m?	collision m?	model m?
P	**with** M_1	no	**with** M_2, M_4	no	**with** M_1	no
M_1	**with** P	**yes**	no	no	**with** P	**yes**
M_2	no	no	**with** P, M_3	**yes**	no	no
M_3	no	no	**with** M_2, M_4	no	no	no
M_4	no	no	**with** P, M_3	**yes**	no	**yes**

	collision suitability	model suitability
L_1	0.6	0.8
L_2	0.2	0.6
L_3	0.6	0.6

If our micro-sample of *four* mutants were representative of all the kinds of errors that may appear in the system P, then we could conclude that the modelling language that provides the best representation accurateness is L_1. Similarly, we can also study sets of systems. If several systems within a given domain, instead of a single system, are considered then the suitability of a modelling language to specify some semantical aspects of these systems can be assessed. Let us note that, in the previous example, we could consider other correctness criterions (e.g., traces inclusion, bisimulation, conformance, etc). Since each criterion focuses on different semantical aspects of systems, each would lead to different suitability results. In the previous example we have considered a straightforward translation from L to each modelling language. Other more elaborated and precise translations could be considered. For example, temporal delays of *different* length could be simulated in L_1 by iterating a tiny delay any non-deterministic number of times. However, such a translation would produce too complex and artificial models, which does not fit into the purpose of a model. Hence, translations must be direct even though details are lost. Finally, let us remind that, depending on the framework, some operations of our methodology could require the participation of a human. For example, checking the equivalence of systems is not decidable if the language is Turing-powerful. The mutant generation function (which applies randomly generated mutations) and the language translation mechanism must be defined by a human, though they can be automatically applied once they are defined.

4 Properties of the Formal Framework

Next we will study the theoretical properties of the concepts presented in the previous section. In particular, we relate the suitability notions presented in Definitions 8 and 9 and we study sufficient conditions to guarantee them. The following results assume the notation criteria introduced in the beginning of Definition 6, that is, $\mathcal{L}_1 = (\alpha_1, \beta_1)$ and $\mathcal{L}_2 = (\alpha_2, \beta_2)$ are two languages, $\mathtt{S}$ is a set of term substitution functions, and M is a mutation function for $\mathcal{L}_1$ and $\mathtt{S}$. Besides, T is a translation function from $\mathcal{L}_1$ to $\mathcal{L}_2$, c_1 and c_2 are correctness criterions for $\mathcal{L}_1$ and $\mathcal{L}_2$, respectively, and $s \in \mathtt{Systems}(\mathcal{L}_1)$.

First, let us note that the relation of collision pairs (see Definition 6) $\updownarrow$ is a symmetric relation.

Lemma 1. For any $m_1, m_2 \in M(s)$ we have that $m_1 \updownarrow m_2$ implies $m_2 \updownarrow m_1$.

Proof. If $m_1 \updownarrow m_2$ then $\mathtt{CMistake}(m_2)$ holds, $m_1, m_2 \in M(s)$, and $T(m_1) = T(m_2)$. Besides, m_1 and m_2 fulfill either the case (a) or the case (b) of Definition 6. In the first case, $\beta_1(m_2) \in c_1(\beta_1(s))$ and $\beta_1(m_1) \notin c_1(\beta_1(s))$. Since $m_1, m_2 \in M(s)$, $T(m_1) = T(m_2)$, and both $\beta_1(m_1) \notin c_1(\beta_1(s))$ and $\beta_1(m_2) \in c_1(\beta_1(s))$ hold, we conclude $m_2 \updownarrow m_1$ holds. The proof for the case where (b) holds is similar. □

Nevertheless, the relation $\updownarrow$ is neither reflexive nor transitive. The next result relates collision mistakes, represented by the relation $\updownarrow$, and model mistakes. If two mutants produce a collision mistake then at least one of them is a model mistake.

Lemma 2. Let $m_1, m_2 \in M(s)$ such that $m_1 \updownarrow m_2$. Then, either $\texttt{MMistake}(m_1)$ or $\texttt{MMistake}(m_2)$.

Proof. If $m_1 \updownarrow m_2$ then $\texttt{CMistake}(m_2)$ holds, $m_1, m_2 \in M(s)$, and $T(m_1) = T(m_2)$. Besides, by Lemma 1, we have $\texttt{CMistake}(m_1)$.

Let us suppose that case (a) of Definition 6 holds (the proof is similar for the case where (b) holds). We have $\beta_1(m_2) \in c_1(\beta_1(s))$ and $\beta_1(m_1) \notin c_1(\beta_1(s))$. Let us show that if $\texttt{MMistake}(m_2)$ is false then we have that $\texttt{MMistake}(m_1)$ holds. If $\texttt{MMistake}(m_2)$ does not hold then, by the condition $\beta_1(m_2) \in c_1(\beta_1(s))$, we infer that $\beta_2(T(m_2)) \in c_2(\beta_2(T(s)))$ necessarily holds. Since $T(m_1) = T(m_2)$, we have that $\beta_2(T(m_1)) \in c_2(\beta_2(T(s)))$ is true as well. Besides, since we also have that $\beta_1(m_1) \notin c_1(\beta_1(s))$, we deduce that $\texttt{MMistake}(m_1)$ holds.

Similarly, we prove that if $\texttt{MMistake}(m_1)$ does not hold then $\texttt{MMistake}(m_2)$ is true. If $\texttt{MMistake}(m_1)$ is false then $\beta_2(T(m_1)) \notin c_2(\beta_2(T(s)))$ because we have $\beta_1(m_1) \notin c_1(\beta_1(s))$. Since $T(m_1) = T(m_2)$ and $\beta_1(m_2) \in c_1(\beta_1(s))$, we conclude $\texttt{MMistake}(m_2)$. □

The previous result allows to relate incomplete collision suitability with incomplete model suitability. If a collision mistake is found while assessing the suitability of a modelling language to represent a set of systems, then the model suitability for this language and set is not full.

Proposition 1. Let $\mathcal{S} = \{(s_1, w_1), \ldots, (s_n, w_n)\} \in \mathcal{P}(\texttt{Systems}(\mathcal{L}_1) \times (0..1])$ with $\sum_{1 \leq i \leq n} w_i = 1$ and such that for all $1 \leq i \leq n$ we have $w_i > 0$. Then,

$$\texttt{CSuit}(\mathcal{S}, \mathcal{L}_2, M, T, c_1) < 1 \Rightarrow \texttt{MSuit}(\mathcal{S}, \mathcal{L}_2, M, T, c_1, c_2) < 1$$

Proof. Taking into account that $\sum_{1 \leq i \leq n} w_i = 1$ and that for all $1 \leq i \leq n$ we have $\texttt{CSuit}(s_i, \mathcal{L}_2, M, T, c_1) \leq 1$, $\texttt{CSuit}(\mathcal{S}, \mathcal{L}_2, M, T, c_1) < 1$ implies that there exists $1 \leq i \leq n$ such that $\texttt{CSuit}(s_i, \mathcal{L}_2, M, T, c_1) < 1$. This implies $\texttt{NumCM}^{M,T,c_1}(s_i) \geq 1$. Then, there exist m_1 and m_2 such that $m_1 \updownarrow^{s_i,M,T,c_1} m_2$.

By Lemma 2, either $\texttt{MMistake}^{s_i,M,T,c_1,c_2}(m_1)$ or $\texttt{MMistake}^{s_i,M,T,c_1,c_2}(m_2)$. In both cases we have $\texttt{NumMM}^{M,T,c_1,c_2}(s_i) \geq 1$. Thus, we also have that the condition $\texttt{MSuit}(s_i, \mathcal{L}_2, M, T, c_1, c_2) < 1$ holds. If we put this disequality together with the fact that for all $1 \leq j \leq n$ we have $\texttt{MSuit}(s_j, \mathcal{L}_2, M, T, c_1, c_2) \leq 1$ and $\sum_{1 \leq j \leq n} w_j = 1$, we infer $\texttt{MSuit}(\mathcal{S}, \mathcal{L}_2, M, T, c_1, c_2) < 1$. □

Let us note that the previous implication holds in spite of c_2 not being considered in the left side, that is, it holds for any correctness criterion c_2. In the following result we use collision mistakes to find as many model mistakes as possible. Since each two mutants involved in a collision mistake induce at least one model mistake, we will group mutants into pairs of collision mistakes. In order to avoid that two pairs provide the same model mistake (this could happen if both pairs share a mutant), we will require that these pairs are *disjoint*. In this way, each pair will add up one model mistake. In the following result, # means *"number of."* Besides, graphs are denoted by pairs (V, E) where V is the set of *vertices* and E is a set such that each of its elements is a set that is comprised of exactly two (distinct) vertices. The elements of E are called *edges*.

Proposition 2. Let $G = (V, E)$ be a graph with $V = M(s)$ and $\{m, m'\} \in E$ iff $m \updownarrow m'$. Let us consider the set of graphs

$$\mathcal{G} = \{G' \mid G' = (V, E') \wedge E' \subseteq E \wedge \text{no path in } G' \text{ traverses 3 different nodes}\}$$

Let $n = \max\{\# \text{ connected components with 2 nodes in } G' \mid G' \in \mathcal{G}\}$. We have that $\texttt{MSuit}(s, \mathcal{L}_2, M, T, c_1, c_2) \leq \frac{|M(s)| - n}{|M(s)|}$.

Proof. Let $G' = (V', E') \in \mathcal{G}$ be the graph where the number of connected components of 2 nodes is maximal. For any edge $\{m, m'\} \in G'$ we have $m \updownarrow m'$. Due to Lemma 2, we have that either $\texttt{MMistake}(m)$ or $\texttt{MMistake}(m')$. Hence, each connected component of 2 nodes in G' provides a model translation mistake. Since the sets of nodes of each connected component in G' are disjoint, mistakes provided by two components do not coincide. Thus, since $\texttt{NumMM}(s) \geq n$, we have $\texttt{MSuit}(s, \mathcal{L}_2, M, T, c_1, c_2) \leq \frac{|M(s)| - n}{|M(s)|}$. □

Next we present a simple sufficient condition to guarantee total collision suitability. Intuitively, if the translation function is injective then two mutants cannot collapse into a single model, so the translation of a correct system and an incorrect system cannot produce correctness inconsistencies. Hence, the collision suitability of a modelling language to represent a set of systems is full.

Proposition 3. Let $\mathcal{S}$ be defined as in Proposition 1. If T is injective then we have $\texttt{CSuit}(\mathcal{S}, \mathcal{L}_2, M, T, c_1) = 1$.

Proof. If T is injective then there do not exist distinct $m, m' \in M(s)$ such that $T(m) = T(m')$. Thus, for all $1 \leq i \leq n$ we have that there does not exist $m \in M(s)$ with $\texttt{CMistake}^{s_i, M, T, c_1}(m)$. Thus, for all $1 \leq i \leq n$ we have $\texttt{NumCM}^{M, T, c_1}(s_i) = 0$, implying that $\texttt{CSuit}(s_i, \mathcal{L}_2, M, T, c_1) = 1$. Since $\sum_{1 \leq i \leq n} w_i = 1$, we conclude that $\texttt{CSuit}(\mathcal{S}, \mathcal{L}_2, M, T, c_1) = 1$. □

The following condition allows to obtain full model suitability. If the correctness criterion of the modelling language assesses a model as correct if and only if this model comes from a correct mutant then we obtain *total* model suitability. Moreover, the reverse claim also holds.

Proposition 4. Let $\mathcal{S}$ be defined as in Proposition 1. For all $1 \leq i \leq n$ let $c_1(\beta_1(s_i)) = B_{1i}$ and let c_2 be such that $c_2(\beta_2(t(s_i))) = B_{2i}$, where

(1) If $b \in B_{1i}$ then for all $m \in M(s_i)$, with $\beta_1(m) = b$, $\beta_2(t(m)) \in B_{2i}$ holds and
(2) If $b \notin B_{1i}$ then for all $m \in M(s_i)$, with $\beta_1(m) = b$, $\beta_2(t(m)) \notin B_{2i}$ holds.

Then, $\texttt{MSuit}(\mathcal{S}, \mathcal{L}_2, M, t, c_1, c_2) = 1$. The reverse implication is also true.

Proof. We prove the left to right implication. First, let us show that if c_2 is defined as above then for all $1 \leq i \leq n$ we have $\texttt{NumMM}^{M, t, c_1, c_2}(s_i) = 0$, that is, there do not exist $1 \leq i \leq n$ and $m' \in M(s_i)$ such that we have $\texttt{MMistake}^{s_i, M, t, c_1, c_2}(m')$. By contrapositive, suppose that there exists $1 \leq i \leq n$ such that $\texttt{MMistake}^{s_i, M, t, c_1, c_2}(m)$ for some $m \in M(s_i)$. On the one hand, let us suppose that $\beta_1(m) \in c_1(\beta_1(s_i))$ and $\beta_2(t(m)) \notin c_2(\beta_2(t(s_i)))$. Then, $\beta_1(m) \in B_{1i}$ and $\beta_2(t(m)) \notin B_{2i}$, which makes a contradiction with the definition of c_2 (condition (1)). On the other hand, if we suppose $\beta_1(m) \notin c_1(\beta_1(s_i))$ and $\beta_2(t(m)) \in c_2(\beta_2(t(s_i)))$ then $\beta_1(m) \notin B_{1i}$ and $\beta_2(t(m)) \in B_{2i}$, which is contradictory with condition (2). Hence, $\texttt{MMistake}^{s_i,M,t,c_1,c_2}(m)$ is not possible for any $1 \leq i \leq n$, so $\texttt{NumMM}^{M,t,c_1,c_2}(s_i) = 0$ for all of them. This implies that for $1 \leq i \leq n$ we have $\texttt{MSuit}(s_i, \mathcal{L}_2, M, t, c_1, c_2) = 1$. Since $\sum_{1 \leq i \leq n} w_i = 1$, we conclude $\texttt{MSuit}(\mathcal{S}, \mathcal{L}_2, M, t, c_1, c_2) = 1$.

Next we prove the right to left implication. $\texttt{MSuit}(\mathcal{S}, \mathcal{L}_2, M, t, c_1, c_2) = 1$ implies that for all $1 \leq i \leq n$ we necessarily have $\texttt{MSuit}(s_i, \mathcal{L}_2, M, t, c_1, c_2) = 1$, since for all of them the condition $w_i > 0$ holds and we have $\sum_{1 \leq i \leq n} w_i = 1$. If $\texttt{MSuit}(s_i, \mathcal{L}_2, M, t, c_1, c_2) = 1$ then $\texttt{NumMM}^{M,t,c_1,c_2}(s_i) = 0$. Now we show that if $\texttt{NumMM}^{M,t,c_1,c_2}(s_i) = 0$ then for all $1 \leq i \leq n$ we have that c_2 follows the previous form. By contrapositive, let us suppose that c_2 does not. Then, for some $1 \leq i \leq n$ either (1) or (2) does not hold. On the one hand, let us suppose that there exist $b \in B_{1i}$ and $m \in M(s_i)$, with $\beta_1(m) = b$, such that $\beta_2(t(m)) \notin B_{2i}$. Then, $\beta_1(m) \in c_1(\beta_1(s_i))$ and $\beta_2(t(m)) \notin c_2(\beta_2(t(s_i)))$. So, $\texttt{MMistake}^{s_i,M,t,c_1,c_2}(m)$ and $\texttt{NumMM}^{M,t,c_1,c_2}(s_i) > 0$, which makes a contradiction. On the other hand, let us suppose that there exist $b \notin B_{1i}$ and $m \in M(s_i)$, with $\beta_1(m) = b$, such that $\beta_2(t(m)) \in B_{2i}$. Then, $\beta_1(m) \notin c_1(\beta_1(s_i))$ and $\beta_2(t(m)) \in c_2(\beta_2(t(s_i)))$. Again, $\texttt{MMistake}^{s_i,M,t,c_1,c_2}(m)$ and $\texttt{NumMM}^{M,t,c_1,c_2}(s_i) > 0$, so we also obtain a contradiction. Hence, for all $1 \leq i \leq n$ we have that c_2 fulfills the conditions (1) and (2). □

Let us note that it is not always possible to construct such a correctness criterion c_2 as required in the previous result. In particular, if there exist $1 \leq i \leq n$ and $m \in M(s_i)$ such that $\texttt{CMistake}^{s_i,M,t,c_1}(m)$ then it will not be possible to fulfill both conditions. Besides, let us note that, even when it is possible, it is not desirable to create *on purpose* a correctness criterion c_2 so that complete suitability is met. On the contrary, the purpose of the correctness criterion is to assess the suitability, so it must be defined *prior to* stating the actual class of systems to be analyzed.

5 Conclusions and Future Work

In this paper we have presented a formal methodology to assess whether a formal modelling language is suitable to represent the *critical* aspects of a system or set of systems. It relies on the idea of creating several alternative systems and using them to exercise the capabilities of the modelling language to distinguish correct/incorrect behaviors. Then, by analyzing whether the modelling language maps correct and incorrect systems as the original language does, we assess the suitability of the modelling language to model these systems.

As future work we plan to apply our methodology to assess the suitability of some well-known modelling languages to represent systems belonging to specific domains. So, our methodology will provide us with an objective and systematic (while heuristic) criterion to compare modelling languages.

References

[AD94] R. Alur and D. Dill. A theory of timed automata. *Theoretical Computer Science*, 126:183–235, 1994.

[AR00] E. Astesiano and G. Reggio. Formalism and method. *Theoretical Computer Science*, 236(1-2):3–34, 2000.

[BG98] M. Bernardo and R. Gorrieri. A tutorial on EMPA: A theory of concurrent processes with nondeterminism, priorities, probabilities and time. *Theoretical Computer Science*, 202:1–54, 1998.

[BG02] M. Bravetti and R. Gorrieri. The theory of interactive generalized semi-Markov processes. *Theoretical Computer Science*, 282(1):5–32, 2002.

[BM99] L. Bottaci and E.S. Mresa. Efficiency of mutation operators and selective mutation strategies: An empirical study. *Software Testing, Verification and Reliability*, 9:205–232, 1999.

[GSS95] R. van Glabbeek, S.A. Smolka, and B. Steffen. Reactive, generative and stratified models of probabilistic processes. *Information and Computation*, 121(1):59–80, 1995.

[Ham77] R.G. Hamlet. Testing programs with the aid of a compiler. *IEEE Transactions on Software Engineering*, 3:279–290, 1977.

[Hil96] J. Hillston. *A Compositional Approach to Performance Modelling*. Cambridge University Press, 1996.

[How82] W.E. Howden. Weak mutation testing and completeness of test sets. *IEEE Transactions on Software Engineering*, 8:371–379, 1982.

[LN01] N. López and M. Núñez. A testing theory for generally distributed stochastic processes. In *CONCUR 2001, LNCS 2154*, pages 321–335. Springer, 2001.

[NS94] X. Nicollin and J. Sifakis. The algebra of timed process, ATP: Theory and application. *Information and Computation*, 114(1):131–178, 1994.

[Zub80] W.M. Zuberek. Timed Petri nets and preliminary performance evaluation. In *7th Annual Symposium on Computer Architecture*, pages 88–96. ACM Press, 1980.

Fork Algebras as a Sufficiently Rich Universal Institution

Carlos G. Lopez Pombo and Marcelo F. Frias*

Department of Computer Science, FCEyN, Universidad de Buenos Aires, Pabellón I, Ciudad Universitaria, Buenos Aires, Argentina (1428) and CONICET
{clpombo, mfrias}@dc.uba.ar

Abstract. Algebraization of computational logics in the theory of fork algebras has been a research topic for a while. This research allowed us to interpret classical first-order logic, several propositional monomodal logics, propositional and first-order dynamic logic, and propositional and first-order linear temporal logic in the theory of fork algebras.

In this paper we formalize these interpretability results as institution representations from the institution of the corresponding logics to that of fork algebra. We also advocate for the institution of fork algebras as a sufficiently rich universal institution into which institutions meaningful in software development can be represented.

1 Introduction

Modeling languages such as the Unified Modeling Language (UML) [1] allow us to model a system through various diagrams. Each diagram provides a view of the system under development. This view-centric approach to software modeling has its advantages and disadvantages. Two advantages are clear:

- Decentralization of the modeling process. Several engineers may be modeling different views of the same system simultaneously.
- Separation of concerns is enforced.

On the other hand, this decentralized process may lead to inconsistencies among different views, or even between different partial models of the same view.

At the same time this modeling process evolved, several results were produced on the interpretability of logics to extensions of the theory of fork algebras [2]. An interpretation of a logic L to fork algebras consists of a mapping $T_L : Sent_L \to Equations_{Fork}$ satisfying the following interpretability condition:

$$\Gamma \models_L \alpha \iff \{T_L(\gamma) : \gamma \in \Gamma\} \vdash_{fork} T_L(\alpha) .$$

Since the language of fork algebras is algebraic, the only predicate is equality. Therefore, formulas are equations and $\vdash_{fork}$ is equational reasoning in the theory

* Research partially funded by Antorchas foundation and project UBACYT X094.

M. Johnson and V. Vene (Eds.): AMAST 2006, LNCS 4019, pp. 235–247, 2006.

of fork algebras (to be introduced later). This makes reasoning in fork algebras simple.

So far, interpretability results have been produced for classical first-order logic with equality [2], monomodal logics [3], propositional dynamic logic [3], first-order dynamic logic [4], propositional linear temporal logic [5] and first-order linear temporal logic [6].

These results constitute the foundations of the $\mathbf{Ar}_{\mathbf{g}}$*entum* Project. $\mathbf{Ar}_{\mathbf{g}}$*entum* is a CASE tool aimed at the analysis of heterogeneous models of software. A system description is a collection of theory presentations coming from different logics, and analysis of the heterogeneous model is achieved by interpreting the presentations to fork algebras and automatically analyzing the resulting fork-algebraic specification.

The idea of having heterogeneous specifications and reasoning across them is not new. A vast amount of work on the subject has been done based on Goguen and Burstall's notion of *institution* [7]. Institutions capture in an abstract way the model theory of a logic. They can be related by means of different kinds of mappings such as institution morphisms [7] and institution representations [8]. These mappings between institutions are extensively discussed by Tarlecki in [9]. The main difference between them being that institution morphisms allow one to build a richer institution from poorer ones, while representations allow us to encode poorer institutions into a richer one. Tarlecki goes even further when he writes:

> "... this suggests that we should strive at a development of a convenient to use proof theory (with support tools!) for a sufficiently rich "universal" institution, and then reuse it for other institutions linked to it by institution representations."

In this paper we pursue this goal by:

1. Introducing the institution of fork algebras.
2. Rephrasing all previous interpretability results, in terms of institution representations in the "universal" institution of fork algebras.

Actually, we will go one step further. Using the foundations of General Logics [8], we provide also an entailment system for fork algebras which happens to be complete. Notice also that tools supporting automatic analysis of specifications in the theory of fork algebras, such as *ReMo* [10], facilitate the search for models or inconsistencies in specifications. Similarly, the extension of the *PVS* semi-automatic theorem prover [11] in order to prove properties in the theory of fork algebras gives us good theorem proving support.

The paper is organized as follows. In Section 2 we introduce the class of full closure fork algebras, as well as their proof calculus, from an algebraic perspective. In Section 3 we present some necessary definitions from the theory of institutions. In Section 4 we present the logic of closure fork algebras from an institutional perspective. In Section 5 we show how theories coming from different logics can be effectively merged in the institution of closure fork algebras. Finally, in Section 6 we draw some conclusions.

2 Full Closure Fork Algebras

Full Closure Fork Algebras with Urelements (denoted by FCFAU) are extensions of relation algebras [12], that is, they possess a relation algebra reduct. In order to introduce this class, we introduce first the class of Pre Proper Closure Fork Algebras with Urelements (denoted by •PCFAU).

Definition 1. *Let* U *be a nonempty set. A* •PCFAU *is a two sorted structure* $\langle 2^{U\times U}, U, \cup, \cap, ^{-}, \emptyset, U\times U, \circ, Id, ^{\smile}, \nabla, ^{\diamond}, ^{*}, \star \rangle$ *such that*

- $\star : U \times U \to U$ *is one to one, but not surjective.*
- Id *is the identity relation on the set* U.
- $\cup$, $\cap$ *and* $^{-}$ *stand for set union, intersection and complement relative to* $U \times U$, *respectively.*
- $x^{\diamond}$ *is the set choice operator defined by the condition:*

$$x^{\diamond} \subseteq x \text{ and } |x^{\diamond}| = 1 \quad \iff \quad x \neq \emptyset .$$

- $\circ$ *is relational composition,* $^{\smile}$ *is transposition, and* * *is reflexive-transitive closure.*
- ∇, *the* fork *operator, is defined by the condition*

$$S \nabla T = \{ \langle x, y \star z \rangle : \langle x, y \rangle \in S \ \wedge \ \langle x, z \rangle \in T \} \ .$$

Notice that $x^{\diamond}$ denotes an arbitrary pair in x. This is why $x^{\diamond}$ is called a *choice* operator. Function $\star$ is used to encode pairs. The fact it is not surjective implies the existence of elements that do not encode pairs. These elements, called *urelements*, will be used to represent the elements from the carriers of the translated logics.

Definition 2. *We define* FCFAU = Rd •PCFAU, *where* Rd *takes reducts to structures of the form* $\langle 2^{U\times U}, \cup, \cap, ^{-}, \emptyset, U\times U, \circ, Id, ^{\smile}, \nabla, ^{\diamond}, ^{*} \rangle$ *(the sort* U *and the function* $\star$ *are forgotten).*

We will refer to the carrier of an algebra $\mathcal{A} \in$ FCFAU as $|\mathcal{A}|$.

The variety generated by FCFAU (the class of Proper Closure Fork Algebras) has a complete ([4, Theorem 1]) equational calculus (the ω-Calculus for Closure Fork Algebras with Urelements — ω-CCFAU) to be introduced next. In order to present the calculus, we provide the grammar for formulas, the axioms of the calculus, and the proof rules. For the sake of simplifying the notation, we will denote the relation $U \times U$ by 1, and the relation $\overline{1 \nabla 1} \cap Id$ by Id_{U}. Relation Id_{U} is the subset of the identity relation that relates the urelements.

Definition 3. *The set of* ω-CCFAU *terms is the smallest set* T *satisfying:*

- $\{ \emptyset, 1, Id \} \subseteq T$,
- *If* $x, y \in T$, *then* $\{ \breve{x}, x^{*}, x^{\diamond}, x \cup y, x \cap y, x \circ y, x \nabla y \} \subseteq T$.

Definition 4. *The set of* ω-CCFAU *formulas is the set of identities* $t_1 = t_2$, *with* $t_1, t_2 \in T$.

Definition 5. *The identities described in formulas (1)–(1) are axioms*[1] *of ω-CCFAU.*

1. *A set of identities axiomatizing the relational calculus [12].*
2. *The following axioms for the fork operator:*

$$x \nabla y = (x \circ (Id \nabla 1)) \cap (y \circ (1 \nabla Id)),$$
$$(x \nabla y) \circ (z \nabla w)^{\smile} = (x \circ \breve{z}) \cap (y \circ \breve{w}),$$
$$(Id \nabla 1)^{\smile} \nabla (1 \nabla Id)^{\smile} \leq Id.$$

3. *The following axioms for the choice operator [13, p. 324]:*

$$x^{\diamond} \circ 1 \circ \breve{x^{\diamond}} \leq Id, \quad \breve{x^{\diamond}} \circ 1 \circ x^{\diamond} \leq Id, \quad 1 \circ (x \cap x^{\diamond}) \circ 1 = 1 \circ x \circ 1.$$

4. *The following axioms for the Kleene star:*

$$x^{*} = Id \cup x \circ x^{*}, \quad x^{*} \circ y \leq y \cup x^{*} \circ (\overline{y} \cap x \circ y).$$

5. *An axiom forcing a nonempty set of urelements.*

$$1 \circ Id_{\mathsf{U}} \circ 1 = 1\ .$$

Definition 6. *The inference rules for the calculus ω-CCFAU are those of equational logic (see for instance [14, p. 94]), extended by adding the following inference rule*[2]*:*

$$\frac{\vdash Id \leq y \qquad x^{i} \leq y \vdash x^{i+1} \leq y}{\vdash x^{*} \leq y} \ (\forall i \in I\!N)$$

The importance of ∇ is twofold; first its inclusion assures the existence of a complete calculus with respect to its class of models (i.e. FCFAU) and second, and most important, it is used to define most of the translations from logical formulas to relational terms in order to interpret a logic in FCFAU.

Notice that only extralogical symbols belong to an equational or first-order signature. Symbols such as $=$ in equational logic, or $\vee$ in first-order logic, have a meaning that is univoquely determined by the carriers and the interpretation of the extralogical symbols. Similarly, once the field of a FCFAU has been fixed, all the operators can be assigned a standard meaning. This gives rise to the following definition of FCFAU signature.

Definition 7. *An* FCFAU *signature is a set of function symbols* $\{f_j\}_{j \in \mathcal{J}}$. *Each function symbol comes equipped with its arity. Notice that since* FCFAU*s have only one sort, the arity is a natural number.*

[1] Since the calculus of relations extends the Boolean calculus, we will denote by $\leq$ the ordering induced by the Boolean calculus in ω-CCFAU. As it is usual, $x \leq y$ is a shorthand for $x \cup y = y$.

[2] Given $i > 0$, by x^{i} we denote the relation inductively defined as follows: $x^{1} = x$, and $x^{i+1} = x \circ x^{i}$.

The set of FCFAU signatures will be denoted as $Sign_{\mathsf{FCFAU}}$. Actually, in order to interpret the logics mentioned in Section 1, constant relational symbols (rather than functions in general) suffice. Since new operators may be necessary in order to interpret new logics in the future, signatures will be allowed to contain functions of arbitrary rank.

In order to extend the definitions of terms (Def. 3) and formulas (Def. 4) to FCFAU signatures, we need to add the following rule.

– If $t_1, \ldots, t_{arity(f_j)} \in T$, then $f_j(t_1, \ldots, t_{arity(f_j)}) \in T$ (for all $j \in \mathcal{J}$).

If $\Sigma \in Sign_{\mathsf{FCFAU}}$, the set of Σ-terms will be denoted as $Term_\Sigma$. In the same way, Sen_Σ will denote the set of equalities between Σ-terms (i.e. the set of Σ-formulas).

Definition 8. *Let* $\Sigma = \{f_j\}_{j \in \mathcal{J}} \in Sign_{\mathsf{FCFAU}}$, *then* $\left\langle \mathcal{P}, \{\underline{f}_j\}_{j \in \mathcal{J}} \right\rangle \in Mod_\Sigma$ *iff* $\mathcal{P} \in \mathsf{FCFAU}$, *and* $\underline{f}_j : |\mathcal{P}|^{arity(f_j)} \to |\mathcal{P}|$, *for all* $j \in \mathcal{J}$.

Definition 9. *Given a signature* $\Sigma = \{f_j\}_{j \in \mathcal{J}} \in Sign_{\mathsf{FCFAU}}$, *and* $M = \left\langle \mathcal{P}, \{\underline{f}_j\}_{j \in \mathcal{J}} \right\rangle \in Mod_\Sigma$, *we denote by* $m_M : Term_\Sigma \to |\mathcal{P}|$ *the function that interprets terms in model* M.

Definition 10. *Let* $\Sigma \in Sign_{\mathsf{FCFAU}}$, *then* $\models^{\Sigma}_{\mathsf{FCFAU}} \subseteq Mod_\Sigma \times Sen_\Sigma$ *is defined as follows:* $M \models^{\Sigma}_{\mathsf{FCFAU}} t_1 = t_2$ *iff* $m_M(t_1) = m_M(t_2)$.

3 Institutions

Burstall and Goguen introduced in [7] the notion of institution as a general and abstract description of the model theory of a logic. This semantic path was then followed by a proof-theoretic approach by Meseguer [8], and Fiadeiro and A. Sernadas [15]. In this section we present the definition of institution, and use the notion of *entailment system* (or π-institution) in order to capture certain proof theoretical aspects of a logic. These concepts are then related by the notion of *logic* [8]. From here on, we assume the reader has a nodding acquaintance with basic concepts from category theory such as the notions of category, functor, natural transformation and colimits. The interested reader is directed to [16] for a gentle introduction to category theory for software engineering.

Definition 11. *A quadruple* $\left\langle \mathsf{Sign}, \mathbf{Sen}, \mathbf{Mod}, \{\models^{\Sigma}\}_{\Sigma \in |\mathsf{Sign}|} \right\rangle$ *is an* institution *if:*

– Sign *is a category of signatures,*
– $\mathbf{Sen} : \mathsf{Sign} \to \mathsf{Set}$ *is a functor (let* $\Sigma \in |\mathsf{Sign}|$, *then* $\mathbf{Sen}(\Sigma)$ *returns the set of* Σ*-sentences),*
– $\mathbf{Mod} : \mathsf{Sign}^{op} \to \mathsf{Cat}$ *is a functor (let* $\Sigma \in |\mathsf{Sign}|$, *then* $\mathbf{Mod}(\Sigma)$ *returns the category of* Σ*-models),*
– $\{\models^{\Sigma}\}_{\Sigma \in |\mathsf{Sign}|}$ *is a family of binary relations* $\models^{\Sigma} \subseteq |\mathbf{Mod}(\Sigma)| \times \mathbf{Sen}(\Sigma)$

and for any signature morphism $\sigma : \Sigma \to \Sigma'$, Σ*-sentence* $\phi \in \mathbf{Sen}(\Sigma)$ *and* Σ'*-model* $M' \in |\mathbf{Mod}(\Sigma)|$ *the following* $\models$*-invariance condition holds*

$$M' \models^{\Sigma'} \mathbf{Sen}(\sigma)(\phi) \;\; \textit{iff} \;\; \mathbf{Mod}(\sigma)(M') \models^{\Sigma} \phi \; .$$

Definition 12. *A triple* $\langle \mathsf{Sign}, \mathbf{Sen}, \{\vdash^{\Sigma}\}_{\Sigma \in |\mathsf{Sign}|} \rangle$ *is an* entailment system *if:*

- Sign *is a category of signatures,*
- $\mathbf{Sen} : \mathsf{Sign} \to \mathsf{Set}$ *is a functor (let* $\Sigma \in |\mathsf{Sign}|$*, then* $\mathbf{Sen}(\Sigma)$ *returns the set of* Σ*-sentences),*
- $\{\vdash^{\Sigma}\}_{\Sigma \in |\mathsf{Sign}|}$ *is a family of binary relations* $\vdash^{\Sigma} \subseteq 2^{\mathbf{Sen}(\Sigma)} \times \mathbf{Sen}(\Sigma)$ *such that for any* $\{\Sigma, \Sigma'\} \subseteq |\mathsf{Sign}|$, $\{\phi\} \cup \{\phi_i\}_{i \in \mathcal{I}} \subseteq \mathbf{Sen}(\Sigma)$, $\{\Gamma, \Gamma'\} \subseteq 2^{\mathbf{Sen}(\Sigma)}$ *the following conditions are satisfied:*
  1. *reflexivity:* $\{\phi\} \vdash^{\Sigma} \phi$,
  2. *monotonicity: if* $\Gamma \vdash^{\Sigma} \phi$ *and* $\Gamma \subseteq \Gamma'$*, then* $\Gamma' \vdash^{\Sigma} \phi$,
  3. *transitivity: if* $\Gamma \vdash^{\Sigma} \phi_i$ *for all* $i \in \mathcal{I}$ *and* $\Gamma \cup \{\phi_i\}_{i \in \mathcal{I}} \vdash^{\Sigma} \phi$*, then* $\Gamma \vdash^{\Sigma} \phi$*, and*
  4. $\vdash$*-translation: if* $\Gamma \vdash^{\Sigma} \phi$*, then for any morphism* $\sigma : \Sigma \to \Sigma'$ *in* Sign, $\mathbf{Sen}(\sigma)(\Gamma) \vdash^{\Sigma'} \mathbf{Sen}(\sigma)(\phi)$.

Definition 13. *Let* $\langle \mathsf{Sign}, \mathbf{Sen}, \{\vdash^{\Sigma}\}_{\Sigma \in |\mathsf{Sign}|} \rangle$ *be an entailment system. Then,* Th*, its category of theories, is a pair* $\langle \mathcal{O}, \mathcal{A} \rangle$ *such that:*

- $\mathcal{O}$ *is the set of pairs* $\langle \Sigma, \Gamma \rangle$ *with* $\Sigma \in |\mathsf{Sign}|$ *and* $\Gamma \subseteq \mathbf{Sen}(\Sigma)$*, and*
- $\mathcal{A}$ *are the theory morphisms, i.e., arrows* $\sigma : \langle \Sigma, \Gamma \rangle \to \langle \Sigma', \Gamma' \rangle$ *in which* $\sigma : \Sigma \to \Sigma'$ *is a signature morphism that satisfies the property:*

$$\textit{for all } \gamma \in \Gamma, \quad \Gamma' \vdash^{\Sigma'} \mathbf{Sen}(\sigma)(\gamma) \; .$$

Definition 14. *A quintuple* $\langle \mathsf{Sign}, \mathbf{Sen}, \mathbf{Mod}, \{\vdash^{\Sigma}\}_{\Sigma \in |\mathsf{Sign}|}, \{\models^{\Sigma}\}_{\Sigma \in |\mathsf{Sign}|} \rangle$ *is a* logic *if:*

- $\langle \mathsf{Sign}, \mathbf{Sen}, \{\vdash^{\Sigma}\}_{\Sigma \in |\mathsf{Sign}|} \rangle$ *is an entailment system,*
- $\langle \mathsf{Sign}, \mathbf{Sen}, \mathbf{Mod}, \{\models^{\Sigma}\}_{\Sigma \in |\mathsf{Sign}|} \rangle$ *is an institution, and*
- *the following* soundness *condition is satisfied: for any* $\Sigma \in |\mathsf{Sign}|$, $\phi \in \mathbf{Sen}(\Sigma)$, $\Gamma \subseteq \mathbf{Sen}(\Sigma)$, $\Gamma \vdash^{\Sigma} \phi \Longrightarrow \Gamma \models^{\Sigma} \phi$.

A logic is complete *if in addition the following condition is also satisfied: for any* $\Sigma \in |\mathsf{Sign}|$, $\phi \in \mathbf{Sen}(\Sigma)$, $\Gamma \subseteq \mathbf{Sen}(\Sigma)$, $\Gamma \models^{\Sigma} \phi \Longrightarrow \Gamma \vdash^{\Sigma} \phi$.

4 The Logic Behind Closure Fork Algebras

In this section we will show how to build a logic (in the sense of Def. 14) on top of closure fork algebras. Since the variety generated by FCFAU is completely characterized by the ω-CCFAU equational calculus, we might consider to relativize the institution (entailment system) of equational logic rather than introducing a new one from scratch. This might work for a while, but there are technical and methodological reasons for presenting the explicit construction.

On the technical side, notice that the actual proof systems for equational logic and ω-CCFAU differ in their proof rules (ω-CCFAU has an extra rule – c.f. Def. 6). This prevents us from modeling the proof calculus [8, Def. 12] ω-CCFAU as a proof subcalculus [8, Def. 14] of equational logic. From the methodological point of view, the categorical construction provides important information to the reader on what operations are part of the logic, how morphisms are defined, etc.

4.1 The Institution Behind Closure Fork Algebras

In this section we will define an institution on top of FCFAU. The section is structured following the order of requirements stated in Def. 11.

Once the definitions for $\mathsf{Sign}_{\mathsf{FCFAU}}$, $\mathbf{Sen}_{\mathsf{FCFAU}}$ or $\mathbf{Mod}_{\mathsf{FCFAU}}$ are precisely stated, proving that indeed $\mathsf{Sign}_{\mathsf{FCFAU}}$ is a category, or that $\mathbf{Sen}_{\mathsf{FCFAU}}$ and $\mathbf{Mod}_{\mathsf{FCFAU}}$ are functors between appropriate categories becomes a simple exercise in category theory (and is therefore left to be proved by the reader). Here lies the beauty of the institution of closure fork algebras. It is a simple, yet very expressive institution, and therefore an appropriate candidate for a "universal" institution.

Lemma 1. *We define* $\mathsf{Sign}_{\mathsf{FCFAU}} = \langle \mathcal{O}, \mathcal{A} \rangle$, *where* $\mathcal{O} = Sign_{\mathsf{FCFAU}}$, *and* $\sigma : \{f_i\}_{i \in \mathcal{I}} \to \{g_j\}_{j \in \mathcal{J}} \in \mathcal{A}$ *whenever* σ *is an arity preserving total function. Then,* $\mathsf{Sign}_{\mathsf{FCFAU}}$ *is a category.*

The intuitive meaning is that an arrow $\sigma : \Sigma \to \Sigma'$ is a translation of Σ-symbols to Σ'-symbols. Since the fork algebra operators are not in the signatures, these cannot be translated.

Definition 15. *Let* $\Sigma \in |\mathsf{Sign}_{\mathsf{FCFAU}}|$, *then* $\mathbf{Term}_{\mathsf{FCFAU}}(\Sigma) = Term_{\Sigma}$.

Let $\sigma : \Sigma \to \Sigma'$ be a FCFAU signature morphism, then σ_{term} is the homomorphic extension of σ to terms of the set $\mathbf{Term}_{\mathsf{FCFAU}}(\Sigma)$. Function σ_{term} translates terms according to the translation of basic symbols induced by σ.

Definition 16. *Let* $\Sigma \in |\mathsf{Sign}_{\mathsf{FCFAU}}|$, *then* $\mathbf{Sen}_{\mathsf{FCFAU}}(\Sigma) = Sen_{\Sigma}$.

Given a signature morphism σ, function σ_{eq} translates FCFAU sentences according to σ. It is defined by the condition

$$\sigma_{eq}(t1 = t2) \stackrel{def}{=} \sigma_{term}(t1) = \sigma_{term}(t2) .$$

Lemma 2. *Let* $\sigma : \Sigma \to \Sigma'$ *be a* $\mathsf{Sign}_{\mathsf{FCFAU}}$ *morphism, then* $\mathbf{Sen}_{\mathsf{FCFAU}} : \mathsf{Sign}_{\mathsf{FCFAU}} \to \mathsf{Set}$ *defined as* $\mathbf{Sen}_{\mathsf{FCFAU}}(\sigma)(S) = \{ \sigma_{eq}(s) : s \in S \}$ *is a functor.*

Lemma 3. *Let* $\Sigma \in |\mathsf{Sign}_{\mathsf{FCFAU}}|$, *then* $\mathbf{Mod}_{\mathsf{FCFAU}}(\Sigma) = \langle \mathcal{O}_{\Sigma}, \mathcal{A}_{\Sigma} \rangle$, *where*

– $\mathcal{O}_{\Sigma} = Mod_{\Sigma}$, $\mathcal{A}_{\Sigma} = \{\gamma : M \to M' : \gamma$ *is a* FCFAU *homomorphism*$\}$,

is a category.

The next definition characterizes the action of $\mathbf{Mod}_{\mathsf{FCFAU}}$ on morphisms in $\mathsf{Sign}_{\mathsf{FCFAU}}$. Morphisms $\sigma : \Sigma \rightarrow \Sigma'$ in $\mathsf{Sign}_{\mathsf{FCFAU}}$ are translations from Σ-symbols to Σ'-symbols. Since $\mathbf{Mod}_{\mathsf{FCFAU}}$ must be contravariant, we will define it as the operation that from Σ'-algebras takes reducts to the signature of Σ-algebras.

Definition 17. *Let* $\Sigma = \{f_i\}_{i\in\mathcal{I}}$ *and* $\Sigma' = \{g_j\}_{j\in\mathcal{J}}$ *be* FCFAU *signatures. Let* $\sigma : \Sigma \rightarrow \Sigma'$ *be a* $\mathsf{Sign}_{\mathsf{FCFAU}}$ *morphism. Let* $M' = \left\langle \mathcal{P}, \{\underline{g}_j\}_{j\in\mathcal{J}} \right\rangle \in |\mathbf{Mod}_{\mathsf{FCFAU}}(\Sigma')|$. *Then,* $M' \restriction_\sigma = \left\langle \mathcal{P}, (\underline{f}_i)_{i\in\mathcal{I}} \right\rangle$, *where* $\underline{f}_i = \underline{\sigma(f_i)}$.

Lemma 4. *Let* $\sigma : \Sigma \rightarrow \Sigma'$ *be a* $\mathsf{Sign}_{\mathsf{FCFAU}}$ *morphism. Then,* $\mathbf{Mod}_{\mathsf{FCFAU}}(\sigma) : \mathbf{Mod}_{\mathsf{FCFAU}}(\Sigma') \rightarrow \mathbf{Mod}_{\mathsf{FCFAU}}(\Sigma)$ *defined by*

$$\mathbf{Mod}_{\mathsf{FCFAU}}(\sigma)(M') = M' \restriction_\sigma, \quad \mathbf{Mod}_{\mathsf{FCFAU}}(\sigma)(\gamma') = \gamma',$$

is a functor.

Lemma 5. $\mathbf{Mod}_{\mathsf{FCFAU}} : {\mathsf{Sign}_{\mathsf{FCFAU}}}^{\mathsf{op}} \rightarrow \mathsf{Cat}$ *is a functor.*

Proof. The proof immediately follows from Lemmas 3 and 4.

Lemma 6. *Let* $\sigma : \Sigma \rightarrow \Sigma'$ *be a* $\mathsf{Sign}_{\mathsf{FCFAU}}$ *morphism. Let* $t_1 = t_2 \in \mathbf{Sen}_{\mathsf{FCFAU}}(\Sigma)$. *Let* $M' \in |\mathbf{Mod}_{\mathsf{FCFAU}}(\Sigma')|$. *Then,*

$$\mathbf{Mod}_{\mathsf{FCFAU}}(\sigma)(M') \models^{\Sigma}_{\mathsf{FCFAU}} t_1 = t_2 \quad \text{iff} \quad M' \models^{\Sigma'}_{\mathsf{FCFAU}} \mathbf{Sen}_{\mathsf{FCFAU}}(\sigma)(t_1 = t_2)\ .$$

Theorem 1. *The structure*

$$\left\langle \mathsf{Sign}_{\mathsf{FCFAU}}, \mathbf{Sen}_{\mathsf{FCFAU}}, \mathbf{Mod}_{\mathsf{FCFAU}}, \{\models^{\Sigma}_{\mathsf{FCFAU}}\}_{\Sigma\in|\mathsf{Sign}_{\mathsf{FCFAU}}|} \right\rangle$$

is an institution.

Proof. The proof follows by Lemmas 1, 2, 5 and 6.

The institution of the closure fork algebras is denoted by I_{FCFAU}.

4.2 The Entailment System Behind Closure Fork Algebras

In this section we use a standard model theoretic construction [8, Prop. 4] in order to build a candidate entailment system. This entailment system, though it defines an entailment relation, does not guarantee the existence of axioms and proof rules implementing the deduction relation. We will address this issue again in Section 4.3.

Definition 18. *Let* $\Sigma \in |\mathsf{Sign}_{\mathsf{FCFAU}}|$. *Let* $\Gamma \subseteq \mathbf{Sen}_{\mathsf{FCFAU}}(\Sigma)$. *We define the category* $\mathbf{Mod}_{\mathsf{FCFAU}}(\Sigma, \Gamma)$ *as the full subcategory of* $\mathbf{Mod}_{\mathsf{FCFAU}}(\Sigma)$ *determined by those models* $M \in |\mathbf{Mod}_{\mathsf{FCFAU}}(\Sigma)|$ *that satisfy all the sentences in* Γ, *i.e.,* $M \models^{\Sigma}_{\mathsf{FCFAU}} \phi$ *for each* $\phi \in \Gamma$.

We also define a relation between sets of sentences and sentences $\Vdash^{\Sigma}_{\mathsf{FCFAU}}$, *as follows:*

$$\Gamma \Vdash^{\Sigma}_{\mathsf{FCFAU}} \phi \quad \text{iff} \quad M \models^{\Sigma}_{\mathsf{FCFAU}} \phi \quad \text{for each } M \in |\mathbf{Mod}_{\mathsf{FCFAU}}(\Sigma, \Gamma)|\ .$$

Since $\mathsf{Sign}_{\mathsf{FCFAU}}$ is a category, and $\mathbf{Sen}_{\mathsf{FCFAU}}$ is a functor. In [8, Prop. 4] it is proved that for each $\Sigma \in |\mathsf{Sign}_{\mathsf{FCFAU}}|$, $\Vdash^{\Sigma}_{\mathsf{FCFAU}}$ satisfies the conditions presented in Def. 12. Thus, the following theorem holds.

Theorem 2. *The structure*

$$\left\langle \mathsf{Sign}_{\mathsf{FCFAU}}, \mathbf{Sen}_{\mathsf{FCFAU}}, \{\Vdash^{\Sigma}_{\mathsf{FCFAU}}\}_{\Sigma \in |\mathsf{Sign}_{\mathsf{FCFAU}}|} \right\rangle$$

is an entailment system.

The entailment system of closure fork algebras is denoted by I^{+}_{FCFAU}.

4.3 The Logic Behind Closure Fork Algebras

Notice that since by definition the relation $\Vdash^{\Sigma}_{\mathsf{FCFAU}}$ is sound and complete with respect to $\models^{\Sigma}_{\mathsf{FCFAU}}$, from Thms. 1 and 2, the following theorem holds.

Theorem 3. *(I_{FCFAU} and I^{+}_{FCFAU} form a logic)*
The structure

$$\left\langle \mathsf{Sign}_{\mathsf{FCFAU}}, \mathbf{Sen}_{\mathsf{FCFAU}}, \mathbf{Mod}_{\mathsf{FCFAU}}, \{\Vdash^{\Sigma}_{\mathsf{FCFAU}}\}_{\Sigma \in |\mathsf{Sign}_{\mathsf{FCFAU}}|}, \{\models^{\Sigma}_{\mathsf{FCFAU}}\}_{\Sigma \in |\mathsf{Sign}_{\mathsf{FCFAU}}|} \right\rangle$$

is a logic.

The logic associated to closure fork algebras will be denoted by L_{FCFAU}.

Having proved the existence of a sound and complete entailment relation $\Vdash^{\Sigma}_{\mathsf{FCFAU}}$ in the way we did, is of little interest. The entailment relation does not give any hints as to how to deduce properties, what would be the axioms, or what are the proof rules employed in order to generate the relation. Actually, it might be the case that no deduction mechanism is available. Fortunately, as the following theorem shows, this is not the case when working with closure fork algebras.

Theorem 4. *Let $\vdash^{\Sigma}_{\mathsf{FCFAU}} \subseteq 2^{\mathbf{Sen}_{\mathsf{FCFAU}}(\Sigma)} \times \mathbf{Sen}_{\mathsf{FCFAU}}(\Sigma)$ be the entailment relation induced by the calculus ω-CCFAU. That is, $\Theta \vdash^{\Sigma}_{\mathsf{FCFAU}} \phi$ iff there is a proof of ϕ from the set of hypotheses Θ. Then, for all $\Sigma \in |\mathsf{Sign}_{\mathsf{FCFAU}}|$, $\vdash^{\Sigma}_{\mathsf{FCFAU}} = \Vdash^{\Sigma}_{\mathsf{FCFAU}}$.*

5 Reasoning Across Logics in Closure Fork Algebras

Once the logic behind FCFAU has been developed, it is possible to review the existing interpretability results in the light of the theory of institutions and institution representations [9]. Moreover, we will present a general technique for building a unique homogeneous theory from heterogeneous ones.

In general, if L is a logic and Σ_L an L-signature, an interpretability result of L in FCFAU is generally presented by resorting to:

- A mapping S from Σ_L to $\Sigma \in |\mathsf{Sign}_{\mathsf{FCFAU}}|$.
- A translation $T_{L \to \mathsf{FCFAU}}$ of Σ_L-formulas to Σ-equations.

– A mapping of Σ_L-models to Σ-models (full closure fork algebras), satisfying:

$$\forall \mathfrak{A} \in \mathsf{FCFAU}, \exists \mathfrak{B}_{\mathfrak{A}} \in |\mathbf{Mod}_L(\Sigma_L)| \\ \left(\mathfrak{B}_{\mathfrak{A}} \models_L^{\Sigma_L} \alpha \iff \mathfrak{A} \models_{\mathsf{FCFAU}}^{\Sigma} T_{L\rightarrow\mathsf{FCFAU}}(\alpha)\right) .$$

– A mapping of Σ-models (full closure fork algebras) to Σ_L-models, satisfying:

$$\forall \mathfrak{B} \in |\mathbf{Mod}_L(\Sigma_L)|, \exists \mathfrak{A}_{\mathfrak{B}} \in \mathsf{FCFAU} \\ \left(\mathfrak{B} \models_L^{\Sigma_L} \alpha \iff \mathfrak{A}_{\mathfrak{B}} \models_{\mathsf{FCFAU}}^{\Sigma} T_{L\rightarrow\mathsf{FCFAU}}(\alpha)\right) .$$

We introduce next institution representations.

Definition 19. *Given institutions* $I = \left\langle \mathsf{Sign}, \mathbf{Sen}, \mathbf{Mod}, \{\models^{\Sigma}\}_{\Sigma\in|\mathsf{Sign}|} \right\rangle$ *and* $I' = \left\langle \mathsf{Sign}', \mathbf{Sen}', \mathbf{Mod}', \{\models^{\Sigma'}\}_{\Sigma\in|\mathsf{Sign}'|} \right\rangle$, *an* institution representation $\rho : I \rightarrow I'$ *consists of a functor* $\rho^{sign} : \mathsf{Sign} \rightarrow \mathsf{Sign}'$, *a natural transformation* $\rho^{sen} : \mathbf{Sen} \Longrightarrow \rho^{sign}; \mathbf{Sen}'$ *and a natural transformation* $\rho^{mod} : (\rho^{sign})^{\mathsf{op}}; \mathbf{Mod}' \Longrightarrow \mathbf{Mod}$ *such that for each* $\sigma \in |\mathsf{Sign}|$, $\phi \in \mathbf{Sen}(\Sigma)$, *and* $M' \in |\mathbf{Mod}(\rho^{sign}(\Sigma))|$ *the following property holds:*

$$M' \models^{\rho^{sign}(\Sigma)} \rho_{\Sigma}^{sen}(\phi) \text{ iff } \rho_{\Sigma}^{mod}(M') \models^{\Sigma} \phi .$$

Notice that the mappings necessary in order to build a representation between institutions can all be trivially obtained from an algebraic interpretation of a logic to closure fork algebras.

Suppose S is a system whose models allow us to retrieve temporal properties (formalized as a linear temporal logic theory $Th_S^{\mathsf{LTL}} = \langle \Sigma_S^{\mathsf{LTL}}, \Gamma_{\mathsf{LTL}} \rangle$), and dynamic properties (formalized as a propositional dynamic logic theory $Th_S^{\mathsf{PDL}} = \langle \Sigma_S^{\mathsf{PDL}}, \Gamma_{\mathsf{PDL}} \rangle$). Using institution representations $\rho_{\mathsf{LTL}\rightarrow\mathsf{FCFAU}} : I_{\mathsf{LTL}} \rightarrow I_{\mathsf{FCFAU}}$ and $\rho_{\mathsf{PDL}\rightarrow\mathsf{FCFAU}} : I_{\mathsf{PDL}} \rightarrow I_{\mathsf{FCFAU}}$ allows us to get new theories that live in the same algebraic world:

$$\begin{array}{ccc} \rho_{\mathsf{LTL}\rightarrow\mathsf{PCFA}}(Th_S^{\mathsf{LTL}}) & \qquad & \rho_{\mathsf{PDL}\rightarrow\mathsf{PCFA}}(Th_S^{\mathsf{PDL}}) \\ \uparrow \rho_{\mathsf{LTL}\rightarrow\mathsf{PCFA}} & & \uparrow \rho_{\mathsf{PDL}\rightarrow\mathsf{PCFA}} \\ Th_S^{\mathsf{LTL}} & & Th_S^{\mathsf{PDL}} \end{array}$$

It is easy to prove that the category $\mathsf{Th}_{\mathsf{FCFAU}}$ is finitely cocomplete, that is, for any finite diagram $\gamma : I \rightarrow \mathsf{Th}_{\mathsf{FCFAU}}$ there exists a colimit of γ (i.e. there exists $T \in |\mathsf{Th}_{\mathsf{FCFAU}}|$ and a commutative cocone with base γ and vertex T).

Notice that if in a diagram we have $T_1 \overset{\sigma_1}{\rightarrow} T_2 \overset{\sigma_2}{\rightarrow} T_3$, ($T_1, T_2, T_3$, theories and σ_1, σ_2 theory morphisms) and $\sigma_1(f_1) = f_2$, $\sigma_2(f_2) = f_3$, then symbols f_1, f_2 and f_3 will be mapped to the same symbol in the colimit.

Using this colimit construction we can compute a FCFAU-theory specifying the whole system. Notice that this theory only equates those symbols that are glued by morphisms from the diagram. A different situation arises when the developer wishes to equate symbols from theories T_1 and T_2 due to system design decisions.

It is seldom the case that T_1 and T_2 will be related by signature morphisms. In [17], Fiadeiro presented, as a tool to solve this synchronization problem, the specification of *channels*. Channels are theories that only contain symbols with no particular behavior (no axioms for the symbols are given). Let us illustrate with an example how channels work. Let $\{P_i\}_{i \in I} \in \Sigma_S^{\mathsf{LTL}}$ and $\{P'_i\}_{i \in I} \in \Sigma_S^{\mathsf{PDL}}$ be proposition symbols. Let us assume they were aimed to represent the same predicates, pairwise (that is, $P_i = P'_i$ for all $i \in I$). The idea is to create a new theory $\langle \{\mathbf{P}_i\}_{i \in I}, \emptyset \rangle$ and theory morphisms σ_{LTL} (mapping $\mathbf{P}_i$ to P_i) and σ_{PDL} (mapping $\mathbf{P}_i$ to P'_i). Then, the symbols P_i and P'_i are adequately equated in the colimit. Unfortunately, this construction is flawed because theories Th_S^{LTL} and Th_S^{PDL} live in different institutions. Fixing this is trivial by introducing a FCFAU theory channel between the representations of the theories, as the following diagram shows:

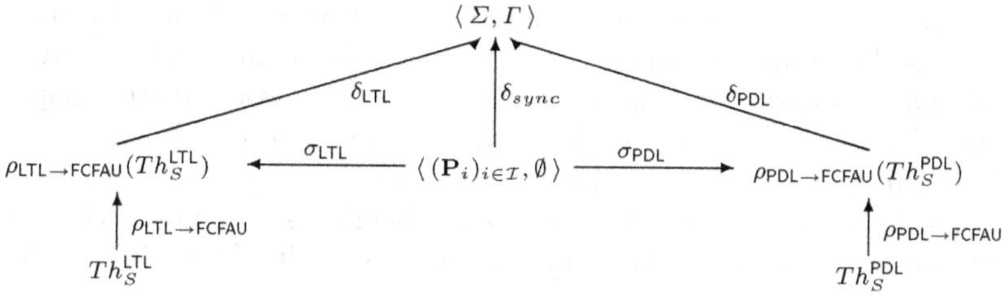

We have successfully equated the proposition symbols. Still there is a major gap between the theories. In effect, the accessibility relation $\mathbf{T}$ induced by the LTL theory is unrelated with the actions $\{\mathbf{A}_j\}_{j \in J}$ in the PDL theory. Notice that it is a very natural decision in the software development process to use PDL for specifying atomic actions and LTL to prescribe the admissible evolutions of the system. Therefore, the relationship $\mathbf{T} = \bigcup_{j \in J} \mathbf{A}_j$ should hold. Since $\mathbf{T}$ is not a symbol in the signature of LTL theories (it appears in the semantics of the logic), a construction along the lines of the channel construction does not seem to work.

Fortunately, the representation of LTL to I_{FCFAU} (see [5] for the mapping of LTL formulas to closure fork algebras) includes symbol $\mathbf{T}$ in signatures, and therefore a construction as the one provided in the following picture, including a theory whose only axiom is $\mathbf{T} = \bigcup_{j \in J} \mathbf{A}_j$, does the job.

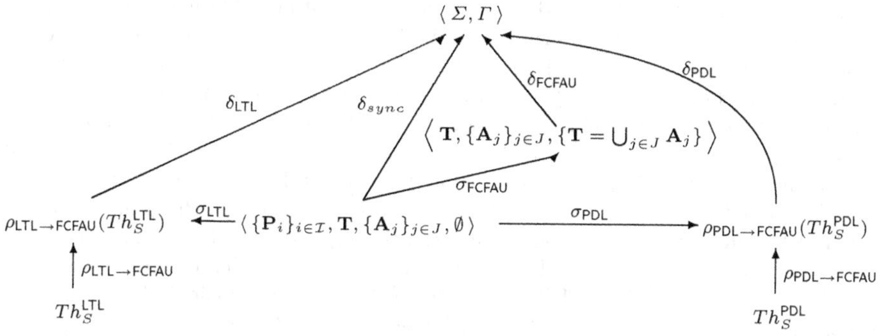

At this point, questions about conservation of properties of the partial specifications might arise. Since equational logic is monotonic, theorems from the particular theories are preserved. Nevertheless, new properties may emerge in the combined theory as a result of the interaction between the properties of the intervening theories.

6 Conclusions

Relations are ubiquitous in logics with applications in software technology. Either in the form of predicates in classical first order logic, as actions in dynamic logic, or as accessibility relations in modal logics, they are everywhere. Therefore, the election of a relational framework to represent logics is fully justified. Many logics have already been interpreted in the theory of closure fork algebras. Reasoning across these logics then becomes feasible by resorting to relational tools. The relational algebraic approach departs from the mainstream work on reasoning across formalisms carried on in the institutional framework. In this paper we have related both approaches and have shown that closure fork algebras are a suitable candidate as a universal logic. The example we have discussed on the relationship between dynamic and temporal theories shows how easy it is to introduce meaningful relationships among theories coming from different logics.

References

1. Booch, G., Rumbaugh, J., Jacobson, I.: The unified modeling language user guide. Addison–Wesley Longman Publishing Co., Inc., Boston, MA, USA (1998)
2. Frias, M.F.: Fork algebras in algebra, logic and computer science. Volume 2 of Advances in logic. World Scientific Publishing Co., Singapore (2002)
3. Frias, M.F., Orlowska, E.: Equational reasoning in non-classical logics. Journal of Applied Non-classical Logics **8** (1998) 27–66
4. Frias, M.F., Baum, G.A., Maibaum, T.S.E.: Interpretability of first-order dynamic logic in a relational calculus. In de Swart, H., ed.: Proceedings of the 6th. Conference on Relational Methods in Computer Science (RelMiCS) - TARSKI. Volume 2561 of Lecture Notes in Computer Science., Oisterwijk, The Netherlands, Springer-Verlag (2002) 66–80
5. Frias, M.F., Lopez Pombo, C.G.: Time is on my side. In Berghammer, R., Möller, B., eds.: Proceedings of the 7th. Conference on Relational Methods in Computer Science (RelMiCS) - 2nd. International Workshop on Applications of Kleene Algebra, Malente, Germany (2003) 105–111
6. Frias, M.F., Lopez Pombo, C.G.: Interpretability of first-order linear temporal logics in fork algebras. Journal of Logic and Algebraic Programming **66** (2006) 161–184
7. Goguen, J.A., Burstall, R.M.: Introducing institutions. In Clarke, E.M., Kozen, D., eds.: Proceedings of the Carnegie Mellon Workshop on Logic of Programs. Volume 184 of Lecture Notes in Computer Science., Springer-Verlag (1984) 221–256
8. Meseguer, J.: General logics. In Ebbinghaus, H.D., Fernandez-Prida, J., Garrido, M., Lascar, D., Artalejo, M.R., eds.: Proceedings of the Logic Colloquium '87. Volume 129., Granada, Spain, North Holland (1989) 275–329

9. Tarlecki, A.: Moving between logical systems. In Haveraaen, M., Owe, O., Dahl, O.J., eds.: Selected papers from the 11th Workshop on Specification of Abstract Data Types Joint with the 8th COMPASS Workshop on Recent Trends in Data Type Specification. Volume 1130 of Lecture Notes in Computer Science., Springer-Verlag (1996) 478–502
10. Frias, M.F., Gamarra, R., Steren, G., Bourg, L.: A strategy for efficient verification of relational specification, based in monotonicity analysis. In Redmiles, D.F., Ellman, T., Zisman, A., eds.: Proceedings of the 20th. IEEE/ACM International Conference on Automated Software Engineering, Long Beach, California, USA, Association for the Computer Machinery and IEEE Computer Society, ACM Press (2005) 305–308
11. Lopez Pombo, C.G., Owre, S., Shankar, N.: A semantic embedding of the $\mathbf{A_g}$ dynamic logic in PVS. Technical Report SRI-CSL-02-04, Computer Science Laboratory, SRI International (2002)
12. Tarski, A.: On the calculus of relations. Journal of Symbolic Logic **6** (1941) 73–89
13. Maddux, R.D.: Finitary algebraic logic. Zeitschrift fur Mathematisch Logik und Grundlagen der Mathematik **35** (1989) 321–332
14. Burris, S., Sankappanavar, H.P.: A course in universal algebra. Graduate Texts in Mathematics. Springer-Verlag, Berlin, Germany (1981)
15. Fiadeiro, J.L., Sernadas, A.: Structuring theories on consequence. In Tarlecki, A., Sannella, D., eds.: Selected papers from the 5th Workshop on Specification of Abstract Data Types. Lecture Notes in Computer Science, Gullane, Scotland, Springer-Verlag (1987) 44–72
16. Fiadeiro, J.L.: Categories for software engineering. Springer-Verlag (2005)
17. Fiadeiro, J.L.: On the emergence of properties in component-based systems. In Wirsing, M., Nivat, M., eds.: Proceedings of the 1996 Algebraic Methodology and Software Technology – AMAST 96. Volume 1101 of Lecture Notes in Computer Science., Munich, Germany, Springer-Verlag (1996)

Realizability Criteria for Compositional MSC*

Arjan Mooij, Judi Romijn, and Wieger Wesselink

Technische Universiteit Eindhoven
Department of Mathematics and Computer Science
P.O. Box 513, 5600 MB Eindhoven, The Netherlands
{a.j.mooij, j.m.t.romijn, j.w.wesselink}@tue.nl

Abstract. Synthesizing a proper implementation for a scenario-based specification is often impossible, due to the distributed nature of implementations. To be able to detect problematic specifications, realizability criteria have been identified, such as non-local choice.

In this work we develop a formal framework to study realizability of compositional MSC [GMP03]. We use it to derive a complete classification of criteria that is closely related to the criteria for MSC from [MGR05]. Comparing specifications and implementations is usually complicated, because different formalisms are used. We treat both of them in terms of a single formalism. Therefore we extend the partial order semantics of [Pra86, KL98] with a way to model deadlocks and with a more sophisticated way to address communication.

1 Introduction

For scenario-based specifications of distributed systems (e.g. in terms of Message Sequence Chart, MSC), it is often impossible to synthesize an implementation with exactly the same behavior. This is caused by the distributed nature of implementations. The best-known phenomenon leading to problems is non-local choice [BAL97], but also other criteria [HJ00, Gen05, MGR05] have been proposed to determine realizability of specifications in practice [MG05]. In this work we develop a formal framework to study such criteria for the MSC extension that is called compositional MSC [GMP03, MM01]. Our work differs from [AEY05], which studies decidability and worst-case complexity of checking whether an MSC specification is realizable, but provides no practical criteria.

Most realizability criteria seem to be tricky formalizations of intuitions about realizability. In contrast, we formally study under what circumstances specifications are trace equivalent to their implementations, and derive a condition that is both necessary and sufficient. From this condition, we derive a complete classification of realizability criteria for compositional MSC. The resulting formal criteria can easily be related to our intuitive criteria in [MGR05].

Several kinds of semantics have been proposed for MSC specifications (e.g. [KL98, Ren99, Hey00, UKM03]), while implementations are typically expressed

* This research is supported by the NWO under project 016.023.015: "Improving the Quality of Protocol Standards".

M. Johnson and V. Vene (Eds.): AMAST 2006, LNCS 4019, pp. 248–262, 2006.

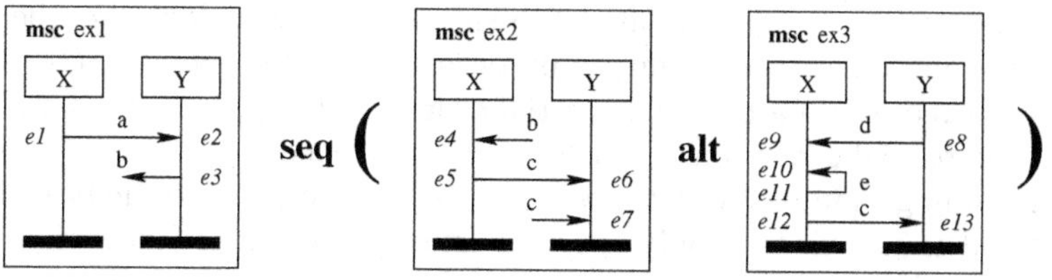

Fig. 1. Running example

in terms of finite state machines. To compare specifications and implementations, two different formalisms must then be related, usually via execution traces (in fact a third formalism). We prefer to use one single formalism for both implementations and specifications, and we want to stay close to the MSC specification formalism. Therefore we use a partial order semantics [Pra86] for our study, and sketch the relation with operational formalisms. In addition to the partial order model in [Pra86, KL98], we introduce a way to model deadlocks and a more sophisticated way to deal with communication.

Overview In Section 2 we introduce our partial order model, which we extend with communication in Section 3. These two sections are rather independent from MSC, but they are the basis of the semantics of compositional MSC in Section 4. In Section 5 we define the typical way of synthesizing an implementation; trace equivalence between specifications and such implementations is studied in Section 6. Finally in Section 7 we classify various realizability criteria. The conclusions and further work are presented in Section 8.

2 Extended Partial Order Model

In this section we define a partial order model and extend it with deadlocks, to make it suitable for studying realizability criteria.

2.1 Running Example

We illustrate our techniques using a running example.

Figure 1 contains a (high-level) MSC consisting of the three basic MSCs ex1, ex2 and ex3. It specifies the behavior of process instances X and Y, such that first the behavior of ex1 occurs, followed by either the behavior of ex2 or the behavior of ex3. For reference purposes we have included arbitrary event names (e_1 to e_{13}) in the basic MSCs.

2.2 LATERs: LAbeled Transitive Event Relations

As a semantic model of behavior, we introduce the notion of a later, which is an acronym for **la**belled **t**ransitive **e**vent **r**elation. A *later* $(E, <, l)$ is a triple that consists of an event set E, a transitive causality relation $<: < \subseteq E \times E$ and a labeling function $l : E \to L$ for a given set of labels L. The behavior of a later is such that any event $e : e \in E$ models a single action with label $l.e$; the event can occur at most once and it may only occur after all events $f : f < e$ have already occurred. Compared to the partial orders in [Pra86], a later is an lposet in which the partial order constraint has been weakened.

In our running example, let laters p_1, p_2 and p_3 correspond to the basic MSCs ex1, ex2 and ex3, such that only the causalities per instance (on each vertical axis) are considered, i.e. without communication. So, $p_1 = (\{e_1, e_2, e_3\}, \{e_2 < e_3\}, l_1)$ and as we will see later on $l_1 = \{e_1 \mapsto !(a, X, Y),\ e_2 \mapsto ?(a, X, Y),\ e_3 \mapsto !(b, Y, X)\}$. The structure of p_1 can be visualized as $\boxed{e_1 \quad e_2 \longrightarrow e_3}$ such that relation $<$ corresponds to the transitive closure of relation $\rightarrow$. In an interleaved execution model where the events are labeled with atomic actions, the maximal behaviors of a partially ordered later are its linearizations; in this example: $e_1 \cdot e_2 \cdot e_3$, $e_2 \cdot e_1 \cdot e_3$ and $e_2 \cdot e_3 \cdot e_1$. Each linearization represents an execution trace, i.e. a sequence of action labels. We prefer to reason about partial orders, because they are better related to MSC and they avoid decomposing each partial order into several over-specific total orders. Another advantage is that true concurrency can be modeled.

The most elementary laters are the empty later, with no events, and the singleton laters, with only one event with a label $k : k \in L$. We introduce the following abbreviations for them:

$$[\epsilon] = (\emptyset, \emptyset, \emptyset)$$
$$[k] = (\{e\}, \emptyset, \{[e \mapsto k]\}) \qquad \text{for } k : k \in L \text{ and arbitrary } e$$

2.3 Isomorphism

The event set of a later is abstract in the sense that a consistent renaming of the events yields a later with the same behavior. This is formalized in the following notion of isomorphism. Laters $(E, <, l)$ and $(E', <', l')$ are *isomorphic*, denoted $(E, <, l) \simeq (E', <', l')$, if there is a bijection $\sim : \sim \subseteq E \times E'$ such that both

- $(\forall e, e' :: e \sim e' \Rightarrow l.e = l'.e')$
- $(\forall e, f, e', f' :: e \sim e' \wedge f \sim f' \Rightarrow (e < f \equiv e' <' f'))$

Relation $\simeq$ is an equivalence relation. In what follows we will hardly mention $\simeq$ explicitly, and implicitly assume that where necessary $\simeq$ has been exploited to obtain suitable laters, e.g. ones that are event disjoint.

2.4 Elementary Later Operators

We often need to relate events to the instance (i.e. computational unit or process) in which they occur. We assume a fixed set of *instance names* I, and a function[1] $\phi : L \rightarrow I$ that maps labels to the instance in which the actions with that label

[1] For a later $(E, <, l)$, [HJ00] uses the slightly different function $\phi' : E \rightarrow I$, which can be obtained from our later-independent ϕ as follows: $\phi'.e = \phi.(l.e)$.

occur. To construct larger laters from the elementary laters, we use the following elementary operators on event disjoint laters (i.e. $E_p \cap E_q = \emptyset$):

$$(E_p, <_p, l_p) \parallel (E_q, <_q, l_q) \;=\; (E_p \cup E_q,\ <_p \cup <_q,\ l_p \cup l_q)$$

$$(E_p, <_p, l_p) \circ_S (E_q, <_q, l_q) \;=\; (E_p \cup E_q,\ <_p \cup <_{\circ_S} \cup <_q,\ l_p \cup l_q)$$
where $<_{\circ_S} = E_p \times E_q$

$$(E_p, <_p, l_p) \circ_W (E_q, <_q, l_q) \;=\; (E_p \cup E_q,\ (<_p \cup <_{\circ_W} \cup <_q)^+,\ l_p \cup l_q)$$
where $<_{\circ_W} = \{(e, f) \mid e, f : e \in E_p \wedge f \in E_q \wedge \phi.(l_p.e) = \phi.(l_q.f)\}$

Operator $\parallel$ denotes parallel composition, and operators $\circ_S$ and $\circ_W$ denote strong and weak sequential composition, respectively. These operators are associative and they have unit element $[\epsilon]$. Since parallel composition is also commutative, we can use $\parallel$ as a quantifier.

In our running example, $\phi.(!(a, X, Y)) = X$ and $\phi.(?(a, X, Y)) = Y$. Let laters p_4 and p_5 be defined as $p_4 = p_1 \circ_W p_2$ and $p_5 = p_1 \circ_W p_3$. The structure of p_5 is visualized as $e_1 \to e_9 \to e_{10} \to e_{11} \to e_{12} \quad e_2 \to e_3 \to e_8 \to e_{13}$.

2.5 Deadlocks

A later $(E, <, l)$ contains a *deadlock* if there is an event $e : e \in E$ such that $e < e$. Conversely, a later is *deadlock-free* if the (transitive) causality relation is a strict partial order (i.e. irreflexive, asymmetric and transitive). These definitions are consistent, since asymmetry implies irreflexivity, and transitivity plus irreflexivity implies asymmetry. In particular, all laters that can be obtained from the elementary laters using the elementary later operators are deadlock-free.

For example, consider later p'_5 (to be defined in Section 3) with the following structure: $e_1 \to e_2 \to e_3 \to e_8 \to e_9 \to e_{10} \leftrightarrows e_{11} \to e_{12} \to e_{13}$. In this later there is a circular dependency between events e_{10} and e_{11}. From the transitivity of relation $<$ it follows that $e_{10} < e_{10}$, hence e_{10} is a deadlock.

The interpretation of the causality relation is such that the set of events "behind any deadlock" cannot occur either. We define the set of deadlocked events Δ for a later $(E, <, l)$ as follows:

$$\Delta.(E, <, l) = \{f \mid e, f : e \in E \wedge f \in E \wedge e < e \wedge e < f\}$$

In our example we obtain $\Delta.p'_5 = \{e_{10}, e_{11}, e_{12}, e_{13}\}$, and hence events e_1, e_2, e_3, e_8 and e_9 are the only events that can occur in later p'_5.

2.6 Prefix

A natural way to compare laters is to compare their possible behaviors. If all possible behaviors of a later p are contained in the possible behaviors of a later q, we call p a prefix of q. To determine whether p is a prefix of q, we only need to consider the deadlock-free part of p. If p is a prefix of q, then (1) p may contain fewer events than q, (2) on this smaller event set, p may contain more causalities than q, (3) q's labeling of events is respected by p, and (4) for each event that is in both p and q, all events that precede the event in q are also in p.

Formally, later p is a *prefix* of later q, denoted $p \preceq q$, if for some laters $(E_p, <_p, l_p) \simeq p$ and $(E_q, <_q, l_q) \simeq q$ the following four conditions hold:

1. $\overline{E_p} \subseteq E_q$
2. $<_q \cap (\overline{E_p} \times \overline{E_p}) \subseteq <_p$
3. $l_p \cap (\overline{E_p} \times L) = l_q \cap (\overline{E_p} \times L)$
4. $(\forall e, f :: e <_q f \wedge f \in \overline{E_p} \Rightarrow e \in \overline{E_p})$

where $\overline{E_p} = E_p \backslash \Delta.(E_p, <_p, l_p)$

In the running example several prefix relations hold, such as $p_1 \preceq p_4$ and $p_1 \preceq p_5$.

As a corollary of $p \preceq q$, we have $\overline{E_p} \subseteq \overline{E_q}$ for $\overline{E_q} = E_q \backslash \Delta.(E_q, <_q, l_q)$. Prefix order $\preceq$ is a pre-order (i.e. reflexive and transitive) with smallest element $[\epsilon]$. Some typical prefixes are $p \preceq p \| q$, $q \preceq p \| q$, $p \preceq p \circ_S q$ and $p \preceq p \circ_W q$. In comparison with [KL98], our definition is more explicit, it can deal with deadlocks and it allows $<_q \cap (\overline{E_p} \times \overline{E_p})$ to be strictly smaller than $<_p$.

Parallel composition is monotonic in both arguments, while both kinds of sequential composition are only monotonic in their second argument (since deadlocks are invisible). A special kind of prefix is a causality extension:

$$< \subseteq <' \Rightarrow (E, <', l) \preceq (E, <, l)$$

As an example consider later p_5', which is a causality extension of later p_5.

2.7 Projection

To restrict the set of events of a later, we define a projection operator π that restricts a later to the events in instance i as follows:

$$\begin{array}{l} \pi_i.(E, <, l) = (F, < \cap (F \times F), l \cap (F \times L)) \\ \quad \text{where } F = \{e \mid e : e \in E \wedge \phi.(l.e) = i\} \end{array}$$

Its relation with parallel composition is $p \preceq (\|i : i \in I : \pi_i.p)$, and it is monotonic with respect to causality extensions:

$$< \subseteq <' \Rightarrow \pi_i.(E, <', l) \preceq \pi_i.(E, <, l)$$

2.8 Sets of Laters

Usually a single later cannot describe all possible behavior of a system. Therefore we study a set of laters (which is the notion of process in [Pra86], and pomset in [KL98]), which represents the set of behaviors of the individual laters. We lift each elementary later operator $\oplus$ and the projection operator π as follows:

$$\begin{array}{ll} P \oplus Q & = \{p \oplus q \mid p, q : p \in P \wedge q \in Q\} \\ \pi_i.P & = \{\pi_i.p \mid p : p \in P\} \end{array}$$

To lift the prefix order $\preceq$, we define order $\sqsubseteq$ as follows:

$$P \sqsubseteq Q \;\equiv\; (\forall p : p \in P : \; (\exists q : q \in Q : \; p \preceq q))$$

Order $\sqsubseteq$ is a pre-order with smallest element $\emptyset$. Like before, parallel composition is monotonic in both arguments, while both kinds of sequential composition are only monotonic in their second argument. Relation $\doteq$ defined as

$$P \doteq Q \;\equiv\; P \sqsubseteq Q \,\wedge\, Q \sqsubseteq P$$

is an equivalence relation. Equivalence $P \doteq Q$ denotes that P and Q have the same sets of deadlock-free prefixes, which means that they are trace equivalent.

3 Asynchronous Communication

In this section we develop an operator that introduces in a later the causalities that correspond to asynchronous message communication. To model distributed systems with communication via message passing, some labels are used to denote sending or receiving a message. The most liberal causalities are obtained by matching sends and receipts in their order of occurrence. This does not require that messages with identical names are communicated in FIFO order.

3.1 Label-Wise Trichotomy

To match events properly, we need to determine the order in which events with identical labels occur. For simplicity reasons, we assume for each label that the events with that label are totally ordered; at least, in the deadlock-free part of the later. Since this deadlock-free part is strict partially ordered, we only need trichotomy (or comparability) for events with identical labels. For notational convenience, we require this property for the whole later and for all labels.

The *label-wise trichotomy* property T is defined as follows:

$$\begin{aligned} &T.P \;\equiv\; (\forall p : p \in P : \; T.p) \\ &T.(E,<,l) \;\equiv\; (\forall e,f :: \; l.e = l.f \;\Rightarrow\; e = f \,\vee\, e < f \,\vee\, f < e) \end{aligned}$$

As we will see in Section 4, this only imposes a few, acceptable restrictions on MSCs. This property is maintained under causality extensions and event restrictions, it holds for the elementary laters, and it is maintained under sequential composition; only for a parallel composition $(E_p, <_p, l_p) \parallel (E_q, <_q, l_q)$ label-disjointness is required, i.e. $(\forall e, f : e \in E_p \,\wedge\, f \in E_q : \; l_p.e \neq l_q.f)$.

3.2 Communication Causalities

We define operator $\Gamma.p$, which introduces the communication causalities in a later p. For compositional MSC, we must also address communication between two sequentially composed laters. Therefore we introduce an extra parameter t to denote the entire preceding behavior of later p in terms of a later.

For each message m, we must ensure that each receipt event (with label $?m$) is preceded by the corresponding/matching send event (with label $!m$). In case there are more receive events than send events, these remaining receipt events are turned into deadlocks. Thus we obtain (provided $T.t$ and $T.P$ hold):

$$\begin{aligned}
&\Gamma^t.P = \{\Gamma^t.p \mid p : p \in P\}\\
&\Gamma^t.(E, <_b, l) = (E, (<_b \cup <_c)^+ \cup <_d, l)\\
&\quad \text{where } <_c = <'_c \cap (E \times E) \text{ and } <_d = <'_d \cap (E \times E)\\
&\quad \text{and } (E', <', l') = t \circ_W (E, <_b, l) \text{ and } \overline{E'} = E' \backslash \Delta.(E', <', l')\\
&\quad \text{and } <'_c = \{(e, f) \mid e, f, m : e \in \overline{E'} \wedge f \in \overline{E'} \wedge l'.e = !m \wedge l'.f = ?m \wedge\\
&\qquad\qquad (\#g :: g <' e \wedge l'.g = !m) = (\#g :: g <' f \wedge l'.g = ?m)\}\\
&\quad \text{and } <'_d = \{(f, f) \mid f, m : f \in \overline{E'} \wedge l'.f = ?m \wedge\\
&\qquad\qquad (\#g :: g \in \overline{E'} \wedge l'.g = !m) \leq (\#g :: g <' f \wedge l'.g = ?m)\}
\end{aligned}$$

In this definition, first an auxiliary later $(E', <', l')$ is computed as the sequential composition of t and $(E, <_b, l)$. Then causalities $<'_c$ are defined for the matching communications, and causalities $<'_d$ are defined for the deadlocked receipt events. Finally, only the causalities on events E (i.e. not on events from previous behavior t) are added to later $(E, <_b, l)$.

For the running example, we define later $p'_4 = \Gamma^{[\epsilon]}.p_4$ and $p'_5 = \Gamma^{[\epsilon]}.p_5$. When visualizing p'_4 and p'_5, we add the additional communication causalities according to $<'_c$ with dashed arrows, and the additional deadlock causality for unmatched receipts ($<'_d$) with a dotted arrow as follows:

p'_4: 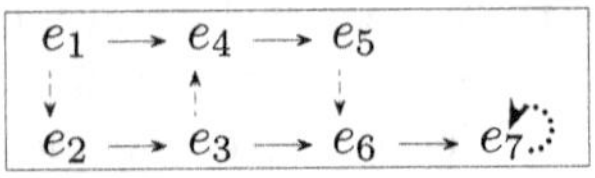

p'_5: 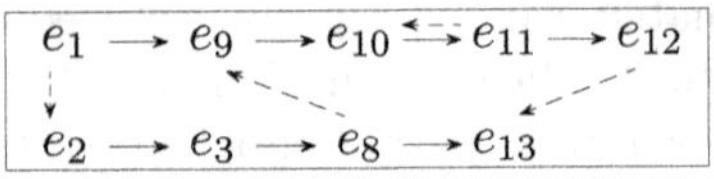

For p'_4 this then boils down to: 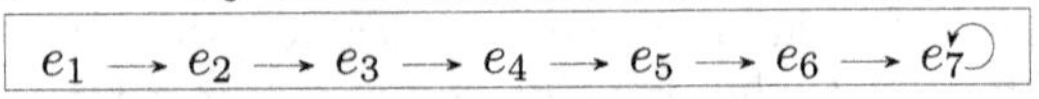

.

For p'_5, the result was already visualized in Section 2.

The role of parameter t of Γ is illustrated in the following important property of sequential composition (see also Section 6):

$$\Gamma^t.(\{p\} \circ_W Q) \;\doteq\; \Gamma^t.(\{p\} \circ_W \Gamma^{t \circ_W p}.Q)$$

Since Γ is a causality extension, it maintains predicate T. However, Γ can introduce deadlocks. The following are some other properties of Γ:

$$\begin{aligned}
&(\textit{shrinking}) && \Gamma^t.p \preceq p\\
&(\textit{idempotence}) && \Gamma^t.p = \Gamma^t.(\Gamma^t.p)\\
&(\textit{monotonicity}) && p \preceq q \Rightarrow \Gamma^t.p \preceq \Gamma^t.q
\end{aligned}$$

These properties can even be generalized to sets of laters.

4 Semantics of Compositional MSC

Using the preceding concepts, we define a semantics of compositional MSC as an extension of the MSC semantics of [KL98]. For simplicity reasons, we delay the introduction of the communication causalities; in Section 6 we will show how they can be introduced earlier (like in [KL98]). We start by giving the semantics of basic MSC, then the semantics of high-level MSC, and finally we complete this semantics by including the communication causalities.

4.1 Basic MSC

The semantics (without communication) of basic MSC B in instance-oriented textual representation [Ren99] is defined as a later $M_{bmsc}[\![B]\!]$ as follows:

$$\begin{aligned}
M_{bmsc}[\![\langle\,\rangle]\!] &= [\epsilon] \\
M_{bmsc}[\![\mathbf{inst}\ i; S\ \mathbf{endinst}; B]\!] &= M_{inst}[\![S]\!](i) \parallel M_{bmsc}[\![B]\!]
\end{aligned}$$

$$\begin{aligned}
M_{inst}[\![\langle\,\rangle]\!](i) &= [\epsilon] \\
M_{inst}[\![a; S]\!](i) &= M_{inst}[\![a]\!](i) \circ_S M_{inst}[\![S]\!](i) \\
M_{inst}[\![\mathbf{in}\ n\ \mathbf{from}\ j]\!](i) &= [?(n,j,i)] \\
M_{inst}[\![\mathbf{out}\ n\ \mathbf{to}\ j]\!](i) &= [!(n,i,j)] \\
M_{inst}[\![\mathbf{local}\ b]\!](i) &= [b(i)] \\
M_{inst}[\![\mathbf{co}\ \langle\,\rangle\ \mathbf{endco}]\!](i) &= [\epsilon] \\
M_{inst}[\![\mathbf{co}\ a; C\ \mathbf{endco}]\!](i) &= M_{inst}[\![a]\!](i) \parallel M_{inst}[\![\mathbf{co}\ C\ \mathbf{endco}]\!](i)
\end{aligned}$$

Function ϕ can then be defined as follows: $\phi.(?(n,j,i)) = i$, $\phi.(!(n,i,j)) = i$ and $\phi.(b(i)) = i$. By construction, each later $M_{bmsc}[\![...]\!]$ is a strict partial order.

To ensure that predicate T is satisfied, we assume that no instance name occurs more than once per bMSC [Ren99], and we require that in each co-region the events are label disjoint. The interest in co-regions is usually very limited (they are completely excluded in [HJ00, GMP03]), so this is no severe restriction. The unrealistic assumption that for each message name there is at most one send event and at most one receipt event per bMSC [KL98], is not required here.

4.2 High-Level MSC

The semantics (without communication) of high-level MSC A in textual representation is defined as a set of laters $M_{hmsc}[\![A]\!]$ as follows:

$$\begin{aligned}
M_{hmsc}[\![\mathbf{empty}]\!] &= \{[\epsilon]\} \\
M_{hmsc}[\![\mathbf{msc}\ name; B\ \mathbf{endmsc}]\!] &= \{M_{bmsc}[\![B]\!]\} \\
M_{hmsc}[\![A\ \mathbf{seq}\ B]\!] &= M_{hmsc}[\![A]\!] \circ_W M_{hmsc}[\![B]\!] \\
M_{hmsc}[\![A\ \mathbf{alt}\ B]\!] &= M_{hmsc}[\![A]\!] \cup M_{hmsc}[\![B]\!]
\end{aligned}$$

By construction, each later in $M_{hmsc}[\![...]\!]$ is a strict partial order, and satisfies predicate T. We do not explicitly address iteration, since it is just repeated sequential composition. Sometimes the parallel composition of high-level MSCs, denoted by **par**, is also considered. Its semantics can easily be expressed in terms of operator $\parallel$ on sets of laters, but we will not consider it in our study.

4.3 MSC

Finally we introduce the causalities imposed by communication:

$$\begin{aligned} M_{msc}[\![A]\!] &= M^{[\epsilon]}_{msc}[\![A]\!] \\ M^{t}_{msc}[\![A]\!] &= \Gamma^{t}.M_{hmsc}[\![A]\!] \end{aligned}$$

This is a proper definition since $M_{hmsc}[\![A]\!]$ satisfies predicate T. By construction, predicate T also holds for $M^{t}_{msc}[\![A]\!]$. Note that the application of Γ^{t} may introduce deadlocks, which violate the strict partial order property. This illustrates one of the reasons for our extended partial order semantics.

Using the example laters from Sections 2 and 3, the semantics of the MSC in Figure 1 corresponds to $\Gamma^{[\epsilon]}.(\{p_1\} \circ_W (\{p_2\} \cup \{p_3\}))$, which simplifies via $\{\Gamma^{[\epsilon]}.(p_1 \circ_W p_2), \Gamma^{[\epsilon]}.(p_1 \circ_W p_3)\}$ into $\{p'_4, p'_5\}$. These two laters represent the possibility of either performing ex1 followed by ex2, or ex1 followed by ex3.

In [GMP03] there is a restriction that receive events in bMSCs may not be matched to send events in future bMSCs. In [MM01] an extension is proposed that drops this restriction. We consider the extension, since the original restriction conflicts with elegant rules, like sequential composition of two bMSCs being equal to simply connecting the instance axis [Ren99].

5 Implementations

In this section we explain how specifications are implemented. The difference between a specification and an implementation is that a specification describes behavior in terms of all instances, while an implementation describes behavior in terms of each individual instance. Thus an implementation for an instance can be represented by a set of laters that contain events of that instance only.

To synthesize an implementation, the specification is decomposed according to the instances. The joint execution behavior of an implementation is obtained by recomposing the instances. We do not consider the unusual implementation with message parameters proposed in [Gen05], which effectively boils down to renaming the messages and shifting the moments of choice. In such an implementation, additional parameters in a request message are sometimes used to fix the choice that should made by the receiver of the request.

5.1 Decomposition

The typical decomposition D of a set of laters M to its instances is:

$$D.M \;=\; \{[i \mapsto \pi_i.M] \mid i : i \in I\}$$

In this set, each instance name is mapped to the corresponding projection of M. Since projection is an event restriction, predicate T is maintained.

For our running example, the decomposition of the laters, $D.\{p'_4, p'_5\}$, yields the following: $\{\, [X \mapsto \{\, \boxed{e_1 \rightarrow e_4 \rightarrow e_5}\,,\ \boxed{e_1 \rightarrow e_9 \rightarrow e_{10} \leftrightarrows e_{11} \rightarrow e_{12}}\ \}]$, $[Y \mapsto \{\, \boxed{e_2 \rightarrow e_3 \rightarrow e_6 \rightarrow e_7 \circlearrowleft}\,,\ \boxed{e_2 \rightarrow e_3 \rightarrow e_8 \rightarrow e_{13}}\ \}]\,\}$.

Let us briefly investigate what might be lost by decomposition. For a singleton set $\{(E, <, l)\}$, note that E and l are partitioned per instance, and hence only the causalities between different instances are lost. For each later in a larger set M, also the link between its projections in the different instances is lost.

5.2 Recomposition

To study the joint execution behavior of the decompositions, the decomposition has to be recomposed. Using the definition from the previous section, the typical recomposition R of a decomposition becomes:

$$R^t.\{[i \mapsto \pi_i.M] \mid i : i \in I\} \;=\; \Gamma^t.(\|i : i \in I : \pi_i.M)$$

This is a proper definition provided $T.M$ holds, since T is maintained under parallel composition with disjoint labels. The projections are label-disjoint, since for each label k all events with that label belong to one instance, viz. $\phi.k$.

We emphasize that $R^t \circ D$, where $\circ$ denotes function composition, is not monotonic with respect to $\sqsubseteq$. For causality extensions like Γ^t, we have:

$$(R^t \circ D).(\Gamma^t.P) \;\sqsubseteq\; (R^t \circ D).P$$

5.3 Implementations in Operational Formalisms

Using our later representation, implementations in operational formalisms can easily be obtained. In an interleaved execution model where the labels denote atomic actions, the maximal behaviors of a single later are the linearizations of the maximal deadlock-free prefix. The set of maximal behaviors of a set of laters is the union of the linearizations of the individual laters. In turn, linearizations can easily be transformed to process algebraic expressions using the delayed choice operator [BM95]. The implementation of our running example corresponds to the following CSP-style implementation:

$$\begin{array}{lll} X: & !a & \cdot\, (?b \cdot !c \;+\; ?d \quad) \\ Y: & ?a \cdot !b & \cdot\, (?c \quad\;\; +\; !d \cdot ?c\,) \end{array}$$

6 Relation Between Specification and Implementation

In this section, we investigate whether compositional MSC specifications are trace equivalent to their implementations, i.e. for all A and t:

$$M^t_{msc}[\![A]\!] \;\doteq\; (R^t \circ D).M^t_{msc}[\![A]\!]$$

For details of the proofs we refer to [MRW06].

6.1 The Implementation Contains the Specification

In this section we show that the specification is contained in the implementation, i.e. for all A and t: $M^t_{msc}[\![A]\!] \sqsubseteq (R^t \circ D).M^t_{msc}[\![A]\!]$. It can be proved as follows:

$$\begin{array}{ll} & (R^t \circ D).M^t_{msc}[\![A]\!] \\ = & \{\text{definition of } R^t \circ D\} \\ & \Gamma^t.(\|i : i \in I : \pi_i.M^t_{msc}[\![A]\!]) \\ \sqsupseteq & \{\text{property of } \pi \text{ and } \|; \text{ monotonicity of } \Gamma\} \\ & \Gamma^t.M^t_{msc}[\![A]\!] \\ = & \{\text{definition of } M^t_{msc}[\![A]\!]; \text{ idempotence of } \Gamma\} \\ & M^t_{msc}[\![A]\!] \end{array}$$

6.2 The Specification Contains the Implementation

In this section we derive conditions under which the implementation is contained in the specification, i.e. for all A and t: $(R^t \circ D).M^t_{msc}[\![A]\!] \sqsubseteq M^t_{msc}[\![A]\!]$. We will set up an inductive argument based on the structure of the high-level MSC. We assume that the following rewrite rules have been applied:

$$\begin{array}{lcl} (\mathbf{empty})\ \mathbf{seq}\ C & \to & C \\ (A\,\mathbf{seq}\,B)\ \mathbf{seq}\ C & \to & A\ \mathbf{seq}\ (B\,\mathbf{seq}\,C) \\ (A\,\mathbf{alt}\,B)\ \mathbf{seq}\ C & \to & (A\,\mathbf{seq}\,C)\ \mathbf{alt}\ (B\,\mathbf{seq}\,C) \end{array}$$

These rules do not change the occurrences of choice, but they ensure that the first argument of sequential composition is just a single bMSC. Using the property of Γ and $\circ_W$ in Section 3, we derive an alternative characterization of $M^t_{msc}[\![...]\!]$ in which communication is addressed earlier (like in [KL98]):

$$\begin{array}{ll} M^t_{msc}[\![\mathbf{msc}\ \mathit{name};\ A\ \mathbf{endmsc}]\!] & = M^t_{msc}[\![\mathbf{msc}\ \mathit{name};\ A\ \mathbf{endmsc\ seq\ empty}]\!] \\ M^t_{msc}[\![\mathbf{empty}]\!] & = \{[\epsilon]\} \\ M^t_{msc}[\![\mathbf{msc}\ \mathit{name};\ A\ \mathbf{endmsc\ seq}\ B]\!] & \doteq \Gamma^t.(\{M_{bmsc}[\![A]\!]\} \circ_W M^{t \circ_W M_{bmsc}[\![A]\!]}_{msc}[\![B]\!]) \\ M^t_{msc}[\![A\ \mathbf{alt}\ B]\!] & = M^t_{msc}[\![A]\!] \cup M^t_{msc}[\![B]\!] \end{array}$$

Empty. For sake of space, we omit the very simple proof of this base case.

Sequential Composition. This inductive case can be proved as follows:

$$\begin{array}{ll} & (R^t \circ D).M^t_{msc}[\![\mathbf{msc}\ \mathit{name};\ A\ \mathbf{endmsc\ seq}\ B]\!] \\ \doteq & \{\text{alternative characterization}\} \\ & (R^t \circ D).(\Gamma^t.(\{M_{bmsc}[\![A]\!]\} \circ_W M^{t \circ_W M_{bmsc}[\![A]\!]}_{msc}[\![B]\!])) \\ \sqsubseteq & \{\text{monotonicity}\} \\ & (R^t \circ D).(\{M_{bmsc}[\![A]\!]\} \circ_W M^{t \circ_W M_{bmsc}[\![A]\!]}_{msc}[\![B]\!]) \\ \doteq & \{\bullet \text{ see below}\} \\ & \Gamma^t.(\{M_{bmsc}[\![A]\!]\} \circ_W (R^{t \circ_W M_{bmsc}[\![A]\!]} \circ D).M^{t \circ_W M_{bmsc}[\![A]\!]}_{msc}[\![B]\!]) \\ \doteq & \{\text{induction hypothesis, monotonicity of } \Gamma \text{ and } \circ_W\} \\ & \Gamma^t.(\{M_{bmsc}[\![A]\!]\} \circ_W M^{t \circ_W M_{bmsc}[\![A]\!]}_{msc}[\![B]\!]) \\ \doteq & \{\text{alternative characterization}\} \\ & M^t_{msc}[\![\mathbf{msc}\ \mathit{name};\ A\ \mathbf{endmsc\ seq}\ B]\!] \end{array}$$

The step marked • follows from the following rule, where m denotes a later that does not order events in different instances, and M denotes a set of laters:

$$(R^t \circ D).(\{m\} \circ_W M) \doteq \Gamma^t.(\{m\} \circ_W (R^{t \circ_W m} \circ D).M)$$

Alternative Composition. This inductive case can be proved as follows:

$$\begin{array}{cl} & (R^t \circ D).M^t_{msc}[\![A \textbf{ alt } B]\!] \\ = & \{\text{alternative characterization}\} \\ & (R^t \circ D).(M^t_{msc}[\![A]\!] \cup M^t_{msc}[\![B]\!]) \\ \sqsubseteq & \{\blacktriangle \text{ see below}\} \\ & (R^t \circ D).M^t_{msc}[\![A]\!] \cup (R^t \circ D).M^t_{msc}[\![B]\!] \\ \doteq & \{\text{induction hypothesis (twice)}\} \\ & M^t_{msc}[\![A]\!] \cup M^t_{msc}[\![B]\!] \\ = & \{\text{alternative characterization}\} \\ & M^t_{msc}[\![A \textbf{ alt } B]\!] \end{array}$$

The step marked ▲ is not only a sufficient condition, but also a necessary one. Since it does not hold for each MSC, we will study it further.

6.3 Sound Choice

Checking condition ▲ is quite involved in practice, since by definition of $R^t \circ D$ arbitrary combinations of projected laters (i.e. from both $M^t_{msc}[\![A]\!]$ and $M^t_{msc}[\![B]\!]$) need to be considered. In Section 7 we will relate various realizability criteria to this condition, but in this section we first strengthen it into a more convenient condition for this purpose; for the details we refer to [MRW06].

We strengthen condition ▲ into what we call the *sound choice* property: there exists an instance k such that for each instance $j : j \neq k$ both

– $\forall\, g :: [I \rightarrow M^t_{msc}[\![A]\!]],\ n :\ n \in \pi_j.M^t_{msc}[\![B]\!] \ \wedge\ \{n\} \not\sqsubseteq \pi_j.M^t_{msc}[\![A]\!]$:

$$\Gamma^t.((\|i : i \neq j :\ \pi_i.g_i) \ \|\ n) \ \preceq\ \Gamma^t.(\|i : i \neq j :\ \pi_i.g_i)$$

– $\forall\, h :: [I \rightarrow M^t_{msc}[\![B]\!]],\ m :\ m \in \pi_j.M^t_{msc}[\![A]\!] \ \wedge\ \{m\} \not\sqsubseteq \pi_j.M^t_{msc}[\![B]\!]$:

$$\Gamma^t.((\|i : i \neq j :\ \pi_i.h_i) \ \|\ m) \ \preceq\ \Gamma^t.(\|i : i \neq j :\ \pi_i.h_i)$$

Here functions g and h represent a chosen later per instance. Later $n : n \in \pi_j.M^t_{msc}[\![B]\!] \wedge \{n\} \not\sqsubseteq \pi_j.M^t_{msc}[\![A]\!]$ denotes a later from MSC B that is no prefix of any later from MSC A. Note that behaviors occurring both in MSC A and MSC B are not problematic for the choice between A and B. The $\preceq$-term expresses that later n (or later m) cannot perform any behavior. Instance k and condition $j \neq k$ ensure that some instance may have initiative.

The choice in our running example is not a sound choice, as can be pointed out by considering both options for k. For $k = X$, we can choose $n = \pi_Y.(\Gamma^{p1}.p3)$ and $g_X = \Gamma^{p1}.p2$, which violate the first $\preceq$ term; and similarly for $k = Y$. We will discuss it in more detail using the non-local choice criterion in Section 7.

Notice that instead of considering arbitrary combinations of projected laters, on the left-hand side of the $\preceq$ in this condition, the combinations of projected laters contain only one later n from B, or only one later m from A respectively. Finally we stress that this condition is stronger than condition ▲.

7 Realizability Criteria

The sound choice property of the previous section implies that the specification and the implementation are trace equivalent; otherwise the specification may not be realizable. In this section we convert the realizability criteria from [MGR05] to high-level MSCs with binary choice, and generalize them to compositional MSC with co-regions. We first depict how the criteria are classified in comparison with sound choice and derived condition ▲ from the previous section:

$$\text{derived condition} \leftarrow \text{sound choice} \leftarrow \begin{cases} \neg \text{ non-local choice} \\ \text{propagating choice} \leftarrow \begin{cases} \neg \text{ non-deterministic choice} \\ \neg \text{ race choice} \end{cases} \end{cases}$$

Example MSCs for (combinations of) these criteria can be found in [MGR05].

7.1 Non-local Choice

A choice between two MSCs is local if at most one instance has initiative in these MSCs; otherwise several instances can independently start executing different MSCs. An instance has initiative in an MSC if some first event of the instance is labeled with either an internal action, or sending a message, or receiving a message that was sent before the choice. The choice in our running example is non-local, since due to events e_4 and e_8 both X and Y have initiative.

Non-local choice follows naturally from sound choice, and in particular from its $\preceq$-terms. Observe that a later n is likely to be problematic if for each label-disjoint later x we have $\Gamma^t.(x\|n) \not\preceq \Gamma^t.x$. This condition follows from $\Gamma^t.n \not\preceq [\epsilon]$, which means that later n contains an initiating event. Due to condition $j \neq k$ in the definition of sound choice, only instance k may have initiative, i.e. no two different instances, say i and j, may have initiative. This leads to the *non-local choice* criterion:

$$\begin{aligned}(\exists i, j, m, n :: \; & i \neq j \wedge m \in \pi_i.M^t_{msc}[\![A]\!] \wedge \{m\} \not\subseteq \pi_i.M^t_{msc}[\![B]\!] \wedge \Gamma^t.m \not\preceq [\epsilon] \\ & \wedge n \in \pi_j.M^t_{msc}[\![B]\!] \wedge \{n\} \not\subseteq \pi_j.M^t_{msc}[\![A]\!] \wedge \Gamma^t.n \not\preceq [\epsilon] \quad)\end{aligned}$$

The difference with other variants of non-local choice in [BAL97, HJ00, MGR05] is in our first two conjuncts on both m and n, where we ensure that sound choice is violated.

7.2 Propagating Choice

Absence of non-local choice is not sufficient to guarantee sound choice. It does guarantee that there is at most one instance that determines the choice, viz. instance k in the definition of sound choice. The other instances j have no initiative and hence their chosen laters n are characterized by $\Gamma^t.n \preceq [\epsilon]$. What remains to guarantee sound choice is that the other instances can resolve the choice, which is characterized by the *propagating choice* property (see also [MGR05]): for each instance j both

– $\forall\, g :: [I \to M^t_{msc}[\![A]\!]],\ n :\ n \in \pi_j.M^t_{msc}[\![B]\!] \ \wedge\ \{n\} \not\subseteq \pi_j.M^t_{msc}[\![A]\!] \ \wedge\ \Gamma^t.n \preceq [\epsilon]$:

$$\Gamma^t.((\|i : i \neq j :\ \pi_i.g_i) \ \|\ n) \ \preceq\ \Gamma^t.(\|i : i \neq j :\ \pi_i.g_i)$$

– $\forall\, h :: [I \to M^t_{msc}[\![B]\!]],\ m :\ m \in \pi_j.M^t_{msc}[\![A]\!] \wedge \{m\} \not\subseteq \pi_j.M^t_{msc}[\![B]\!] \wedge \Gamma^t.m \preceq [\epsilon]$:

$$\Gamma^t.((\|i : i \neq j :\ \pi_i.h_i) \ \|\ m) \ \preceq\ \Gamma^t.(\|i : i \neq j :\ \pi_i.h_i)$$

7.3 Non-deterministic Choice

Propagating choice is an important property, but it is not easy to apply. A simple case that violates it is when the MSCs contain behaviors m and n that are different, although they share a common prefix p, i.e. $p \preceq m$ and $p \preceq n$. In case such a prefix p starts with a receipt behavior, instance j cannot resolve the choice using one of its initial events. This is characterized by the *non-deterministic choice* criterion (see also [MGR05]):

$$\begin{array}{l}(\exists j, m, n, p ::\ p \preceq m \ \wedge\ p \preceq n \ \wedge \\ \qquad m \in \pi_j.M^t_{msc}[\![A]\!] \ \wedge\ \{m\} \not\subseteq \pi_j.M^t_{msc}[\![B]\!] \ \wedge\ \Gamma^t.m \preceq [\epsilon] \\ \qquad \wedge\, n \in \pi_j.M^t_{msc}[\![B]\!] \ \wedge\ \{n\} \not\subseteq \pi_j.M^t_{msc}[\![A]\!] \ \wedge\ \Gamma^t.n \preceq [\epsilon] \\ \qquad \wedge\, (\exists g, h :\ g :: [I \to M^t_{msc}[\![A]\!]] \ \wedge\ h :: [I \to M^t_{msc}[\![B]\!]] : \\ \qquad\quad (\Gamma^t.((\|i : i \neq j :\ \pi_i.g_i) \ \|\ p) \ \not\preceq\ \Gamma^t.(\|i : i \neq j :\ \pi_i.g_i) \\ \qquad\quad \vee\, \Gamma^t.((\|i : i \neq j :\ \pi_i.h_i) \ \|\ p) \ \not\preceq\ \Gamma^t.(\|i : i \neq j :\ \pi_i.h_i))\,)\,)\end{array}$$

This criterion can be made more syntactic by weakening the inner existential quantification into condition $p \not\preceq [\epsilon]$. Although non-deterministic choice violates sound choice, it does not guarantee that the condition ▲ in Section 6 is violated; so sound choice has been a real strengthening.

7.4 Race Choice

Absence of non-deterministic choice is not sufficient to guarantee propagating choice. It does guarantee that the choice can be resolved when no initiating receipt event can end up receiving a message intended for a non-initial receipt event in another MSC. The other cases are characterized by the *race choice* criterion (see also [MGR05]). Its formal definition is very similar to the definitions of propagating choice and non-deterministic choice, see also [MRW06].

In [HJ00] the reconstructible choice criterion is proposed in order to guarantee realizability. However, this claim contradicts their example of a reconstructible MSC (see Figure 15 in [HJ00]). In terms of our classification, it suffers from race choice: if instance A sends message $m1$ before message $m5$, instance B may receive message $m6$ (related to $m5$) before message $m3$ (related to $m1$).

8 Conclusions and Further Work

We have developed a denotational semantics for compositional MSC through our extension of pomsets with deadlocks. In this formalism we have studied realizability, especially of the choice construct. We have discussed various proposed realizability criteria and shown completeness of our classification in [MGR05].

Realizability problems can also be detected by verifying the implementation [UKM03]. However, it is far more effective to have criteria for specifications, and to develop ways to make specifications realizable [HJ00]. For the latter, we plan to evaluate our proposals in [MG05, MGR05] using the current framework, and to automate them.

A possible extension is to explore other realizability criteria, especially since sound choice is a real strengthening. In addition, more syntactical criteria would better allow automation. Also the realizability of other MSC constructs may be studied, of which parallel composition is a challenging one.

References

[AEY05] R. Alur, K. Etessami, and M. Yannakakis. Realizability and verification of MSC graphs. *Theoretical Computer Science*, 331:97–114, 2005.

[BAL97] H. Ben-Abdallah and S. Leue. Syntactic detection of process divergence and non-local choice in Message Sequence Charts. In *Proceedings of TACAS'97*, volume 1217 of *LNCS*, pages 259–274. Springer, 1997.

[BM95] J.C.M. Baeten and S. Mauw. Delayed choice: an operator for joining Message Sequence Charts. In *Formal Description Techniques*, pages 340–354, 1995.

[Gen05] B. Genest. Compositional Message Sequence Charts (CMSCs) are better to implement than MSCs. In *Proceedings of TACAS'05*, volume 3440 of *LNCS*, pages 429–440. Springer, 2005.

[GMP03] E.L. Gunter, A. Muscholl, and D.A. Peled. Compositional Message Sequence Charts. *International Journal on Software Tools for Technology Transfer*, 5(1):78–89, November 2003. An earlier version appeared at TACAS'01.

[Hey00] S. Heymer. A semantics for MSC based on Petri-Net components. In *Proceedings of SAM'00: 2nd Workshop on SDL and MSC*, 2000.

[HJ00] L. Hélouët and C. Jard. Conditions for synthesis of communicating automata from HMSCs. In *Proceedings of 5th FMICS Workshop*, 2000.

[KL98] J.-P. Katoen and L. Lambert. Pomsets for Message Sequence Charts. In *Proceedings of SAM'98: 1st Workshop on SDL and MSC*, 1998.

[MG05] A.J. Mooij and N. Goga. Dealing with non-local choice in IEEE 1073.2's standard for remote control. In *Proceedings of SAM'04: 4th Workshop on SDL and MSC*, volume 3319 of *LNCS*, pages 257–270. Springer, 2005.

[MGR05] A.J. Mooij, N. Goga, and J.M.T. Romijn. Non-local choice and beyond: Intricacies of MSC choice nodes. In *Proceedings of FASE'05*, volume 3442 of *LNCS*, pages 273–288. Springer, 2005.

[MM01] P. Madhusudan and B. Meenakshi. Beyond message sequence graphs. In *Proceedings of FASE'01*, LNCS 2245, pages 256–267. Springer, 2001.

[MRW06] A.J. Mooij, J.M.T. Romijn, and J.W. Wesselink. Realizability criteria for compositional MSC. Computer Science Report 06-11, Technische Universiteit Eindhoven, March 2006.

[Pra86] V. Pratt. Modelling concurrency with partial orders. *International Journal of Parallel Programming*, 15(1):33–71, 1986.

[Ren99] M.A. Reniers. *Message Sequence Chart: Syntax and Semantics.* PhD thesis, Technische Universiteit Eindhoven, June 1999.

[UKM03] S. Uchitel, J. Kramer, and J. Magee. Synthesis of behavioral models from scenarios. *IEEE Transactions on Software Engineering*, 29(2):99–115, 2003.

Quantales and Temporal Logics

Bernhard Möller[1], Peter Höfner[1,*], and Georg Struth[2]

[1] Institut für Informatik, Universität Augsburg
D-86135 Augsburg, Germany
{moeller, hoefner}@informatik.uni-augsburg.de

[2] Department of Computer Science, University of Sheffield
Sheffield S1 4DP, UK
G.Struth@dcs.shef.ac.uk

Abstract. We propose an algebraic semantics for the temporal logic CTL* and simplify it for its sublogics CTL and LTL. We abstractly represent state and path formulas over transition systems in Boolean left quantales. These are complete lattices with a multiplication that preserves arbitrary joins in its left argument and is isotone in its right argument. Over these quantales, the semantics of CTL* formulas can be encoded via finite and infinite iteration operators; the CTL and LTL operators can be related to domain operators. This yields interesting new connections between representations as known from the modal μ-calculus and Kleene/ω-algebra.

1 Introduction

The temporal logic CTL* and its sublogics CTL and LTL are prominent tools in the analysis of concurrent and reactive systems. Although they are by now well-understood, one rarely finds algebraic treatments of their semantics. First results along these lines were obtained by von Karger and Berghammer [23, 24]. But the semantic operators involved were characterised only implicitly. For LTL compact closed expressions could be obtained by Desharnais, Möller and Struth in [5] and, in the framework of fork algebras, by Frías and Lopez Pombo [10].

In the present paper we provide compact closed semantic expressions for CTL and LTL by using modal operators in combination with finite and infinite iteration. This is achieved in two steps. First we provide an algebraic semantics for the more expressive logic CTL* on the basis of quantales, i.e., complete lattices with an operation of multiplication that preserves arbitrary joins in its left and non-empty joins in its right argument. In quantales, sets of states and hence the semantics of state formulas can be represented as test elements in the sense of Kozen [15], while general elements represent the semantics of path formulas.

We define suitable mappings that, for the CTL and LTL formulas, transform their general CTL* semantics into simplified versions in ω-regular form. This yields interesting new connections between representations as known from the modal μ-calculus [12] and Kleene/ω-algebra. Our reasoning is purely semantical;

* This research was partially supported by DFG (German Research Foundation).

M. Johnson and V. Vene (Eds.): AMAST 2006, LNCS 4019, pp. 263–277, 2006.

we do not intend to provide something like an interpretation between logical theories.

The remainder of this paper is organised as follows. Section 2 briefly recapitulates the standard semantics of CTL* and gives a set-based view of it that prepares the algebraic semantics. In Section 3 we present the algebraic framework of quantales enriched by tests, modal operators and iteration. Section 4 gives an algebraic semantics of full CTL* that abstracts a set-based view of the standard semantics. The next section discusses the algebraic properties of the semantic element that models the next-time operator. Section 6 shows that the denotations of state formulas are in one-to-one correspondence with tests, i.e., abstract representations of sets of states. This prepares the simplified semantics for CTL and LTL that are derived from the full semantics in Sections 7 and 8. It turns out that much weaker requirements on the underlying algebras now suffice: modal Kleene algebra with a convergence operator in the case of CTL and plain modal Kleene algebra for LTL. A brief conclusion is presented in Section 9.

2 Modelling CTL*

The language Ψ of *CTL* formulas* (see e.g. [9]) over a set Φ of atomic propositions is defined by the grammar

$$\Psi \ ::= \ \bot \mid \Phi \mid \Psi \to \Psi \mid \mathsf{X}\Psi \mid \Psi \,\mathsf{U}\, \Psi \mid \mathsf{E}\Psi,$$

where X and U are the next-time and until operators and E is the existential quantifier on paths. The logical connectives $\neg, \wedge, \vee, \mathsf{A}$ are defined, as usual, by $\neg\varphi =_{df} \varphi \to \bot$, $\varphi\wedge\psi =_{df} \neg(\varphi \to \neg\psi)$, $\varphi\vee\psi =_{df} \neg\varphi \to \psi$ and $\mathsf{A}\varphi =_{df} \neg\,\mathsf{E}\neg\varphi$. The sublanguages Σ of *state formulas* that denote sets of computation traces and Π of *path formulas* that denote sets of states are given by

$$\Sigma ::= \bot \mid \Phi \mid \Sigma \to \Sigma \mid \mathsf{E}\Pi,$$
$$\Pi ::= \Sigma \mid \Pi \to \Pi \mid \mathsf{X}\Pi \mid \Pi \,\mathsf{U}\, \Pi.$$

To motivate our algebraic semantics, we briefly recapitulate the standard CTL* semantics formulas. Its basic objects are traces σ from S^+ or S^ω, the sets of finite non-empty or infinite words over some set S of states. The i-th element of σ (indices starting with 0) is denoted σ_i, and σ^i is the trace that results from σ by removing its first i elements.

Each atomic proposition $\pi \in \Phi$ is associated with the set $S_\pi \subseteq S$ of states for which π is true. The relation $\sigma \models \varphi$ of *satisfaction* of a formula φ by a trace is defined inductively (see e.g. [9]) by

$$\begin{array}{ll}
\sigma \not\models \bot, & \\
\sigma \models \pi & \text{iff } \sigma_0 \in S_\pi, \\
\sigma \models \varphi \to \psi & \text{iff } \sigma \models \varphi \text{ implies } \sigma \models \psi, \\
\sigma \models \mathsf{X}\varphi & \text{iff } \sigma^1 \models \varphi, \\
\sigma \models \varphi \,\mathsf{U}\, \psi & \text{iff } \exists\, j \geq 0 \,.\, \sigma^j \models \psi \text{ and } \forall\, k < j \,.\, \sigma^k \models \varphi, \\
\sigma \models \mathsf{E}\varphi & \text{iff } \exists\, \tau \,.\, \tau_0 = \sigma_0 \text{ and } \tau \models \varphi.
\end{array}$$

In particular, $\sigma \models \neg\varphi$ iff $\sigma \not\models \varphi$.

From this semantics one can extract a set-based one by assigning to each formula φ the set $[\![\varphi]\!] =_{df} \{\sigma \mid \sigma \models \varphi\}$ of paths that satisfy it. This is the basis of the algebraic semantics in Section 4.

We quickly repeat the proof of validity of the CTL^* axiom

$$\neg \mathsf{X}\varphi \leftrightarrow \mathsf{X}\neg\varphi, \tag{1}$$

since this will be crucial for the algebraic representation of X in Section 4:

$$\sigma \models \neg\mathsf{X}\varphi \Leftrightarrow \sigma \not\models \mathsf{X}\varphi \Leftrightarrow \sigma^1 \not\models \varphi \Leftrightarrow \sigma^1 \models \neg\varphi \Leftrightarrow \sigma \models \mathsf{X}\neg\varphi\,.$$

3 Quantales, Modal Operators and Iteration

W now prepare the algebraic setting. A *left quantale* is a structure $(S, \leq, 0, \cdot, 1)$ where $(S, \leq)$ is a complete lattice with least element 0 and an associative multiplication (to model sequential composition) that preserves arbitrary joins in its left and non-empty joins in its right argument. Moreover, 1 is required to be neutral w.r.t. multiplication, playing the role of inaction. The meet and join of two elements $a, b \in S$ are denoted by $a \sqcap b$ and $a + b$, resp. Both operators have equal binding power, which is lower than that of multiplication. The greatest element of S is denoted by $\top$. The definition implies that $\cdot$ is *left-strict*, i.e., that $0 \cdot a = 0$ for all $a \in S$.

A *right quantale* is defined symmetrically. Finally, $(S, \leq, 0, \cdot, 1)$ is a *quantale* [20] if it is both a left and right one. In a (right) quantale multiplication is right-strict, i.e., $a \cdot 0 = 0$ for all $a \in S$. The notion of a quantale is equivalent to that of a *standard Kleene algebra* [3].

A (left) quantale is called *Boolean* if its underlying lattice is distributive and complemented, whence a Boolean algebra. An important Boolean quantale is REL(M), the algebra of binary relations over a set M under set union and relational composition; further examples will be presented below.

General quantale elements abstractly represent sets of paths, i.e., the semantics of path formulas. To model state formulas we use tests as introduced into Kleene algebras by Kozen [15]. In REL(M) a set of elements can be modelled as a subset of the identity relation; meet and join of such partial identities coincide with their composition and union. Generalising this, a *test* in a (left) quantale is an element $p \leq 1$ that has a complement q relative to 1, i.e., $p + q = 1$ and $p \cdot q = 0 = q \cdot p$. The set of all tests of a quantale S is denoted by $\mathsf{test}(S)$. It is not hard to show that $\mathsf{test}(S)$ is closed under $+$ and $\cdot$ and has 0 and 1 as its least and greatest elements. Moreover, the complement $\neg p$ of a test p is uniquely determined. Hence $\mathsf{test}(S)$ forms a Boolean algebra. If S itself is Boolean then $\mathsf{test}(S)$ coincides with the set of all elements below 1. We will consistently write $a, b, c \ldots$ for arbitrary semiring elements and $p, q, r, \ldots$ for tests. Also, we will freely use the standard Boolean operations on $\mathsf{test}(S)$, for instance implication $p \rightarrow q = \neg p + q$, with their usual laws.

With the above definition of tests we deviate slightly from [15], in that we do not allow an arbitrary Boolean algebra of subidentities as $\mathsf{test}(S)$ but only the maximal complemented one. The reason is that the axiomatisation of domain to be presented below will force this maximality anyway (see [6]).

A set of states will now be represented abstractly by a test. Left and right multiplication by a test correspond to restricting an element on the input and output side, resp. This allows us to represent the set of all possible paths that start with a state in set p by the *test ideal* $p \cdot \top$.

Example 3.1. We now introduce two further important Boolean left test quantales. Both are based on finite and infinite words over an alphabet A. Next to their classical interpretation as characters, the elements of A may e.g. be interpreted as states in a computation system, or, in connection with graph algorithms, as nodes in a graph. So words over A can be used to model paths in a transition system. As usual, A^* is the set of all finite words over A including the empty word ε. Moreover, A^ω is the set of all infinite words over A. We set $A^\infty =_{df} A^* \cup A^\omega$. Concatenation is denoted by juxtaposition, where $st =_{df} s$ if $s \in A^\omega$.

A *language* over A is a subset of A^∞. As usual, we identify a singleton language with its only element. For a language $U \subseteq A^\infty$ we define its infinite and finite parts by

$$\mathsf{inf}\, U =_{df} U \cap A^\omega, \qquad \mathsf{fin}\, U =_{df} U - \mathsf{inf}\, U \;.$$

The left Boolean quantale $\mathrm{WOR}(A) = (\mathcal{P}(A^\infty), \subseteq, \emptyset, \cdot, \{\varepsilon\})$ is obtained by extending concatenation to languages in the following way:

$$U \cdot V =_{df} \mathsf{inf}\, U \cup (\mathsf{fin}\, U)\, V \;.$$

Note that in general $U \cdot V \neq ST$; for $V = \emptyset$ one has $ST = \emptyset$, whereas $U \cdot V = \mathsf{inf}\, U$. It is straightforward to show that $\mathrm{WOR}(A)$ is indeed a left quantale. This algebra is well-known from the classical theory of ω-languages (see e.g. [22] for a survey). However, its neutral element is $\{\varepsilon\}$ and therefore its test algebra $\mathsf{test}(\mathrm{WOR}(A)) = \{\emptyset, \{\varepsilon\}\}$ is rather trivial and not suitable for our purposes.

Therefore, besides this model we use a second one with a more refined view of multiplication and hence a richer and more useful test algebra. It uses non-empty words and the *fusion product* $\bowtie$ of words as a language-valued multiplication operation. For $s \in A^+$, $t \in A^\omega$, $u \in A^\infty - \varepsilon$ and $x, y \in A$,

$$sx \bowtie xu =_{df} sxu \;, \qquad sx \bowtie yu =_{df} \emptyset \;\text{ if } x \neq y \;, \qquad t \bowtie u =_{df} t \;.$$

Informally, a finite non-empty word s can be fused with a non-empty word t iff the last letter of s coincides with the first one of t; only one copy of that letter is kept in the fused word.

Since we view the infinite words as streams of computations, we call the left Boolean quantale based on this multiplication operation $\mathrm{STR}(A)$ and define it by $\mathrm{STR}(A) =_{df} (\mathcal{P}(A^\infty - \varepsilon), \subseteq, \emptyset, \bowtie, A)$, where $\bowtie$ is extended to languages in the following way:

$$U \bowtie V =_{df} \mathsf{inf}\, U \cup \{s \bowtie t : s \in \mathsf{fin}\, U \wedge t \in V\} \;.$$

This operation has the language A as its neutral element. Moreover, as above, we have $U \bowtie \emptyset = \inf U$ and hence $U \bowtie \emptyset = \emptyset$ iff $\inf U = \emptyset$. A transition relation over a state set A can be modelled in STR(A) as a set R of words of length 2. The powers R^i of R then consist of the words (or paths) of length $i+1$ that are generated by R-transitions.

The multiplicative identity A has exactly the subsets of A as its subobjects, so that in this quantale the tests faithfully represent sets of states. □

Over a left Boolean quantale S the *domain operation* $\ulcorner\_ : S \to \mathsf{test}(S)$ returns, for a set of paths represented by an element $a \in S$, the set of their starting states. It is axiomatised by the Galois connection

$$\ulcorner a \leq p \Leftrightarrow a \leq p \cdot \top \ .$$

This is well defined, since in a Boolean left quantale $\cdot$ preserves arbitrary meets of tests in its left argument [4], and hence in left Boolean quantales domain always exists. By general properties of Galois connections, domain preserves arbitrary joins. For further domain properties see [6].

We list a number of important properties of tests, test ideals and domain; for the proofs see [17].

Lemma 3.2. *Assume a left Boolean quantale.*

(a) $\ulcorner(p \cdot \top) = p$.
(b) $p \leq q \Leftrightarrow p \cdot \top \leq q \cdot \top$.
(c) *If the meet* $a \sqcap b$ *exists then* $p \cdot a \sqcap b = p \cdot (a \sqcap b)$.
Hence also $p \cdot \top \sqcap a = p \cdot a$ *and* $p \cdot (a \sqcap b) = p \cdot a \sqcap p \cdot b$.
(d) $p \cdot a \sqcap q \cdot a = p \cdot q \cdot a$.
(e) $\neg p \cdot \top = \overline{p \cdot \top}$.

By (b) the set of test ideals is isomorphic to the set of tests. To use the above properties freely, we assume for the remainder that S is a Boolean left quantale.

Using domain we define (forward) modal operators. For $a \in S$, $q \in \mathsf{test}(S)$,

$$\langle a \rangle q =_{df} \ulcorner(a \cdot q) \ , \qquad [a]q =_{df} \neg\langle a \rangle \neg q \ .$$

The diamond is an abstract inverse-image operator, whereas box generalises the notion of the weakest liberal precondition wlp to Boolean left quantales. If we view a as the transition relation of a command then the test $[a]q$ characterises those states from which no transition under a is possible or the execution of a is guaranteed to end up in a final state that satisfies test q. Both operators are isotone in their test argument. Hence in a Boolean quantale we have the full power of the modal μ-calculus [12] available.

In particular, the *convergence* $\Delta a \in \mathsf{test}(S)$ of an element a, defined by

$$\Delta a =_{df} \mu x \,.\, [a]x \ ,$$

characterises the set of states from which no infinite transition paths emerge.

To make the modal operators well-behaved w.r.t. composition we need to assume that the underlying quantale satisfies

$$\ulcorner(a \cdot b) = \ulcorner(a \cdot \ulcorner b), \tag{2}$$

since then $\langle a \cdot b \rangle = \langle a \rangle \circ \langle b \rangle$ and $[a \cdot b] = [a] \circ [b]$, where $\circ$ is composition of modal operators. Therefore we call a (left) quantale with this property *modal*. Both WOR(A) and STR(A) are modal.

We will also need finite iteration a^* and infinite iteration a^ω of quantale elements. They are defined as usual by

$$a^* =_{df} \mu x \,.\, 1 + a \cdot x \ , \qquad a^\omega =_{df} \nu x \,.\, a \cdot x \ ,$$

where μ and ν are the least and greatest fixpoint operators, resp. If, like in a Boolean quantale, $+$ is completely conjunctive then, as shown in [1], these operations satisfy the axioms of a left Kleene/ω-algebra [14, 2]. The two operations are connected as follows (see e.g. [1]):

$$a^* \cdot b = \mu x \,.\, b + a \cdot x \ , \qquad a^\omega + a^* \cdot b = \nu x \,.\, b + a \cdot x \ . \tag{3}$$

In a modal left quantale, star, convergence and box interact according to the following induction and coinduction laws [6, 7]:

$$x \le p \cdot [a]x \Rightarrow x \le [a^*]p, \tag{4}$$

$$\triangle a \cdot [a^*]p = \mu x \,.\, p \cdot [a]x. \tag{5}$$

Dual laws hold for the diamond operator.

Modal quantales (and, more generally, modal ω/convergence algebras) offer additional flexibility compared to PDL [12] and the μ-calculus, since the modal operators are defined for ω-regular expressions, not only for atomic actions.

4 Algebraic Semantics of CTL*

We now give our algebraic interpretation of CTL* over a Boolean modal quantale S. To save some notation we set $\Phi = \mathsf{test}(S)$. Moreover, we fix an element n (n standing for "next") that represents the transition system underlying the logic. The precise requirements for n will be discussed in Section 5. Then the concrete semantics above generalises to a function $[\![\_]\!] : \Psi \to S$, where $[\![\varphi]\!]$ abstractly represents the set of paths satisfying formula φ:

$$\begin{aligned}
[\![\bot]\!] &= 0, \\
[\![p]\!] &= p \cdot \top, \\
[\![\varphi \to \psi]\!] &= \overline{[\![\varphi]\!]} + [\![\psi]\!], \\
[\![\mathsf{X}\,\varphi]\!] &= \mathsf{n} \cdot [\![\varphi]\!], \\
[\![\varphi \,\mathsf{U}\, \psi]\!] &= \bigsqcup_{j \ge 0} (\mathsf{n}^j \cdot [\![\psi]\!] \sqcap \bigsqcap_{k<j} \mathsf{n}^k \cdot [\![\varphi]\!]), \\
[\![\mathsf{E}\varphi]\!] &= \ulcorner[\![\varphi]\!] \cdot \top.
\end{aligned}$$

Using these definitions, it is straightforward to check that

$$[\![\varphi \vee \psi]\!] = [\![\varphi]\!] + [\![\psi]\!], \qquad [\![\varphi \wedge \psi]\!] = [\![\varphi]\!] \sqcap [\![\psi]\!], \qquad [\![\neg\varphi]\!] = \overline{[\![\varphi]\!]}.$$

Given a set A of states, over the left quantale STR(A) (see Example 3.1) this semantics coincides with that of Section 2. Another important check of the adequacy of our definitions is provided by the following theorem. The restriction on n mentioned in the assumption will be discussed in the next section.

Theorem 4.1. *Assume that left multiplication with* n *distributes through meets. Then the element* $[\![\varphi \,\mathsf{U}\, \psi]\!]$ *is the least fixpoint* μf *of the function* $f(y) =_{df} [\![\psi]\!] + ([\![\varphi]\!] \sqcap \mathsf{n} \cdot y)$.

Proof. Since in a Boolean quantale multiplication and binary meet preserve arbitrary joins, f preserves arbitrary joins, too, and hence is continuous. So by Kleene's fixpoint theorem $\mu f = \bigsqcup_{j \geq 0} f^j(0)$. A straightforward induction shows that

$$f^i(0) = \bigsqcup_{j \leq i} (\mathsf{n}^j \cdot [\![\psi]\!] \sqcap \bigsqcap_{k<j} \mathsf{n}^k \cdot [\![\varphi]\!]),$$

from which the claim is immediate. □

We define the usual abbreviations:

$$\mathsf{A}\varphi =_{df} \neg\mathsf{E}\neg\varphi, \qquad \mathsf{F}\varphi =_{df} \top\mathsf{U}\varphi, \qquad \mathsf{G}\varphi =_{df} \neg\mathsf{F}\neg\varphi.$$

Theorem 4.1 and (3) yield the following closed representation of F:

Corollary 4.2. $[\![\mathsf{F}\varphi]\!] = \mathsf{n}^* \cdot [\![\varphi]\!]$.

5 The Next-Time Operator

We now want to find suitable requirements on n by considering axiom (1) in the algebraic setting. To satisfy it, we need to have for all formulas φ and their semantical values $b =_{df} [\![\varphi]\!]$,

$$\overline{\mathsf{n} \cdot b} = [\![\neg\mathsf{X}\varphi]\!] = [\![\mathsf{X}\neg\varphi]\!] = \mathsf{n} \cdot \overline{b}. \tag{6}$$

This semantic property can equivalently be characterised as follows (property (a) was already shown in [4]).

Lemma 5.1. *Consider a Boolean left quantale* S *and* $\mathsf{n} \in S$ *such that* $\mathsf{n} \cdot 0 = 0$.

(a) $\forall b \in S : \mathsf{n} \cdot \overline{b} \leq \overline{\mathsf{n} \cdot b} \Leftrightarrow \forall b, c \in S : \mathsf{n} \cdot (b \sqcap c) = \mathsf{n} \cdot b \sqcap \mathsf{n} \cdot c$.
(b) $\forall b \in S : \overline{\mathsf{n} \cdot b} \leq \mathsf{n} \cdot \overline{b} \Leftrightarrow \mathsf{n} \cdot \top = \top \Leftrightarrow \mathsf{n}^\omega = \top$.

Proof. (a) ($\Rightarrow$) It suffices to show ($\geq$), since the reverse inequality follows by isotony. By shunting, the assumption $\mathsf{n} \cdot \overline{b} \leq \overline{\mathsf{n} \cdot b}$, distributivity, Boolean algebra, and lattice algebra:

$$\mathsf{n}\cdot b \sqcap \mathsf{n}\cdot c \le \mathsf{n}\cdot(b\sqcap c) \Leftrightarrow \mathsf{n}\cdot b \le \overline{\mathsf{n}\cdot c} + \mathsf{n}\cdot(b\sqcap c) \Leftarrow \mathsf{n}\cdot b \le \mathsf{n}\cdot\overline{c} + \mathsf{n}\cdot(b\sqcap c)$$
$$\Leftrightarrow \mathsf{n}\cdot b \le \mathsf{n}\cdot(\overline{c} + (b\sqcap c)) \Leftrightarrow \mathsf{n}\cdot b \le \mathsf{n}\cdot(\overline{c}+b) \Leftrightarrow \mathsf{TRUE}.$$

($\Leftarrow$) We calculate, using the assumption in the third step:

$$0 = \mathsf{n}\cdot 0 = \mathsf{n}\cdot(b\sqcap\overline{b}) = \mathsf{n}\cdot b \sqcap \mathsf{n}\cdot\overline{b}.$$

Now the claim is immediate by shunting.

(b) By shunting, distributivity, complement, greatest element, and $\mathsf{n}^\omega = \nu y\,.\,\mathsf{n}\cdot y$:

$$\overline{\mathsf{n}\cdot b} \le \mathsf{n}\cdot\overline{b} \Leftrightarrow \top \le \mathsf{n}\cdot b + \mathsf{n}\cdot\overline{b} \Leftrightarrow \top \le \mathsf{n}\cdot(b+\overline{b}) \Leftrightarrow \top \le \mathsf{n}\cdot\top \Leftrightarrow \top = \mathsf{n}\cdot\top \Leftrightarrow \mathsf{n}^\omega = \top.$$

□

In relation algebra, the special case $\mathsf{n}\cdot\overline{1} \le \overline{\mathsf{n}}$ of the property in (a) characterises n as a partial function and is equivalent to the full property [21]. But in general quantales the special and the general case are not equivalent [4]. Moreover, again from [4], we know that in quantales such as WOR and STR an element n is left-distributive over meet iff it is prefix-free, i.e. if no member of n is a prefix of another member. This holds in particular if all words in n have equal length, which is the case if n models a transition relation and hence consists only of words of length 2. The equivalent condition $\forall b\,.\,\mathsf{n}\cdot b \sqcap \mathsf{n}\cdot\overline{b} = 0$ was used in the computation calculus of R.M. Dijkstra [8].

But what about property (b)? Only rarely will a quantale be "generated" by an element n in the sense that $\mathsf{n}^\omega = \top$. The solution is to choose a left-distributive element n and restrict the set of semantical values to the subset $\mathrm{SEM}(\mathsf{n}) =_{df} \{b : b \le \mathsf{n}^\omega\}$, taking complements relative to n^ω. This set is clearly closed under $+$ and $\sqcap$ and under prefixing by n, since by isotony

$$\mathsf{n}\cdot b \le \mathsf{n}\cdot\mathsf{n}^\omega = \mathsf{n}^\omega\ .$$

Finally, it also contains all elements $p\cdot\mathsf{n}^\omega$ with $p \in \mathsf{test}(S)$, since $p \le 1$. Hence the above semantics is well-defined in $\mathrm{SEM}(\mathsf{n})$ if we replace $\top$ by n^ω.

6 The Semantics of State Formulas

In this section we show, next to some other properties, that the semantics of each state formula has the special form of a test ideal and hence directly corresponds to a test, i.e., an abstract representation of a set of states. This will be the key to the simplified CTL semantics in Section 7.

Theorem 6.1. *Let φ be a state formula of* CTL^*.

(a) $[\![\varphi]\!]$ *is a test ideal, and hence, by Lemma 3.2(a),* $[\![\varphi]\!] = \ulcorner[\![\varphi]\!]\cdot\top$.

(b) $[\![\mathsf{E}\varphi]\!] = [\![\varphi]\!]$.

(c) $[\![\mathsf{A}\varphi]\!] = \neg\ulcorner(\overline{[\![\varphi]\!]})\cdot\top$.

Proof. (a) The proof is by induction on the structure of φ.

- For $\bot$ and $p \in \mathsf{test}(S)$ this is immediate from the definition.
- Assume that the claim already holds for state formulas φ and ψ. We calculate, using the definitions, the induction hypothesis, Lemma 3.2(e), distributivity and the definitions again,

$$\begin{aligned} \llbracket \varphi \to \psi \rrbracket &= \overline{\llbracket \varphi \rrbracket} + \llbracket \psi \rrbracket = \overline{\ulcorner \llbracket \varphi \rrbracket \cdot \top} + \ulcorner \llbracket \psi \rrbracket \cdot \top = \neg \ulcorner \llbracket \varphi \rrbracket \cdot \top + \ulcorner \llbracket \psi \rrbracket \cdot \top \\ &= (\neg \ulcorner \llbracket \varphi \rrbracket + \ulcorner \llbracket \psi \rrbracket) \cdot \top = (\ulcorner \llbracket \varphi \rrbracket \to \ulcorner \llbracket \psi \rrbracket) \cdot \top . \end{aligned}$$

- For $\mathsf{E}\varphi$ the claim is immediate from the definition.

(b) Immediate from (a) and the definition of $\llbracket \mathsf{E}\varphi \rrbracket$.
(c) Similar to (b). □

Moreover, state formulas are closed under $\neg, \wedge, \vee$ and A.

Next, we derive some properties of U and its relatives for state formulas. For this we use knowledge about dual functions and their fixpoints. The *(de Morgan) dual* f° of a function $f : S \to S$ over a Boolean quantale is, as usual, defined by $f^\circ(y) =_{df} \overline{f(\overline{y})}$. Then $\mu f = \overline{\nu f^\circ}$ and $\nu f = \overline{\mu f^\circ}$.

Lemma 6.2. *Let φ, ψ be state formulas of CTL^* and $p \cdot \top =_{df} \llbracket \varphi \rrbracket, q \cdot \top =_{df} \llbracket \psi \rrbracket$.*
(a) $\llbracket \varphi \mathsf{U} \psi \rrbracket = (p \cdot \mathsf{n})^* \cdot q \cdot \top = (\llbracket \varphi \rrbracket \sqcap \mathsf{n})^* \cdot \llbracket \psi \rrbracket$.
(b) $\llbracket \mathsf{G} \varphi \rrbracket = (p \cdot \mathsf{n})^\omega = (\llbracket \varphi \rrbracket \sqcap \mathsf{n})^\omega$.
Hence we have the shunting rule $(p \cdot \mathsf{n})^\omega = \overline{\mathsf{n}^* \cdot \neg p \cdot \top}$.

Proof. (a) Using Theorem 4.1 and Lemma 3.2(c) we calculate

$$\llbracket \varphi \mathsf{U} \psi \rrbracket = \mu y \,.\, q \cdot \top + (p \cdot \top \sqcap \mathsf{n} \cdot y) = \mu y \,.\, q \cdot \top + p \cdot \mathsf{n} \cdot y ,$$

and the claim follows by (3).
(b) Since $\llbracket \mathsf{F} \varphi \rrbracket = \mu f_p$ where $f_p(y) = p \cdot \top + \mathsf{n} \cdot y$, we have, by Lemma 3.2(e), $\llbracket \mathsf{G} \varphi \rrbracket = \llbracket \neg \mathsf{F} \neg \varphi \rrbracket = \nu f^\circ_{\neg p}$, where, again by Lemma 3.2(e) and by (6),

$$f^\circ_{\neg p}(y) = \overline{\neg p \cdot \top + \mathsf{n} \cdot \overline{y}} = \overline{\neg p \cdot \top} \sqcap \overline{\mathsf{n} \cdot \overline{y}} = p \cdot \top \sqcap \mathsf{n} \cdot y = p \cdot \mathsf{n} \cdot y .$$

Hence the claim follows by the definition of ω. □

The case $p = 1$ yields again Corollary 4.2. Now we deal with E.

Lemma 6.3. $\llbracket \mathsf{EX} \varphi \rrbracket = \llbracket \mathsf{EXE} \varphi \rrbracket$.

Proof. By the definitions, properties of domain, (2) and the definitions again,

$$\llbracket \mathsf{EXE} \varphi \rrbracket = \ulcorner (\mathsf{n} \cdot \ulcorner \llbracket \varphi \rrbracket \cdot \top) \cdot \top = \ulcorner (\mathsf{n} \cdot \ulcorner \llbracket \varphi \rrbracket) \cdot \top = \ulcorner (\mathsf{n} \cdot \llbracket \varphi \rrbracket) \cdot \top = \llbracket \mathsf{EX} \varphi \rrbracket . \quad \square$$

Next, we collect a number of properties of A. The proofs are straightforward calculations.

Lemma 6.4. *For atomic proposition $p \in \mathsf{test}(S)$,*

$$\begin{aligned} \llbracket \mathsf{A} \bot \rrbracket &= 0, & \llbracket \mathsf{A} \top \rrbracket &= \top, \\ \llbracket \mathsf{A}(p \vee \varphi) \rrbracket &= p + \llbracket \mathsf{A} \varphi \rrbracket, & \llbracket \mathsf{A}(p \wedge \varphi) \rrbracket &= p \cdot \llbracket \mathsf{A} \varphi \rrbracket . \end{aligned}$$

Moreover, for the axiom $\mathsf{EX}\top$ we obtain

Lemma 6.5. $[\![\mathsf{EX}\top]\!] = \top \Leftrightarrow {}^{\ulcorner}\mathsf{n} = 1 \Leftrightarrow \mathsf{n}$ *total.*

Proof. This follows by Lemma 3.2(b), since $[\![\mathsf{EX}\top]\!] = {}^{\ulcorner}(\mathsf{n} \cdot \top) \cdot \top = {}^{\ulcorner}\mathsf{n} \cdot \top$. □

We conclude this section by noting that EX and AX are de Morgan duals; again the proof is a straightforward calculation.

Lemma 6.6. $[\![\mathsf{AX}\varphi]\!] = [\![\neg\mathsf{EX}\neg\varphi]\!]$.

From this and Lemma 6.3 we obtain

Corollary 6.7. $[\![\mathsf{AX}\varphi]\!] = [\![\mathsf{AXA}\varphi]\!]$.

7 From CTL* to CTL

For a number of applications the sublogic CTL of CTL* suffices. We will see that it can be modelled in plain Kleene/convergence algebra. Syntactically, CTL consists of those CTL* state formulas that only use path formulas of the restricted form $\Pi ::= \mathsf{X}\Sigma \mid \Sigma \mathsf{U} \Sigma$.

From the previous section we already know that the semantics of every CTL formula is a test ideal t, from which, by Theorem 6.1(a), we can extract the corresponding test (or state set) as ${}^{\ulcorner}t$. This is reflected by the simplified semantics

$$[\![\varphi]\!]_d =_{df} {}^{\ulcorner}[\![\varphi]\!].$$

This enables us to calculate solely with tests.

First, for the Boolean connectives we obtain by disjunctivity of domain and Lemma 3.2,

$$[\![\varphi \vee \psi]\!]_d = [\![\varphi]\!]_d + [\![\psi]\!]_d, \qquad [\![\varphi \wedge \psi]\!]_d = [\![\varphi]\!]_d \cdot [\![\psi]\!]_d, \qquad [\![\neg\varphi]\!]_d = \neg[\![\varphi]\!]_d.$$

Next, we transfer the properties of A from Lemma 6.4 to the simplified semantics. Again the proofs are straightforward calculations.

Lemma 7.1. *For atomic proposition* $p \in \mathsf{test}(S)$,

$$\begin{aligned} [\![\mathsf{A}\bot]\!]_d &= 0, & [\![\mathsf{A}\top]\!]_d &= 1, \\ [\![\mathsf{A}(p \vee \varphi)]\!]_d &= p + [\![\mathsf{A}\varphi]\!]_d, & [\![\mathsf{A}(p \wedge \varphi)]\!]_d &= p \cdot [\![\mathsf{A}\varphi]\!]_d. \end{aligned}$$

Now we can calculate the inductive behaviour of $[\![\_]\!]_d$ for all CTL formulas.

Theorem 7.2

(a) $[\![\bot]\!]_d = 0$,

(b) $[\![p]\!]_d = p$,

(c) $[\![\varphi \rightarrow \psi]\!]_d = [\![\varphi]\!]_d \rightarrow [\![\psi]\!]_d$,

(d) $[\![\mathsf{EX}\varphi]\!]_d = \langle \mathsf{n} \rangle [\![\varphi]\!]_d$,

(e) $[\![\mathsf{AX}\varphi]\!]_d = [\mathsf{n}][\![\varphi]\!]_d = [\![\mathsf{AXA}\varphi]\!]_d$,

(f) $[\![\mathsf{AF}\varphi]\!]_d = \neg\,\overline{{}^{\ulcorner}\mathsf{n}^* \cdot [\![\varphi]\!]_d \cdot \top} = \neg{}^{\ulcorner}(\neg[\![\varphi]\!]_d \cdot \mathsf{n})^{\omega}$,

(g) $[\![\mathsf{E}(\varphi \mathsf{U} \psi)]\!]_d = \langle ([\![\varphi]\!]_d \cdot \mathsf{n})^* \rangle [\![\psi]\!]_d$,

(h) $[\![\mathsf{A}(\varphi \mathsf{U} \psi)]\!]_d = [\![\mathsf{AF}\varphi]\!]_d \cdot [b^*]([\![\varphi]\!]_d + [\![\psi]\!]_d)$ *where* $b =_{df} \neg[\![\varphi]\!]_d \cdot \mathsf{n}$.

The lengthy proof by induction on the structure of the state formulas can be found in the Appendix. This theorem shows that the sublogic CTL needs fewer algebraic concepts than full CTL^*: general joins and complementation (and therefore also general meet) are not needed. For the CTL semantics a modal left omega algebra [17] is sufficient.

To complete the picture, we show the validity of the usual least-fixpoint characterisation of $\mathsf{A}(u)$, where $u = [\![\varphi \mathsf{U} \psi]\!]$ for state formulas φ and ψ. Then, by Lemma 4.1, the definition of f, Lemma 6.4 twice and Corollary 6.7, we obtain

$$\mathsf{A}(u) = \mathsf{A}(f(u)) = \mathsf{A}(q \cdot \top + p \cdot \mathsf{n} \cdot u) = q \cdot \top + p \cdot \mathsf{A}(\mathsf{n} \cdot u) = q \cdot \top + p \cdot \mathsf{A}(\mathsf{n} \cdot \mathsf{A}(u)).$$

In general quantales, however, $\mathsf{A}(u)$ need not be the least fixpoint of the associated function. We need an additional assumption on the underlying quantale S, namely that unlimited finite iteration can be extended to infinite iteration in the following sense:

$$\forall b \in S : \bigsqcap_{i \in \mathbb{N}} \ulcorner(b^i) \leq \ulcorner(b^\omega). \tag{7}$$

In particular, S must have "enough" infinite elements to make $b^\omega \neq 0$ if all $b^i \neq 0$. This property is violated in the subquantale LAN of WOR in which only languages of finite words are allowed, because in LAN finite languages may be iterated indefinitely, but no infinite "limits" exist.

Now we can show the desired leastness of A.

Theorem 7.3. *Assume (7).*

(a) $\neg\ulcorner(b^\omega) = \Delta b$.
(b) *If b is total, i.e., $\ulcorner b = 1$ then also $\ulcorner(b^\omega) = 1$.*
(c) *If $[\![\varphi]\!] = p \cdot \top$ then $[\![\mathsf{AF}\varphi]\!]_d = \Delta(\neg p \cdot a)$*
(d) $[\![\varphi \mathsf{U} \psi]\!]_d = \mu h$, *where* $h(y) =_{df} q + p \cdot [\mathsf{n}]y$.

Proof. (a) First, $\neg\ulcorner(b^\omega)$ is a fixpoint of $[b]$:

$$\neg\ulcorner(b^\omega) = \neg\ulcorner(b \cdot (b^\omega)) = \neg\ulcorner(b \cdot \neg\neg(b^\omega)) = [b](\neg\ulcorner(b^\omega)).$$

Hence $\Delta b = \mu[b] \leq \neg\ulcorner(b^\omega)$. For the converse inequation we calculate By shunting, (7), and the definition of meet:

$$\neg\ulcorner(b^\omega) \leq \Delta b \Leftrightarrow \neg\Delta b \leq \ulcorner(b^\omega) \Leftarrow \neg\Delta b \leq \bigsqcap_{i \in \mathbb{N}} \ulcorner(b^i) \Leftarrow \forall i \in \mathbb{N} : \neg\Delta b \leq \ulcorner(b^i).$$

Using $\neg\Delta b \leq 1$, isotony of domain, the definition of box and that Δb is a fixpoint of $[b]$, we have indeed $\ulcorner(b^i) \geq \ulcorner(b^i \cdot \neg\Delta b) = \neg[b^i]\Delta b = \neg\Delta b$.

(b) By the assumption (2) of modality multiplication preserves totality: if $\ulcorner a = \ulcorner b = 1$ then $\ulcorner(a \cdot b) = \ulcorner(a \cdot \ulcorner b) = \ulcorner(a \cdot 1) = \ulcorner a = 1$. Now an easy induction shows $\ulcorner b = 1 \Rightarrow \forall i : \ulcorner b^i = 1$ and assumption (7) immediately implies the claim.

(c) Immediate from Theorem 7.2(f) and (a).

(d) From the definition of h we get by Boolean algebra

$$h(y) = (q + p) \cdot (q + [\mathsf{n}]y).$$

Now the claim follows from (5), Theorem 7.2(h) and (b). □

This result shows that for CTL we can even do without omega iteration and need only a convergence algebra. Recently it has been shown [13] that property (a) is equivalent to validity of the coinduction rule

$$p \leq \ulcorner(q + a \cdot p) \Rightarrow p \leq \ulcorner(a^\omega + a^* \cdot q) \ .$$

8 From CTL* to LTL

The logic LTL is the fragment of CTL* in which only A may occur, once and outermost only, as path quantifier. More precisely, the LTL path formulas are given by

$$\Pi ::= \Phi \mid \bot \mid \Pi \rightarrow \Pi \mid \mathsf{X}\,\Pi \mid \Pi\,\mathsf{U}\,\Pi.$$

The LTL semantics is embedded into the CTL* one by assigning to $\varphi \in \Pi$ the semantic value $[\![\mathsf{A}\varphi]\!]$.

Unfortunately, except for the cases $[\![\mathsf{AX}\varphi]\!] = [\mathsf{n}][\![\mathsf{A}\varphi]\!]$ and $[\![\mathsf{AG}\varphi]\!] = [\mathsf{n}^*][\![\mathsf{A}\varphi]\!]$ the semantics does not propagate nicely in an inductive way into the subformulas, and so a simplified semantics cannot be obtained directly from the CTL* one.

However, by a slight change of view we can still achieve our goal. In the considerations based on the concrete quantales WOR and STR, the semantic element n representing X "glued" transitions to the front of traces. However, as is frequently done, one can also interpret n as a relation that maps a trace σ to its tail σ^1. This is the basis for a simplified semantics of LTL over the Boolean quantale REL(A^ω) (since standard LTL considers only infinite traces) for some set A of states.

What are the tests involved? Obviously, they now correspond to sets of paths, since they are subrelations of the identity relation on traces. So in this view the semantics of LTL formulas is again given by test ideals, only in a different algebra.

Therefore we can re-use the simplified CTL semantics. In particular, we set

$$[\![\mathsf{X}\varphi]\!]_\mathsf{L} =_{df} \langle\mathsf{n}\rangle[\![\varphi]\!]_\mathsf{L}.$$

This means that $[\![\mathsf{X}\varphi]\!]_\mathsf{L}$ is the inverse image of $[\![\varphi]\!]_\mathsf{L}$ under the tail relation; hence the standard LTL semantics is captured faithfully.

What does axiom (1) mean in this interpretation? It is equivalent to the equation $\langle\mathsf{n}\rangle = [\mathsf{n}]$ which characterises $\langle\mathsf{n}\rangle$ as a total function. This holds indeed for the tail relation on A^ω.

The semantics of $\bot$ and $\rightarrow$ are as before. It remains to work out the semantics of U. With $p =_{df} [\![\varphi]\!]_\mathsf{L}$ and $q =_{df} [\![\psi]\!]_\mathsf{L}$, we want $[\![\varphi\mathsf{U}\psi]\!]_\mathsf{L}$ to be the least fixpoint of the function $h(y) =_{df} q + p \cdot \langle\mathsf{n}\rangle y$, which by the dual of box induction (5) is $\langle(p \cdot \mathsf{n})^*\rangle q$. By this, the semantics of $\mathsf{F}\psi$ and $\mathsf{G}\psi$ work out to $\langle\mathsf{n}^*\rangle q$ and $[\mathsf{n}^*]q$.

Summarising, our LTL semantics now reads (see also [5])

$$\begin{aligned}
[\![\bot]\!]_{\mathsf{L}} &= 0,\\
[\![p]\!]_{\mathsf{L}} &= p,\\
[\![\varphi \to \psi]\!]_{\mathsf{L}} &= [\![\varphi]\!]_{\mathsf{L}} \to [\![\psi]\!]_{\mathsf{L}},\\
[\![\mathsf{X}\varphi]\!]_{\mathsf{L}} &= \langle \mathsf{n} \rangle [\![\varphi]\!]_{\mathsf{L}},\\
[\![\varphi \,\mathsf{U}\, \psi]\!]_{\mathsf{L}} &= \langle ([\![\varphi]\!]_{\mathsf{L}} \cdot \mathsf{n})^* \rangle [\![\psi]\!]_{\mathsf{L}},\\
[\![\mathsf{F}\psi]\!]_{\mathsf{L}} &= \langle \mathsf{n}^* \rangle [\![\psi]\!]_{\mathsf{L}},\\
[\![\mathsf{G}\psi]\!]_{\mathsf{L}} &= [\mathsf{n}^*] [\![\psi]\!]_{\mathsf{L}}.
\end{aligned}$$

This shows that for LTL we can weaken the requirements on the underlying semantic algebra even further, viz. to that of a modal Kleene algebra.

9 Conclusion

We have provided a compact algebraic semantics for full CTL^* in the framework of modal quantales and shown that for the two sublogics CTL and LTL the semantics can be mapped to closed expressions using modal operators as well as Kleene star and ω-iteration or the convergence operator. Compared with representations of CTL^* in the modal μ-calculus the compactness is achieved, since in quantales the modal operators are defined for ω-regular expressions (and even more generally), not only for atomic actions. Moreover, we have shown that for CTL and LTL the requirements on the semantic algebra can be relaxed to that of a modal omega or convergence algebra an even just a modal Kleene algebra, resp.

Future research will concern use of the algebraic semantics for concrete calculations in case studies as well the extension from the current propositional case to the first-order one; for this Tarskian frames as introduced in [16] seem a promising candidate.

Acknowledgements. We are grateful to the anonymous referees and to Kim Solin for valuable comments and remarks.

References

1. R. C. Backhouse et al.: Fixed point calculus. Inform. Proc. Letters, 53:131–136 (1995)
2. E. Cohen: Separation and reduction. In R. Backhouse and J.N. Oliveira (eds.): Mathematics of Program Construction. LNCS 1837. Springer 2000, 45–59
3. J.H. Conway: Regular algebra and finite machines. London: Chapman and Hall 1971
4. J. Desharnais, B. Möller: Characterizing determinacy in Kleene algebra. Special Issue on Relational Methods in Computer Science, Information Sciences — An International Journal 139, 253–273 (2001)
5. J. Desharnais, B. Möller, G. Struth: Modal Kleene algebra and applications — a survey. J. Relational Methods in Computer Science 1, 93–131 (2004)
6. J. Desharnais, B. Möller, G. Struth: Kleene algebra with domain. ACM Transactions on Computational Logic 2006 (to appear)

7. J. Desharnais, B. Möller, G. Struth: Termination in modal Kleene algebra. In J.-J. Lévy, E. Mayr, and J. Mitchell, editors, Exploring new frontiers of theoretical informatics. IFIP International Federation for Information Processing Series 155. Kluwer 2004, 653–666
8. R.M. Dijkstra: Computation calculus bridging a formalisation gap. Science of Computer Programming **37**, 3-36 (2000)
9. E.A. Emerson: Temporal and modal logic. In J. van Leeuwen (ed.): Handbook of theoretical computer science. Vol. B: Formal models and semantics. Elsevier 1991, 995–1072
10. M.F. Frías and C. Lopez Pombo. Interpretability of linear time temporal logic in fork algebra. Journal of Logic and Algebraic Programming, 66(2):161-184 (2006)
11. V. Goranko: Temporal logics of computations. Introductory course, 12th European summer School in Logic, Language and Information, Birmingham, 6–18 August 2000
12. D. Harel, D. Kozen, J. Tiuryn: Dynamic Logic. MIT Press 2000
13. P. Höfner, B. Möller, K. Solin: Omega Algebra, demonic refinement algebra and commands. Proc. 9th RelMiCS/4th AKA 2006 (to appear)
14. D. Kozen: A completeness theorem for Kleene algebras and the algebra of regular events. Information and Computation **110:2**, 366–390 (1994)
15. D. Kozen: Kleene algebras with tests. ACM TOPLAS **19**, 427–443 (1997)
16. D. Kozen: Some results in dynamic model theory. Science of Computer Programming 51, 3–22 (2004)
17. B. Möller: Kleene getting lazy. Science of Computer Programming, Special issue on MPC 2004 (to appear). Previous version: B. Möller: Lazy Kleene algebra. In D. Kozen (ed.): Mathematics of program construction. LNCS 3125. Springer 2004, 252–273
18. B. Möller, G. Struth: Algebras of Modal Operators and Partial Correctness Theoretical Computer Science 351, 221-239 (2006)
19. B. Möller, G. Struth: wp is wlp. In W. MacCaull, M. Winter and I. Duentsch (eds.): Relational Methods in Computer Science. LNCS 3929, Springer 2006, 200–211
20. K.I. Rosenthal: Quantales and their applications. Pitman Research Notes in Mathematics Series, Vol. 234. Longman Scientific&Technical 1990
21. G. Schmidt, T. Ströhlein: Relations and Graphs — Discrete Mathematics for Computer Scientists. EATCS Monographs on Theoretical Computer Science. Springer 1993
22. L. Staiger: Omega languages. In G. Rozenberg, A. Salomaa (eds.): Handbook of formal languages, Vol. 3. Springer 1997, 339–387
23. B. von Karger: Temporal algebra. Mathematical Structures in Computer Science 8:277–320, 1998
24. B. von Karger, R. Berghammer: A relational model for temporal logic. Logic Journal of the IGPL 6, 157–173, 1998

Appendix: Proof of Theorem 7.2

The proof is again by induction on the structure of the state formulas. The cases (a)–(c) of $\bot$, p and $\varphi \rightarrow \psi$ have already been covered in the proof of Theorem 6.1.

(d) Using again Theorem 6.1, the definition of $[\![\ ]\!]$, (2) and the definitions again, we calculate $[\![\mathsf{EX}\varphi]\!]_d = \ulcorner[\![\mathsf{X}\varphi]\!] = \ulcorner(\mathsf{n}\cdot[\![\varphi]\!]) = \ulcorner(\mathsf{n}\cdot\ulcorner[\![\varphi]\!]) = \langle\mathsf{n}\rangle[\![\varphi]\!]_d$.

(e) By Theorem 6.1(c) and Lemma 3.2(b), definition and Theorem 6.1, by (6), by Lemma 3.2(b), domain property, and the definition:

$$[\![\mathsf{AX}\varphi]\!]_d = \neg\ulcorner\overline{[\![\mathsf{X}\varphi]\!]} = \neg\ulcorner\overline{\mathsf{n}\cdot[\![\varphi]\!]_d\cdot\top} = \neg\ulcorner(\mathsf{n}\cdot\overline{[\![\varphi]\!]_d\cdot\top}) = \neg\ulcorner(\mathsf{n}\cdot\neg[\![\varphi]\!]_d\cdot\top) = \\ \neg\ulcorner(\mathsf{n}\cdot\neg[\![\varphi]\!]_d) = [\mathsf{n}][\![\varphi]\!]_d.$$

Moreover, $[\![\varphi]\!]_d = [\![\mathsf{A}\varphi]\!]_d$ follows from Lemma 7.1.

(f) Assume $[\![\varphi]\!] = p\cdot\top$. By the definition of A and the explicit representation of F from Corollary 4.2 we obtain $[\![\mathsf{AF}\varphi]\!] = \neg\ulcorner\overline{\mathsf{n}^*\cdot p\cdot\top}\cdot\top$. Now the claim follows from the shunting rule of Lemma 6.2(b) and the definition of $[\![\ ]\!]_d$.

(g) For $[\![\mathsf{E}(\varphi\mathsf{U}\psi)]\!]$ we use the principle of *least-fixpoint fusion* [1]: If h preserves arbitrary joins and $h\circ f = g\circ h$ then $h(\mu f) = \mu g$.
Set, for abbreviation, $p =_{df} [\![\varphi]\!]_d$ and $q =_{df} [\![\psi]\!]_d$. Then, by Lemma 4.1 and Lemma 3.2(c), $u =_{df} [\![\varphi\mathsf{U}\psi]\!] = \mu f$ where $f(y) =_{df} q\cdot\top + (p\cdot\mathsf{n}\cdot y)$. Second, by Theorem 6.1 and (5), $\langle(p\cdot\mathsf{n})^*\rangle = \mu g$ where $g(p) =_{df} q + \langle(p\cdot\mathsf{n})\rangle p$. We need to show $\ulcorner(\mu f) = \mu g$. By the principle of least-fixpoint fusion this is implied by $\ulcorner\circ f = g\circ\ulcorner$, since $\ulcorner$ preserves arbitrary joins. We calculate: By definition f, additivity of domain, Lemma 3.2(a), by (2), definition diamond, and definition g:

$$\ulcorner(f(y)) = \ulcorner(q\cdot\top + (p\cdot\mathsf{n}\cdot y)) = \ulcorner(q\cdot\top) + \ulcorner(p\cdot\mathsf{n}\cdot y)) = q + \ulcorner(p\cdot\mathsf{n}\cdot y) = \\ q + \ulcorner(p\cdot\mathsf{n}\cdot\ulcorner y) = q + \langle p\cdot\mathsf{n}\rangle\cdot\ulcorner y = g(\ulcorner y).$$

(h) For $r =_{df} [\![\mathsf{A}(\varphi\mathsf{U}\psi)]\!]$ we use that, by Theorem 6.1(c), $r = \neg\ulcorner\overline{u}$, where $u =_{df} [\![\varphi\mathsf{U}\psi]\!]$. Let, for abbreviation, $p\cdot\top =_{df} [\![\varphi]\!]$ and $q\cdot\top =_{df} [\![\psi]\!]$. Since $u = \mu f$ where $f(y) = q\cdot\top + p\cdot\mathsf{n}\cdot y$, we have $\overline{u} = \nu f^\circ$. By the definitions, de Morgan, Lemma 3.2(e), Lemma 3.2(c) and de Morgan, Lemma 3.2(e) and (6), complement, distributivity, and de Morgan:

$$f^\circ(y) = \overline{q\cdot\top + p\cdot\mathsf{n}\cdot\overline{y}} = \overline{q\cdot\top}\sqcap\overline{p\cdot\mathsf{n}\cdot\overline{y}} = \neg q\cdot\top\sqcap\overline{p\cdot\top\sqcap\mathsf{n}\cdot\overline{y}} \\ = \neg q\cdot(\overline{p\cdot\top} + \overline{\mathsf{n}\cdot\overline{y}}) = \neg q\cdot(\neg p\cdot\top + \overline{\overline{\mathsf{n}\cdot y}}) = \neg q\cdot(\neg p\cdot\top + \mathsf{n}\cdot y) \\ = \neg q\cdot\neg p\cdot\top + \neg q\cdot\mathsf{n}\cdot y = \neg(p+q)\cdot\top + \neg q\cdot\mathsf{n}\cdot y.$$

By the above, (3), distributivity and de Morgan, Lemma 6.2 (b) and a domain property, Theorem 6.1(c) and definition of box, and Lemma 4.2:

$$\begin{aligned} & r \\ = {} & \neg\ulcorner(\nu f^\circ) \\ = {} & \neg\ulcorner((\neg q\cdot\mathsf{n})^\omega + (\neg q\cdot\mathsf{n})^*\cdot\neg(p+q)\cdot\top) \\ = {} & \neg\ulcorner((\neg q\cdot\mathsf{n})^\omega)\cdot\neg\ulcorner((\neg q\cdot\mathsf{n})^*\cdot\neg(p+q)\cdot\top) \\ = {} & \neg\ulcorner\overline{(\mathsf{n}^*\cdot q\cdot\top)}\cdot\neg\ulcorner((\neg q\cdot\mathsf{n})^*\cdot\neg(p+q)) \\ = {} & \mathsf{A}(\mathsf{n}^*\cdot q\cdot\top)\cdot[(\neg q\cdot\mathsf{n})^*](p+q) \\ = {} & (\mathsf{AF}q)\cdot[(\neg q\cdot\mathsf{n})^*](p+q). \end{aligned}$$

Fractional Semantics*

Härmel Nestra

Institute of Computer Science, University of Tartu
J. Liivi 2, 50409 Tartu, Estonia
harmel.nestra@ut.ee

Abstract. Transfinite semantics have been argued to be a proper framework for reasoning about correctness of certain program transformation techniques, e.g. program slicing. But transfinite semantics work fine only for non-recursive programs because of infinity being "one-way".

This paper presents transfinite trace semantics in a different form which we call fractional semantics. The components of traces are indexed with rational numbers rather than ordinals. Rational numbers form both infinite ascending and infinite descending chains, so the principal obstacle of handling recursion disappears.

Although we have not yet found a fractional semantics appropriate for all cases of recursion, the approach seems to be promising. Another contribution achieved with help of fractional semantics is presenting both standard and transfinite trace semantics uniformly using fixpoints.

1 Introduction

Standard semantics consider computations all of whose proper initial parts are finite. States of computation traces are indexed with natural numbers.

By *transfinite semantics*, one means a semantics according to which computation may continue after an infinite number of steps from some limit state determined somehow by the infinite computation performed. States of traces are indexed with ordinals. Transfinite semantics have turned out to be useful for constructing an adequate theoretical setting for some applications.

For example, the first study of transfinite semantics has been done for functional programming, see Kennaway et al. [5]. Later, Giacobazzi and Mastroeni [4] introduced transfinite semantics for a simple imperative language with the aim to overcome the so-called semantic anomaly of program slicing. The idea was proposed already by Cousot [2]. We have followed up this approach in [6, 7].

Program slicing is a kind of program transformation where the aim is to find an executable part of a program which is responsible for computing all the values important to the user. An introduction can be found in Binkley and Gallagher [1]. A semantic anomaly arises when the slices found via data-flow analysis being definitely perfect for practical purposes are not correct w.r.t. standard semantics. The discrepancy occurs when infinite loops are sliced away and running the slice causes execution of code not reached during the run of the original program. For

* Partially supported by Estonian Science Foundation under grant no. 6713.

M. Johnson and V. Vene (Eds.): AMAST 2006, LNCS 4019, pp. 278–292, 2006.

this reason, a semantics for giving a theoretical framework to program slicing must not fatally distinguish between termination and non-termination.

Transfinite semantics may be called *lazy* trace semantics since they allow recovery from non-termination. The term "lazy" comes from Danicic et al. [3] where a lazy denotational semantics was proposed with the same aim. "Semantic anomaly" can arise also due to runtime errors if one slices away a statement like assignment of `1 / 0` to a variable with no influence to important variables via data-flow. Lazy semantics can recover also from such errors by treating only the assigned variable as infected by the error rather than the whole computation.

In this paper, we propose *fractional semantics* as a uniform framework for both strict and lazy trace semantics. In fractional semantics, states of traces are indexed with rational numbers. Our design of these semantics leads to parts of the code being in fact statically associated with certain intervals of rationals.

For example, the execution trace of `z := x ; (x := y ; y := z)` at state $(\mathtt{x} \mapsto 1, \mathtt{y} \mapsto 2, \mathtt{z} \mapsto 0)$ is

$$
\begin{array}{ll}
0 & \mapsto \langle(\mathtt{x} \mapsto 1, \mathtt{y} \mapsto 2, \mathtt{z} \mapsto 0) \mid [\mathtt{z\ :=\ x\,;(x\ :=\ y\,;y\ :=\ z)}]\rangle\ , \\
\frac{1}{2} & \mapsto \langle(\mathtt{x} \mapsto 1, \mathtt{y} \mapsto 2, \mathtt{z} \mapsto 1) \mid [\mathtt{x\ :=\ y\,;y\ :=\ z}]\rangle\ , \\
\frac{3}{4} & \mapsto \langle(\mathtt{x} \mapsto 2, \mathtt{y} \mapsto 2, \mathtt{z} \mapsto 1) \mid [\mathtt{y\ :=\ z}]\rangle\ , \\
1 & \mapsto \langle(\mathtt{x} \mapsto 2, \mathtt{y} \mapsto 1, \mathtt{z} \mapsto 1) \mid [\varepsilon]\rangle\ .
\end{array}
$$

If the 2nd assignment is replaced with W = **while** `z > 0` **do** `z := z - 1` then the placing of runs of other parts on the trace remains unchanged:

$$
\begin{array}{ll}
0 & \mapsto \langle(\mathtt{x} \mapsto 1, \mathtt{y} \mapsto 2, \mathtt{z} \mapsto 0) \mid [\mathtt{z\ :=\ x\,;}(W\mathtt{\,;y\ :=\ z})]\rangle\ , \\
\frac{1}{2} & \mapsto \langle(\mathtt{x} \mapsto 1, \mathtt{y} \mapsto 2, \mathtt{z} \mapsto 1) \mid [W\mathtt{\,;y\ :=\ z}]\rangle\ , \\
\frac{5}{8} & \mapsto \langle(\mathtt{x} \mapsto 1, \mathtt{y} \mapsto 2, \mathtt{z} \mapsto 1) \mid [(\mathtt{z\ :=\ z\ -\ 1\,;}W)\mathtt{;y\ :=\ z}]\rangle\ , \\
\frac{11}{16} & \mapsto \langle(\mathtt{x} \mapsto 1, \mathtt{y} \mapsto 2, \mathtt{z} \mapsto 0) \mid [W\mathtt{\,;y\ :=\ z}]\rangle\ , \\
\frac{3}{4} & \mapsto \langle(\mathtt{x} \mapsto 1, \mathtt{y} \mapsto 2, \mathtt{z} \mapsto 0) \mid [\mathtt{y\ :=\ z}]\rangle\ , \\
1 & \mapsto \langle(\mathtt{x} \mapsto 1, \mathtt{y} \mapsto 0, \mathtt{z} \mapsto 0) \mid [\varepsilon]\rangle\ .
\end{array}
$$

This would be the case even if the initial state was changed. As every countable set of ordinals can be order-preservingly mapped into the set of all rationals, this framework enables accommodating transfinite execution traces as easily.

The lazy variants of our semantics defined later are not directly corresponding to the denotational lazy semantics of Danicic et al. [3]. The essential difference is that when a branching predicate takes an erroneous value then the semantics of [3] merges both branches but our semantics simply choose the false-branch. This treatment is enough for applications to program slicing.

In Sect. 2, we define the base language of our work together with a standard denotational semantics. In Sect. 3, we present trace semantics definition schemata for our language capturing both strict and lazy fractional semantics. The schemata refer to abstract fixpoints in the case of looping constructs. Defining these fixpoints appropriately is the topic of the following two sections. Section 4 does this for while-loop. Section 5 discusses what can happen if the same fixpoint specification is carried over to the case of recursion. Section 6 concludes.

Syntactic categories.

$$\begin{array}{ll} \mathit{Var} & \text{— the set of all variables} \\ \mathit{Proc} & \text{— the set of all procedure identifiers} \\ \mathit{Expr} & \text{— the set of all expressions} \\ \mathit{Stmt} & \text{— the set of all statements} \\ \mathit{Decl} & \text{— the set of all procedure declarations} \end{array}$$

Grammar.

$$\begin{array}{rcl} \mathit{Stmt} & \to & \mathit{Var}\ \texttt{:=}\ \mathit{Expr} \\ & | & \textbf{if}\ \mathit{Expr}\ \textbf{then}\ \mathit{Stmt}\ \textbf{else}\ \mathit{Stmt} \\ & | & \textbf{while}\ \mathit{Expr}\ \textbf{do}\ \mathit{Stmt} \\ & | & \textbf{call}\ \mathit{Proc}\texttt{(}\mathit{Expr}\texttt{,}\ \ldots\texttt{,}\ \mathit{Expr}\texttt{)} \\ & | & \mathit{Stmt}\ ;\ \mathit{Stmt} \\ & | & \varepsilon \end{array}$$

$$\begin{array}{rcl} \mathit{Decl} & \to & \textbf{proc}\ \mathit{Proc}\texttt{(}\mathit{Var}\texttt{,}\ \ldots\texttt{,}\ \mathit{Var}\texttt{)}\ \textbf{is}\ \mathit{Stmt} \\ & | & \mathit{Decl}\ ;\ \mathit{Decl} \\ & | & \varepsilon \end{array}$$

Laws.

$$\forall S \in \mathit{Stmt}\ (\varepsilon\ ;\ S = S = S\ ;\ \varepsilon)$$
$$\forall D \in \mathit{Decl}\ (\varepsilon\ ;\ D = D = D\ ;\ \varepsilon)$$

Fig. 1. Abstract syntax of **Proc**

2 Language Proc and Its Denotational Semantics

We are going to work on the language **Proc** — a simple programming language with procedures. Its abstract syntax is given in Fig. 1.

The inner structure of expressions is irrelevant for our purposes and we leave it unspecified. There are two kinds of looping constructs in **Proc**: while-loops and procedure recursion. Explicit return-statement is not included; procedures returning a value can be equivalently reformulated using an extra variable.

Note that ; is used to denote both statement and declaration succession and ε is used to denote both empty statement and empty declaration. In both cases, ε is assumed to be the unit of ; at syntactic level.

Let *Val* be the set of all values variables can take whereby it contains at least the truth values *tt*, *ff*. Let $\mathit{State} = \mathit{Var} \to \mathit{Val}$. All semantic categories throughout this paper are assumed to contain a special value $\perp$ used in the cases of failure or undefinedness. In the definitions, we will not show the attendancy of $\perp$ explicitly. Functions need not be strict, i.e., if applied to $\perp$, they not necessarily give $\perp$ as value. But if $\perp$ is applied to something, the result will be $\perp$.

For a function f, its potential argument x and value v, write $f[x \mapsto v]$ for the function working like f except for on x where it takes value v. If there is a family $x_1, \ldots, x_n$ of arguments and a corresponding family $v_1, \ldots, v_n$ of values, write $f[\forall i\, x_i \mapsto v_i]$ for the function working like f except for on arguments x_i where

Types.

$$\begin{array}{ll} Act = State \to State & \text{— program actions on states} \\ Env = Proc \to (Val^* \to Act) & \text{— procedure environments} \end{array}$$

Signatures.

$$\begin{array}{ll} \mathcal{W}^{E,T} \in Act \to (Act \to Act) & \text{— auxiliary} \\ \mathcal{P}^{D} \in (Env \to Env) \to (Env \to Env) & \text{— auxiliary} \end{array}$$

$$\begin{array}{ll} \mathcal{E} \in Expr \to (State \to Val) & \text{— expression evaluations} \\ \mathcal{S} \in Stmt \to (Env \to Act) & \text{— statement actions} \\ \mathcal{C} \in Decl \to (Env \to Env) & \text{— small-step environment changes} \\ \mathcal{D} \in Decl \to (Env \to Env) & \text{— big-step environment changes} \end{array}$$

Definitions.

$$\mathcal{W}^{E,T}(h)(g)(s) = \begin{cases} g(h(s)) & \text{if } \mathcal{E}(E)(s) = tt \\ s & \text{otherwise} \end{cases}$$

$$\mathcal{P}^{D}(g) = g \,;\, \mathcal{C}(D)$$

$$\mathcal{S}(X \texttt{ := } E)(e)(s) = s[X \mapsto \mathcal{E}(E)(s)]$$

$$\mathcal{S}(\texttt{if } E \texttt{ then } T_1 \texttt{ else } T_2)(e)(s) = \begin{cases} \mathcal{S}(T_1)(e)(s) & \text{if } \mathcal{E}(E)(s) = tt \\ \mathcal{S}(T_2)(e)(s) & \text{otherwise} \end{cases}$$

$$\mathcal{S}(\texttt{while } E \texttt{ do } T)(e) = \mathrm{fix}(\mathcal{W}^{E,T}(\mathcal{S}(T)(e)))$$

$$\mathcal{S}(\texttt{call } P(E_1, \ldots, E_n))(e)(s) = e(P)(\mathcal{E}(E_1)(s), \ldots, \mathcal{E}(E_n)(s))(s)$$

$$\mathcal{S}(T_1 \,;\, T_2)(e) = \mathcal{S}(T_1)(e) \,;\, \mathcal{S}(T_2)(e)$$

$$\mathcal{S}(\varepsilon)(e) = \mathrm{id}$$

$$\mathcal{C}(\texttt{proc } P(X_1, \ldots, X_n) \texttt{ is } S)(e)$$
$$= e[P \mapsto \lambda(v_1, \ldots, v_n).\, \lambda s.\, \mathcal{S}(S)(e)(s[\forall i\, X_i \mapsto v_i])[\forall i\, X_i \mapsto s(X_i)]]$$

$$\mathcal{C}(C_1 \,;\, C_2) = \mathcal{C}(C_1) \,;\, \mathcal{C}(C_2)$$

$$\mathcal{C}(\varepsilon) = \mathrm{id}$$

$$\mathcal{D}(D) = \mathrm{fix}\, \mathcal{P}^{D}$$

Fig. 2. Denotational semantics for **Proc**

it takes values v_i, respectively. We often write non-$\bot$ functions using λ-notation. Function composition is denoted by ; (the left-hand function is applied first); id stands for identity functions. All these conventions hold throughout the paper.

Fig. 2 presents a standard denotational semantics for **Proc**. It is given in fixpoint form like it is common for denotational semantics. We will present our trace semantics also in similar form.

Expression evaluation function $\mathcal{E}$ is unspecified here since also the set of expressions remained unspecified. Assume expressions having no side-effects. Due to the missing return-statements, $\mathcal{E}$ does not depend on procedure environment.

The value of operator fix on an argument function $\mathcal{G}$ is a fixpoint of $\mathcal{G}$. The classical approach is to choose the least fixpoint w.r.t. the order in which $\bot$ is less than "normal" states, among which all are pairwise incomparable. It is a classical result that the operators being the arguments of fix in the definitions of Fig. 2 possess least fixpoint, see Nielson and Nielson [8] for proof techniques.

We omit from our semantics everything concerning static errors like type errors or argument vector mismatch on procedure call. All programs we deal with are assumed statically correct.

In figure 2, we have omitted even everything concerning runtime errors because this semantics has been presented for a guiding example only where the peripheral details are of secondary importance. In the definitions of semantics in the rest of the paper, runtime errors are taken into account.

3 Various Kinds of Trace Semantics of Proc

We parametrize items of our general semantics schema with semantics kind κ. To clearly distinguish between strict and lazy semantics, we use symbol $\top$ for failure cases in lazy semantics and $\bot$ in strict semantics. This does not hint at any order relation; difference of $\bot$ and $\top$ is used to avoid repeating constructs like "it is one way for strict κ but otherwise for lazy κ".

So assume $\mathit{State} = \mathit{Var} \to \mathit{Val}$ like before whereby $\mathit{Val} \ni \top$. Variables can take value $\top$ either because of erroneous value assignment or, like in transfinite semantics, after an infinite loop during which the value of the variable does not stabilize. Denoting failures with different elements in different kinds of semantics is the only reason why expression evaluation function $\mathcal{E}_\kappa$ is also parametrized.

The components of traces are configurations consisting of the current state and the rest of the code. The latter is also called *program point*. Let Conf be the set of all configurations and PP the set of all program points; then $\mathit{Conf} = \mathit{State} \times \mathit{PP}$. Denote the configuration of state s and program point p by $\langle s \mid p \rangle$. Define $\mathrm{st}\langle s \mid p \rangle = s$, $\mathrm{st}\,\bot = \bot$ and $\mathrm{pp}\langle s \mid p \rangle = p$, $\mathrm{pp}\,\bot = \bot$. Write $c \sim \bot$ iff $\mathrm{st}\,c = \bot$ or $\mathrm{pp}\,c = \bot$. Call a trace *failing* iff it ends with a configuration $c \sim \bot$.

Program point as the part of the program to be run yet must entail the current call-stack, including remainders of every pendent procedure. Basing on this observation, take $\mathit{PP} = \mathrm{List}\,\mathit{Stmt}$ where $\mathrm{List}\,A$ is the set of all lists (either finite or infinite but not transfinite) with elements from A. Let nil denote the empty list and, for $a \in A$ and $l \in \mathrm{List}\,A$, let $a : l$ denote the list starting with a and continuing with elements of l in the same order; for strict version of $:$ (giving $\bot$ on argument $\bot$), use $\bar{:}$. Let $\mathrm{hd} \in \mathrm{List}\,A \to A$ be the operator giving the first element of any non-empty argument list as value and $\bot$ if the argument list is either empty or $\bot$. We will often denote finite or infinite lists by describing their contents exhaustively in brackets. By $|l|$, denote the length of list l. If $p \in \mathit{PP}$ and $T \in \mathit{Stmt}$, denote by $p \,;\, T$ the program point obtained by supplementing the first element of p, if it exists, with T using $;$, i.e.

$$p \,;\, T = \begin{cases} (S \,;\, T) : q & \text{if } p = S : q \\ p & \text{otherwise} \end{cases} .$$

Figures 3 and 4 together present trace semantics schema for **Proc** parametrically over the kind of semantics κ. Fig. 3 contains precisely all items that are to be specified differently for different kinds κ.

Types.

$$\mathit{Trace}_\kappa \text{ — the set of all traces}$$

Signatures.

single_κ	$\in \mathit{Conf} \to \mathit{Trace}_\kappa$	— trace of one configuration
cons_κ	$\in \mathit{Conf} \times \mathit{Trace}_\kappa \to \mathit{Trace}_\kappa$	— list constructor
snoc_κ	$\in \mathit{Trace}_\kappa \times \mathit{Conf} \to \mathit{Trace}_\kappa$	— dual list constructor
hd_κ	$\in \mathit{Trace}_\kappa \to \mathit{Conf}$	— first configuration on trace
dh_κ	$\in \mathit{Trace}_\kappa \to \mathit{Conf}$	— last configuration on trace
brz_κ	$\in \mathit{Trace}_\kappa \times \mathit{Trace}_\kappa \to \mathit{Trace}_\kappa$	— brazing concatenation
map_κ	$\in (\mathit{Conf} \to \mathit{Conf}) \to (\mathit{Trace}_\kappa \to \mathit{Trace}_\kappa)$	— mapping of each element

Fig. 3. Varying part of trace semantics of **Proc**

The trace concatenation operator is called *brazing* to express the idea that the last element on the first trace and the first element on the second trace are fused together into one element.

Auxiliary operators wrap_κ^L play the role of guards permitting a computation if some expressions do not produce $\perp$ and outputting a trivial trace with program point $\perp$ otherwise. Operator lift_κ applies its argument action to the last state of given trace and concatenates the result with the given trace. This operator is used in situations where two actions have to be composed (statement succession, while-loop). Operators $\mathcal{W}_\kappa^{E,T}$ concatenate a single step of execution of a while-loop with a given action. Similarly, $\mathcal{P}_\kappa^D$ composes a single step of interpretation of a declaration with a given environment transformation.

The call-stack is actually reversed in the program point: the procedure called latest appears as the last element in the list (the bottom of the stack).

For standard trace semantics, the unspecified items of Fig. 3 are naturally defined as shown in Fig. 5. The first equality reflects that a trace is never empty. The operator snoc_σ adds one element to the end of a given trace if it is finite and unfailing and gives the original trace back otherwise. The operator brz_σ gives non-$\perp$ result only if either the first argument ends with the element with which the second starts or the first argument is infinite or failing. In the former case, it concatenates the two traces, fusing the double configuration at the hook-up place into one. In the latter case, the first list is given as the result. Function map_σ operates as `map` known from functional languages.

The definitions of items of Fig. 3 for fractional semantics are shown in Fig. 6; a few new symbols are used which we define below. The principles behind the definitions of the operators are similar to that for standard semantics.

Denote the set of all rational numbers by $\mathbb{Q}$. For arbitrary $a, b \in \mathbb{Q}$ with $a \leqslant b$, denote by $[a;b]$ the set of all $r \in \mathbb{Q}$ for which $a \leqslant r \leqslant b$. By $A \dashrightarrow B$, denote the set of all partial functions from A to B; by $\operatorname{dom} f$, denote the set of elements on which f is defined.

Types.

$$\begin{array}{ll} Act_\kappa = State \to Trace_\kappa & \text{— program actions} \\ Env_\kappa = Proc \to (Val^* \to Act_\kappa) & \text{— procedure environments} \end{array}$$

Signatures.

$$\begin{array}{lll} \mathrm{wrap}^L_\kappa & \in State \to (Trace_\kappa \to Trace_\kappa) & \text{— auxiliary} \\ \mathrm{lift}_\kappa & \in Act_\kappa \to (Trace_\kappa \to Trace_\kappa) & \text{— auxiliary} \\ \mathcal{W}^{E,T}_\kappa & \in Act_\kappa \to (Act_\kappa \to Act_\kappa) & \text{— auxiliary} \\ \mathcal{P}^D_\kappa & \in (Env_\kappa \to Env_\kappa) \to (Env_\kappa \to Env_\kappa) & \text{— auxiliary} \end{array}$$

$$\begin{array}{ll} \mathcal{E}_\kappa \in Expr \to (State \to Val) & \text{— expression evaluations} \\ \mathcal{S}_\kappa \in Stmt \to (Env_\kappa \to Act_\kappa) & \text{— computations according to statements} \\ \mathcal{C}_\kappa \in Decl \to (Env_\kappa \to Env_\kappa) & \text{— small-step environment changes} \\ \mathcal{D}_\kappa \in Decl \to (Env_\kappa \to Env_\kappa) & \text{— big-step environment changes} \end{array}$$

Definitions.

$$\mathrm{wrap}^{(E_1,\ldots,E_n)}_\kappa\, s\, l = \left\{\begin{array}{ll} l & \text{if } \forall i\,(\mathcal{E}_\kappa(E_i)(s) \neq \bot) \\ \mathrm{single}_\kappa\langle s \mid \bot\rangle & \text{otherwise} \end{array}\right\}$$

$$\mathrm{lift}_\kappa\, g\, l$$
$$= \mathrm{brz}_\kappa(\mathrm{map}_\kappa(\lambda\langle u \mid p\rangle.\,\langle u \mid p\,;\,\mathrm{hd}(\mathrm{pp}(\mathrm{hd}_\kappa(g(\mathrm{st}(\mathrm{dh}_\kappa\, l))))))\rangle)\, l, g(\mathrm{st}(\mathrm{dh}_\kappa\, l)))$$

$$\mathcal{W}^{E,T}_\kappa(h)(g)(s)$$
$$= \mathrm{cons}_\kappa(\langle s \mid [\mathtt{while}\ E\ \mathtt{do}\ T]\rangle, \mathrm{wrap}^{(E)}_\kappa\, s \left\{\begin{array}{ll} \mathrm{lift}_\kappa\, g(h(s)) & \text{if } \mathcal{E}_\kappa(E)(s) = tt \\ \mathrm{single}_\kappa\langle s \mid [\varepsilon]\rangle & \text{otherwise} \end{array}\right\})$$

$$\mathcal{P}^D_\kappa(g) = g\,;\,\mathcal{C}_\kappa(D)$$

$$\mathcal{S}_\kappa(X\ \mathtt{:=}\ E)(e)(s)$$
$$= \mathrm{cons}_\kappa(\langle s \mid [X\ \mathtt{:=}\ E]\rangle, \mathrm{wrap}^{(E)}_\kappa\, s(\mathrm{single}_\kappa\langle s[X \mapsto \mathcal{E}_\kappa(E)(s)] \mid [\varepsilon]\rangle))$$

$$\mathcal{S}_\kappa(\mathtt{if}\ E\ \mathtt{then}\ T_1\ \mathtt{else}\ T_2)(e)(s)$$
$$= \mathrm{cons}_\kappa(\langle s \mid [\mathtt{if}\ E\ \mathtt{then}\ T_1\ \mathtt{else}\ T_2]\rangle,$$
$$\mathrm{wrap}^{(E)}_\kappa\, s \left\{\begin{array}{ll} \mathcal{S}_\kappa(T_1)(e)(S) & \text{if } \mathcal{E}_\kappa(E)(s) = tt \\ \mathcal{S}_\kappa(T_2)(e)(S) & \text{otherwise} \end{array}\right\})$$

$$\mathcal{S}_\kappa(\mathtt{while}\ E\ \mathtt{do}\ T)(e) = \mathrm{fix}(\mathcal{W}^{E,T}_\kappa(\mathcal{S}_\kappa(T)(e)))$$

$$\mathcal{S}_\kappa(\mathtt{call}\ P(E_1,\ \ldots,\ E_n))(e)(s)$$
$$= \mathrm{cons}_\kappa(\langle s \mid [\mathtt{call}\ P(E_1,\ \ldots,\ E_n)]\rangle,$$
$$\mathrm{map}_\kappa(\lambda\langle u \mid p\rangle.\,\langle u \mid \varepsilon \mathbin{\bar{:}} p\rangle)(\mathrm{wrap}^{(E_1,\ldots,E_n)}_\kappa\, s(e(P)(\mathcal{E}_\kappa(E_1)(s),\ldots,\mathcal{E}_\kappa(E_n)(s)))(s)))$$

$$\mathcal{S}_\kappa(T_1\,;\,T_2)(e) = \mathcal{S}_\kappa(T_1)(e)\,;\,\mathrm{lift}_\kappa(\mathcal{S}_\kappa(T_2)(e))$$

$$\mathcal{S}_\kappa(\varepsilon)(e)(s) = \mathrm{single}_\kappa\langle s \mid [\varepsilon]\rangle$$

$$\mathcal{C}_\kappa(\mathtt{proc}\ P(X_1,\ \ldots,\ X_n)\ \mathtt{is}\ S)(e)$$
$$= e[P \mapsto \lambda(v_1,\ldots,v_n).\,\lambda s.\,\mathrm{snoc}_\kappa(t, \langle \mathrm{st}(\mathrm{dh}_\kappa\, t)[\forall i\, X_i \mapsto s(X_i)] \mid \mathrm{nil}\rangle)]$$
$$(\text{where } t = \mathcal{S}_\kappa(S)(e)(s[\forall i\, X_i \mapsto v_i]))$$

$$\mathcal{C}_\kappa(C_1\,;\,C_2) = \mathcal{C}_\kappa(C_1)\,;\,\mathcal{C}_\kappa(C_2)$$

$$\mathcal{C}_\kappa(\varepsilon) = \mathrm{id}$$

$$\mathcal{D}_\kappa(D) = \mathrm{fix}\,\mathcal{P}^D_\kappa$$

Fig. 4. Parametric specifications common to all trace semantics of **Proc**

$$Trace_\sigma = \text{List}\,Conf \setminus \{\text{nil}\}$$

$$\text{single}_\sigma\, c = [c] \qquad \text{hd}_\sigma\, l = \text{hd}\, l$$

$$\text{cons}_\sigma(c,l) = \begin{Bmatrix} c : l & \text{if } c \not\sim \bot \\ [c] & \text{otherwise} \end{Bmatrix} \qquad \text{dh}_\sigma\, l = \begin{Bmatrix} d_n & \text{if } l = [d_1,\ldots,d_n] \\ \bot & \text{otherwise} \end{Bmatrix}$$

$$\text{snoc}_\sigma(l,c) = \begin{Bmatrix} [d_1,\ldots,d_n,c] & \text{if } l = [d_1,\ldots,d_n],\ d_n \not\sim \bot \\ l & \text{otherwise} \end{Bmatrix}$$

$$\text{brz}_\sigma(l_1,l_2) = \begin{Bmatrix} \begin{Bmatrix} d_1 : \ldots : d_{n-1} : l_2 & \text{if } \text{dh}_\sigma\, l_1 = \text{hd}_\sigma\, l_2 \\ \bot & \text{otherwise} \end{Bmatrix} & \text{if } l_1 = [d_1,\ldots,d_n],\ d_n \not\sim \bot \\ l_1 & \text{otherwise} \end{Bmatrix}$$

$$\text{map}_\sigma\, f(d_1 : \ldots : d_n : l) = f(d_1) : \ldots : f(d_n) : l \text{ if } l \in \{\text{nil},\bot\}$$

$$\text{map}_\sigma\, f[d_1,d_2,\ldots] = [f(d_1), f(d_2),\ldots]$$

Fig. 5. Underlying elements of standard trace semantics

$$Trace_\varphi = \{l \in [0;1] \dashrightarrow Conf \mid 0 \in \text{dom}\, l \ \wedge\ ((\exists r > 0\,(r \in \text{dom}\, l)) \Rightarrow\ 1 \in \text{dom}\, l)\}$$

$$\text{single}_\varphi\, c = (0 \mapsto c) \quad \text{cons}_\varphi(c,l) = (0 \mapsto c) \cup \text{right}\, l \quad \text{hd}_\varphi\, l = l(0)$$

$$\text{snoc}_\varphi(l,c) = \text{left}\, l \cup (1 \mapsto c) \qquad \text{dh}_\varphi\, l = \begin{Bmatrix} l(1) & \text{if } 1 \in \text{dom}\, l \\ l(0) & \text{otherwise} \end{Bmatrix}$$

$$\text{brz}_\varphi(l_1,l_2) = \begin{Bmatrix} \text{left}\, l_1 \cup \text{right}\, l_2 & \text{if } 1 \in l_1 \ \wedge\ 1 \in l_2 \\ l_1 \cup \text{right}\, l_2 & \text{if } 1 \in l_1 \ \wedge\ 1 \notin l_2 \\ l_1 \cup l_2 & \text{otherwise} \end{Bmatrix} \qquad \text{map}_\varphi\, f\, l = l\, ;\, f$$

Fig. 6. Underlying elements of fractional trace semantics

A partial function $l \in [0;1] \dashrightarrow Conf$ can be envisioned as a trace as follows: for every $r \in \text{dom}\, l$, write $l(r)$ at point r on the number axis, and read the configurations from left to right.

The condition imposed on functions belonging to $Trace_\varphi$ tells that every trace has a first component at point 0 and, whenever there are more components, the trace has a last component at point 1. Obviously, if the trace consists of one component, it is its last component, so, consequently, every trace has a last component. It is useful to define also generalized traces where the endpoints are not necessarily at 0 or 1. For all $a, b \in \mathbb{Q}$ such that $a \leqslant b$, denote $Trace_\varphi^{a,b} = \{f \in [a;b] \dashrightarrow Conf \mid a, b \in \text{dom}\, f\}$. So $Trace_\varphi = Trace_\varphi^{0,1} \cup Trace_\varphi^{0,0}$.

Expression $(a \mapsto x)$ denotes the partial function in $Trace_\varphi^{a,a}$ whose value on a is x. Operators left, right transform partial functions $f \in [a;b] \dashrightarrow Conf$ with $\text{dom}\, f \neq \varnothing$ according to rules

$$\text{left}\, f = \begin{Bmatrix} (0 \mapsto \text{elem}\, f) & \text{if } |\text{dom}\, f| = 1 \\ (\lambda r.\, 2r)\, ;\, f & \text{otherwise} \end{Bmatrix},$$

$$\text{right}\, f = \begin{Bmatrix} (1 \mapsto \text{elem}\, f) & \text{if } |\text{dom}\, f| = 1 \\ (\lambda r.\, 2r - 1)\, ;\, f & \text{otherwise} \end{Bmatrix}$$

where elem f denotes the only value taken by partial function f with one-element domain. The domains of left f and right f for $f \in Trace_\varphi$ are contained in $[0;\frac{1}{2}]$

and $[\frac{1}{2};1]$, respectively. The idea behind both operators is compressing traces to twice smaller space.

Let $a,b,c,d \in \mathbb{Q}$, $0 \leqslant a \leqslant b \leqslant c \leqslant d \leqslant 1$ and let $f \in \mathit{Trace}_{\varphi}^{a,b}$, $g \in \mathit{Trace}_{\varphi}^{c,d}$. Assume that $f(b) \sim \bot$ or $b \neq c$ or $f(b) = g(c)$. Then define $f \cup g \in \mathit{Trace}_{\varphi}^{a,d} \cup \mathit{Trace}_{\varphi}^{a,a}$ as follows:

$$\mathrm{dom}(f \cup g) = \left\{ \begin{array}{ll} \mathrm{dom}\, f \cup \mathrm{dom}\, g & \text{if } f(b) \not\sim \bot \\ (\mathrm{dom}\, f \setminus \{b\}) \cup \{d\} & \text{if } f(b) \sim \bot \wedge a \neq b \\ \{a\} & \text{otherwise} \end{array} \right\} \subseteq \mathrm{dom}\, f \cup \mathrm{dom}\, g \ ,$$

$$\forall r \in \mathrm{dom}(f \cup g) \left((f \cup g)(r) = \left\{ \begin{array}{ll} f(r) & \text{if } r \in \mathrm{dom}\, f \\ g(r) & \text{if } r \notin \mathrm{dom}\, f \wedge f(b) \not\sim \bot \\ f(b) & \text{otherwise} \end{array} \right\} \right) .$$

The idea is to join partial functions as sets of argument-value pairs if f is unfailing and rearrange the values of f to possibly wider domain $[a;d]$ if f is failing.

Two examples illustrating fractional semantics were given already in Sect. 1. To demonstrate the behaviour of strict fractional semantics in the case of errors, replace the second assignment in the first example of Sect. 1 with `x := 1 / 0`. The fractional execution trace on initial state $(\mathtt{x} \mapsto 1, \mathtt{y} \mapsto 2, \mathtt{z} \mapsto 0)$ is as follows:

$$\begin{array}{l} 0 \mapsto \langle(\mathtt{x} \mapsto 1, \mathtt{y} \mapsto 2, \mathtt{z} \mapsto 0) \mid [\mathtt{z\ :=\ x\ ;(x\ :=\ 1\ /\ 0\ ;y\ :=\ z)}]\rangle \ , \\ \frac{1}{2} \mapsto \langle(\mathtt{x} \mapsto 1, \mathtt{y} \mapsto 2, \mathtt{z} \mapsto 1) \mid [\mathtt{x\ :=\ 1\ /\ 0\ ;y\ :=\ z}]\rangle \ , \\ 1 \mapsto \langle(\mathtt{x} \mapsto 1, \mathtt{y} \mapsto 2, \mathtt{z} \mapsto 1) \mid \bot\rangle \ . \end{array}$$

As the second assignment fails, it leads the execution to program point $\bot$. The configuration after this assignment is indexed by 1 rather than $\frac{3}{4}$ since, according to the definition of $\cup$, final erroneous configurations are transferred to the rightmost point of the second argument trace.

Correctness of the definition of the semantics is not clear before specifying precisely the meaning of fix in two places in Fig. 4. The intention is that they should stand for fixpoint operators. We return to this in the following sections.

Another concern is the well-definedness of $\cup$ for all its uses. For that, we must restrict the assortment of actions.

Definition 3.1. *If* $G = \{g \in \mathit{Act}_{\varphi} \mid \mathrm{st}(\mathrm{hd}_{\varphi}(g(s))) = s \wedge |\mathrm{pp}(\mathrm{hd}_{\varphi}(g(s)))| = 1\}$, *denote*

$$\begin{array}{rl} \mathit{Act}_{\varphi}^{+} &= \{g \in G \mid \mathrm{dh}_{\varphi}(g(s)) \not\sim \bot \Rightarrow \mathrm{pp}(\mathrm{dh}_{\varphi}(g(s))) = [\varepsilon]\} \ , \\ \mathit{Env}_{\varphi}^{\circ} &= \mathit{Proc} \to (\mathit{Val}^{*} \to \{g \in G \mid \mathrm{dh}_{\varphi}(g(s)) \not\sim \bot \Rightarrow \mathrm{pp}(\mathrm{dh}_{\varphi}(g(s))) = \mathrm{nil}\}) \ . \end{array}$$

In the whole definition of fractional semantics, we may take $\mathit{Act}_{\varphi}^{+}$ at place of Act_{φ} and $\mathit{Env}_{\varphi}^{\circ}$ at place of Env_{φ}. The only suspicion about $\cup$ possible being used incorrectly can be when they arise due to uses of lift_{φ} since, in other cases, the domains of the arguments of $\cup$ do not intersect. The following proposition dismisses all doubt.

Proposition 3.2. *If* $g, h \in \mathit{Act}_{\varphi}^{+}$ *and* $s \in \mathit{State}$ *then* $\mathrm{lift}_{\varphi}\, g(h(s))$ *is well-defined.*

$$\mathit{Trace}_\delta = \mathit{Conf}$$

$$\text{single}_\delta\, c = c \quad \text{cons}_\delta(c,d) = \text{snoc}_\delta(c,d) = \text{brz}_\delta(c,d) = \left\{ \begin{matrix} d & \text{if } c \not\sim \bot \\ c & \text{otherwise} \end{matrix} \right\}$$
$$\text{hd}_\delta\, c = \text{dh}_\delta\, c = c \quad \text{map}_\delta\, f\, c = f(c)$$

Fig. 7. Underlying elements of denotational degenerate trace semantics

Proof. Let

$$t = \text{st}(\text{dh}_\varphi(\hbar(s))) \ , \quad k = \text{map}(\lambda\langle u \mid p\rangle.\, \langle u \mid p\, ;\text{hd}(\text{pp}(\text{hd}_\varphi(g(t))))\rangle)(\hbar(s)) \ .$$

It suffices to show that $\text{dh}_\varphi\, k = \text{hd}_\varphi(g(t))$ whenever $\text{dh}_\varphi\, k \not\sim \bot$. We have

$$\begin{aligned} \bot \not\sim \text{dh}_\varphi\, k &= (\lambda\langle u \mid p\rangle.\, \langle u \mid p\, ; \text{hd}(\text{pp}(\text{hd}_\varphi(g(t))))\rangle)(\text{dh}_\varphi(\hbar(s))) \\ &= \langle t \mid \text{pp}(\text{dh}_\varphi(\hbar(s)))\, ; \text{hd}(\text{pp}(\text{hd}_\varphi(g(t))))\rangle \ . \end{aligned}$$

Thus $t \neq \bot$ and $\text{dh}_\varphi(\hbar(s)) \not\sim \bot$. By $g \in \mathit{Act}^+_\varphi$, we have $\text{pp}(\text{hd}_\varphi(g(t))) = [S]$ for some $S \in \mathit{Stmt}$. By $\hbar \in \mathit{Act}^+_\varphi$, we get

$$\text{pp}(\text{dh}_\varphi\, k) = \text{pp}(\text{dh}_\varphi(\hbar(s)))\, ; \text{hd}(\text{pp}(\text{hd}_\varphi(g(t)))) = [\varepsilon]\, ; S = [S] \ .$$

Therefore $\text{dh}_\varphi(k) = \langle t \mid [S]\rangle = \text{hd}_\varphi(g(t))$. □

Lastly, it is worth to note that if we define the items of Fig. 3 as shown in Fig. 7 then we in principle obtain the denotational semantics schema in Fig. 2 with error handling added. The difference that the original denotational semantics specifies final states while the semantics obtained from Figures 4 and 7 gives final configurations with trivial program point is purely formal.

4 Fractional Semantics of While-Loop

In this section, we establish a way to define fix for fractional semantics of while-loop. Due to limited space, we present proof sketches only.

A preliminary study of the definition of $\mathcal{W}^{E,T}_\varphi$ would show that, for arbitrary fixpoints x, y of $\mathcal{W}^{E,T}_\varphi(\hbar)$ and state s, traces $x(s)$ and $y(s)$ coincide everywhere except for possibly at point 1. So there is not much freedom anyway.

A classical way to construct fixpoints for semantics is taking the least upper bound of monotonic chains w.r.t. some ordering where the monotonic chains are constructed by iteration of a continuous operator on some trivial value. We nearly follow this schema; we have abstractly stabilizing sequences at place of monotonic chains and a limit operator defined via stabilization at place of least upper bound. Both notions are defined below.

Definition 4.1. *Call a sequence* $(c_i : i \in \mathbb{N})$ stabilizing to v *iff* $\forall i \geqslant n\, (c_i = v)$ *for some* $n \in \mathbb{N}$. *Call a sequence* stabilizing *iff it is stabilizing to some* v.

For every set $C \ni \bot$, let σ_C be defined by $\sigma_C(c) = tt$ for all $c \in C \setminus \{\bot\}$ and $\sigma_C(\bot) = ff$.

Definition 4.2.

(i) *Call a function* $f \in C \to D$ *an* abstract constant *iff* $f \,;\, \sigma_D$ *is a constant function. Call* f *an* abstract identity *iff* $f \,;\, \sigma_D = \sigma_C$.

(ii) *Depending on the construction of the domain containing the elements of sequence, define* abstract stabilization *as follows:*

1. *for* $(v_i : i \in \mathbb{N})$ *with* $v_i \in \mathit{Val}$, *the sequence is abstractly stabilizing if* $(\sigma_{\mathsf{Val}}(v_i) : i \in \mathbb{N})$ *is stabilizing;*
2. *for* $(p_i : i \in \mathbb{N})$ *with* $p_i \in \mathit{PP}$, *abstract stabilization and stabilization coincide;*
3. *if abstract stabilization is defined for domain* B *then, for* $(f_i : i \in \mathbb{N})$ *with* $f_i \in A \to B$, *call the sequence abstractly stabilizing iff either it is stabilizing to* $\bot$ *or it is componentwise abstractly stabilizing. Among others, this applies to sequences with elements from* $\mathsf{State} = \mathsf{Var} \to \mathit{Val}$*;*
4. *if abstract stabilization is defined for domains* A *and* B *then, for* $(p_i : i \in \mathbb{N})$ *with* $p_i \in A \times B$, *call the sequence abstractly stabilizing iff either it is stabilizing to* $\bot$ *or it is componentwise abstractly stabilizing. Among others, this applies to sequences with elements from* $\mathsf{Conf} = \mathsf{State} \times \mathit{PP}$*;*
5. *for* $((a_i, b_i) : i \in \mathbb{N})$ *and* $(f_i : i \in \mathbb{N})$ *where* $f_i \in \mathsf{Trace}_{\varphi}^{a_i, b_i}$, *call the sequence* $(f_i : i \in \mathbb{N})$ *abstractly stabilizing iff* $((a_i, b_i) : i \in \mathbb{N})$ *is stabilizing and, for each* $r \in \mathbb{Q}$, *there exists* $n \in \mathbb{N}$ *such that either* $r \notin \operatorname{dom} f_i$ *for each* $i \geqslant n$ *or* $r \in \operatorname{dom} f_i$ *for each* $i \geqslant n$ *and* $(f_i(r) : i \geqslant n)$ *is abstractly stabilizing; in other words, abstract stabilization is defined componentwise where the bounds* (a_i, b_i) *and the definedness/undefinedness status are considered as components on which abstract stabilization and stabilization coincide.*

Definition 4.3.

(i) *Let* C *be a set where abstract stabilization is defined. Depending on the construction of the domain* C, *define limit operator* lim *on every abstractly stabilizing sequence* $(c_i : i \in \mathbb{N})$ *such that all* $c_i \in C$ *as follows:*

1. *if* $C = \mathit{Val}$ *or* $C = \mathit{PP}$ *then*

$$\lim(c_i : i \in \mathbb{N}) = \begin{cases} d & \text{if } (c_i : i \in \mathbb{N}) \text{ stabilizes to } d \\ \top & \text{otherwise} \end{cases} ;$$

2. *for other domains, define* lim *componentwise together with the clause that if the sequence stabilizes to* $\bot$ *then the limit is also* $\bot$.

(ii) *Let* C, D *be sets where abstract stabilization is defined. A function* $f : C \to D$ *is called* continuous *iff, for every abstractly stabilizing sequence* $(c_i : i \in \mathbb{N})$ *with all* $c_i \in C$, *the sequence* $(f(c_i) : i \in \mathbb{N})$ *is abstractly stabilizing whereby* $f(\lim(c_i : i \in \mathbb{N})) = \lim(f(c_i) : i \in \mathbb{N})$.

The operator lim is well-defined on Val and PP since $\mathit{Val} \ni \top$ and all abstractly stabilizing program point sequences are stabilizing. On State, the definition gives the same limit operator as used in transfinite semantics papers [4, 6, 7].

Define operator fix for trace semantics of while-loop as follows.

Definition 4.4. *Let κ be a kind of trace semantics and $\lambda s.\, s'$ be a fixed state transformation. For $\mathcal{G} \in Act_\kappa \to Act_\kappa$, choose*

$$\mathrm{fix}\,\mathcal{G} = \lim(\mathcal{G}^i(\lambda s.\ \mathrm{cons}_\kappa(\langle s \mid [\varepsilon]\rangle, \mathrm{single}_\kappa\langle s' \mid [\varepsilon]\rangle)) : i \in \mathbb{N})\ .$$

Obviously, fix depends on both κ and $\lambda s.\, s'$. As it depends also on the argument type (there is another fix for operators in $(Env_\kappa \to Env_\kappa) \to (Env_\kappa \to Env_\kappa)$ in Fig. 4) and the conditions under which the definition of fix is sound heavily depend on κ, we preferred to treat fix as an informal denotation hinting at some kind of construction rather than an operator and omitted the indices.

The choice of $\lambda s.\, s'$ starts playing a role when the loop does not terminate. The state into which the computation falls after executing the body for infinitely many times equals to the limit of the $'$-images of states occurring after any finite number of body iterations.

To obtain strict treatment of infinite loops, define $s' = \bot$ for every s. Then the limit is also $\bot$ and the configuration c after infinite run satisfies $c \sim \bot$ which means that this configuration is the last one irrespective of the possibly non-empty rest of code. Therefore a non-terminating loop is never followed by any computation, hence the semantics is essentially standard. But taking $s' = s$ for every s leads to transfinite semantics like the second one given by us in [6].

Definition 4.4 is sound only if the sequence of the iterations is abstractly stabilizing. This is what Theorem 4.5 states for the while-loop case of fractional semantics. Furthermore, it is easy to see that if, in addition, $\mathcal{G}$ is continuous then fix $\mathcal{G}$ is indeed a fixpoint of $\mathcal{G}$.

As $\lambda s.\, \bot$ is an abstract constant and $\lambda s.\, s$ is an abstract identity, $\lambda s.\, s'$ meets the requirements of Theorem 4.5 for both variants considered above.

Theorem 4.5. *Let $E \in Expr$, $T \in Stmt$ and $\hbar \in Act^+_\varphi$. Let $\lambda s.\, s'$ be either an abstract constant or an abstract identity.*

(i) *$\mathcal{W}^{E,T}_\varphi(\hbar)$ is continuous.*

(ii) *The sequence*

$$((\mathcal{W}^{E,T}_\varphi(\hbar))^i(\lambda s.\ \mathrm{cons}_\varphi(\langle s \mid [\varepsilon]\rangle, \mathrm{single}_\varphi\langle s' \mid [\varepsilon]\rangle)) : i \in \mathbb{N})$$

is abstractly stabilizing.

Therefore $\mathrm{fix}(\mathcal{W}^{E,T}_\varphi(\hbar))$ is well-defined.

Proof. Denote $\mathcal{G} = \mathcal{W}^{E,T}_\varphi(\hbar)$ and $a = \lambda s.\ \mathrm{cons}_\varphi(\langle s \mid [\varepsilon]\rangle, \mathrm{single}_\varphi\langle s' \mid [\varepsilon]\rangle)$.

(i) Prove that operators left and right are continuous. Prove that $\cup$ is continuous in its both arguments. Prove that hd_φ is continuous, brz_φ is continuous in its both arguments, and map_φ is continuous. Prove that $\mathrm{wrap}^L_\varphi(s)$ for any $s \in State$ and lift_φ are continuous. Then the claim follows.

(ii) The case $s = \bot$ is simple since $\mathcal{G}^i(a)(\bot) = (0 \mapsto \langle \bot \mid [\texttt{while}\ E\ \texttt{do}\ T]\rangle)$ for $i > 0$. Assume now $s \neq \bot$.

First show that $\mathcal{G}^i(a)(s) \in Trace^{0,1}_\varphi$ for every $i \in \mathbb{N}$. It is easy to see that $0 \in \mathrm{dom}(\mathcal{G}^i(a)(s))$; for $1 \in \mathrm{dom}(\mathcal{G}^i(a)(s))$, it requires a little case-study.

It remains to show that $(\mathcal{G}^i(a)(s) : i \in \mathbb{N})$ is componentwise abstractly stabilizing. This is done in two parts: the first is to prove it for all rationals in $[0;1]$ except for 1, the second is proving the abstract stabilization of $(\mathcal{G}^i(a)(s)(1) : i \in \mathbb{N})$.

For the former, take arbitrary $r \in [0;1] \setminus \{1\}$, choose $n \in \mathbb{N}$ so that $1 - \frac{1}{2^n} \leqslant r < 1 - \frac{1}{2^{n+1}}$ (there is precisely one such n), and argue by induction on n.

Proving the abstract stabilization of $(\mathcal{G}^i(a)(s)(1) : i \in \mathbb{N})$ goes in two parts again. If there exists an $n \in \mathbb{N}$ such that either $\mathrm{dh}_\varphi(\hbar((\hbar\,;\,\mathrm{dh}_\varphi\,;\,\mathrm{st})^n(s))) \sim \perp$ or $\mathcal{E}_\varphi(E)((\hbar\,;\,\mathrm{dh}_\varphi\,;\,\mathrm{st})^n(s)) \neq tt$ then argue by induction on n. Otherwise, prove by induction on n that $\mathcal{G}^n(a)(s)(1) = \langle((\hbar\,;\,\mathrm{dh}_\varphi\,;\,\mathrm{st})^n(s))' \mid [\varepsilon]\rangle$ and use the assumption about $\lambda s.\,s'$. □

Corollary 4.6. *For every $S \in \mathit{Stmt}$ and $e \in \mathit{Env}^\circ_\varphi$, Figures 4 and 6 soundly specify $\boldsymbol{\mathcal{S}}_\varphi(S)(e) \in \mathit{Act}^+_\varphi$.*

Proof. By induction on the structure of S. □

We already mentioned that taking $s' = s$ for every $s \in \mathit{State}$ leads to a transfinite semantics observed in [6]. Stating this connection in a precise form would be cumbersome since the syntax of the language was a bit different there. Anyway, the following holds; it implies that $\boldsymbol{\mathcal{S}}_\varphi$ with lazy φ is in principle transfinite semantics provided no illegal orderings are introduced by environment.

Theorem 4.7. *Let $e \in \mathit{Env}^\circ_\varphi$ such that $\mathrm{dom}(e(P)(v_1,\ldots,v_n)(s))$ is well-ordered for every procedure identifier P, values $v_1,\ldots,v_n$, and state s. Define fix as in Definition 4.4 where $\lambda s.\,s' = \mathrm{id}$. Then $\mathrm{dom}(\boldsymbol{\mathcal{S}}_\varphi(S)(e)(s))$ is well-ordered for every $S \in \mathit{Stmt}$ and $s \in \mathit{State}$ whereby the type of the well-order is a successor ordinal.*

Proof. By induction on the structure of S, while-loop being the main case. □

For example, the components of the execution trace of **while** `true` **do** ε are numbered by ordinals $0, 1, 2, 3, \ldots$ and ω in transfinite semantics and by $0, \frac{1}{2}, \frac{3}{4}, \frac{7}{8}, \ldots$ and 1 in lazy fractional semantics; the components of double infinite loop **while** `true` **do** **while** `true` **do** ε are indexed by ordinals from 0 to ω^2 in transfinite semantics and as shown in Fig. 8 in lazy fractional semantics.

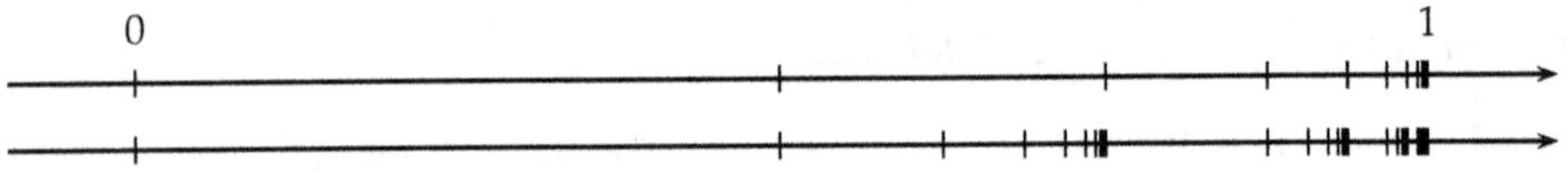

Fig. 8. The domains of lazy fractional semantics of statements **while** `true` **do** ε (above) and **while** `true` **do** **while** `true` **do** ε

5 Fractional Semantics for Recursion

To handle recursion lazily, unloading infinitely deep recursion must be enabled. This involves chains with no least element, so transfinite semantics do not qualify.

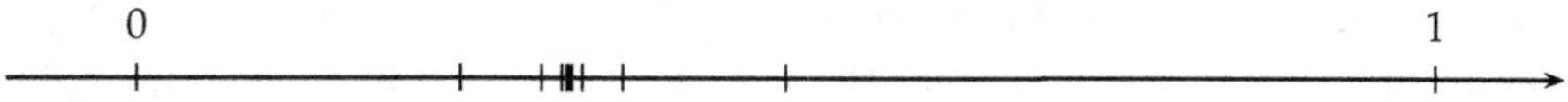

Fig. 9. The domain of lazy fractional semantics of p in context of declaration (1)

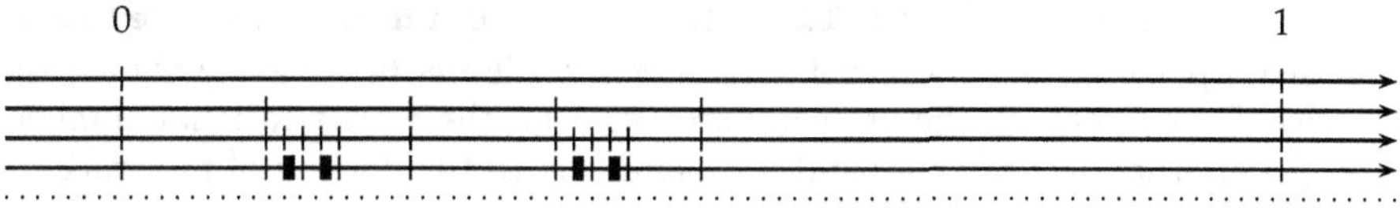

Fig. 10. Construction of the domain of lazy fractional semantics of q in context of declaration (2)

The simplest example is obtained by declaration

$$\textbf{proc}\ \texttt{p()}\ \textbf{is}\ \textbf{call}\ \texttt{p()}\ . \tag{1}$$

Originating from the assumption that $\text{fix}\, \mathcal{P}_{\varphi}^{D}$ is a fixpoint of $\mathcal{P}_{\varphi}^{D}$ in Fig. 4, we obtain the lazy trace depicted in Fig. 9. Two infinite sequences — one ascending and another descending — are both converging to $\frac{1}{3}$.

The trace domain of the fixpoint in the case of declaration

$$\textbf{proc}\ \texttt{q()}\ \textbf{is}\ (\textbf{call}\ \texttt{q()}\,;\textbf{call}\ \texttt{q()}) \tag{2}$$

is built step-by-step in Fig. 10. Each step adds twice more points than the previous since the number of uninterpreted calls doubles every level. The fixpoint domain forms a fractal structure. A rational number between 0 and 1 belongs to it iff its octal representation is finite and each its digit after octal point is either 1 or 3 except for the last one which can be also 2 or 4. The set of all possible limits of converging sequences of rationals in this set is uncountable.

The next interesting issue is the way how to define states of variables for configurations on such traces. Constructing the fixpoint similarly to the case of while-loop which seems a natural approach results in the following definition.

Definition 5.1. *Let κ be a kind of trace semantics and $\lambda s.\, s'$ be a fixed state transformation. For $\mathcal{G} \in (\mathit{Env}_{\kappa} \to \mathit{Env}_{\kappa}) \to (\mathit{Env}_{\kappa} \to \mathit{Env}_{\kappa})$, choose*

$$\text{fix}\, \mathcal{G} = \lim(\mathcal{G}^{i}(\lambda e.\, \lambda P.\, \lambda v_1, \ldots, v_n.\, \lambda s.\ \text{cons}_{\kappa}(\langle s \mid [\varepsilon]\rangle, \text{single}_{\kappa}\langle s' \mid \text{nil}\rangle)) : i \in \mathbb{N})\ .$$

Unfortunately, this approach has severe flaws. We finish the section by pointing them out; a way to define fix appropriately for all cases is yet to be found.

Firstly, $\text{fix}\, \mathcal{P}_{\varphi}^{D}$ in lazy fractional semantics turns out not to be necessarily a fixpoint of $\mathcal{P}_{\varphi}^{D}$. For seeing that, consider declaration

$$D = \textbf{proc}\ \texttt{d()}\ \textbf{is}\ (\texttt{x}\ \textbf{:=}\ \texttt{-x}\,;\textbf{call}\ \texttt{d()}\,;\texttt{z}\ \textbf{:=}\ \texttt{x * x})\ .$$

In the sequence in Definition 5.1, the value of x at $\frac{3}{8}$ (the place of the very last assignment to z) in the case of a fixed initial state is alternating between x

and $-x$ (where x is the initial value of x), hence the value assigned to z at $\frac{3}{8}$ is stabilizing to x^2. In the limit trace therefore x has value $\top$ at $\frac{3}{8}$ but z has value x^2. In the result of applying $\mathcal{P}_{\varphi}^{D}$ to the limit, z is assigned $\top \neq x^2$.

Secondly, the sequence in Definition 5.1 whose limit has to be found is even not necessarily abstractly stabilizing. Namely, the predicate in declaration **proc** c() **is** (x **:=** -x;**call** c();**if** x > 0 **then** z **:=** 0 **else** ε) can take opposite values every two consecutive levels of iteration. Since the lengths of statements in the branches are different, there exist rational numbers (e.g. $\frac{7}{16}$) alternately belonging and not belonging to the domain of the trace.

6 Conclusion

In the paper, we proposed fractional semantics to replace the less powerful transfinite semantics for defining program slicing soundly. We gave a general definition schema for both strict and lazy variants of this semantics, the latter subsuming transfinite semantics. Schemata of both variants differ very little, the fixpoint operator for while-loop has a parameter whose value determines the variant.

In the related papers [4, 6, 7], transfinite semantics may seem a bit ad hoc. Our parametric representation shows artlessness of this choice. It also turned out that the rules producing transfinite semantics in the case of while-loop lead to fractal structures in the case of recursion. However, a precise way to define fixpoints for recursion case was missing, this remains an open problem.

References

1. Binkley, D. W., Gallagher, K. B.: Program Slicing. Advances in Computers **43** (1996) 1–50
2. Cousot, P.: Constructive Design of a Hierarchy of Semantics of a Transition System by Abstract Interpretation. Electronic Notes in Theoretical Computer Science, **6** (1997) 25 pp.
3. Danicic, S., Harman, M., Howroyd, J., Ouarbya, L.: A Lazy Semantics for Program Slicing. In *Proceedings of the 1st International Workshop on Programming Language Interference and Dependence*. Available at `http://profs.sci.univr.it/~mastroen/download/PLID/Proceedings/Proceedings.html` (2004)
4. Giacobazzi, R., Mastroeni, I.: Non-Standard Semantics for Program Slicing. Higher-Order Symbolic Computation, **16** (2003) 297–339
5. Kennaway, R., Klop, J. W., Sleep, R., Vries, F.-J. de.: Transfinite Reductions in Orthogonal Term Rewriting Systems. Information and Computation, **119**(1) (1995) 18–38
6. Nestra, H.: Transfinite Corecursion. Nordic Journal of Computing **12**(2) (2005) 133–156
7. Nestra, H.: Transfinite Semantics in Program Slicing. Proceedings of the Estonian Academy of Sciences. Engineering **11**(4) (2005) 313–328
8. Nielson, H. R., Nielson, F.: Semantics with Applications: A Formal Introduction. Wiley Professional Computing, Wiley (1992)

Reasoning About Data-Parallel Pointer Programs in a Modal Extension of Separation Logic*

Susumu Nishimura

Department of Mathematics, Faculty of Science, Kyoto University
Sakyo-ku, Kyoto 606-8502, Japan
susumu@math.kyoto-u.ac.jp

Abstract. This paper proposes a modal extension of Separation Logic [1, 2] for reasoning about data-parallel programs that manipulate heap allocated linked data structures. Separation Logic provides a formal means for expressing allocation of disjoint substructures, which are to be processed in parallel. A modal operator is also introduced to relate the global property of a parallel operation with the local property of each sequential execution running in parallel. The effectiveness of the logic is demonstrated through a formal reasoning on the parallel list scan algorithm featuring the pointer jumping technique.

1 Introduction

Parallel prefix or *scan* on arrays is a fundamental collective operation in parallel computing [3, 4]. For example, the parallel prefix sum algorithm computes the sums of prefix subsequences of an integer array. The prefix computation can be efficiently implemented on parallel computers, where the basic technique is simultaneous addition of array elements at indices of exponentially increasing intervals. The same technique applies to implement a range of parallel algorithms including parallel sorting, maximum segment sum, etc. A *data-parallel* programming paradigm is best suited for the implementation, where the same sequential program processes every different array element simultaneously, as attributed by the SPMD (single program, multiple data-stream) execution scheme [5].

A similar but more sophisticated programming technique can even implement parallel collective operations on linked data structures, *e.g.*, lists and trees. The program in Figure 1 implements a data-parallel scan operation on integer list that computes the sum of every sublist, where the integer list is expressed by a linked structure. Each cons cell allocates an integer value in the head position and a pointer to the successor cell (or a special value nil when there is no successor cell) in the tail position.

The programming technique employed in the program is called *pointer jumping* [3, 6]. As depicted in Figure 1, in each step of parallel execution, the tail position in every cons cell is updated with the value contained in the tail position of the successor cell and also the integer value in the head position is added with that in the successor

* A preliminary short summary of this paper was presented at the 3rd Workshop on Semantics, Program Analysis, and Computing Environments for Memory Management (SPACE 2006), Charleston, SC, January 2006.

M. Johnson and V. Vene (Eds.): AMAST 2006, LNCS 4019, pp. 293–307, 2006.

```
(* p is a pointer to the initial cell of the list *)
q := [p+1];
while q ≠ nil do
  begin
    forall x ↦ n,t in ⟨allocation addresses of list cells⟩ do
      begin var m;
        if t ≠ nil then m := [t]; t := [t+1]; [x] := n+m; [x+1] := t
      end;
    q := [p+1]
  end
```

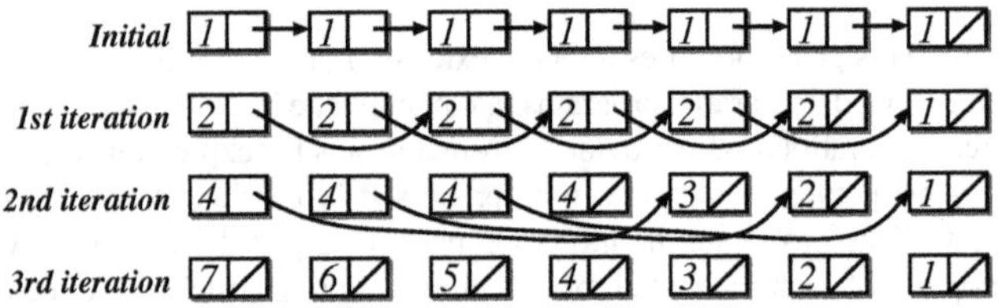

Fig. 1. Data-parallel list scan algorithm

cell. Every single step of parallel execution is expressed in the program by a data-parallel primitive, the **forall** command. In the program, the **forall** command executes the command body for every different cons cell in parallel, with x bound to the allocation address of the cons cell and n and t bound to values stored in the head part and the tail part of the cell, respectively. In the command body, $[p]$ stands for dereferencing of a pointer p, $[p+1]$ for dereferencing with displacement 1, and the command $[e] := e'$ updates the address e with the value of e'. The iteration is terminated as soon as the tail position of the initial cell has been set to nil. It is easy to see that the iteration requires only a logarithmic number of steps, since the length of pointers is doubled per each iteration.

This paper proposes a program logic, in the style of Hoare, for verifying such data-parallel programs implementing a collective operation on linked data structure, on top of a formal semantics of a suitable data-parallel programming language.

So far several formal semantics have been proposed for data-parallel programming languages (*e.g.*, [7, 8, 9]). The most prospective for our purpose would be the assertional approach by Bougé et al. [7]. However, they only deal with arrays as the primary data structure for parallel processing. Thus pointer jumping is expressed with indirection of interpreting pointers as array indices. This is not merely a notational issue. More significantly, it obfuscates the logical structure of the program, by having the formal reasoning stick to particular properties of integer arithmetics.

Our formal proof system for verification deals with a data-parallel programming language in which pointers are first-class citizens. Pointer operations are notoriously hard to reason about, even in sequential programs. We solve the difficulty by adopting Separation Logic [1, 2], which has been recently developed for compositional reasoning on pointer manipulating programs.

As we shall discuss later, parallel list scan instantiates the divide-and-conquer strategy, where a data set is decomposed into disjoint subcomponents that are subject to successive parallel processing. The parallel processing of subcomponents is guaranteed safe, as disjointness implies non-interference among parallel threads of sequential execution. Separation Logic allows us to express this property of the program in a notably elegant way.

In addition to the standard features of Separation Logic, we extend it with a modal operator, expressing modal possibility up to alteration of the heap contents. (Although our modal operator is much alike Berger, Honda, and Yoshida's content quantification [10], ours has a fundamental difference from theirs: theirs takes care of a single address, while ours mentions about the entire set of allocation addresses of the current heap.) The modal operator provides a logical means for relating the global property of the execution of a **forall** command with the property local to every sequential execution running in parallel.

The rest of the paper is organized as follows. Section 2 introduces Separation Logic with a modal extension. Section 3 informally explains the logical structure of parallel list scan algorithm and interprets it into a formal specification in Separation Logic. In Section 4, we give a formal definition of the data-parallel programming language and a sound proof system for it. Section 5 gives a formal correctness proof for the parallel list scan. Finally, Section 6 concludes the paper.

2 The Assertion Language of Separation Logic with Modality

This section gives the formal definition of the syntax and semantics of the assertion language of Separation Logic extended with modality. Separation logic extends first-order logic with assertions for mentioning about heap storage. We assume a shared memory model for the heap storage, where a single global memory space is shared among all parallel execution instances.

Given a partial function f, let us write $f[x_1 \mapsto v_1, \ldots, x_n \mapsto v_n]$ for a partial function g such that $dom(g) = dom(f) \cup \{x_1, \ldots, x_n\}$, $g(x_i) = v_i$ for every i, and $g(y) = f(y)$ for every $y \in dom(f) \setminus \{x_1, \ldots, x_n\}$. We write $f \sharp g$ to mean that f and g have disjoint domains. We also write $f * g$ for the disjoint union of the two functions, if $f \sharp g$; $f * g$ is undefined otherwise. Sometimes we represent a partial function f by its *graph* $\{(x, f(x)) \mid x \in dom(f)\}$. The notation $f \restriction A$ expresses a restriction of f to a set A, *i.e.*, $f \restriction A = \{(x, f(x)) \mid x \in dom(f) \cap A\}$.

Let *Var* be a set of of variables. We define the set *Val* of values as the set of integers and the set *Addr* of addresses as the set of non-negative integers, hence $Addr \subseteq Val$.

Our formal model of storage consists of two semantic domains, *store* and *heap*:

$$Store = Var \to Val \qquad Heap = Addr \to_{fin} Val \qquad State = Store \times Heap.$$

A store is a total mapping from variables to values. A heap is a finite partial mapping that associates each address with its content. The state of the entire storage is represented by a pair of a store and a heap.

The syntax of the assertion language is given below.

$$e ::= \langle integers \rangle \mid \mathsf{nil} \mid x \mid e + e \qquad \text{(expressions)}$$

$$\begin{aligned} P ::= &\ \mathbf{true} \mid \mathbf{false} \mid e = e \mid \neg P \mid P \vee P \mid P \wedge P \mid \exists x.P \mid \forall x.P \mid P \Rightarrow P \\ &\mid \mathbf{emp} \mid e \mapsto e \mid P * P \mid P -\!\!* P \mid \Diamond P \end{aligned} \qquad \text{(assertions)}$$

An expression e is either an integer value, a constant symbol nil, a variable, or a sum of integers. We assume the symbol nil stands for a non-address value. In addition to the standard connectives of first-order logic, the assertion language provides several separating connectives and a modality. We write $FV(P)$ for the set of free variables, *i.e.*, variables whose occurrence in P is not in the scope of any enclosing quantifier.

Given $(s,h) \in State$, the interpretation of assertions is defined as below. Let us write $[\![e]\!]s$ to denote the value of expression e interpreted under the store s. We define the judgment $s,h \models P$ as follows. (The interpretation of the remaining connectives is standard and is omitted for the lack of space.)

$s,h \models \mathbf{emp}$	iff $dom(h) = \emptyset$.
$s,h \models e \mapsto e'$	iff $[\![e]\!]s \in Addr$ and $h = \{([\![e]\!]s, [\![e']\!]s)\}$.
$s,h \models P * Q$	iff $h = h_1 * h_2$, $s,h_1 \models P$, and $s,h_2 \models Q$ for some h_1, h_2.
$s,h \models P \mathrel{-\!\!*} Q$	iff $h \sharp h'$ and $s,h' \models P$ implies $s,h*h' \models Q$, for all h'.
$s,h \models \Diamond P$	iff $dom(h) = dom(h')$ and $s,h' \models P$ for some h'.

The assertion **emp** indicates an empty heap that allocates no contents yet. Points-to relation $e \mapsto e'$ indicates a singleton heap, which allocates a single content e' at the address e. Separating conjunction $P * Q$ holds for the current heap h iff there exists a disjoint separation of the heap $h = h_1 * h_2$ such that P holds for h_1 and Q holds for h_2. Separating implication $P \mathrel{-\!\!*} Q$ says that Q holds up to any expansion of the current heap that satisfies P. The modal operator is new to this paper. The assertion $\Diamond P$ means that the assertion P can be made true by appropriately changing currently allocated values.

Here we introduce some notational conventions. We write $e \mapsto e_1, e_2, \ldots, e_n$ for $(e \mapsto e_1) * (e+1 \mapsto e_2) * \cdots * (e+n-1 \mapsto e_n)$ $(n \geq 1)$, namely, a block of size n allocating values $e_1, \ldots, e_n$ at consecutive addresses starting from e. Inexact variant of points-to relation $e \hookrightarrow e_1, \ldots, e_n$ abbreviates $e \mapsto e_1, \ldots, e_n * \mathbf{true}$, which indicates that the current heap *at least* contains the allocation as expressed by $e \mapsto e_1, \ldots, e_n$. A symbol $-$ in the right hand side of $\mapsto$ or $\hookrightarrow$ stands for an existentially quantified variable, *e.g.*, $x \mapsto 1, -, -$ for $\exists yz.x \mapsto 1, y, z$. Throughout the paper, we assume that $*$ binds more tightly than $\wedge$; we follow the usual convention on the precedence of bindings for other connectives.

We also consider the following subclasses of assertions.

- An assertion P is called *pure* if it is independent to heap, that is, $s,h \models P$ implies $s,h' \models P$ for any h'. Syntactically, any assertion is pure if it is free from assertions and connectives that are affected by heap, *i.e.*, **emp**, $*$, $\mathrel{-\!\!*}$, $\mapsto$, and $\Diamond$.
- An assertion P is called *precise* if, for any store s and heap h, there exists at most one subheap h' such that $h' \subseteq h$ and $s,h' \models P$. Assertions which are built from logical expressions only using **emp**, $*$, $\Diamond$, $e \mapsto e'$, and $e \mapsto -$ form a conservative subclass of precise assertions.
- An assertion P is called *strictly exact* if P determines at most one heap, that is, for any store s and heaps h, h', $s,h \models P$ and $s,h' \models P$ implies $h = h'$. Any assertion that is built from logical expressions only using **emp**, $*$, and $e \mapsto e'$ is strictly exact.
  Obviously any strictly exact assertion is precise.

Some logical properties of the assertion language will be given later in Section 5.

3 Specifying Parallel List Scan

A large body of parallel algorithms can be explained as an instance of the divide-and-conquer strategy, where a problem is divided into subproblems of smaller sizes and solutions to the subproblems later combine to give the final solution. The strategy merits parallel implementation, as disjoint subproblems can be safely solved in parallel.

3.1 An Informal Description of the Correctness of Parallel Scan

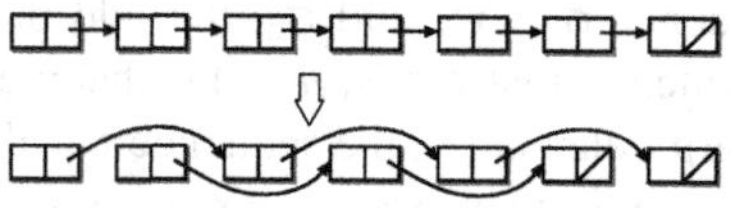
Fig. 2a. Odd-even list partitioning

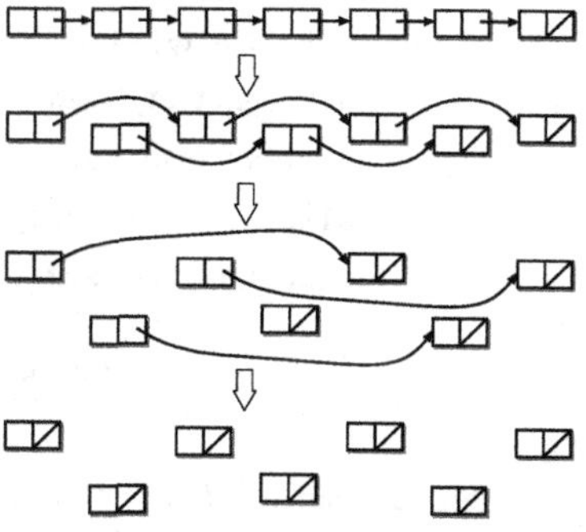
Fig. 2b. Iterated partitioning

Let us show that the data-parallel list scan algorithm is another instance of divide-and conquer. Figure 2a gives a graphical presentation of a single step of pointer jumping on a linked list. The figure indicates that a single step of pointer jumping corresponds to an odd-even partitioning of the list: the parallel operation splits the list into two disjoint sublists, one consisting of cells sitting at odd positions of the original list and the other consisting of cells sitting at even positions. Successive execution of parallel pointer jumping simultaneously operates on the partitioned sublists, further splitting each sublist into two disjoint smaller sublists. The iteration continues until the original list is decomposed into a set of singleton lists (Figure 2b).

Note that the iteration respects the following loop invariant: "Each iteration preserves the sum of integers reachable from every cons cell." This invariant implies that, when the program terminates, every cons cell holds the sum of all the integer values reachable from that cell in the original list. Thus, we obtain the list scan.

3.2 Formal Specification in Separation Logic

In the rest of this section, we express a formal specification of the properties discussed above in Separation Logic. We give a *specification* of program in Hoare's partial correctness assertion form, written $\{P\}\,C\{Q\}$, which means that, for any state satisfying the precondition P, the program C safely executes without errors such as memory faults and, if it ever terminates, it ends up with a final state satisfying the postcondition Q.

Let us write $[]$ to stand for an empty sequence of values and $a :: \ell$ for a sequence beginning with a value a followed by a sequence ℓ. We also abbreviate $a_1 :: (a_2 :: (\cdots :: [])\cdots)$ by $[a_1, a_2, \ldots, a_n]$. In what follows, we use meta-variables $\tau, \tau', \ldots$ to range over sequences of integers and $\sigma, \sigma', \ldots$ over sequences of addresses.

Let us define a few predicates for sequences.

- $\mathtt{part}(\sigma,\sigma_1,\sigma_2)$ iff either $\sigma = \sigma_1 = \sigma_2 = []$ or $\sigma = [p_1, p_2, \ldots, p_k]$, $\sigma_1 = [p_1, p_3, \ldots, p_{2\lfloor\frac{k-1}{2}\rfloor+1}]$, and $\sigma_2 = [p_2, p_4, \ldots, p_{2\lfloor\frac{k}{2}\rfloor}]$ $(k \geq 1)$.
- $\mathtt{alts}(\tau,\tau_1,\tau_2)$ iff either $\tau = \tau_1 = \tau_2 = []$ or $\tau = [n_1, n_2, \ldots, n_k]$, $\tau_1 = [\rho(1), \rho(3), \ldots, \rho(2\lfloor\frac{k-1}{2}\rfloor+1)]$, and $\tau_2 = [\rho(2), \rho(4), \ldots, \rho(2\lfloor\frac{k}{2}\rfloor)]$, where $\rho(k) = n_k$ and $\rho(i) = n_i + n_{i+1}$ when $0 \leq i \leq k-1$ $(k \geq 1)$.
- $\mathtt{sum}(\tau, m)$ iff $\tau = [n_1, n_2, \ldots, n_k]$ and $m = \sum_{i=1}^{k} n_i$ $(k \geq 0)$.

The predicate $\mathtt{part}(\sigma,\sigma_1,\sigma_2)$ gives the result of odd-even partitioning of the sequence σ in σ_1 and σ_2, the odd part and the even part, respectively; $\mathtt{alts}(\tau,\tau_1,\tau_2)$ indicates that τ_1 (τ_2, *resp.*) is the integer sequence obtained by adding every odd (even, *resp.*) element in the sequence τ with its successor element; $\mathtt{sum}(\tau,n)$ gives the sum of integer sequence τ in n. Figure 3 lists some properties relevant to these predicates.

$$\mathtt{part}(\sigma,[],\sigma_2) \Leftrightarrow \sigma = \sigma_2 = [] \tag{3.1}$$
$$\mathtt{alts}(\tau,[],\tau_2) \Leftrightarrow \tau = \tau_2 = [] \tag{3.2}$$
$$\mathtt{alts}([n],\tau_1,\tau_2) \Leftrightarrow \tau_1 = [n] \wedge \tau_2 = [] \tag{3.3}$$
$$\mathtt{part}(n :: \sigma, m :: \sigma_1, \sigma_2) \Leftrightarrow n = m \wedge \mathtt{part}(\sigma,\sigma_2,\sigma_1) \tag{3.4}$$
$$\mathtt{alts}(n :: m :: \tau, n' :: \tau_1, \tau_2) \Leftrightarrow n' = n + m \wedge \mathtt{alts}(m :: \tau, \tau_2, \tau_1) \tag{3.5}$$
$$\mathtt{alts}(\tau,\tau_1,\tau_2) \wedge \mathtt{sum}(\tau,n) \Leftrightarrow \mathtt{alts}(\tau,\tau_1,\tau_2) \wedge \mathtt{sum}(\tau_1,n) \tag{3.6}$$

Fig. 3. Properties of sequences

We define assertion $R(i,p,\sigma,\tau)$ to indicate the heap state when the i-th iteration of pointer jumping has just finished, where p is the pointer to the initial cell of the list and σ and τ are the sequences of allocation addresses and integer values of the original list, respectively. The inductive definition[1] of the assertion is given below.

$$\begin{aligned}
R(i,p,[],[]) &\triangleq i \geq 0 \wedge p = \mathsf{nil} \wedge \mathbf{emp} \\
R(0,p,r :: \sigma, n :: \tau) &\triangleq p = r \wedge \exists q.(p \mapsto n, q * R(0,q,\sigma,\tau)) \\
R(i,p,r :: \sigma, n :: \tau) &\triangleq i > 0 \wedge p = r \wedge (R(i-1,p,\sigma_1,\tau_1) * \exists q.R(i-1,q,\sigma_2,\tau_2)) \\
&\qquad \text{where } \mathtt{part}(r :: \sigma,\sigma_1,\sigma_2) \text{ and } \mathtt{alts}(n :: \tau,\tau_1,\tau_2).
\end{aligned}$$

When $i = 0$, the assertion represents a non-circular, heap allocated linked list structure [1]. When $i > 0$, the assertion indicates that the heap allocates two disjoint sublists, the odd part $R(i-1,p,\sigma_1,\tau_1)$ and the even part $R(i-1,q,\sigma_2,\tau_2)$, where each sublist is further partitioned $i-1$ times more.

We also define assertion $\Pi^n(\sigma)$ $(n \geq 1)$, whose inductive definition is given by:

$$\Pi^n([]) \triangleq \mathbf{emp} \qquad \Pi^n(p :: \sigma) \triangleq \exists x_1 \cdots x_n. p \mapsto x_1, \ldots, x_n * \Pi^n(\sigma).$$

[1] Though the present assertion language does not formally include inductive definitions, it would be extended with fixed point operators as in [11].

The assertion $\Pi^n(\sigma)$ indicates that the heap allocates non-overlapping blocks of the same size n at addresses as listed in σ.

The properties about the program given in Figure 1 that we have informally discussed above can be specified as follows.

Proposition 1. *Let $C_{\mathbf{forall}}$ and $C_{\mathbf{while}}$ denote the command bodies of the* **forall** *and* **while** *command of the program in Figure 1, respectively. Then the following properties hold.*

(a) $\{R(i,p,\sigma,\tau)\}$ **forall** $x \mapsto n,t$ **in** σ **do** $C_{\mathbf{forall}}$ $\{R(i+1,p,\sigma,\tau)\}$
(b) $\{\exists i.R(i,p,\sigma,w) \wedge p+1 \hookrightarrow q\}$ **while** $q \neq \mathsf{nil}$ **do** $C_{\mathbf{while}}$ $\{\exists i.R(i,p,\sigma,w) \wedge p+1 \hookrightarrow q \wedge q = \mathsf{nil}\}$
(c) $R(i,p,p_1 :: \sigma, n_1 :: \tau) \wedge p+1 \hookrightarrow \mathsf{nil} \Rightarrow \exists q.(R(i,q,\sigma,\tau) \wedge (q = \mathsf{nil} \vee q+1 \hookrightarrow \mathsf{nil})) * (p = p_1 \wedge p \hookrightarrow n, \mathsf{nil})$, *whenever* $\mathtt{sum}(n_1 :: \tau, n)$.

The specification (a) indicates that every single execution of the **forall** command corresponds to a single step of parallel pointer jumping. The specification (b) identifies $\exists i.R(i,p,\sigma,w) \wedge p+1 \hookrightarrow q$ as the loop invariant. The implication (c) indicates that the invariant property on the sum of integers discussed earlier is intrinsic to the definition of $R(i,p,\sigma,\tau)$.

4 Program Logic for a Data-Parallel Programming Language

We consider a simple data-parallel programming language as given below:

$$b ::= \mathbf{true} \mid \mathbf{false} \mid e = e \mid \neg b \mid b \vee b \mid b \wedge b$$

$$C ::= x := e \mid x := [e] \mid [e] := e' \mid \mathbf{skip} \mid C;C \mid \mathbf{begin\ var}\ x;\ C\ \mathbf{end}$$
$$\mid \mathbf{if}\ b\ \mathbf{then}\ C\ \mathbf{else}\ C \mid \mathbf{while}\ b\ \mathbf{do}\ C \mid \mathbf{forall}\ x \mapsto y_1,\ldots,y_n\ \mathbf{in}\ \sigma\ \mathbf{do}\ C\ (n \geq 0)$$

where meta-variable e ranges over the set of expressions and σ in the **forall** command over the set of non-empty sequences of address constants.

The language consists of the following components. Assignment command $x := e$ updates the variable x by the value of the expression e; Lookup command $x := [e]$ assigns x with the dereferenced value of the address e; Mutation $[e] := e'$ updates the content at the address e by the value of e'; The command **skip** does no operation. These atomic commands are composed via sequencing C_1; C_2, block structures **begin** $\cdots$ **end** with local variable declaration, conditionals, while loops, and the **forall** primitive for parallel execution. As usual, **if** b **then** C abbreviates **if** b **then** C **else skip**.

We write $FV(C)$ for the set of free variables in C, *i.e.*, variables which are not in the scope of local variable declaration of a block structure or a parallel command. We also write $MD(C)$ to denote the set of free variables which can be updated by a variable assignment, *i.e.*, variables that have a free occurrence of the form $x := \cdots$ in C. We assume that, for every parallel command **forall** $x \mapsto y_1,\ldots,y_n$ **in** σ **do** C, $MD(C) \subseteq \{x,y_1,\ldots,y_n\}$.

We give the formal semantics of this language in the style of big-step operational semantics that derives an evaluation relation either of the form $(s,h),C \rightsquigarrow (s',h')$ or $(s,h),C \rightsquigarrow \mathsf{abort}$. The former indicates that the command C with initial state (s,h)

$$\frac{}{(s,h),x := e \rightsquigarrow (s[x \mapsto [\![e]\!]s],h)} \qquad \frac{[\![e]\!]s \in dom(h)}{(s,h),x := [e] \rightsquigarrow (s[x \mapsto h([\![e]\!]s)],h)}$$

$$\frac{[\![e]\!]s \in dom(h)}{(s,h),[e] := e' \rightsquigarrow (s,h[[\![e]\!]s \mapsto [\![e']\!]s])} \qquad \frac{(s,h),C \rightsquigarrow (s',h')}{(s,h),\textbf{begin var } x; C \textbf{ end} \rightsquigarrow (s'[x \mapsto s(x)],h')}$$

$$\frac{\begin{array}{c} MD(C) \subseteq \{x,y_1,\ldots,y_n\} \qquad \sigma = [p_1,\ldots,p_m]\ (m \geq 1) \qquad h = h' * h'' \\ h' = h'_1 * \cdots * h'_m \qquad dom(h'_i) = \{p_i + d \mid 0 \leq d \leq n-1\} \text{ for every } i \in \{1,\ldots,m\} \\ (s[x \mapsto p_i, y_1 \mapsto h'(p_i),\ldots,y_n \mapsto h'(p_i+n-1)],h'),C \rightsquigarrow (s_i,h''_i) \text{ for every } i \in \{1,\ldots,m\} \end{array}}{(s,h),\textbf{forall } x \mapsto y_1,\ldots,y_n \textbf{ in } \sigma \textbf{ do } C \rightsquigarrow (s,\bigcup_{i=1}^{m}\{(p_i+d,h''_i(p_i+d)) \mid 0 \leq d \leq n-1\} * h'')}$$

Fig. 4. Operational semantics of the data-parallel programming language

ends up with a final state (s',h'), while the latter stands for an execution aborted by an error.

Figure 4 gives derivation rules for a part of commands, where the notation $[\![e]\!]s$ denotes the result of evaluating the expression e under the store s. We omitted standard rules for other commands and the rules that lead to abort. A program can be aborted either by a memory fault (via a lookup or a mutation into a non-allocated address) or by an underallocation in the execution of **forall**, *i.e.*, the heap allocates fewer addresses than the required set σ of addresses designated in the **forall** command.

In the execution of the command **forall** $x \mapsto y_1,\ldots,y_n$ **in** σ **do** C, the command body C is simultaneously evaluated for every different address x taken from a fixed finite set of heap addresses explicitly given by σ. (In practice, σ could be automatically derived from, say, a reference name to a collection of heap allocated data, in a suitable extension of the present language.) Every different execution of the command body is in charge of updating the contents allocated in a contiguous block of size n starting from the address x, with $y_1,\ldots,y_n$ bound to the values stored in the block. The condition $h' = h'_1 * \cdots * h'_m$ in the premise of the operational semantics requires that the allocation addresses of different blocks do not overlap. The variables $x,y_1,\ldots,y_n$ are local to each execution instance; their variable assignments will be restored to the original ones, upon termination of the parallel execution.

In order to avoid inconsistencies that may arise by concurrent writes to the shared heap memory, we assume that every execution instance of a parallel command operates on its own local copy of the entire store and the heap blocks subject to the parallel processing. Every instance is allowed to read and write the local copy of the storage, except that global variables are read only. However, the effect of updates is only reflected to the local copy and is not accessible from other instances during parallel execution. Upon termination of the parallel execution, the store is restored to the original one and the contents of every heap block are updated to those of the local copy of the corresponding block held in the associated execution instance; the contents of irrelevant blocks in the local copy of each execution instance are discarded.

To summarize, an update to the heap is meaningful only if the updated address belongs to the block associated with the instance that executes the update. We also note that no execution instance can change the heap domain, as the language does not include commands for heap allocation and deallocation. There are many data-parallel programs that do not adhere to this limited set of heap operations, of course. However, the primary goal of this paper is to develop a formal proof system that gives a clear logical account for pointer jumping, a common programming technique in parallel computing. We presumed this reduced pattern of heap operations for the sake of a simpler specification of the parallel command.

4.1 Hoare Logic for Data-Parallelism

We define a specification $\{P\}\,C\,\{Q\}$ is valid iff $(s,h),C \rightsquigarrow (s,h')$ implies $s',h' \models Q$, for all $(s,h),(s',h') \in \mathit{State}$ satisfying $s,h \models P$,

The inference rules for deriving valid specifications are given in Figure 5. The upper half of the figure consists of the rules for commands. Most of the rules are standard except for LKP, MUT, and PAR. LKP and MUT give a weakest (liberal) precondition for the lookup and mutation command, respectively, where logically equivalent variant rules can substitute for them [1,2].

The inference rule PAR is explained as follows. The condition $P \Rightarrow \Pi^n(\sigma)$ in the premise indicates that the heap mentioned by the precondition P has to disjointly allocate a contiguous block of size n at every address listed in σ. The conjunct $(\Pi^1(\sigma) \wedge x \hookrightarrow -) * \mathbf{true}$ of the precondition implies that in any execution instance x must be bound to the initial address of a separate block. The rest of conjuncts $P \wedge x \hookrightarrow y_1,\ldots,y_n$ indicates that $y_1,\ldots,y_n$ must be bound to the contents allocated in the block.

Every different execution of the command body C is intended to update a contiguous block (referred to by x) with contents as mentioned by the postcondition Q. However, simply putting $\exists y_1 \cdots y_n.(Q \wedge x \hookrightarrow y_1,\ldots,y_n)$ overspecifies the postcondition of the command body, since every single execution instance of the command body is only in charge of updating the block allocated at the address x and does not care about other blocks. (If it ever updates other blocks, the effect will be canceled by other parallel execution instances that are in charge of updating those blocks.)

Here we utilize the modal operator as follows in order to mention about such a limited portion of heap addresses out of the entire set of addresses:

$$\exists y_1,\cdots,y_n.(x \hookrightarrow y_1,\cdots,y_n \wedge \Diamond(Q \wedge x \hookrightarrow y_1,\cdots,y_n)).$$

The first conjunct indicates that the heap allocates contents $y_1,\ldots,y_n$ at the address x; The second conjunct requires the contents $y_1,\ldots,y_n$ to respect the assertion Q but leaves those contents allocated at other addresses unspecified.

The inference rule comes with a few side conditions for technical reasons. In the precondition of the premise, the variable x is aliased to a fresh variable z, because the denotation of x may be altered by an assignment. Variables $x,y_1,\ldots,y_n$ are local to every sequential execution of the subcommand C and hence they are assumed not free in P or Q. Finally, we require that Q be a strictly exact assertion.

The condition that Q be a strictly exact assertion is vital for the soundness of the inference rule. When we reason about, somewhat informally, a data-parallel execution, we resort to the assumption that every different parallel execution of the same command ends up with a single unique result as specified by the postcondition. If Q is strictly exact, this uniqueness is guaranteed. Otherwise, inconsistencies may arise. For instance,

$$\text{ASGN}\ \frac{}{\{P[e/x]\}\,x := e\,\{P\}} \qquad \text{LKP}\ \frac{}{\{\exists z.(P[z/x] \land e \hookrightarrow z)\}\,x := [e]\,\{P\}}\ z \notin FV(e) \cup FV(P)$$

$$\text{MUT}\ \frac{}{\{e \mapsto - * (e \mapsto e' \mathrel{-\!\!*} P)\}\,e := e'\,\{P\}} \qquad \text{SKIP}\ \frac{}{\{P\}\,\mathbf{skip}\,\{P\}}$$

$$\text{SEQ}\ \frac{\{P\}\,C\,\{Q'\} \quad \{Q'\}\,C'\,\{Q\}}{\{P\}\,C;C'\,\{Q\}} \qquad \text{BLK}\ \frac{\{P\}\,C\,\{Q\}}{\{P\}\,\mathbf{begin\ var}\ x;\ C\ \mathbf{end}\,\{Q\}}\ x \notin FV(P) \cup FV(Q)$$

$$\text{IF}\ \frac{\{P \land Q\}\,C\,\{P'\} \quad \{P \land \neg Q\}\,C'\,\{P'\}}{\{P\}\,\mathbf{if}\ Q\ \mathbf{then}\ C\ \mathbf{else}\ C'\,\{P'\}} \qquad \text{WHILE}\ \frac{\{P \land Q\}\,C\,\{P\}}{\{P\}\,\mathbf{while}\ Q\ \mathbf{do}\ C\,\{P \land \neg Q\}}$$

$$\text{PAR}\ \frac{P \Rightarrow \Pi^n(\sigma) \qquad \left\{ \begin{array}{c} ((\Pi^1(\sigma) \land x \hookrightarrow -) * \mathbf{true}) \land \\ P \land x \hookrightarrow y_1, \cdots, y_n \land x = z \end{array} \right\} C \left\{ \begin{array}{c} \exists y_1 \cdots y_n.(z \hookrightarrow y_1, \cdots, y_n \land \\ \Diamond(Q \land z \hookrightarrow y_1, \cdots, y_n)) \end{array} \right\}}{\{P\}\,\mathbf{forall}\ x \mapsto y_1, \cdots, y_n\ \mathbf{in}\ \sigma\ \mathbf{do}\ C\,\{Q\}}$$

$x, y_1, \ldots, y_n, z \notin FV(P) \cup FV(Q)$, $z \notin FV(C)$, and Q is strictly exact.

$$\text{CONSEQ}\ \frac{P \Rightarrow P' \quad \{P'\}\,C\,\{Q'\} \quad Q' \Rightarrow Q}{\{P\}\,C\,\{Q\}} \qquad \text{EXQ}\ \frac{\{P\}\,C\,\{Q\}}{\{\exists x.P\}\,C\,\{\exists x.Q\}}\ x \notin FV(C)$$

$$\text{DISJ}\ \frac{\{P_1\}\,C\,\{Q_1\} \quad \{P_2\}\,C\,\{Q_2\}}{\{P_1 \lor P_2\}\,C\,\{Q_1 \lor Q_2\}} \qquad \text{FRAME}\ \frac{\{P\}\,C\,\{Q\}}{\{P * R\}\,C\,\{Q * R\}}\ MD(C) \cap FV(R) = \emptyset$$

Fig. 5. Inference rules for Hoare triples

consider a derivation for a parallel command $\{P\}\,\mathbf{forall}\ x \mapsto y\ \mathbf{in}\ [1001, 1002]\ \mathbf{do}\ C\,\{Q\}$ with $Q \triangleq (1001 \mapsto 1 * 1002 \mapsto 3) \lor (1001 \mapsto 4 * 1002 \mapsto 6)$, which is not a strictly exact assertion. Then the execution of the parallel command can result in a heap, say, $h = \{(1001, 1), (1002, 6)\}$, as this adheres to $\exists y.(\Diamond(Q \land z \hookrightarrow y) \land z \hookrightarrow y)$. However h does not satisfy Q and therefore the inference is not sound.

Theorem 1 (Soundness). *If $\{P\}\,C\,\{Q\}$ is derivable, then $\{P\}\,C\,\{Q\}$ is valid.*

Proof. The soundness of the FRAME rule follows from the safety monotonicity and the frame property [12], which are proved by induction on the size of derivation of evaluation relations.

To show the soundness of the PAR rule, suppose we have a derivation that ends up with a conclusion $\{P\}\,\mathbf{forall}\ x \mapsto y_1, \cdots, y_n\ \mathbf{in}\ \sigma\ \mathbf{do}\ C\ \{Q\}$, with $\sigma = [p_1, \ldots, p_m]$ ($m \geq 1$). Let (s, h) be any state such that $s, h \models P$. By the premise $P \Rightarrow \Pi^n(\sigma)$ of the inference rule, we have $h = h_1 * \cdots * h_m$ where $h_i = h \restriction \{p_i + d \mid 0 \leq d \leq n-1\}$ ($i \in \{1, \cdots, m\}$). Let $s_i = s[x \mapsto p_i, y_1 \mapsto h(p_i), \ldots, y_n \mapsto h(p_i + n - 1)]$. Since $x, y_1, \ldots, y_n, z \notin FV(P)$, we have $s_i[z \mapsto p_i], h \models ((\Pi^1(\sigma) \land x \hookrightarrow -) * \mathbf{true}) \land P \land x \hookrightarrow y_1, \cdots, y_n \land x = z$. Hence, by induction hypothesis, for every i, we have $(s_i[z \mapsto p_i], h), C \rightsquigarrow (s_i', h_i')$, for some $(s_i', h_i') \in$ *State* satisfying $dom(h_i') = dom(h)$ and

$$s_i', h_i' \models \exists y_1 \cdots y_n.(z \hookrightarrow y_1, \cdots, y_n \land \Diamond(Q \land z \hookrightarrow y_1, \cdots, y_n)). \qquad (\dagger)$$

Here we notice that $s'_i[z \mapsto s(z)]$ agrees with s for any variable other than $MD(C)$. Hence, it follows from $x, y_1, \ldots, y_n, z \notin FV(Q)$ and (†) that $s, h''_i \models Q$ for some h''_i such that $h''_i(q) = h'_i(q)$ for every $i \in \{1, \ldots, m\}$ and $q \in \{p_i + d \mid 0 \le d \le n-1\}$. Since Q is strictly exact, we have $h''_i = h''_j$ for every $i, j \in \{1, \cdots, m\}$. This implies that $s, h' \models Q$ holds, where $h' = \bigcup_{i=1}^{m} \{(p_i + d, h'_i(p_i + d)) \mid 0 \le d \le n-1\}$.

The proof of other rules is rather a routine and is omitted. □

5 The Correctness Proof for Parallel List Scan

This section gives a proof for Proposition 1 and shows the correctness of the parallel list scan program given in Figure 1.

In the proof, we will exploit the properties of assertions listed in Figure 6. In addition to general properties of the classical BI logic [1], the list contains those specific to connectives $\mapsto$, $\hookrightarrow$ and also to the subclasses of pure and precise assertions. Note that the given set of axioms and inference rules are by no means complete; neither they are not minimal in the sense that some properties are derivable from others. In the subsequent formal derivation, the rules in Figure 6 that have no reference number will be used in the subsequent proof without explicitly mentioned.

Lemma 1. *Let n be any positive integer. Then $R(i, p, \sigma, \tau)$ is strictly exact, $\Pi^n(\sigma)$ is precise. Also, the following formulas are all valid.*

(a) $\Pi^n(\sigma) \Rightarrow \Pi^1(\sigma) * \mathbf{true}$
(b) $\Pi^n(\sigma) \Leftrightarrow \Pi^n(\sigma_1) * \Pi^n(\sigma_2)$, *whenever* $\mathtt{part}(\sigma, \sigma_1, \sigma_2)$.
(c) $R(i, p, p' :: \sigma, n :: \tau) \Rightarrow p \hookrightarrow n, - \wedge p = p'$
(d) $R(i, p, [p'], [n]) \Leftrightarrow p' \mapsto n, \mathsf{nil} \wedge p = p' \wedge i \ge 0$
(e) $\Diamond R(i, p, \sigma, \tau) \Leftrightarrow \Diamond R(i+1, p, \sigma, \tau) \wedge i \ge 0$
(f) $(\Pi^1(\sigma) \wedge p \hookrightarrow -, -) * \mathbf{true} \wedge R(0, p, \sigma, \tau) \Rightarrow \mathbf{false}$
(g) $R(i, p_1, p_1 :: \sigma_1, \tau_1) * (R(i, p_2, \sigma_2, \tau_2) \wedge p_2 + 1 \hookrightarrow \mathsf{nil}) \Rightarrow p_1 + 1 \hookrightarrow \mathsf{nil}$, *whenever* $\mathtt{part}(\sigma, p_1 :: \sigma_1, \sigma_2)$ *and* $\mathtt{alts}(\tau, \tau_1, \tau_2)$ *for some* σ *and* τ.

We omit the proof, which is by routine induction on i and the length of σ.

Proof of Proposition 1(a). By the inference rule PAR, it is sufficient to show the derivability of the specification:

$$\left\{ \begin{array}{c} ((\Pi^1(\sigma) \wedge x \hookrightarrow -) * \mathbf{true}) \wedge \\ R(i, p, \sigma, \tau) \wedge x \hookrightarrow n, t \wedge x = z \end{array} \right\} C_{\mathbf{forall}} \left\{ \begin{array}{l} \exists nt.(z \hookrightarrow n, t \wedge \\ \quad \Diamond(R(i+1, p, \sigma, \tau) \wedge z \hookrightarrow n, t)) \end{array} \right\}. \quad (5.11)$$

Proof is by induction on i and the length of σ (and τ).

Case $\sigma = \tau = []$. It vacuously holds by the rule CONSEQ.

Case $\sigma = [p']$ and $\tau = [n']$. We have $R(i, p, \sigma, \tau) \wedge x \hookrightarrow n, t \wedge x = z \Leftrightarrow i \ge 0 \wedge x \mapsto n, t \wedge n = n' \wedge x = z \wedge z = p \wedge p = p' \wedge t = \mathsf{nil}$ by lemma 1(d), (5.5), and (5.1). Since the conjunction of this formula and $t \ne \mathsf{nil}$ leads to absurdity, by the rules IF and CONSEQ, (5.11) is derived from the implication $x \mapsto n, t \wedge n = n' \wedge x = z \wedge z = p \wedge p = p' \wedge t = \mathsf{nil} \Rightarrow \exists nt.(z \mapsto n, t \wedge \Diamond(p' \mapsto n', \mathsf{nil} \wedge p = p' \wedge z \hookrightarrow n, t))$, which follows from (5.1), (5.5), and (5.6).

$$(P*Q)*R \Leftrightarrow P*(Q*R) \qquad P*Q \Leftrightarrow Q*P \qquad P*\mathbf{emp} \Leftrightarrow P$$

$$(P_1 \vee P_2)*Q \Leftrightarrow (P_1*Q) \vee (P_2*Q) \qquad (P_1 \wedge P_2)*Q \Rightarrow (P_1*Q) \wedge (P_2*Q)$$

$$(\exists x.P)*Q \Leftrightarrow \exists x.(P*Q) \quad \text{and} \quad (\forall x.P)*Q \Rightarrow \forall x.(P*Q) \quad \text{where } x \text{ is not free in } Q$$

$$\frac{P_1 \Rightarrow Q_1 \qquad P_2 \Rightarrow Q_2}{P_1*P_2 \Rightarrow Q_1*Q_2} \qquad \frac{P*Q \Rightarrow R}{P \Rightarrow (Q -\!\!* R)} \qquad \frac{P \Rightarrow (Q -\!\!* R)}{P*Q \Rightarrow R} \qquad \frac{P \Rightarrow Q}{\Diamond P \Rightarrow \Diamond Q}$$

$$P \Rightarrow \Diamond P \qquad \Diamond(P \vee Q) \Leftrightarrow \Diamond P \vee \Diamond Q \qquad \Diamond(P \wedge Q) \Rightarrow \Diamond P \wedge \Diamond Q$$

$$\exists x.\Diamond P \Leftrightarrow \Diamond \exists x.P \qquad \Diamond(P*Q) \Leftrightarrow \Diamond P * \Diamond Q$$

$$x \mapsto y_1,\cdots,y_n \wedge x' \mapsto y'_1,\cdots,y'_n \Leftrightarrow x \mapsto y_1,\cdots,y_n \wedge x = x' \wedge \textstyle\bigwedge_{i=1}^n y_i = y'_i \tag{5.1}$$

$$x = x' \wedge x \hookrightarrow y \wedge x' \hookrightarrow y' \Rightarrow y = y' \tag{5.2}$$

$$x \mapsto - * x' \mapsto - * \mathbf{true} \Rightarrow x \neq x' \tag{5.3}$$

$$\textstyle\bigwedge_{i=1}^n (x+i-1 \hookrightarrow y_i) \Leftrightarrow x \hookrightarrow y_1,\ldots,y_n \tag{5.4}$$

$$x \mapsto y_1,\cdots,y_n \wedge x' \mapsto y'_1,\cdots,y'_n \Leftrightarrow x \mapsto y_1,\cdots,y_n \wedge x' \hookrightarrow y'_1,\cdots,y'_n \tag{5.5}$$

$$\Diamond(x \mapsto y) \Leftrightarrow x \mapsto - \tag{5.6}$$

$$(P_1*P_2) \wedge x \hookrightarrow y \Leftrightarrow (P_1 \wedge x \hookrightarrow y)*P_2 \vee P_1*(P_2 \wedge x \hookrightarrow y) \tag{5.7}$$

For any pure assertions P, P_1, P_2,

$$P_1 \wedge P_2 \Leftrightarrow P_1*P_2 \quad (P \wedge Q)*R \Leftrightarrow P \wedge (Q*R) \quad P \Leftrightarrow \Diamond P \quad \Diamond(P \wedge Q) \Leftrightarrow P \wedge \Diamond Q$$

$$((P \wedge Q) -\!\!* (P \wedge R)) \wedge P \Rightarrow Q -\!\!* (P \wedge R) \tag{5.8}$$

For any precise assertions P, P_1, P_2,

$$(P*Q_1) \wedge (P*Q_2) \Leftrightarrow P*(Q_1 \wedge Q_2) \tag{5.9}$$

$$(P_1 \wedge Q_1)*P_2 \wedge P_1*(P_2 \wedge Q_2) \Leftrightarrow (P_1 \wedge Q_1)*(P_2 \wedge Q_2) \tag{5.10}$$

Fig. 6. Properties of assertions

Case $\sigma = p_1 :: p_2 :: \sigma'$, $\tau = n_1 :: n_2 :: \tau'$, and $i = 0$. We prove by case analysis on the equality of p and z.

First consider the case $p = z$. Below we show the proof outline for the **then** clause of the conditional in $C_{\mathbf{forall}}$:

$\{(p_1 \mapsto n_1, p_2 \wedge x \hookrightarrow n, t) * p_2 \mapsto n_2, q \wedge p = p_1 \wedge x = z\}$

$\{(x \mapsto -,- * p_2 \mapsto -,-) \wedge t \hookrightarrow n_2, q \wedge p = p_1 \wedge p_1 = z \wedge n_1 = n \wedge x = z\}$ by (5.5)

$\quad m := [t]; t := [t+1]$

$\{(x \mapsto -,- * p_2 \mapsto -,-) \wedge p = p_1 \wedge p_1 = z \wedge n_1 = n \wedge n_2 = m \wedge q = t \wedge x = z\}$

$\{(x \mapsto -,- * p_2 \mapsto -,-) * (x \mapsto (n+m), t -\!\!* x \mapsto (n_1+n_2), q)$ — (*)

$\quad \wedge p = p_1 \wedge p_1 = z \wedge x = z\}$

$\quad [x] := n+m; [x+1] := t$

$\{(x \mapsto (n_1+n_2), q * p_2 \mapsto -,-) \wedge p = p_1 \wedge p_1 = z \wedge x = z\}$

$\{(z \mapsto (n_1+n_2), q * p_2 \mapsto -,-) \wedge p = p_1 \wedge p_1 = z\}$.

The proof step $(*)$ is derived as follows. Let $P_0 \triangleq n_1 = n \wedge n_2 = m \wedge q = t$. Then we have $\mathbf{emp} \wedge P_0 \Rightarrow (x \mapsto (n_1+n_2), q \wedge P_0 \multimap x \mapsto (n_1+n_2), q \wedge P_0) \wedge P_0 \Rightarrow x \mapsto (n+m), t \multimap x \mapsto (n_1+n_2), q$ by $\mathbf{emp} \Rightarrow Q \multimap Q$, (5.8) and (5.1).

Thus we have the proof outline for $C_{\mathbf{forall}}$ as below, where σ_1', σ_2', τ_1', τ_2' denote sequences satisfying $\texttt{part}(p_1 :: p_2 :: \sigma', p_1 :: \sigma_1', p_2 :: \sigma_2')$ and $\texttt{alts}(n_1 :: n_2 :: \tau', (n_1+n_2) :: \tau_1', \tau_2')$.

$\{p = z \wedge (\Pi^1(\sigma) \wedge x \hookrightarrow -) * \mathbf{true} \wedge R(0,p,\sigma,\tau) \wedge x \hookrightarrow n,t \wedge x = z\}$
$\{(p_1 \mapsto n_1, p_2 \wedge x \hookrightarrow n,t) * (p_2 \mapsto n_2, q) * R(0,q,\sigma',\tau') \wedge p = p_1 \wedge x = z\}$
$C_{\mathbf{forall}}$ by FRAME,EXQ, (5.7), (5.3)
$\{(z \mapsto (n_1+n_2), q * p_2 \mapsto -,-) * R(0,q,\sigma',\tau') \wedge p = p_1 \wedge p_1 = z\}$
$\{z \hookrightarrow (n_1+n_2), q \wedge \Diamond(p = p_1 \wedge p_1 = z \wedge (p_1 \mapsto (n_1+n_2), q \wedge z \hookrightarrow (n_1+n_2), q) * R(0,q,\sigma_1',\tau_1') * R(0,p_2 :: \sigma_2',\tau_2'))\}$ by (5.1), (5.6), and lemma 1(e)
$\{\exists nt.(z \hookrightarrow n,t \wedge \Diamond(R(1,p,\sigma,\tau) \wedge z \hookrightarrow n,t))\}$ by (5.7) and EXQ.

Let $\sigma_1', \sigma_2', \tau_1', \tau_2'$ be defined as above. The other case $p \neq z$ is proved as follows.
$\{p \neq z \wedge (\Pi^1(\sigma) \wedge x \hookrightarrow -) * \mathbf{true} \wedge R(0,p,\sigma,\tau) \wedge x \hookrightarrow n,t \wedge x = z\}$
$\{p_1 \mapsto - * (\Pi^1(p_2 :: \sigma') \wedge x \hookrightarrow -) * \mathbf{true} \wedge p_1 \mapsto n_1, p_2 * (R(0,p_2,p_2 :: \sigma', n_2 :: \tau') \wedge x \hookrightarrow n,t) \wedge p = p_1 \wedge x = z\}$ by EXQ, (5.7), (5.3)
$\{(p = p_1 \wedge p_1 \mapsto n_1, p_2) * (\Pi^1(p_2 :: \sigma') \wedge x \hookrightarrow -) * \mathbf{true} \wedge R(0,p_2,p_2 :: \sigma', n_2 :: \tau') \wedge x \hookrightarrow n,t \wedge x = z\}$ by (5.5), (5.1), (5.9), (5.7), (5.4), lemma 1(f)
$C_{\mathbf{forall}}$
$\{(p = p_1 \wedge p_1 \mapsto n_1, p_2) * \exists nt.(z \hookrightarrow n,t \wedge \Diamond(R(1,p_2,p_2 :: \sigma, n_2 :: \tau) \wedge z \hookrightarrow n,t))\}$ by I.H. and FRAME
$\{\exists nt.(z \hookrightarrow n,t \wedge \Diamond(p_1 \mapsto (n_1+n_2), q * R(0,q,\sigma_1',\tau_1') * R(0,p_2 :: \sigma_2',\tau_2') \wedge p = p_1 \wedge z \hookrightarrow n,t)\}$ by (5.6)
$\{\exists nt.(z \hookrightarrow n,t \wedge \Diamond(R(1,p,\sigma,\tau) \wedge z \hookrightarrow n,t))\}$. by EXQ

Combining the two cases, we derive the specification (5.11) by the rule DISJ.

Case $\sigma = p_1 :: p_2 :: \sigma'$, $\tau = n_1 :: n_2 :: \tau'$, and $i > 0$. Let σ_1', σ_2', τ_1, τ_2 be sequences satisfying $\texttt{part}(\sigma', \sigma_1', \sigma_2')$ and $\texttt{alts}(\tau, \tau_1, \tau_2)$, and also define $R_j(i) \triangleq R(i, p_j, p_j :: \sigma_j', \tau_j)$ for every $j \in \{1,2\}$. Then for every $j \in \{1,2\}$ we have:
$\{p = p_1 \wedge (\Pi^1(\sigma) \wedge x \hookrightarrow -) * \mathbf{true} \wedge R_j(i-1) * (R_{3-j}(i-1) \wedge x \hookrightarrow n,t) \wedge x = z\}$
$\{(p = p_1 \wedge R_j(i-1)) * ((\Pi^1(\sigma_{3-j}) \wedge x \hookrightarrow -) * \mathbf{true} \wedge R_{3-j}(i-1) \wedge x \hookrightarrow n,t \wedge x = z)\}$ by lemma 1(a) and (b), (5.7), (5.10), (5.3)
$C_{\mathbf{forall}}$
$\{(p = p_1 \wedge R_j(i-1)) * \exists nt.(z \hookrightarrow n,t \wedge \Diamond(R_{3-j}(i) \wedge z \hookrightarrow n,t))\}$ by I.H. and FRAME
$\{\exists nt.(z \hookrightarrow n,t \wedge \Diamond(p = p_1 \wedge R_j(i) * R_{3-j}(i) \wedge z \hookrightarrow n,t))\}$ by lemma 1(e) and (5.7).

Therefore (5.11) follows from $R(i,p,\sigma,\tau) \Leftrightarrow p = p_1 \wedge R_1(i-1) * R_2(i-1)$ and (5.7) by the rule DISJ. □

Proof of Proposition 1(b). When $\sigma = []$, the assertion vacuously holds. Suppose $\sigma \neq []$. We have $\{R(i,p,\sigma,w)\}\, C_{\mathbf{forall}}\, \{R(i+1,p,\sigma,w)\}$ by proposition 1(a) and $\{\exists z.(R(i+1,p,\sigma,w) \wedge p+1 \hookrightarrow z)\}\; q := [p+1]\; \{R(i+1,p,\sigma,w) \wedge p+1 \hookrightarrow q\}$ by the rule LKP. Since $R(i+1,p,\sigma,w) \Rightarrow \exists z.(R(i+1,p,\sigma,w) \wedge p+1 \hookrightarrow z$ by lemma 1(c), it follows that $\{R(i,p,\sigma,w)\}\, C_{\mathbf{forall}};\, q := [p+1]\; \{R(i+1,p,\sigma,w) \wedge p+1 \hookrightarrow q\}$ by the rules SEQ and CONSEQ. Thus the assertion follows by the rules EXQ and WHILE. □

Proof of Proposition 1(c). By induction on i and the length of σ (and τ).

When $i=0$, we have $R(0,p,p_1 :: \sigma, n_1 :: \tau) \wedge p+1 \hookrightarrow \mathsf{nil} \Rightarrow \exists q.(p \hookrightarrow n_1, q * R(0,q,\sigma',\tau')) \wedge p = p_1 \wedge p+1 \hookrightarrow \mathsf{nil} \Rightarrow \exists q.(R(0,q,\sigma,\tau) \wedge q = \mathsf{nil}) * (p = p_1 \wedge p \hookrightarrow n_1, \mathsf{nil})$ by (5.2). This indicates that $\sigma = \tau = []$ and thus $\mathtt{sum}(n_1 :: \tau, n_1)$.

Suppose $i > 0$. The case $\sigma = \tau = []$ is likewise proved. Let $\sigma', \sigma'_1, \sigma'_2, \tau', \tau'_1, \tau'_2$ be sequences satisfying $\sigma = p_2 :: \sigma'$, $\tau = n_2 :: \tau'$, $\mathtt{part}(\sigma', \sigma'_1, \sigma'_2)$, and $\mathtt{alts}(n_2 :: \tau', n_2 :: \tau'_2, \tau'_1)$, and also let n' be an integer such that $\mathtt{alts}(n_1 :: \tau, n')$. Then we have:

$$
\begin{array}{lr}
R(i,p,p_1 :: \sigma, n_1 :: \tau) \wedge p+1 \hookrightarrow \mathsf{nil} & \\
\Rightarrow (R(i-1,p_1,p_1 :: \sigma'_1,(n_1+n_2) :: \tau'_1) \wedge p_1+1 \hookrightarrow \mathsf{nil}) & \\
\quad * R(i-1,p_2,p_2 :: \sigma'_2, n_2 :: \tau'_2) \wedge p = p_1 & \text{by (5.7), (5.3)} \\
\Rightarrow \exists q.(R(i-1,q,\sigma'_1,\tau'_1) \wedge (q = \mathsf{nil} \vee q+1 \hookrightarrow \mathsf{nil})) & \\
\quad * R(i-1,p_2,p_2 :: \sigma'_2, n_2 :: \tau'_2) * (p = p_1 \wedge p_1 \hookrightarrow n', \mathsf{nil}) & \text{by induction hypothesis} \\
\Rightarrow (R(i-1,p_2,p_2 :: \sigma'_2, n_2 :: \tau'_2) \wedge p_2+1 \hookrightarrow \mathsf{nil}) & \\
\quad * \exists q.R(i-1,q,\sigma'_1,\tau'_1) * (p = p_1 \wedge p_1 \hookrightarrow n', \mathsf{nil}) & \text{— (**)} \\
\Rightarrow (R(i,p_2,\sigma,\tau) \wedge p_2+1 \hookrightarrow \mathsf{nil}) * (p = p_1 \wedge p_1 \hookrightarrow n', \mathsf{nil}) & \text{by (3.4), (3.5), (5.7)}
\end{array}
$$

The implication (**) is derived as follows. If $q = \mathsf{nil}$, then $R(i-1,q,\sigma'_1,\tau'_1)$ implies $\sigma'_1 = []$ and thus $\sigma'_2 = []$ by (3.1). Hence $R(i-1,p_2,p_2 :: \sigma'_2, n_2 :: \tau'_2) \Rightarrow p_2+1 \hookrightarrow \mathsf{nil}$. Otherwise, we have $(R(i-1,q,\sigma'_1,\tau'_1) \wedge q+1 \hookrightarrow \mathsf{nil}) * R(i-1,p_2,p_2 :: \sigma'_2, n_2 :: \tau'_2) \Rightarrow p_2+1 \hookrightarrow \mathsf{nil}$ by lemma 1(g). □

Theorem 2. *Let $C_{\mathbf{scan}}$ denote the program in Figure 1. Then the following specification is valid.*

$$\{R(0,p,\sigma,\tau) \wedge p \neq \mathsf{nil}\}\, C_{\mathbf{scan}}\, \{\exists i.R(i,p,\sigma,\tau) \wedge p+1 \hookrightarrow \mathsf{nil}\}$$

This theorem follows from proposition 1(b). As a corollary, we can deduce from proposition 1(c) that the program computes the list scan, *i.e.*, when the program terminates, every cons cell holds the sum of the corresponding sublist.

6 Conclusion

We have proposed a program logic for reasoning about data-parallel programs. We have worked out a formal correctness proof for the parallel list scan algorithm that employs the pointer jumping, a common method for parallel processing of linked data structures. The proof system adopts Separation Logic as a formal means to represent disjoint partitioning of linked data structures and further extends it with modality to provide a sound specification for the data-parallel command **forall**. This enables us to formally present the parallel list scan algorithm as another instance of the divide-and-conquer strategy.

We believe that our program logic can also apply to other variants of parallel algorithms based on parallel prefix or scan [4]. However, in the present paper, it is assumed that each parallel execution instance can only update heap contents owned by the instance itself. In some parallel algorithms, it is vital that every execution instance is possible to update heap contents owned by other instances. Such algorithms are more difficult to verify, because of possible race conditions caused by concurrent writes to the heap storage. It would be an interesting future topic to refine the present proof system for allowing concurrent writes. We hope that the notion of ownership transfer [13] might give a relevant solution to this issue.

Acknowledgment. I thank anonymous reviewers for their helpful comments.

References

1. Ishtiaq, S., O'Hearn, P.W.: BI as an assertion language for mutable data structures. In: Proceedings of the 28th ACM SIGPLAN-SIGACT symposium on Principles of programming languages, ACM Press (2001) 14–26
2. Reynolds, J.C.: Separation logic: A logic for shared mutable data structures. In: 17th IEEE Symposium on Logic in Computer Science (LICS 2002), IEEE Computer Society (2002) 55–74
3. Hillis, W., Steele Jr., G.L.: Data parallel algorithms. Communications of ACM **29**(12) (1986) 1170–1183
4. Blelloch, G.E.: Prefix sums and their applications. Technical Report CMU-CS-90-190, Carnegie Mellon University (1990)
5. Hatcher, P., Quinn, M.: Data-Parallel Programming on MIMD Computers. MIT Press (1991)
6. Gibbons, A., Rytter, W.: Efficient Parallel Algorithms. Cambridge University Press (1988)
7. Bougé, L., Cachera, D., Guyadec, Y.L., Utard, G., Virot, B.: Formal validation of data-parallel programs: A two-component assertional proof system for a simple language. Theoretical Computer Science **189**(1-2) (1997) 71–107
8. Blelloch, G.E., Greiner, J.: A provable time and space efficient implementation of NESL. In: ACM SIGPLAN International Conference on Functional Programming (ICFP '96), ACM Press (1996) 213–225
9. Nishimura, S., Ohori, A.: Parallel functional programming via data-parallel recursion. Journal of Functional Programming **9**(4) (1999) 427–463
10. Berger, M., Honda, K., Yoshida, N.: A logical analysis of aliasing in imperative higher-order functions. In: International Conference on Functional Programming (ICFP 2005), ACM Press (2005) 280–293
11. Sims, É.J.: Extending separation logic with fixpoints and postponed substitution. In: Algebraic Methodology and Software Technology, 10th International Conference, AMAST 2004. Volume 3116 of LNCS., Springer (2004) 475–490
12. Yang, H., O'Hearn, P.: A semantic basis for local reasoning. In: Foundations of Software Science and Computation Structures, 5th International Conference, FOSSACS 2002. Volume 2303 of LNCS., Springer (2002) 402–416
13. O'Hearn, P.: Resources, concurrency and local reasoning. In: CONCUR 2004 — Concurrency Theory, 15th International Conference. Volume 3170 of LNCS., Springer (2004) 49–67

Testing Semantics: Connecting Processes and Process Logics

Dusko Pavlovic[1,⋆], Michael Mislove[2,⋆], and James B. Worrell[3]

[1] Kestrel Institute, Palo Alto, CA
[2] Tulane University, New Orleans, LA
[3] Oxford University, Oxford, UK

Abstract. We propose a methodology based on testing as a framework to capture the interactions of a machine represented in a denotational model and the data it manipulates. Using a connection that models machines on the one hand, and the data they manipulate on the other, testing is used to capture the interactions of each with the objects on the other side: just as the data that are input into a machine can be viewed as tests that the machine can be subjected to, the machine can be viewed as a test that can be used to distinguish data. This approach is based on generalizing from duality theories that now are common in semantics to logical connections, which are simply contravariant adjunctions. In the process, it accomplishes much more than simply moving from one side of a duality to the other; it faithfully represents the interactions that embody what is happening as the computation proceeds.

Our basic philosophy is that tests can be used as a basis for modeling interactions, as well as processes and the data on which they operate. In more abstract terms, tests can be viewed as formulas of process logics, and testing semantics connects processes and process logics, and assigns computational meanings to both.

1 Introduction: The Problem of Testing

Testing a family Ξ of systems by a family Θ of tests, or process logic formulas, is a map

$$\Xi \times \Theta \xrightarrow{\mathbb{T}} \Omega$$

where Ω is the type of observations, or truth values. The simplest case is $\Omega = \{0,1\}$, where 1 represents "accept", or "succeed", or "truth", and 0 is "reject", or "fail", or "diverge", or "false". A richer semantics can be achieved if one replaces the truth values $\{0,1\}$ by the interval $[0,1]$, and interprets the result of a test as the probability a process passes it. But the problem with either approach is that once the test is performed, we have only the result. Making tests more dynamic requires taking a slightly different view.

⋆ The support of the NSF and the US Office of Naval Research is gratefully acknowledged.

M. Johnson and V. Vene (Eds.): AMAST 2006, LNCS 4019, pp. 308–322, 2006.

The goal of testing is to find bugs, which distinguish an implemented, real system $R \in \Xi$ from an ideal reference system $S \in \Xi$, or to demonstrate that they are indistinguishable. A bug can be construed as a test $b \in \Theta$, which leads to an observation $R \models b$, different from the observation $S \models b$. On the other hand, if $(R \models t) = (S \models t)$ for all tests $t \in \Theta$, then the systems are computationally indistinguishable, modulo testing equivalence

$$R \sim S \iff \forall t \in \Theta.\ (R \models t) = (S \models t)$$

The basic methods of studying computation in terms of tests on automata go back to the 1950s and E.P. Moore's seminal paper [1]. Moore introduced distinguishing sequences of tests, as well as testing equivalence, and several other fundamental ideas, which later led to a broad range of methods of *conformance testing*, which is the discipline of proving that an implementation R conforms to a standard S. Other problems resolved through testing include determining the current or the final state of a given automaton, or characterizing an unknown automaton.[1] One of Moore's most interesting contributions was the method of extracting minimal automata, i.e. the canonical representatives of computational behaviors, from equivalence classes of states modulo testing equivalence.

The starting point of the present work is a small modification of Moore's idea: we represent equivalent states, which form a state of a minimal automaton, not as equivalence classes of states, but as the maps from tests to observations that they induce: two states are equivalent if and only if they induce the same map. Either way, the computational behaviors arise as the elements in the image L of the semantic map, in the form

$$\begin{array}{ccc} \Xi & \xrightarrow{\quad\models\quad} & \Omega^{\Theta} \\ & \searrow \qquad \nearrow & \\ & L & \end{array}$$

The choice of representatives, of course, does not matter for abstract theory, but it turns out to make a lot of difference when it comes to analyzing state-based systems which arise in the design of reactive and embedded systems, involving stochastic, continuous, temporal or hybrid dynamics. The study of labelled Markov processes [4] provides a striking example. On the other hand, a generic categorical framework where states are represented as truth assignments of logical formulas has been used in [5, 6, 7]. In this paper, we will confine our presentation to the possibilistic setting, leaving the probabilistic setting for further work. For this setting the categorical trace semantics of finite state automata [8] and context-free languages [9] are clear examples, and are close conceptual predecessors of testing semantics. What appears to be new is our ability to bring Turing machines into the same setting.

[1] Excellent surveys of testing methodologies (albeit a bit outdated in applications) are [2, 3].

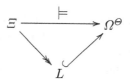

2 Logical Connections

A *logical connection* is a contravariant adjunction $M^{op} \dashv P : \mathcal{S}^{op} \longrightarrow \mathcal{T}$ between a category of "spaces" and a category of "types" or "theories". In one direction, a space X is mapped to the type PX of "predicates" over it; in the other direction, a type A is mapped to the space MA of its models. Among the many dualities that are examples of logical connections, we mention just a few:

Self-duality of sets: $\wp^{op} \dashv \wp : \mathsf{Set}^{op} \longrightarrow \mathsf{Set}$, which can be viewed as duality of discrete spaces and complete atomic Boolean algebras (the category of which is equivalent to Set^{op}). In more detail, the functor associates to a set A the family $\wp(A)$ of all subsets of A, and to a mapping $f\colon A \longrightarrow B$ between sets, the mapping $f^{-1}\colon \wp(B) \longrightarrow \wp(A)$. The power set of a set is a complete, atomic Boolean algebra, and the mapping f^{-1} preserves all unions, intersections and complements. Thus our duality identifies each set with the complete atomic Boolean algebra it generates, and to each algebra, its set of atoms.

Here are some other notable connections, many of them dualities:

Stone duality: More generally, if we let $\mathcal{T}$ be Boolean algebras (viewed as propositional theories), then $\mathcal{S}$ becomes Stone spaces (whose points are the ultrafilters, i.e. models of Boolean propositional theories),

Topological spaces and complete Heyting algebras: Generalizing to intuitionistic logic, we can let $pt \dashv \mathcal{O} : \mathsf{Esp}^{op} \longrightarrow \mathsf{Frm}$ [10], At this level, we get a logical connection; to obtain a duality, a restriction to sober spaces and spatial frames, respectively, is needed, but that is not required for our results,

Various spectral correspondences: $C \dashv S : \mathsf{Esp}^{op} \longrightarrow \mathsf{Rng}$, connecting topological spaces and rings (and leading to significant extensions of the notion of a logical theory),

Denotational semantics: and connections of domains and spaces with program logics [11].

The Schizophrenic object. The power set of a set can equally be represented as the family of functions from the set to the two-point set, $2 = \{0,1\}$, where one identifies a subset with its characteristic function. Dually, 2 is a Boolean algebra, and the set of atoms of a complete, atomic Boolean algebra B is in one-to-one correspondence with the Boolean algebra maps from B to 2. Thus, 2 is a primary example of a *schizophrenic object*, one which lives in both categories and that gives rise to a duality using the morphisms of the category. In general, when $\mathcal{S}$ and $\mathcal{T}$ have enough limits and colimits, and in particular a final object 1, then a connection between them can be viewed as homming into a "schizophrenic object" Ω, that lives in both categories, as the type $P1$ and space $M1$. Indeed, it is easy to see that these two objects have the same underlying set $Obs = |P1| = |M1|$.[2] For every space X we also have the canonical maps

[2] We write $|C| = \mathcal{C}(1, C)$ for any object C of a category $\mathcal{C}$.

$$\frac{\coprod_{|X|} 1 \longrightarrow X}{PX \longrightarrow P(\coprod_{|X|} 1) \xrightarrow{\sim} \prod_{|X|} P1}$$

where the isomorphism arises from the fact that $P : \mathcal{S}^{op} \longrightarrow \mathcal{T}$ is a right adjoint. Similarly, for every type A there is a canonical map $MA \longrightarrow \prod_{|A|} M1$. These maps are usually monic, which means that Ω is a cogenerator[3] both in $\mathcal{S}$ and in $\mathcal{T}$. Abusing notation, we define the functors $\Omega^X = \prod_{|X|} P1$ and $\Omega^A = \prod_{|A|} M1$, and arrive at monic natural transformations

$$PX \rightarrowtail \Omega^X \qquad \text{and} \qquad MA \rightarrowtail \Omega^A.$$

3 Process Logics as Test Algebras

Process logics are modal logics for describing the behavior of computational processes. Process formulas can be viewed as tests: a process *satisfies* a formula if and only if it passes the test that the formula represents.

The first and probably best known process logic is Hennessy-Milner logic [12], which will be presented in section 6.2. In fact, computational traces can be viewed as degenerate process formulas, with no logical operations, only modalities. On the other hand, dynamic logics can be viewed as a natural extension of process logics, where modalities are generated over arbitrary programs, and not just atomic actions.

In this work, process modalities are generated over a given alphabet Σ, representing atomic actions. Sometimes we distinguish the input alphabet Σ and the output alphabet Γ; or Σ represents the external actions (terminal symbols), and Γ the internal ones.

Besides modalities, process formulas are generated by various logical signatures, i.e. sets of logical connectors represented by the theory monad $T : \mathcal{T} \longrightarrow \mathcal{T}$. If a type $A \in \mathcal{T}$ is thought of as a set of propositional letters, then the type TA is the free propositional theory, containing all formulas generated by A in the given signature. E.g., if the only logical connector is conjunction, then TA is the free semilattice over A; but it has proven useful to also consider free commutative groups, rings, and even C^*-algebras of a certain type, as "logical" theories, generating tests for certain process behaviors. In all cases, the considered algebraic theories have a distinguished constant, denoting "truth", represented by a natural transformation $1 \xrightarrow{\top} T$.

Assumption: Ω is T-algebra. It is assumed that the schizophrenic object Ω comes equipped with a canonical algebraic structure $T\Omega \longrightarrow \Omega$, which lifts to all $TPX \longrightarrow PX$ along the inclusion $PX \rightarrowtail \Omega^X$.

[3] In fact, the duality of $\mathcal{S}$ and $\mathcal{T}$ is usually built by restricting them to the parts injectively cogenerated by the object Ω, embodying their connection.

3.1 Test Theories

Test theories are obtained by extending T-algebras ("propositional theories") by the modal operators generated by Σ. For example, if T is the power set functor, then we generate the so-called modal Boolean algebras by lifting actions of Σ on a transition system to modal operators for the power set of its state space. In general, a test theory is a (weak) algebra for either of the functors

$$F_0 X = TX + \Sigma \times X \qquad \text{or} \qquad F_1 X = T(\Sigma \times X)$$

In both cases the universal test theory is obtained as the initial weak algebra

$$\Theta_i = \mu X.\ F_i X$$

Tests are thus generated by the grammars

$$t_0 ::= \top \mid f(t_0, \ldots, t_0) \mid a.t_0 \qquad \text{and} \qquad t_1 ::= \top \mid f(a.t_1, \ldots, a.t_1)$$

where f a logical connector from the signature of T. By pre-composing with the monad T, we see that the weak F_0-algebra Θ_0 is a weak algebra for the functor T, while Θ_1 is just the free T-algebra for the monad T generated by Σ. In fact, Θ_1 is the initial *action algebra:*

Definition 1. *An* action algebra *for a monad $T : \mathcal{T} \longrightarrow \mathcal{T}$ and alphabet Σ is an algebra $TA \xrightarrow{\alpha} A$ for the monad T, together with a map $\Sigma \times A \xrightarrow{\cdot} A$, called* prefixing. *An action algebra homomorphism is a T-algebra homomorphism which also preserves prefixing.*

Proposition 1. *The free action algebra for the monad T and the alphabet Σ generated by B is the initial weak algebra $\Theta_B = \mu X.\ T(\Sigma \times (B + X))$.*

4 Automata and Processes as Coalgebras

Nondeterminism and more recently probabilistic choice are staples of computation. The constructors for choice operators are represented by a monad $S : \mathcal{S} \longrightarrow \mathcal{S}$.

Definition 2. *A (state)* machine *with inputs from Σ, outputs from Γ and final states predicated over Υ is represented by*

- *a coalgebra $X \longrightarrow GX$ where $GX = \Upsilon \times (S(\Gamma \times X))^{\Sigma}$*
- *an initial state $x \in X$.*

A process *is a machine where any state may be final, i.e. $\Upsilon = 1$. A process thus boils down to a coalgebra $\partial : X \longrightarrow (S(\Gamma \times X))^{\Sigma}$ and the initial state $x \in X$. A machine where $\Upsilon \neq 1$ is often called an* automaton. *When the coalgebra $X \longrightarrow GX$ is clear from the context, we speak of the automaton or process $x \in X$.*

A coalgebra structure of a machine consists of a pair $X \xrightarrow{\langle \Phi, \partial \rangle} \Upsilon \times (S(\Gamma \times X))^{\Sigma}$, where $\Phi : X \longrightarrow \Upsilon$ is the characteristic function of the final states, and $\partial : X \longrightarrow (S(\Gamma \times X))^{\Sigma}$ assigns to each state a choice of an output and a next state.[4] Final states are usually evaluated in the type of truth values $\Upsilon = L$. For the possibilistic automata, $\Upsilon = 2$, and $\Phi : X \longrightarrow 2$ is just the characteristic function of the set of final states. In general, Υ may be different from L, e.g. an arbitrary semiring [13].

The computational differences between *reactive* (or reading) machines, where $\Gamma = 1$, and *generating* (or writing) machines are discussed in [14]. Coalgebras $X \longrightarrow (S(\Gamma \times X))^{\Sigma}$ thus represent processes that both read and write, which is perhaps clearer in the transposed form $\Sigma \times X \longrightarrow S(\Gamma \times X)$.

Initially we focus on reactive processes, which are represented by the final weak coalgebra $\Xi = \nu X.\ (SX)^{\Sigma}$.

Assumption: Ω is S-algebra. It is assumed that Ω comes equipped with a canonical algebraic structure $S\Omega \longrightarrow \Omega$, which lifts to all $SMA \longrightarrow MA$ along the inclusion $MA \rightarrowtail \Omega^A$.

5 Testing Semantics

The behaviors of processes from Ξ are captured by testing whether they satisfy formulas from Θ and observing the results in Ω via $\Xi \times \Theta \xrightarrow{\mathbb{T}} \Omega$. However, since Ξ and Θ generally live in the different universes $\mathcal{S}$ and $\mathcal{T}$, respectively, their interaction can only be observed using the connection between these universes, in one of the two forms:

$$\frac{\Xi \xrightarrow{\models} \Omega^{\Theta}}{\Theta \xrightarrow{=\!|} \Omega^{\Xi}}$$

In general, given a coalgebra $X \longrightarrow GX$, and an algebra $A \longleftarrow FA$, we define two semantic maps

$$\frac{X \xrightarrow{\models} MA}{A \xrightarrow{=\!|} PX}$$

connected by the adjunction. Each state $x \in X$ induces a map $x \models (-) : A \longrightarrow \Omega$ which maps each piece of data $a \in A$ to the observation $(x \models a) \in \Omega$ in which the computation of x on a will result. Dually, each piece of data $a \in A$ induces a map

$$a \mathrel{=\!|} (-) \in PX \longrightarrow \Omega^X$$

[4] Anticipating semantics, we point out that the execution is always allowed to continue beyond a final state. This is in contrast with the *deadlock* states, which are represented by a choice functor G of the form $G = 1 + G'$. The deadlock states of a coalgebra $X \longrightarrow 1 + G'X$ are those that get mapped into 1.

which gives for each state $x \in X$ the observation $a \dashv x$. Theorem 1 below describes how these various views of semantics transform the algebraic structure of tests and the coalgebraic structure of processes.

5.1 Connecting Algebras and Coalgebras: Representation Theorem

Logical view. The logical operation of negation can be viewed as a very special case of a connection: if $\mathbb{A}$ is a pseudocomplemented lattice (Heyting algebra), then $\neg^{op} \vdash \neg : \mathbb{A}^{op} \longrightarrow \mathbb{A}$ is clearly a connection. Indeed, for every $\omega \in \mathbb{A}$, the operation $(-) \Rightarrow \omega : \mathbb{A}^{op} \longrightarrow \mathbb{A}$ is self adjoint. In posets and lattices, functors $F, G : \mathbb{A} \longrightarrow \mathbb{A}$ are monotone operators, algebras are super-fixpoints $a \geq Fa$, and colagebras are sub-fixpoints $a \leq Ga$; the initial algebra $\mu x.Fx$ is the least fixpoint, and the final coalgebra $\nu x.Gx$ is the greatest fixpoint.

For logical intuition, connections can be thought of as generalisations of negation. From that perspective, the following theorem can be viewed as a categorical elaboration of the fact that

$$\frac{Gx \leq \neg F \neg x}{\nu x.Gx \leq \nu x.\neg F \neg x \leq \neg \mu a.Fa}$$

What is the relevance of this fact? As explained in the introduction, the goal of this work is to explore the interplay of algebra and coalgebra in the theory of processes and in the practice of system specification. In practice, the behavior of a system is often specified as a quotient of a final coalgebra $\nu X.GX$ of processes using an initial algebra $\mu A.FA$ of tests. The connection $M^{op} \dashv P : \mathcal{S}^{op} \longrightarrow \mathcal{T}$ now allows deriving the semantics $\nu X.GX \xrightarrow{\models} M\mu X.FX$ if there is a distributive law $FP \longrightarrow PG$, i.e.

$$\frac{G \longrightarrow MFP}{\nu X.GX \longrightarrow \nu X.MFPX \longrightarrow M\mu A.FA}$$

The specified behavior is then the MFP-coalgebra L which is the image of $\nu X.GX$ in $\nu X.MFPX$. Furthermore, the carrier L can be conveniently represented as a subobject of $M\mu A.FA$. Informally, this is the content of the next theorem.

Relating a MFP-coalgebra and a M-image of a F-algebra requires a homomorphism which is consistent with the algebra and coalgebra structures both on the covariant and on the contravariant side of the correspondence (i.e., the "negation"). This is captured by the notion of *twisted* coalgebra homomorphisms, defined in the statement of the theorem.

Theorem 1. [5] *For a connection $M^{op} \dashv P : \mathcal{S}^{op} \longrightarrow \mathcal{T}$, endofunctors $G : \mathcal{S} \longrightarrow \mathcal{S}$ and $F : \mathcal{T} \longrightarrow \mathcal{T}$, and a distributive law $\lambda : FP \longrightarrow PG$ the following hold.*

[5] For simplicity and generality of the statement of the theorem, we avoid the finality and the initiality requirements, and spell out just the relations of F-algebras, and $G-$ and MFP-coalgebras.

(a) The predicate functor $P : \mathcal{S}^{op} \longrightarrow \mathcal{T}$ *lifts to* $\widehat{P} : (\mathcal{S}_G)^{op} \longrightarrow {}_F\mathcal{T}$, *mapping*

$$\frac{X \xrightarrow{\partial} GX}{\widehat{P}\partial : FPX \xrightarrow{\lambda} PGX \xrightarrow{P\partial} PX}$$

(b) $\widehat{P}$ *does not generally have an adjoint, but there is a correspondence of algebra homomorphisms and of* twisted *coalgebra homomorphisms*

$$\frac{\alpha \longrightarrow \widehat{P}\partial}{\Lambda\partial \longrightarrow M\alpha}$$

where $\Lambda : \mathcal{S}_G \longrightarrow \mathcal{S}_{MFP}$ *is the functor mapping the coalgebra* $X \xrightarrow{\partial} GX$ *to* $X \xrightarrow{\partial} GX \xrightarrow{\lambda'} MFPX$.

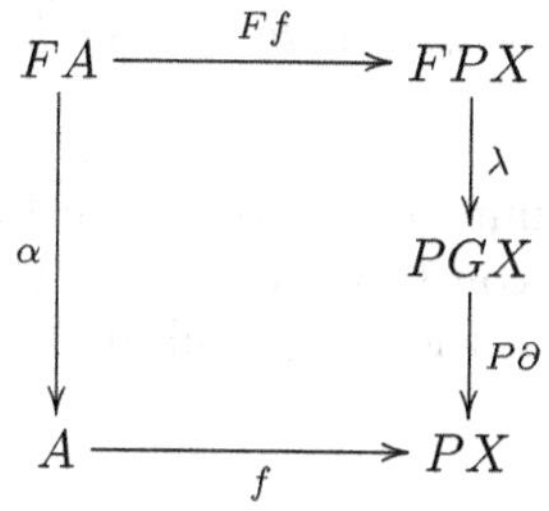

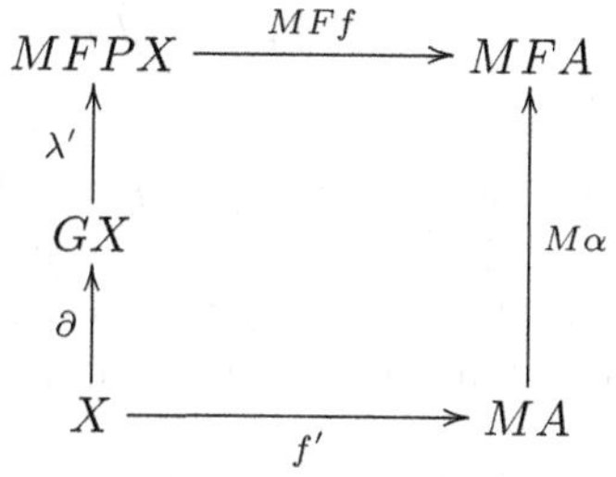

(c) If $\mathcal{T}$ *is a regular category, and* $F : \mathcal{T} \longrightarrow \mathcal{T}$ *preserves reflective coequalizers, then* ${}_F\mathcal{T}$ *is a regular category. In particular, every* F*-algebra homomorphism* $\alpha \xrightarrow{f} \widehat{P}\partial$ *has a regular epi-mono factorisation.*

(d) If $\mathcal{S}$ *is a regular category, and* MFP *preserves weak pullbacks, then every twisted coalgebra homomorphism* $\Lambda\partial \xrightarrow{f'} M\alpha$ *has a regular epi-mono factorisation, which induces a coalgebra* $\ell : L \longrightarrow MFPL$ *as the image of* $\Lambda\partial$.

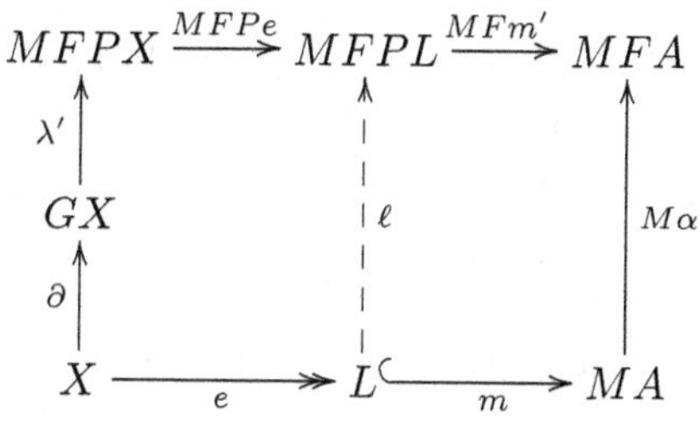

(e) If the coalgebra $X \longrightarrow GX$ *is final, then the coalgebra* $L \longrightarrow MFPL$ *is final if and only if the functor* $\Lambda : \mathcal{S}_G \longrightarrow \mathcal{S}_{MFP}$ *is essentially surjective.*

Comment. The correspondence $\mathcal{T}/P \cong \mathcal{S}/M^{op}$ thus lifts to ${}_F\mathcal{T}/\widehat{P} \cong \Lambda/M$, where the last denotes the comma construction for twisted homomorphisms. Abstractly, this does not seem like a very natural construction; the examples show that this is the ubiquitous framework where quotienting of the G-coalgebras ∂ induced by testing semantics takes place.

Definition 3. *A* duality *is a connection where the functors M and P are equivalences.*

Corollary 1. *Suppose that the connection $M^{op} \dashv P : \mathcal{S}^{op} \longrightarrow \mathcal{T}$ is a duality. Then the following are true.*
(f) The algebra $\alpha : FA \longrightarrow A$ is initial if and only if the coalgebra $M(\alpha \circ \varepsilon) : MA \longrightarrow MFPMA$ is final. When that is the case, then the behavior $\ell : L \longrightarrow MFPL$ is a subcoalgebra of the final MFP-coalgebra.

(g) If $F \cong PGM$ (equivalently $G \cong MFP$), then the behavior $\ell : L \longrightarrow MFPL$, constructed in Theorem 1, is isomorphic to the coalgebra $\partial : X \longrightarrow GX$. If $\partial : X \longrightarrow GX$ is final and $\alpha : FA \longrightarrow A$ is initial, then $\partial \cong M\alpha$ and $\alpha = P\partial$.

In many cases, the functor $F = PGM$ has a simpler representation than G, and the initial algebra $\Theta = \mu A.FA$ is easier to construct in $\mathcal{T}$ than the final coalgebra of $\Xi = \nu X.GX$ is in $\mathcal{S}$. In such cases, the isomorphism $\Xi = M\Theta$ offers significant technical advantages [4].

5.2 Specifying Semantics

Given a coalgebra $X \xrightarrow{\partial} GX$ and the initial test algebra $F\Theta \xrightarrow{\varrho} \Theta$, we define a semantics $X \xrightarrow{\models} M\Theta$ by induction over Θ, using the fact that Ω is a T-algebra in $\mathcal{T}$ and an S-algebra in $\mathcal{S}$ — i.e. that each PX is a T-algebra in $\mathcal{T}$, whereas each $M\Theta$ is an S-algebra in $\mathcal{S}$. Given an initial state x of a machine X, we define a map $x \models (-) : \Theta \longrightarrow \Omega$.

Loose tests. Since an element of $\Theta_0 = \mu X.\, TX + \Sigma \times X$ is in the form

$$t ::= \top \mid f(t_0 \ldots t_n) \mid a.t$$

where $\top$ is the distinguished constant of the algebraic theory of the monad T, and f is an operation from that theory

$$(x \models \top) = \top \quad (1)$$

$$(x \models f(t_0 \ldots t_n)) = f((x \models t_0) \ldots (x \models t_n)) \quad (2)$$

$$(x \models a.t) = (\delta(x, a) \models t) \quad (3)$$

where $\delta : X \times \Sigma \longrightarrow SX$ is the transpose of $X \xrightarrow{\partial} GX = (SX)^{\Sigma}$, and $\models$ extends along $X \xrightarrow{\models} SX \longrightarrow M\Theta_0 \longrightarrow \Omega^{\Theta_0}$.

Remark. Clauses (1) and (2) say that $x \models (-) : A \longrightarrow \Omega$ is a T-algebra homomorphism. Clause (3) extends $x \models (-)$ beyond $T\Theta_0$ to $\Sigma \times \Theta_0$, using the fact that Ω (viz $M\Theta_0$) is an S-algebra, and extending $X \xrightarrow{\models} M\Theta_0$ to an S-algebra homomorphism $SX \xrightarrow{\models} M\Theta_0$.

Tight tests. Since an element of $\Theta_1 = \mu X.\ T(\Sigma \times X)$ is in the form

$$t ::= \top \mid f(a_0.t_0 \ldots a_n.t_n)$$

the semantics retains clause 1, deletes clause 3, and replaces clause 2 with

$$(x \models f(a_0.t_0 \ldots a_n.t_n)) = f\big((\delta(x, a_0) \models t_0) \ldots (\delta(x, a_n) \models t_n)\big)$$

Note further that testing a coalgebra $X \xrightarrow{\langle \Phi, \partial \rangle} \Omega \times GX$, where $\Phi : X \longrightarrow \Omega$ denotes the final states, changes the base clause of semantics to $\big(x \models \top\big) = \Phi(x)$.

6 Possibilistic Semantics

Possibilistic semantics is evaluated in $\Omega = \{0, 1\}$. In the simplest case, both state spaces and data types are modeled in the same universe $\mathcal{S} = \mathcal{T} = \mathsf{Set}$ of sets and functions. The contravariant powerset functor is self-adjoint $\wp^{op} \dashv \wp :$ $\mathsf{Set}^{op} \longrightarrow \mathsf{Set}$, and maps a state to the type of predicates over it, and a type to the space of its models.[6]

Possibilistic systems. Possibilistic nondeterminism means that there can be several possible transitions from a state $x \in X$, for a given action $a \in \Sigma$. The choice monad is thus based on the (covariant) finite powerset functor $S = \wp_f :$ $\mathcal{S} \longrightarrow \mathcal{S}$. Simple processes are thus coalgebras in the form $X \longrightarrow (\wp_f X)^{\Sigma}$, or $X \longrightarrow \wp_f(\Sigma \times X)$.

6.1 Linear Semantics: Trace Testing

A trace semantics describes computations over strings of symbols. The tests are thus pure modal formulas, with no logical operations except the constant $\top$. The logic monad is thus the smallest possible: $TA = \top$, for all $A \in \mathcal{T} = \mathsf{Set}$. The loose and the tight semantics for it coincide, and the test algebra Θ is initial for $FA = 1 + \Sigma \times A$, i.e. the free monoid Σ^*. Trace semantics have been investigated as an extension of coalgebraic methods in [8, 9]. We describe three examples.

[6] In the Hennessy-De Nicola [15] style testing semantics, tests are a special class of processes. In our testing framework, this means that tests and processes live in the same universe $\mathcal{S} = \mathcal{T}$, and moreover that the test algebra $F\Theta \longrightarrow \Theta$ is contained in (can be completed to) a choice coalgebra $\Xi \longrightarrow G\Xi$. Indeed, the trace algebra $\Theta = \Sigma^*$ is a coalgebra $\Sigma^* \longrightarrow \Sigma \times \Sigma^* \longrightarrow \wp_f(\Sigma \times \Sigma^*)$.

Finite state automata. Possibilistic automata are coalgebras in the form $X \xrightarrow{\langle \Phi, \partial \rangle} 2 \times \wp_f(\Sigma \times X)$. The trace semantics of finite state automata is obtained by instantiating (1-3)

$$(x \models \top) = \Phi(x) \tag{4}$$

$$(x \models a.t) = \bigvee_{x \xrightarrow{a} y} (y \models t) \tag{5}$$

where $x \xrightarrow{a} y$ means that $y \in \delta(x, a)$. Note that (4) says that $x \models (-) : \Theta \longrightarrow \Omega$ only preserves $\top$ where Φ holds. The final states Φ are an explicit relativisation of the $\top$-preservation requirement; the semantics $x \models (-) : \Theta \longrightarrow \Omega$ is a T-algebra homomorphism up to Φ.

Let Aut denote bisimulation classes of finite-state automata, let $GX = 2 \times \wp_f(\Sigma \times X) \cong \wp_f(1 + \Sigma \times X)$, and let $\mathsf{Aut} \longrightarrow G(\mathsf{Aut})$ be final for all *finite* G-coalgebras. Then the trace semantics $\mathsf{Aut} \xrightarrow{\models} \wp\Sigma^*$ maps each automaton $x \in \mathsf{Aut}$ to the language $L_x = \{\sigma \in \Sigma^* \mid x \models \sigma\}$.

Pushdown automata. While finite state automata behaviors were obtained by structuring the alphabet Σ, pushdown automata are obtained by structuring the state spaces X. Fix a set Γ, to be used as "non-terminal" symbols, and extend each state space X by the free monoid action to $X \times \Gamma^*$. A pushdown automaton is a coalgebra for the functor $G : \mathcal{S} \longrightarrow \mathcal{S}$, defined

$$GX = 2 \times \wp_f(X \times \Gamma^*)^{\Sigma+1}$$

where the "blank" symbol $\sqcup \in 1$ allows pure non-terminal rewrites. A start non-terminal symbol $Z_0 \in \Gamma$ is assumed to be distinguished, or freely added. The test algebra is still the same, $\Theta = \Sigma^*$.

Turing Machines. Turing machines act on tapes. The obvious idea is to view the contents of a tape as a test. The problem is that the essential property of the tape is that it can be extended in both directions, so at the first sight, the Turing machine interaction does not seem not fit naturally into the inductive testing framework.

Another look at the acceptance condition for Turing machines offers a solution. A Turing machine X accepts a word $t \in \Sigma^*$ if and only if reaches a final state, in any configuration, after having started a computation with the head just to the left of the word t, presented on the tape. — So the accepted words initially extend to the right of the head. The left part of the tape is only used for intermediary computation.

A Turing machine can thus be modeled following the idea of a pushdown automaton: the tape to the left of the head can be viewed as a *stack*, and treated as a part of the state; the tape to the right of the head can be construed as another stack, containing the actual test. Unlike a pushdown automaton, a Turing machine allows words in the same alphabet in both stacks. A pushdown automaton had two disjoint alphabets, Γ and Σ for the left and the right stack, respectively.

Moreover, the right "stack" of a pushdown automaton is not a real stack, since it only allows popping.

A Turing machine can thus be viewed as a machine with two real stacks, representing the two parts of its tape, on the two sides of the head. Just like a pushdown automaton, besides the alphabet Σ, it may allow non-terminal symbols, at least $\sqcup$, used in computation, but not in the tested words.

A nondeterministic Turing machine is thus a coalgebra $X \xrightarrow{\langle \Phi, \partial \rangle} 2 \times \wp_f(X \times \Gamma \times \{\lhd, \rhd\})^{\Gamma}$, where $\Gamma \supseteq \Sigma + \{\sqcup\}$. As before, the component $X \xrightarrow{\Phi} 2$ marks the final states, whereas the transition function $X \times \Gamma \xrightarrow{\delta} \wp_f(X \times \Gamma \times \{\lhd, \rhd\})$ assigns to each state and each input the possible next states, outputs, and the direction for the move of the head. We represent the move of the head by popping a symbol from one stack and pushing it onto the other.

6.2 Branching Semantics: Set-Tree Testing

Here not only are the universes $\mathcal{S}$ and $\mathcal{T}$ identical, but we also take the logic monad T to be the same as the choice monad S: they are both the finite powerset $\wp_f : \mathsf{Set} \longrightarrow \mathsf{Set}$. So both the space of the choices $\wp_f X$ and the logic of tests $\wp_f A$ are free semilattices. But the two lattices will be used differently: the former as a join semilattice (because the process can continue with this computation *or* with that computation...), and the latter as a meet semilattice (because the testing formula is a conjunction).

Remark. The same class of computational behaviors could be formalized by taking either of the monads T and S, or both of them, to be the diagonal functor $\Delta X = X \times X$. This would just mean that nondeterministic branching would always be binary, and that tests would be just binary conjunctions. Associativity, commutativity and idempotence of these operations would be imposed later. The intermediary options would be to take the functor $\wp_{\leq 2} X = \{x_0, x_1\}$ of (at most) two-element subsets, imposing commutativity and idempotence, and leaving out associativity.

Two-way simulation. In the simplest case $TA = \wp_f A$. The tests are thus in the form

$$t ::= \top \mid t \wedge t \cdots \wedge t \mid a.t$$

where $\wedge$ is an associative, commutative, idempotent operation with unit $\top$. The semantics (1-3) becomes

$$(x \models \top) = \top$$
$$(x \models \bigwedge_{i=1}^{n} t_i) = \bigwedge_{i=1}^{n} (x \models t_i)$$
$$(x \models a.t) = \bigvee_{y \in \delta_\kappa(x,a)} (y \models t)$$

The functors generating data and processes are thus

$$FA = \wp_f A + \Sigma \times A$$
$$GX = \wp_f(\Sigma \times X)$$

Proposition 2. *Let Θ be the initial F-algebra and Ξ the final G-coalgebra. Let the partial order on X be defined by*

$$x \leq y \iff \forall t \in \Theta.\ (x \models t) \leq (y \models t) \tag{6}$$

Then the process x can be simulated by the process y if and only if $x \leq y$, i.e.

$$x \leq y \iff \forall a \in \Sigma \forall x' \in \delta(x,a) \exists y' \in \delta(y,a).\ x' \leq y' \tag{7}$$

Bisimulation. Adding negation to the logic

$$t ::= \top \mid t \wedge t \cdots \wedge t \mid \neg t \mid a.t$$

i.e. testing by

$$FA = \wp_f A + A + \Sigma \times A$$

the semantics is extended by the clause

$$(x \models \neg t) = \neg(x \models t)$$

This gives an interesting strengthening of the testing power.

Proposition 3. *The equivalence*

$$x \sim y \iff \forall t \in \Theta.\ (x \models t) = (y \models t)$$

means just that the processes x and y are bisimilar

$$\begin{aligned} x \sim y \iff & \forall a \in \Sigma \\ & (\forall x' \in \delta(x,a) \exists y' \in \delta(y,a).\ x' \sim y') \wedge \\ & (\forall y' \in \delta(y,a) \exists x' \in \delta(y,a).\ x' \sim y') \end{aligned}$$

Strong Bisimulation by Stone Duality. Strong bisimilarity is classified by the final coalgebra $\Xi \longrightarrow \wp_f(\Sigma \times \Xi)$. Using the restriction of the Stone duality $S \dashv C : \mathsf{Set}^{op} \longrightarrow \mathsf{caBa}$ from Stone spaces and Boolean algebras to sets (discrete spaces) and complete atomic Boolean algebras, allows applying Corollary 1. Setting $FA = C\wp_f(\Sigma \times SA)$ allows a representation of the bisimulation classes as characters of the Boolean algebra $\Theta = \mu A.FA$.

7 Summary, Related and Future Work

We have proposed combining coalgebras as models of processes with algebras as models of their testing regimes via logical connections between the two, in order to realize models that capture the interactions of processes with the data on which they operate, as well as to model the processes themselves. Our Main Thereom 1 is the key to this approach. In addition, we have illustrated our approach with standard examples: finite automata, pushdown automata and Turing machines, as well as results that show how our approach captures simulation and strong bisimulation of processes in concurrency.

There is a wealth of work using duality theory – especially Stone Duality – emanating from the seminal work of Abramsky [16] and the work on coalgebras exemplified by Rutten's [17] and the work of Plotkin and Turi [18]. None of these works has the same aims as our work. Indeed, the work along this line has been aimed at achieving a setting in which both operational and denotational models of the same language or process algebra could be presented and related.

The closest work to what we have presented is that of Kupke, Kurz and Pattinson [19] and of Bonsangue and Kurz [20]. The former uses similar theoretical machinery – but restricted to duality theories, and applies it to the study of finitary modal logics as specification languages for Set-coalgebras. The latter also models transition systems as coalgebras, and then uses a duality to arrive at a logic for the transition system. Their main result is the soundness, completeness and expressiveness of the logic. They also extend to the setting of Vietoris coalgebras on topological spaces, and apply it to derive adequate logics on posets, sets, spectral spaces and Stone spaces. The logics in these works employ the usual modalities, possibility and necessity. In addition, the results rely heavily on the duality theory to transfer initial algebras to final coalgebras and back.

As we stated in earlier, our approach has a rather different goal, and employs weaker assumptions. Our goal is to understand the interactions of a state machine and the data on which it operates during computation. These are fundamentally different objects – programs are executed, but data are not. Our work is based on logical connections, and does not require a duality theory. In fact, one could argue that our results begin when the connection used is not a duality. In addition, we are dealing with process logics where process formulas are a possible interpretation of tests: a process modality $\langle a \rangle$ is a test assigned to the action $a \in \Sigma$. That said, we believe the present paper has just begun to scratch the surface, and a lot remains to do. A primary goal is to present probabilistic systems from this perspective, and, in particular to apply it to extend the work in [4], as well as to explore the relationship between probability and nondeterminism, as presented, e.g., in [21].

References

1. Moore, E.F.: Gedanken experiments on sequential machines. In: Automata Studies, Princeton (1956) 129–153
2. Holzmann, G.J.: Design and Validation of Computer Protocols. Software Series. Prentice Hall, London (1991)
3. Lee, D., Yannakakis, M.: Principles and methods of testing finite state machines - A survey. In: Proceedings of the IEEE. Volume 84. (1996) 1090–1126

4. Mislove, M., Ouaknine, J., Pavlovic, D., Worrell, J.: Duality for labelled Markov processes. In Walukiewicz, I., ed.: Proceedings of FoSSaCS 2004. Volume 2987 of Lecture Notes in Computer Science., Springer Verlag (2004) 393–407
5. Pavlovic, D., Smith, D.R.: Composition and refinement of behavioral specifications. In: Automated Software Engineering 2001. The Sixteenth International Conference on Automated Software Engineering, IEEE (2001)
6. Pavlovic, D., Smith, D.R.: Guarded transitions in evolving specifications. In Kirchner, H., Ringeissen, C., eds.: Proceedings of AMAST 2002. Volume 2422 of Lecture Notes in Computer Science., Springer Verlag (2002) 411–425
7. Pavlovic, D., Pepper, P., Smith, D.R.: Colimits for concurrent collectors. In Dershowitz, N., ed.: Verification — Theory and Practice. Essays Dedicated to Zohar Mana on the Occasion of His 64th Birthday. Volume 2772 of Lecture Notes in Computer Science., Springer Verlag (2003) 568–597
8. Jacobs, B.: Trace semantics for coalgebras. In: Coalgebraic Methods in Computer Science (CMCS) 2004. Volume 106 of Electr. Notes in Theor. Comp. Sci. (2004)
9. Hasuo, I., Jacobs, B.: Context-free languages via coalgebraic trace semantics. In Fiadeiro, J., Harman, N., Roggenbach, M., Rutten, J., eds.: Algebra and Coalgebra in Computer Science (CALCO'05). Volume 3629 of Lecture Notes in Computer Science., Berlin, Springer-Verlag (2005) 213–231
10. Johnstone, P.: Stone Spaces. Number 3 in Cambridge Studies in Advanced Mathematics. Cambridge University Press (1982)
11. Abramsky, S.: Domain theory in logical form. Annals of Pure and Applied Logic **51** (1991) 1–77
12. Milner, R.: Communication and concurrency. International Series in Computer Science. Prentice Hall, London (1989)
13. Rutten, J.: Behavioral differential equations: a coinductive calculus of streams, automata and power series. Theor. Comp. Sci. **308** (2003) 1–53
14. Glabbeek, R., Smolka, S.A., Steffen, B.: Reactive, generative, and stratified models of probabilistic processes. Inf. Comput. **121** (1995) 59–80
15. DeNicola, R., Hennessy, M.: Testing equivalences for processes. Theoretical Computer Science **34** (1984) :83–133
16. Abramsky, S.: A domain equation for bisimulation. Information and Computation **92** (1991) 161–218
17. Rutten, J.: Universal coalgebra: a theory of systems. Theor. Comp. Sci. **249** (2000) 3–80
18. Plotkin, G.D., Turi, D.: Towards a mathematical operational semantics. (In: Proceedings of LICS)
19. C. Kupke, A.K., Pattinson, D.: Ultrafilter extensions for coalgebras. In: Proceedings of CALCO. Volume 3629 of Lecture Notes in Computer Science. (2005) 263–277
20. Bonsangue, M., Kurz, A.: Duality for logics of transition systems. In: Proceedings of FoSSaCS. Volume 3441 of Lecture Notes in Computer Science. (2005) 455–469
21. Varacca, D.: The powerdomain of indexed valuations. In: Proceedings 17th IEEE Symposium on Logic in Computer Science, IEEE Press (2002)

Tableaux for Lattices

Georg Struth

Department of Computer Science, The University of Sheffield
Regent Court, 211 Portobello Street, Sheffield, S1 4DP, UK
g.struth@dcs.shef.ac.uk

Abstract. We formally derive tableau calculi for various lattices. They solve the word problem for the free algebra in the respective class. They are developed in and integrated into the ordered resolution theorem proving framework as special-purpose procedures. Theory-specific and procedural information is included by rewriting techniques and by imposing the subformula property on the ordering constraints. Intended applications include modal logic and the automated proof support for set-based formal methods. Our algebraic study also contributes to the foundations of tableau and sequent calculi, explaining the connection of distributivity with the data-structure of sequents and with cut-elimination.

1 Introduction

The development of focused algebraic calculi, the integration of theory-specific knowledge, is an important, but difficult task in automated deduction. For complex theories, when axiomatic reasoning is hopeless, it is indispensable to derive theory-specific inference rules in a systematic way from a given axiomatisation. Recently, such a derivation method has been proposed for the ordered resolution framework [12] and applied to algebraic structures of considerable complexity [13, 14, 15]. This includes lattice-based calculi for sets with applications to formal methods like B or Z. While these calculi treat elementary theories, the question of integrating special-purpose (decision) procedures for equational reasoning into the ordered resolution framework is also very interesting.

A second, more technical question is the following. Using the derivation method, resolution-like calculi at the lattice-level have been constructed within resolution at the logical meta-level by encoding multisets—the natural data-structure for clauses—at the lattice-level in the ordering constraints for resolution [13, 14]. Therefore, can one in a similar way enforce the construction of tableaux within resolution by encoding the subformula property?

A third, more foundational question follows from the observation that a cut-like rule for lattices characterises precisely distributivity of the lattice (cf. [11]). So what is the connection with the sequent calculus? How is distributivity handled there? What is the algebraic role of the structural rules and of cut-elimination in the context of lattices?

The present paper answers these questions. First, we show how to integrate tableau-like decision procedures for semilattices and distributive lattices into the

M. Johnson and V. Vene (Eds.): AMAST 2006, LNCS 4019, pp. 323–337, 2006.

ordered-resolution framework. Second, we extend these procedures to further interesting cases: to Boolean lattices, to certain modal logics and to operational reasoning with sets. Third, we show that the derivation method supports a natural synthesis of tableaux by integrating rewriting techniques and by encoding the subformula property in the ordering constraints of resolution. The trick is to develop tableaux using negative literals. This allows one to capture also demonic choice, which is essential for tableaux but not expressible by rewriting. Fourth, we provide new insights into the correspondence between algebra and logic by comparing word problems for lattices with sequent calculi and tableaux. Our analysis of the distributivity law explains the algebraic role of the data-structure of sequents and the cut rule in the sequent calculus. Our derivation yields not only an algebraic reconstruction, but also a novel algebraic completeness proof of propositional tableaux. In particular, this formally demonstrates that the rules of the sequent calculus are independent in a strictly formal sense. Finally, the paper is not only reconstructive. The integration of tableaux into the ordered resolution framework makes strong redunancy elimination techniques available to tableaux. This may be a considerable benefit in applications.

Here, we can only sketch some proofs. An extended version can be found at the author's web-site. We also assume knowledge on tableau and sequent calculi (we identify tableaux with cut-free sequent calculi). See [10, 3] for introductions.

2 Resolution and Redundancy

This section revisits some well-known results, most of them originating in [9]. Let $T_\Sigma(X)$ be a set of terms with signature Σ and variables in X. A term is *ground* if no variable occurs in it. An *atomic formula* is an expression $p(t_1,\dots,t_n)$, where p is an n-ary predicate symbol and $t_1,\dots,t_m \in T_\Sigma(X)$. A *literal* is an atomic formula ϕ (*positive literal*) or its negation $\neg\phi$ (*negative literal*). A *clause* is a finite multiset of literals. A clause is positive (negative) if it consists solely of positive (negative) literals or if it is empty. A *Horn clause* contains at most one positive literal. A *clause set* is a set of clauses. If Γ is a clause and ϕ a literal, we write Γ,ϕ instead of $\Gamma \cup \{\phi\}$. We denote clauses by $\Gamma \longrightarrow \Delta$, where Γ (Δ) is a multiset of negative (positive) literals.

We consider calculi constrained by syntactic orderings. This may considerably narrow the search space. *Term* and a *literal orderings* $\prec$ are well-founded total orderings on the respective ground expressions. They are lifted to non-ground expressions by stipulating $e_1 \prec e_2$ iff $e_1\sigma \prec e_2\sigma$ for all ground substitutions σ. A literal l is *maximal* with respect to a multiset Γ of literals if $l \nprec l'$ for all $l' \in \Gamma$. It is *strictly maximal* with respect to Γ if $l \npreceq l'$ for all $l' \in \Gamma$. The non-ground orderings are still well-founded, but need no longer be total.

Literal orderings are extended to clauses, measuring clauses as multisets of literals and comparing them via the multiset extension of the literal ordering. A literal is assigned greater weight when it is negative than when it is positive. See Section 5 for more details. A clause ordering inherits totality and well-foundedness from the literal ordering. Again, the non-ground extension need not be total. We usually denote all syntactic orderings by $\prec$.

Definition 1 (Ordered Resolution Calculus). *Let $\prec$ be a literal ordering. The* ordered resolution calculus OR *consists of the following deduction inference rules. The* ordered resolution rule

$$\frac{\Gamma \longrightarrow \Delta, \phi \qquad \Gamma', \psi \longrightarrow \Delta'}{\Gamma\sigma, \Gamma'\sigma \longrightarrow \Delta\sigma, \Delta'\sigma}, \tag{Res}$$

where σ is a most general unifier of ϕ and ψ, $\phi\sigma$ is strictly maximal with respect to $\Gamma\sigma, \Delta\sigma$ and maximal with respect to $\Gamma'\sigma, \Delta'\sigma$. The ordered factoring rule

$$\frac{\Gamma \longrightarrow \Delta, \phi, \psi}{\Gamma\sigma \longrightarrow \Delta\sigma, \phi\sigma}, \tag{Fact}$$

where σ is a most general unifier of ϕ and ψ and $\phi\sigma$ is strictly maximal with respect to $\Gamma\sigma$ and maximal with respect to $\Delta\sigma$.

In all inference rules, *side formulas* are the parts of clauses denoted by capital Greek letters. Literals occurring explicitly in the premises are called *minor formulas*, those in the conclusion *principal formulas.*

Let S be a clause set and $\prec$ a clause ordering. A clause Γ is *$\prec$-redundant* or simply redundant in S if it is a semantic consequence of instances from S which are all smaller than Γ with respect to $\prec$. A ground inference is *redundant* in S if either the maximal premise is redundant or else its conclusion is a semantic consequence of instances from S which are all smaller than the maximal premise with respect to $\prec$. An inference is *redundant* if all its ground instances are. Closing S under OR up to redundant inferences and eliminating redundant clauses on the fly transforms S into an *ordered resolution basis* (an *orb*).

As usual, an OR-proof is a finite tree whose nodes are labelled by clauses and whose edges are determined by OR-inferences. An OR-*refutation* from a clause set S is an OR-proof with all leaves in S and with the empty clause as root.

Proposition 1

(i) Orbs of inconsistent clause sets contain the empty clause.
(ii) Fair OR*-implementations refute inconsistent clause sets in finite time.*
(iii) For every inconsistent clauses set containing an orb there is a refutation in which no OR*-inference has both premises from the orb.*

3 The Derivation Method

We now recall the derivation method for focused calculi with theory-specific inference rules from [12]. It is of general interest for compiling algebraic knowledge into resolution-based theorem proving. It has a syntactic and a semantic side.

At the syntactic side, consider a partition of some clause set into a set T of *theory clauses* and a set S of *non-theory* clauses. We intend to internalise T into a set of derived inference rules in refutations. The (ground) chaining rule

$$\frac{\Gamma \rightarrow \Delta, a \leq b \qquad \Gamma' \rightarrow \Delta', b \leq c}{\Gamma, \Gamma' \rightarrow \Delta, \Delta', a \leq c},$$

for instance, internalises the instance $a \leq b, b \leq c \rightarrow a \leq c$ of the transitivity law in a two-step resolution proof. In general, this internalisation is possible if there exists an OR-refutation of $S \cup T$ in which theory clauses are "sufficiently separated" to admit proof patterns in which all but one literal of a theory clause can successively be consumed by non-theory clauses. While in the case of non-ordered resolution such theory resolution rules can quite easily be established, here, the permutation-invariance of refutations is strongly restricted by the ordering constraints. This makes the extraction of such patterns non-trivial. See for instance [14] for a general discussion. Here, however, due to our special problem structure, we need only assume that the theory clauses form an orb. Then, by Proposition 1, they are sufficiently separated.

These observations suggest the following three-step scenario: For a given theory specification T, (1)construct an orb of T, (2) extract focused inference rules from the interaction of non-theory clauses with the orb and (3) lift ground inference rules to the non-ground level.

An essential feature is the modularity of orb constructions. Incrementing a theory specification, an orb need not be recompiled. Only the effect of the new clauses on the orb must be determined.

At the semantic side of the derivation method, we use two ways to integrate declarative and procedural algebraic knowledge. First, by selecting an appropriate theory specification. Here, in particular, by a characterisation of distributivity in terms of a cut-rule. Second, by choosing the syntactic orderings $\prec$. We will essentially enforce tableau rules by encoding the subformula property.

4 Lattices

We study word problems for free lattices. Our signature is $\Sigma = \{\sqcup, \sqcap\}$. Its elements are varyadic operation symbols denoting the lattice join and meet operations. Besides equality, $\leq$ is the only (binary) predicate symbol of our language. It denotes a partial ordering. As usual, a *join semilattice* is a poset closed under finite least upper bounds or joins. Dually, a *meet semilattice* is closed under finite greatest lower bounds or meets. (In lattice theory, the *dual* of a statement is obtained by interchanging joins and meets and converting the ordering.) A *lattice* is a poset that is both a join and a meet semilattice. It is *distributive* if (cut) holds (see below). See [11] for further discussion, including the relevance of (cut) to lattice-word problems and resolution. A similar non-standard axiomatisation of distributivity has been used earlier in [7].

The antisymmetry law of a poset can be given a special treatment. We only need it for decomposing an identity $s \approx t$ between lattice terms into the equivalent expression $s \leq t \wedge t \leq s$. This can be done as a preprocessing, for instance during the transformation to clause normal form. We thus eliminate the predicate symbol $\approx$ from our language and use only the pre-ordering axioms of reflexivity and transitivity. Consider the following clausal axioms.

$$\longrightarrow x \leq x, \tag{ref}$$

$$x \leq y, y \leq z \longrightarrow x \leq z, \tag{trans}$$

$$\longrightarrow x \sqcap y \leq x, \qquad \longrightarrow x \sqcap y \leq y, \tag{lb}$$

$$x \leq y, x \leq z \longrightarrow x \leq y \sqcap z, \tag{glb}$$

$$\longrightarrow x \leq x \sqcup y, \qquad \longrightarrow y \leq x \sqcup y, \tag{ub}$$

$$x \leq z, y \leq z \longrightarrow x \sqcup y \leq z, \tag{lub}$$

$$x_1 \leq y_1 \sqcup z, x_2 \sqcap z \leq y_2 \longrightarrow x_1 \sqcap x_2 \leq y_1 \sqcup y_2. \tag{cut}$$

The class of join semilattices, meet semilattices, lattices and distributive lattices are axiomatised by the sets

$$J = \{(\text{ref}), (\text{trans}), (\text{lub}), (\text{ub})\}, \qquad M = \{(\text{ref}), (\text{trans}), (\text{glb}), (\text{lb})\},$$
$$L = J \cup M, \qquad D = L \cup \{(\text{cut})\}.$$

Joins and meets are associative, commutative, idempotent ($x \sqcap x = x = x \sqcup x$) and isotone in the associated ordering. We will henceforth consider all inequalities modulo associativity and commutativity. The similarities between the rules in J, M, D and those of the sequent or tableau calculus are already quite apparent. (glb) and (lub) are similar to the right conjunction and left disjunction rule, the similarity between (cut) and the cut rule is evident. (glb) and (lub) will later be transformed into equivalent, but more meaningful rules that correspond to the left conjunction and right disjunction rules.

Let K be some variety of lattices. The *word problem* for K is the following: Determine if an identity $s \approx t$ over some set of constants (or generators) in the language for K holds in K or, equivalently, in the free algebra in K. This is the case iff s and t are congruent modulo the equational axioms of K. A *solution* to the word problem is an algorithm that decides the problem for all inputs. Since $s \leq t \Leftrightarrow s \sqcup t = t \Leftrightarrow s \sqcap t = s$ holds, respectively, for join and meet semilattices, we identify word and reachability problems as well as inequalities and identities.

In the context of ordered resolution, instead of solving some set of positive identities, we attempt to refute a set of negative ones. Now, when the class K is axiomatised by an orb consisting of a set of Horn clauses, the resolution process has a particularly simple structure. First, ordered factoring is never applicable. Second, the theory clauses serve as (independent) closure rules that generate new negative clauses from old ones. Third, because of this particular structure, the orb must contain a positive literal whenever the process generates the empty clause. To obtain a decision procedure, it suffices that the input clause is maximal with respect to the syntactic ordering in the closure induced by the process and that the number of ground clauses smaller than the input clause is finite. It is well-known that the free semilattice, distributive lattice and Boolean lattice generated by a finite set of constants is finite (cf [1]). Obviously, therefore, one should try to construct a syntactic ordering that is compatible with these conditions. This is the purpose of the following section.

5 Syntactic Orderings

There is a natural syntactic ordering for the sequent calculus: any ordering enforcing the subformula property. Here, any AC-compatible simplification ordering will serve our purposes. Roughly, an ordering is AC-compatible if it respects AC-equivalence classes. Orderings that are appropriate for our purposes exist [2]. A simplification ordering contains in particular the subterm ordering: Every term is greater than each of its subterms. One can for instance choose an AC-compatible ordering with a precedence in which the join and meet operation are maximal and identical. Let $\prec$ be such an ordering. Like in Section 2, we now extend this ordering to literals and clauses.

Let $\mathbb{B}$ be the two-element Boolean algebra with ordering $<_{\mathbb{B}}$. Let $M = G \times \mathbb{B} \times \mathbb{B} \times G$, where G denotes a multiset of generators. Let A be a set of atoms occurring in some clause $C = \Gamma \longrightarrow \Delta$. The ordering $\prec_1 \subseteq M \times M$ is the lexicographic combination of $\prec$ for the first and last component of M and $<_{\mathbb{B}}$ for the others. A ground *literal measure* (for clause C) is the mapping $\mu_C : A \longrightarrow M$ defined by $\mu_C : \phi \mapsto (t_\nu(\phi), p(\phi), s(\phi), t_\mu(\phi))$ for each (ground) literal $\phi \in A$ occurring in C. Hereby $t_\nu(\phi)$ $(t_\mu(\phi))$ denotes the maximal (minimal) term with respect to $\prec$ in ϕ. $p(\phi) = 1$ $(p(\phi) = 0)$ if ϕ occurs in Γ (in Δ). $s(\phi) = 1$ $(s(\phi) = 0)$ if $\phi = s < t$ and $s \succeq t$ $(s \prec t)$. The (ground) *literal ordering* $\prec_2 \subseteq A \times A$ is defined by $\phi \prec_2 \psi$ iff $\mu_C(\phi) \prec_1 \mu_C(\psi)$ for $\phi, \psi \in A$. Hence $\prec_2$ is embedded in $\prec_1$ via the literal measure. The ordering $\prec_1$ is total and well-founded by construction. Via the embedding, $\prec_2$ inherits these properties. See [12] for a motivation of the components arising in a similar ordering. Intuitively, the syntactic ordering enforces that all non-theory clauses are split into clauses containing only subterms by the clauses in D. This yields the subformula property of the sequent calculus.

All orderings are extended to the non-ground case and to clauses as described in Section 2. In unambiguous situations we will denote them all by $\prec$.

In [13, 14], a similar construction enforces resolution-like calculi at the lattice level within ordered resolution at the meta-level. The specific difference is that there the term ordering encodes multisets as the natural data-structure of clauses, after transforming all lattice terms to such a format. This yields resolution-like rules. Consequently, the derivation method is quite flexible. The procedural behaviour of the algebraic calculi under construction can directly be influenced by the choice of the syntactic ordering.

6 Tableau Calculi

We now present tableau calculi as solutions to the word problems for free semilattices and distributive lattices. Using standard techniques, these calculi can be lifted to the non-ground case. As usual in resolution-based theorem proving, queries are given as negative literals and then refuted by the calculi. The calculi are formally derived in the next two sections.

Definition 2 (Distributive Lattice Tableau). *Let $\prec$ be the literal and clause ordering of Section 5. The* tableau calculus for (finite) distributive lattices DT *consists of the following inference rules.*

$$\frac{\Gamma, x \leq x \longrightarrow}{\Gamma \longrightarrow}, \qquad \text{(Ref)}$$

$$\frac{\Gamma, x \leq y \sqcap z \longrightarrow}{\Gamma, x \leq y, x \leq z \longrightarrow}, \quad \text{(MR)} \qquad \frac{\Gamma, x \sqcup y \leq z \longrightarrow}{\Gamma, x \leq z, y \leq z \longrightarrow}, \quad \text{(JL)}$$

$$\frac{\Gamma, x \leq w \sqcup (y \sqcap z) \longrightarrow}{\Gamma, x \leq w \sqcup y, x \leq w \sqcup z \longrightarrow}, \quad \text{(EMR)} \qquad \frac{\Gamma, w \sqcap (x \sqcup y) \leq z \longrightarrow}{\Gamma, w \sqcap x \leq z, w \sqcap y \leq z \longrightarrow}, \quad \text{(EJL)}$$

$$\frac{\Gamma, x \sqcap y \leq z \longrightarrow}{\Gamma, x \leq z \longrightarrow}, \quad \text{(ML)} \qquad \frac{\Gamma, x \leq y \sqcup z \longrightarrow}{\Gamma, x \leq z \longrightarrow}. \quad \text{(JR)}$$

In all rules, the minor formula is maximal in the premise. All rules are meant modulo associativity, commutativity and idempotence.

(Ref) stands for *reflexivity*, (MR) for *meet right*, (EMR) for *extended meet right*, (ML) for *meet left*, (JL) for *join left*, (EJL) for *extended join left*, (JR) for *join right.* The respective join and meet rules are completely dual. There is no variant of a cut rule (cf. Section 8 for an explanation). Note also the correspondence with tableau or sequent calculus rules. See finally Section 10 for a discussion of the role of (EJL) and (EMR).

Definition 3 (Semilattice Tableaux). *Under the conditions of Definition 2, the inference rules of the* tableau calculi JT and MT for join and meet semilattices *are restrictions of the* DT*-rules to join and meet semilattice terms.*

Thus in particular, JT consists solely of variants of the rules (JL) and (JR), MT of variants of (ML) and (MR). JT and MT are dual and of course JT can be used also for the meet semilattice, dualising meet semilattice identities.

7 Constructing the Orb

We now perform the first step of the derivation of the tableau calculi. Our input specifications are J, M and D. With the orderings of Section 5, we compute the respective orbs; the OR-closures modulo redundancy elimination. Appealing to duality prevents us from repetitions.

We first index clauses according to their orientation under $\prec$: i (increasing) if the antecedent is smaller than the succedent, d (decreasing) in the converse case and ? if orientation must be instance-wise. Note that all clauses in J, M and D are indexed by i, except (trans) and (cut), which are indexed by ?.

Consider now the Horn clauses

$$x \leq w \sqcup y, x \leq w \sqcup z \longrightarrow_i x \leq w \sqcup (y \sqcap z), \tag{emr}$$
$$x \leq y \sqcap z \longrightarrow_d x \leq y, \tag{imr}$$
$$x \leq w \sqcup (y \sqcap z) \longrightarrow_d x \leq w \sqcup y, \tag{eimr}$$
$$x \leq z \longrightarrow_i x \sqcap y \leq z \tag{ml}$$

and their duals (ejl), (ijl), (eijl) and (jr). Let

$$J' = \{(\text{ref}), (\text{lub}), (\text{imr}), (\text{ml}), (\text{cut})\}, \qquad M' = \{(\text{ref}), (\text{glb}), (ijl), (jr), (\text{cut})\},$$
$$D' = J' \cup M' \cup \{(\text{emr}), (\text{eimr}), (ejl), (eijl), (\text{cut})\}.$$

In particular, restricted variants of (cut), for instance $x_1 \leq z, x_2 \sqcap z \leq y_2 \longrightarrow x_1 \sqcap x_2 \leq y_2$ occur in M' and J'. Moreover, (trans) is a restriction of these (cut) rules, forgetting the lattice term structure. Now—up to the extended rules (ejl), (eijl), (emr) and (eimr)—all rules are reminiscent to those in the sequent calculus. The inverse rules (imr) and (ijl) also hold in the sequent calculus by the inversion lemma (they are derivable with and admissible without the cut rule [10]). Consequently, all rules in J' and M' immediately correspond to rules of the sequent calculus. The extended rules are combinations of the non-extended rules and isotonicity of join and meet. They also encode the effect of distributivity in absence of sequents. See Section 10 for further discussion.

The following lemma shows that the members of J', M' and D' are independent in a strictly formal sense. By the above correspondence, also the rules of tableau and sequent calculi are therefore independent.

Lemma 1. *Let $\prec$ be the literal ordering defined in Section 5.*

(i) M' is an orb for meet semilattices.
(ii) J' is an orb for join semilattices.
(iii) D' is an orb for distributive lattices.

We always implicitly normalise with respect to idempotence of join and meet and consider terms modulo associativity and commutativity.

Proof. The proofs consist of three steps. First, we orient the rules in J, M and D with respect to $\prec$. Second, we derive the rules in J', M' and D' in OR. Third, we show that all conclusions of (ground instances of) theory/theory inferences in OR with respect to J', M' and D' are redundant. Here, we show only some inferences. The complete case analysis is included in an appendix.

Orientation has already been described and the derivation of the rules of J', M' and D' from those of J, M and D is straightforward. So it remains to show that these sets do indeed form orbs.

(ad i). As an example, consider the inference between (ml) and (imr). It is depicted in the following diagram.

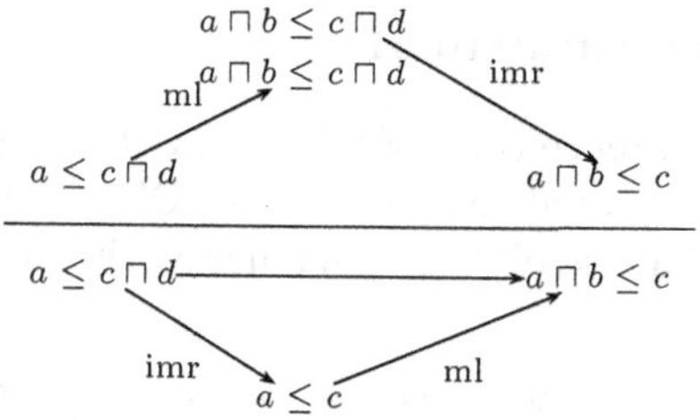

The upper part of the diagram is the resolution step, the lower part shows a smaller proof of the resolvent using (imr) and (ml).

As a second example, consider the inference

$$\frac{\longrightarrow_i a \sqcap c \leq a \sqcap c \quad a \leq b, c \sqcap b \leq c \sqcap b \longrightarrow_d a \sqcap c \leq b \sqcap c}{a \leq b \longrightarrow_i a \sqcap c \leq b \sqcap c}$$

between (ref) and (cut), that yields isotonicity of meet. But (ml) yields $\rightarrow a \sqcap c \leq b$ and $\rightarrow a \sqcap c \leq c$ from $\rightarrow a \leq b$ and from $\rightarrow c \leq c$, that is (ref). Using (glb), we obtain $a \sqcap c \leq b \sqcap c$ from these rules. Thus we can prove the isotonicity clause already using (ml), (glb) and (ref). All these rules are indexed also with i. Therefore in every proof, an inference using the isotonicity rule can be replaced by a proof with members of M', whence the isotonicity clause is entailed by smaller instances from M' and therefore redundant. The analysis of the other resolution steps arising from M is similar.

(ad ii) This follows by duality from (i).

(ad iii) As an example, consider the resolution step between (emr) and (eimr). It is shown in the following diagram.

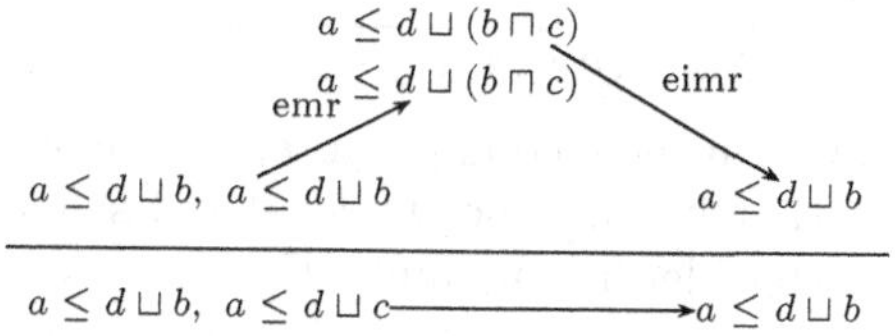

The resolvent is a tautology. The remaining steps are similar. □

Proposition 1(iii) and Lemma 1 immediately imply the following consequence of independence, which is essential for the arguments in the following section.

Corollary 1. *For every inconsistent clause set containing J', M' or D' there exists a refutation in which no* OR*-inference has both premises from this set.*

Continuing our discussion from Section 4 and Section 5, we still have no solution to the respective word problem, since resolution inferences with (cut) may introduce new variables (remind that (cut) is indexed with ?) and lead to non-isotone proofs. This is analogous to the sequent calculus, where propositional decidability depends on cut elimination. We will now show an algebraic variant.

8 Deriving the Tableau Rules

We now derive the inference rules of DT from OR-derivations with D'. Our main assumptions are refutational completeness of OR and the fact that our ordering constraints rule out all resolution inference with both premises from the orb (Corollary 1).

Theorem 1. *The tableau calculus* DT *solves the word problem for the free distributive lattice: For every input* $s \leq t \rightarrow$, *such that* $s \leq t$ *holds in the free distributive lattice generated by the constants in* s *and* t, *there exists a refutation in* DT.

Proof. Consider a refutation of an input $s \leq t \rightarrow$ in presence of the members of D'. By Corollary 1, there are no inferences for which both premises are from D'. Moreover, by our discussion in Section 4 all clauses that are generated in the resolution process are again negative (or empty). We can therefore restrict our attention to non-theory clauses of this form. Also, since D' contains solely Horn clauses, the refutation does not contain factoring steps.

We therefore need only consider ordered resolution inferences between negative non-theory clauses and members of D'. The extraction of inference rules is very simple, but there are some cases to inspect.

(case i) Resolution of a clause $\Gamma, a \leq a \longrightarrow$ and (ref) is

$$\frac{\longrightarrow a \leq a \qquad \Gamma, a \leq a \longrightarrow}{\Gamma \longrightarrow},$$

where, due to the constraints of ordered resolution, the identity $a \leq a$ majorises Γ. Internalising (ref) immediately yields the rule (Ref).

(case ii) Resolution of a clause $\Gamma, a \leq b \sqcap c \longrightarrow$ and (glb) is

$$\frac{a \leq b, a \leq c \longrightarrow a \leq b \sqcap c \qquad \Gamma, a \leq b \sqcap c \longrightarrow}{\Gamma, a \leq b, a \leq c \longrightarrow},$$

where $a \leq b \sqcap c$ is maximal in the second premise. Internalising (glb) immediately yields (MR). The fact that in this rule the left-hand side of an identity is split shows the necessity to consider a non-empty Γ.

(case iii) Resolution of a clause $\Gamma, a \leq b \sqcup (c \sqcap d) \longrightarrow$ and (emr) is

$$\frac{a \leq b \sqcup c, a \leq b \sqcup d \longrightarrow a \leq b \sqcup (c \sqcap d) \qquad \Gamma, a \leq b \sqcup (c \sqcap d) \longrightarrow}{\Gamma, a \leq b \sqcup c, a \leq b \sqcup d \longrightarrow},$$

where $a \leq b \sqcup (c \sqcap d)$ is maximal in the second premise. This yields (EMR).

(case iv) The antecedent of (imr) is greater than the succedent according to $\prec$ and never satisfies the ordering constraints of ordered resolution with a clause with empty succedent. Therefore it does not contribute an inference rule.

(case v) For (eimr), the situation is analogous to (case iv).

(case vi) Resolution of a clause $\Gamma, a \sqcap b \leq c \longrightarrow$ and (ml) is

$$\frac{a \leq c \longrightarrow_S a \sqcap b \leq c \qquad \Gamma, a \sqcap b \leq c \longrightarrow}{\Gamma, a \leq c \longrightarrow},$$

where $a \sqcap b \leq c$ is maximal in the second premise. This yields (ML).

(case vii) to (case xi), yielding the inference rules (JL), (EJL) and (JR) from the clauses (lub), (ejl), (ijl), (eijl) and (jr) are dual to (case ii) to (case vi).

(case xii) Resolution of a query $\Gamma, a \sqcap b \leq c \sqcup d \longrightarrow$ with (cut) is

$$\frac{a \leq c \sqcup e, b \sqcap e \leq d \longrightarrow a \sqcap b \leq c \sqcup d \qquad \Gamma, a \sqcap b \leq c \sqcup d \longrightarrow}{\Gamma, a \leq c \sqcup e, b \sqcap e \leq d \longrightarrow}.$$

We show by induction on the distance from such an inference to the empty clause and the *cut rank* of the lattice term, that is the size of the minor term which is cut out, that this inference is not needed. In proof-theoretic terms we show a version of cut elimination. Since the proof is standard we give only a sketch and refer to [6, 10] for details. In particular, for simplicity, we assume that $c = 0$.

(case α) Let e be a generator. Then $a \leq e$ must be of the form $a' \sqcap e \leq e$ and $b \sqcap e \leq d$ must either be of the form $b' \sqcap d \sqcap e \leq d$ or $d = e$ such that $b \sqcap d \leq d$ in order to eliminate both identities from the conclusion. So also $a \sqcap b \leq d$ either is of the form $a' \sqcap b \sqcap d \leq d$ or of the form $a \sqcap b' \sqcap d \leq d$ and already the minor formula of the right-hand premise can be eliminated using (ML) and (Ref).

(case β) Let $e = e_1 \sqcap e_2$. Then we may assume that (MR) has been applied to the identityy $a \leq e_1 \sqcap e_2$, which transforms the conclusion of the above inference into $\Gamma, a \leq e_1, a \leq e_2, b \sqcap e_1 \wedge e_2 \leq c \longrightarrow$. Using the induction hypothesis we can then argue that this sequent has been obtained from the right-hand premise of the above inference by two smaller cuts, respecting the ordering constraints. Hence in any case the above inference is not needed.

Since we have considered all clauses from D' and all these clauses produce conclusions with empty succedent, we have compiled a refutationally complete set of inference rules for a negative input from the theory specification. The inference rules yield a decision procedure, since, as a simple inspection shows, all conclusions are smaller than the (maximal) premise and the number of terms that is generated by the procedure is bounded by the subterms of the input. Thus our calculus has the subterm property. □

It is obvious that due to the copying by the extended rules, the procedure runs at least in exponential time with respect to the size of the input. This is what can be expected for distributive lattices.

Corollary 2. *The tableau calculi* JT *and* MT *solve the word problem for the free join and meet semilattice.*

Again, the procedures run in exponential time, which is sub-optimal, since the word problem for lattices can be solved in polynomial time (cf. [5]). They will, however, become polynomial when renaming techniques and data-structures like union-find are used. Note that also Whitman's algorithm [16] for the word problem for free lattices is exponential without such refinements.

9 Three Extensions

Our previous results are the basis for interesting applications. We now sketch three simple extensions of the tableau calculus for distributive lattices. First, an extension to Boolean lattices, second to certain basic modal logics and third to operational reasoning with sets.

The extension to Boolean lattices is straightforward.

Corollary 3. *In a lattice with* 0 *and* 1*, let x' denote the complement of x, that is $x' \sqcup x = 1$ and $x' \wedge x = 0$. The rules of* DT *together with the rules*

$$\frac{\Gamma, x \wedge y' \leq z \longrightarrow}{\Gamma, x \leq y \sqcup z \longrightarrow}, \qquad \frac{\Gamma, x \leq y' \sqcup z \longrightarrow}{\Gamma, x \wedge y \leq z \longrightarrow}, \qquad \frac{\Gamma, 0 \leq x \rightarrow}{\Gamma \rightarrow}, \qquad \frac{\Gamma, x \leq 1 \rightarrow}{\Gamma \rightarrow},$$

for eliminating complements and axiomatising zero and one solve the word problem for the free Boolean lattice[1].

The complement rules just encode the usual Galois connections $x \sqcap y' \leq z \Leftrightarrow x \leq y \sqcup z$ and $x \leq y' \sqcup z \Leftrightarrow x \sqcap y \leq z$ as inference rules.

Another interesting case is the extension to distributive and Boolean lattices with operators. Operators are mappings $h : L \rightarrow L$, where L is a distributive or Boolean lattice, that are strict join-homomorphisms, that is, they satisfy $h(0) = 0$ and $h(x \sqcup y) = h(x) \sqcup h(y)$. Alternatively, we could also consider co-strict meet-homomorphisms, that is mappings satisfying $h(1) = 1$ and $h(x \sqcap y) = h(x) \sqcap h(y)$. It is well-known that the first kind of mappings corresponds to modal diamond operators and the second one to modal box operators. There is also a strong connection with isotone predicate transformers, since both kinds of homomorphisms are isotone. Operators can be integrated in our tableaux by standard techniques. The main idea is to push them to the leaves of lattice terms, using the laws for homomorphisms and strictness. Terms of the form $h(s \sqcap t)$, when h is a join-homomorphism, can simply be renamed, since $\phi(s) \Leftrightarrow \exists x.(x = s \wedge \phi(x)$ holds in first-order logic. Thus an expression $\Gamma, h(r \sqcap s) \leq t \rightarrow$, for instance, can be replaced by the clause $\Gamma, c \leq r \sqcap s, r \sqcap s \leq c, h(c) \leq t \rightarrow$, where c is a new constant.

Atomic distributive lattices may serve as a calculus for operational reasoning with sets [14, 15]. Atoms are those elements of a lattice that are situated immediately above the zero. In every atomic distributive lattice with zero, atoms can be axiomatised by the identities $\alpha \not\leq 0$ and $\alpha \leq x \sqcup y \Leftrightarrow \alpha \leq x \vee \alpha \leq y$, where α denotes an atom. In the set-theoretic model of atomic distributive lattices, atoms correspond to singleton sets, whence to elements of a set. The expression $\alpha \leq x$ then reads as "α is an element of the set x". A lattice with zero is atomic iff every element can be expresses as a least upper bound of some set of atoms. In distributive lattices with zero, atomicity is equivalent to $x \not\leq y \Leftrightarrow \exists \alpha.(\alpha \leq x \wedge \alpha \sqcap y \leq 0)$. This equivalence can be used for hypothesis elimination, that is to transform arbitrary input clauses into positive ones (which are then negated for refutation). The second axiom for atoms yields a tableau rule that further splits identities. In particular, every finite Boolean lattice is atomic. The atomicity axiom introduces Skolem functions, but in inferences, the respective variables are always instantiated by subterms of terms occurring the input clause. This yields novel

[1] Here we assume that terms containing zeroes and ones are implicitly simplified during preprocessing, reducing, for instance, $s \leq t \sqcap 0$ to $s \leq 0$ or $s \sqcap 1 \leq t$ to $s \leq t$.

tableau-based decision procedures for the elementary theories of finite atomic distributive and Boolean lattices, whence for operational element-wise reasoning with sets. See [14] for a deeper discussion on the connection between atomic distributive lattices and sets and for related resolution-like calculi. A full formal treatment of atomic distributive lattices remains beyond the scope of this paper.

10 Discussion

Our solution to the word problem for the free distributive lattice implements tableaux via rewriting within the ordered resolution framework. This may surprise, since rewriting can only express angelic non-determinism, whereas tableaux and sequent calculi use also its demonic counterpart. In the sequent calculus, demonic non-determinism is implemented by shifting terms to the sequent level. Here, in contrast, the trick is to develop the tableau in the antecedent of clauses, where the comma means conjunction. Moreover, the extended rules essentially handle distributivity. In contrast, sequent calculi handle distributivity by lifting expressions to the sequent level. Consider, for instance, the following derivation in some variant of the cut-free sequent calculus. $\leq$ is now replaced by the sequent-arrow $\longrightarrow$ (both are pre-orderings); x, y and z are propositional constants.

$$\dfrac{\dfrac{x, y \vee z \longrightarrow x, z \qquad \dfrac{x, y \longrightarrow y, z \quad x, z \longrightarrow y, z}{x, y \vee z \longrightarrow y, z}}{x, y \vee z \longrightarrow x \wedge y, z}}{x \wedge (y \vee z) \longrightarrow (x \wedge y) \vee z}$$

Lifting formulas to sequents, the distributivity law is implicitly applied to multiply out terms and make the invertible conjunctive rules applicable, whereas the comma model the disjunctive ones. For a comparison, a proof in DT is

$$\dfrac{\dfrac{\dfrac{\dfrac{x \wedge (y \sqcup z) \leq (x \wedge y) \sqcup z \longrightarrow}{x \wedge (y \sqcup z) \leq x \sqcup z, x \wedge (y \sqcup z) \leq y \sqcup z \longrightarrow}\,(\mathrm{EMR})}{x \wedge y \leq x \sqcup z, x \wedge z \leq x \sqcup z, x \wedge y \leq y \sqcup z, x \wedge z \leq y \sqcup z \longrightarrow}\,(\mathrm{EJL})}{x \leq x, x \leq x, y \leq y, z \leq z \longrightarrow}\,(\mathrm{ML})(\mathrm{JR})}{\longrightarrow}\,(\mathrm{Ref})$$

Obviously, the sequent proof synthesises the goal formula, whereas the resolution proof analyses it. But reading the sequent proof backwards it turns out that both proofs are essentially the same.

Our algebraic analysis shows that the cut rule of the sequent calculus essentially encodes distributivity, (see also [11] for further discussion). But in the sequent calculus, distributivity is already included via the shift to sequents. In the free case, in absence of further relations between generators, there is no need to derive further consequences of relations (by analytic cut) or even to invent new constants (by non-analytic cut). Algebraically, therefore, admissibility of cut in the sequent calculus is very natural. Our reconstruction supports this intuition with a formal argument. In presence of relations between generators, of course, further consequences of these relations must be computed, possibly using cut: In case of finitely presented distributive lattices, when further relations between generators exist, resolution steps using (cut) cannot in general be circumvented. They can even be turned into the central ingredient of the calculus, as the chaining calculi for distributive lattices show [13]. This has a correspondence in the sequent calculus, where in presence of assumptions, cut is often unavoidable.

Our tableau calculi are more focused than mere derivations with the axioms in J, M or D. For instance, a resolution inference with two instances of (trans) may eagerly introduce new variables that might be meaningless for a refutation. In unordered resolution, strategies to avoid such kind of reasoning are well-known; for instance set of support or theory resolution. But the transfer of these strategies to ordered resolution is non-trivial, as we have seen. Here, the main advantage of JT, MT or DT over plain ordered resolution with the orbs J', M' and D' is cut elimination. The remaining tableau rules offer no additional efficiency over forbidding theory/theory inferences a priori (for instance by colouring clauses) instead of explicitly testing for redundancy. In general, however, the specific inference rules can be much more effective than plain resolution with orbs (cf. [12, 13, 14]). In particular, it follows from our orb construction that the tableau and sequent rules are independent in a strictly formal sense: they are elements of an irredundant irreducible basis. This construction is reminiscent to that of an orthogonal basis of a vector space. Last, but not least, a particular benefit of the integration of tableaux into the ordered resolution framework is that the full power of redundancy elimination rules of ordered resolution becomes available to tableaux.

11 Conclusion

We have used the derivation method from [12] to integrate tableau calculi that solve lattice-theoretic word problems as special-purpose procedures into the ordered resolution framework. In contrast to standard tableau or sequent calculi, distributivity has not been included by introducing an additional data-structure of sequents, but by allowing certain splittings below contexts. We have seen that cut-rules arise naturally in lattice theory in presence of distributivity. They can be eliminated in the free case, in absence of relations between generators.

Alternative, but different, proof systems are the Genzen systems for distributive lattices by Font and Verdú [4] and the natural-decuction-style proof systems for (non-distributive) lattices of Negri and von Plato [8]. In particular, Negri and Plato show that the axiom of transitivity is not needed in their calculus; a special case of our cut-elimination property related to distributivity. However, none of these proof systems have been formally derived and none of them seems compatible with the ordered resolution framework.

The derivation of tableaux is only one of several applications of the derivation method in lattice theory. There are also focused chaining calculi for transitive relations, pre-orderings, semilattices, distributive lattices, Boolean lattices, atomic distributive lattices and atomic Boolean lattices [12, 13].

The extensions of our basic procedures to simple modal logics and to reasoning with sets demonstrate the potential and modularity of the approach. In contrast to the usual semantic translation to first-order logic, our novel technique specifically exploits the underlying algebra. The consideration of more complex applications, for instance temporal and dynamic algebras and logics, is left for future work.

References

1. G. Birkhoff. *Lattice Theory*, volume 25 of *Colloquium Publications*. American Mathematical Society, 1984. Reprint.
2. C. Delor and L. Puel. Extension of the associative path ordering to a chain of associative-commutative symbols. In C. Kirchner, editor, *5th Conference on Rewrite Techniques and Applications*, volume 690 of *LNCS*, pages 389–404. Springer-Verlag, 1993.
3. M. Fitting. *First-Order Logic and Automated Theorem Proving*. Springer-Verlag, second edition, 1996.
4. J. M. Font and V. Verdú. Algebraic logic for classical conjunction and disjunction. *Studia Logica*, 50:391–419, 1991.
5. R. Freese, J. Ježek, and J.B. Nation. *Free Lattices*, volume 42 of *Surveys and Monographs*. American Mathematical Society, 1995.
6. G. Gentzen. Untersuchungen über das logische Schließen. *Mathematische Zeitschrift*, 39:176–210, 405–431, 1935.
7. P. Lorenzen. Algebraische und logistische Untersuchungen über freie Verbände. *The Journal of Symbolic Logic*, 16(2):81–106, 1951.
8. S. Negri and J. von Plato. Proof systems for lattice theory. *Mathematical Structures in Computer Science*, 14(4):507–526, 2004.
9. M. Rusinowitch. *Démonstration Automatique: Techniques de Réecriture*. Science Informatique. InterEditions, Paris, 1989.
10. H. Schwichtenberg. Proof theory: Some applications of cut-elimination. In J. Barwise, editor, *Handbook of Mathematical Logic*, pages 867–895. North-Holland, 1977.
11. G. Struth. An algebra of resolution. In L. Bachmair, editor, *Rewriting Techniques and Applications, 11th International Conference*, volume 1833 of *LNCS*, pages 214–228. Springer-Verlag, 2000.
12. G. Struth. Deriving focused calculi for transitive relations. In A. Middeldorp, editor, *Rewriting Techniques and Applications, 12th International Conference*, volume 2051 of *LNCS*, pages 291–305. Springer-Verlag, 2001.
13. G. Struth. Deriving focused lattice calculi. In S. Tison, editor, *Rewriting Techniques and Applicaions, 13th International Conference*, volume 2378 of *LNCS*, pages 83–97. Springer-Verlag, 2002.
14. G. Struth. A calculus for set-based program development. In J. S. Dong and J. Woodcock, editors, *Formal Methods and Software Engineering: 5th International Conference on Formal Engineering Methods*, volume 2885 of *LNCS*, pages 541–559. Springer-Verlag, 2003.
15. G. Struth. Automated element-wise reasoning with sets. In J. R. Cuellar and Z. Liu, editors, *2nd International Conference on Software Engineering and Formal Methods*, pages 320–329. IEEE Computer Society, 2004.
16. Ph.M. Whitman. Free lattices. *Ann. of Math.*, 42(2):325–330, 1941.

Accelerated Modal Abstractions of Labelled Transition Systems

Miguel Valero Espada[1] and Jaco van de Pol[2]

[1] Universidad Complutense de Madrid, Spain
mvaleroe@pdi.ucm.es
[2] Centrum voor Wiskunde en Informatica, Amsterdam, The Netherlands
Jaco.van.de.Pol@cwi.nl

Abstract. Modal Labelled Transition Systems (*Modal*-LTSs) can be used to specify system behaviour. They distinguish between required behaviour and allowed behaviour. This makes *Modal*-LTSs a suitable formalism to specify abstractions of a system by over- and under-approximations. This paper studies an extension to *Modal*-LTSs by allowing *accelerated*-transitions, i.e. transitions labelled with regular expressions. This permits to represent that a process can reach a state by executing some sequence of actions, abstracting away the intermediate states. We show how accelerated transitions improve the expressiveness of abstractions. Consequently, more *liveness* properties can be checked.

1 Introduction

Automatic verification techniques, such as model checking, normally require the exploration of the state space corresponding to a formal specification. These techniques are quite limited by the size of the state spaces, which may be too large or even infinite. Abstraction is being widely used to reduce the complexity of the analysed systems. The main idea is to prove properties for a (small) abstract system and to infer their satisfaction or refutation in the (large) concrete original system.

Abstract Model Checking usually integrates the following steps. First, we depart from a concrete specification, whose state space is too large to generate or infinite. From the formal specification we extract an abstract state space, for example by interpreting the concrete operations of the model on smaller data domains (see, for example [2]). Then we apply model checking on the abstract system. The results of the abstract model checking can be inferred to the concrete system following some specific rules. Applying abstraction causes some loss of information, so it is not always possible to prove the satisfaction or refutation of some properties. In theses cases, it is necessary to refine the abstractions.

Modal Labelled Transition Systems (or *Modal*-LTSs) have been used to describe abstract state spaces. Basically, the transition systems contain two kinds of transitions or modalities *may* and *must*. The *may*-transitions are used to represent the possible behaviours of the system, in other words, the behaviours that can appear in the refinements. *Must* transitions represent the necessary

M. Johnson and V. Vene (Eds.): AMAST 2006, LNCS 4019, pp. 338–352, 2006.

behaviours, the ones that have to appear in all the refinements. The set of possible behaviours constitutes an over-approximation of the concrete system and the necessary behaviours an under-approximation. Therefore, abstractions described by *Modal*-LTSs doubly approximate the concrete systems. The seminal idea is due to Larsen and Thomsen [11].

Informally, we can distinguish two sets of properties: *safety* and *liveness*. A *safety* property expresses that "something bad never happens." Typical safety properties are those forbidding "bad" execution sequences on the *Modal*-LTSs. Therefore, in order to prove a *safety* property on an abstract system we would have to check the possible executions that are related to the *may* transitions. A *liveness* property expresses that "something good eventually happens." So, typical *liveness* properties assert the existence of some desired executions, therefore its satisfaction will be related by the presence of the suitable *must* transitions in the *Modal* transition system.

Abstraction has been successfully used to prove mainly safety properties. Even though some frameworks allow the inference of liveness properties their verification remains one of the major challenges of abstraction theories [14]. The problem comes from the lack of guaranteed (required) behaviours. The number of *necessary* behaviours reduces due to the non-determinism added by abstractions. This fact makes it difficult to prove liveness properties.

To deal with this problem, we propose a new formalism to represent abstractions. We enhance *Modal*-LTSs by allowing *must*-transitions to match sequences of actions, which captures the idea that in a finite computation a state can be reached from a given one. This extension will allow to capture more accurately abstract systems and therefore to infer stronger liveness properties.

In the next section, we present the definition of the new type of structure *Accelerated Modal Labelled Transition Systems* in which *must*-transitions are labelled with regular expressions built over the action labels and the usual operators. This new type of transition allows to represent finite computations by single transitions. A motivating example of *accelerated*-steps is given in Section 2. Then, we will present how to generate abstractions with *accelerations* and we give the preservation results. In fact, we have proved that the framework is sound and complete for Propositional Dynamic Logic (PDL [8]); we use a three-valued interpretation in the sense of [6].

Besides the theoretical foundation of using *Accelerated Modal*-LTS for specifying abstractions and approximations of behaviour, we present a model checking algorithm. This algorithm checks whether a PDL formula holds (necessarily or possibly) for an *Accelerated Modal*-LTS. So the non-trivial model-checking problem for PDL on *Accelerated Modal*-LTSs is decidable.

Our approach to obtain more accurate specifications is novel. The model of *Accelerated Modal*-LTS is new as well, which makes the model checking algorithm quite different from usual PDL model checking. Note that our approach is complementary to approaches that try to eliminate spurious may-behaviour, like e.g. [10]. The current paper is based on the work presented in [19], which also contains the full proofs.

2 Accelerated Modal Transition Systems

Labelled Transition Systems are common structures to define the semantics of system specifications. Let S be a non-empty set of states and Act a non-empty set of transition labels, then:

Definition 1. *We define a* Labelled Transition System *(LTS) as a tuple* $(S, Act, \rightarrow, s_0)$, $\rightarrow$ *is a possibly infinite set of transitions and* s_0 *in* S *is the initial state. A transition is a triple* $s \xrightarrow{a} s'$ *with* $a \in Act$ *and* $s,s' \in S$.

Figure 1 presents a simple *resettable* counter that, starting from some value bigger than zero, decreases until it arrives to zero and then ends by executing an action that informs about its expiration. The counter may be reset to its initial value at any time (if it has not expired).

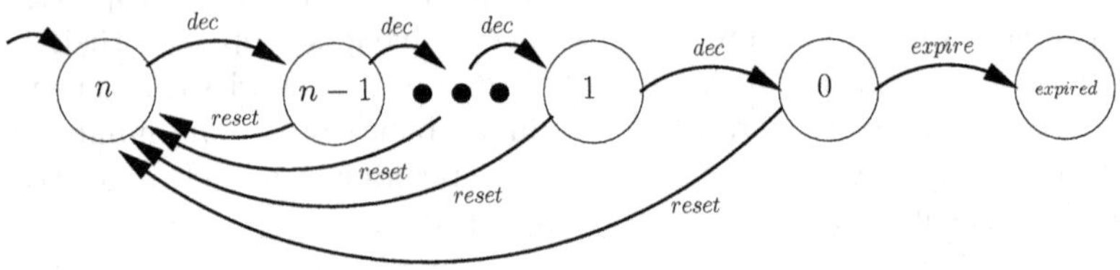

Fig. 1. Concrete *resettable* counter

As we have mentioned in the introduction, *Modal* Labelled Transition Systems are more suitable structures to model abstraction, because they allow to describe incomplete or underspecified systems. Let us recall the definition:

Definition 2. *A* Modal Labelled Transition System *(*Modal-*LTS) is a tuple* (S, Act, $\rightarrow_\Diamond, \rightarrow_\Box, s_0$) *where* S*,* Act *and* s_0 *are as above and* $\rightarrow_\Diamond, \rightarrow_\Box$ *are possibly infinite sets of (may or must) transitions of the form* $s \xrightarrow{a}_x s'$ *with* $s,s' \in S$, $a \in Act$ *and* $x \in \{\Diamond, \Box\}$. *We require that every* must*-transition is a* may*-transition* $(\xrightarrow{a}_\Box \subseteq \xrightarrow{a}_\Diamond)$.

The requirement $\xrightarrow{a}_\Box \subseteq \xrightarrow{a}_\Diamond$ is needed to satisfy the intuitive property that every *necessary* behaviour is also *possible* (see, for example [16]). Note, that every LTS corresponds to a trivially equivalent *Modal*-LTS in which $\rightarrow_\Diamond = \rightarrow_\Box$, we call this subclass concrete *Modal*-LTSs.

We will formally describe how to generate abstractions in section 3. We now present as illustration one possible abstraction of the previous example. Figure 2 represents[1] a simple modal abstraction of the *resettable* counter in which the values greater or equal to 1 are collapsed to a single abstract state, '+'. The relationship between Figure 1 and 2 will be made precise by definition 4.

[1] In figures, *must*-transitions are represented by solid lines and *may*-transitions by dashed ones. For clarity, when there is a *must* transition we do not include the corresponding *may* one.

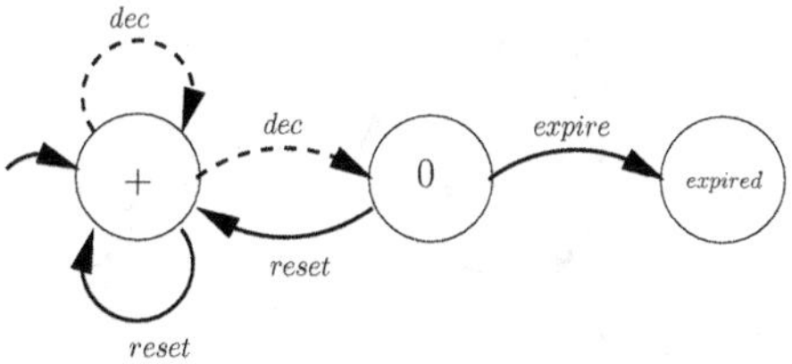

Fig. 2. Modal Abstraction of a *resettable* counter

Note that in Figure 2 there is no guaranteed path from '+' to 'expired', because from state '+' there is no *must-dec* step. The intuitive reason is that abstract state '+' captures both the concrete state '1' (where *dec* would lead to '0') as well as states '> 1', where *dec* would lead to '+'. So, although the concrete system in Figure 1 has a path to 'expired', this path is lost in the *Modal*-LTS abstraction.

In order to improve this situation, we enhance *Modal*-LTSs by changing the definition of *must*-transitions. We call *accelerated must*-transitions, those transitions that condense a sequence of steps into a single one. *Accelerated must*-transitions will be labelled by regular expressions σ built over the alphabet *Act* and the usual operators $\cdot, *$ and $|$, where '$\cdot$' stands for the concatenation operator, '$|$' is the choice operator, '$*$' is the transitive and reflexive closure operator. By $[\![\sigma]\!]$ we denote the language generated from σ. Let us see the definition:

Definition 3. *An* Accelerated Modal-*LTS is a tuple* $(S, Act, \rightarrow_\diamond, \rightarrow_\boxplus, s_0)$ *where* S, *Act and* s_0, $\rightarrow_\diamond$ *are as in the previous definition, and* $\rightarrow_\boxplus$ *is a possibly infinite set of* accelerated must-*transitions of the form* $s \xrightarrow{\sigma}_\boxplus s'$ *with* $s,s' \in S$, *and* σ *is a regular expression. Furthermore, we require:*

- *Every* accelerated must-*transition corresponds to a finite sequence of* may-*transitions:* $s \xrightarrow{\sigma}_\boxplus s' \implies \exists\, a_0, ..., a_i.\, s \xrightarrow{a_0}_\diamond ... \xrightarrow{a_i}_\diamond s' \wedge [a_0 \cdots a_i] \in [\![\sigma]\!]$

Basically, the new definition allows *must*-transitions to be labelled with arbitrary regular expressions. Examples of correct *accelerated*-transitions are $s \xrightarrow{a|b}_\boxplus s'$, $s \xrightarrow{a\cdot a*}_\boxplus s'$ (which can be abbreviated by $s \xrightarrow{a+}_\boxplus s'$) and $s \xrightarrow{a\cdot b*\cdot a}_\boxplus s'$.

A trivial result is that every *Modal*-LTS is an *Accelerated Modal*-LTS . It follows from the fact that every *must* transition is an *accelerated must*-transition in which σ is equal to a single action label. The condition that every *must*-transition corresponds to a sequence of *may*-transitions, is similar to the one imposed in the *Modal*-LTS and, as we will see in section 4, it will help to define a consistent logical characterisation of the abstractions.

Figure 3 presents a modal abstraction of the *resettable* counter with *accelerated*-transitions. Note that the difference with the simple *Modal*-abstraction (Figure 2) is that we can capture the fact that from the state '+' we can reach the state '0' in an indeterminate but finite number of steps. This is represented by the *accelerated*-transition $+ \xrightarrow{dec+}_\boxplus 0$. This fact was not possible to be expressed

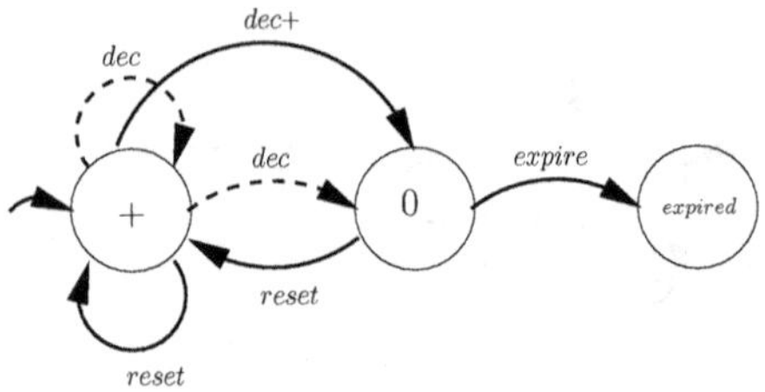

Fig. 3. Resettable counter with one *accelerated must*-transition

by the simple *Modal*-LTS. The relationship between Figures 1 and 3 will be made precise by definition 6 and 7, in the next section.

3 Accelerated Modal Abstractions

From a concrete system described by an LTS or a *Modal*-LTS, we can generate an abstraction by relating concrete states with abstract states. A widely applied approach uses homomorphic functions to define abstractions. This idea was suggested by Clarke, Grumberg and Long [2]. An alternative approach proposed by S. Graf etal. and Dams [13, 3] is based on Galois Connections, which allow concrete states to be related to more than one abstract state. The Galois Connection framework gives, in general, more accurate abstractions than the homomorphic approach, but the latter is conceptually simpler. In this paper, we are going to use simple mappings (homomorphisms), but we believe that the extension to more complicated relations can be done following the ideas of [15].

First we are going to present the *Modal*-LTS based abstraction and then we will present the extension to transition systems with *accelerations*. Let us assume a set of abstract states $\widehat{S}$ and a total and surjective function from concrete states to abstract ones, $h : S \to \widehat{S}$. An abstract value $\widehat{s}$ corresponds to all the states s for which $h(s) = \widehat{s}$. Then, we can generate an abstraction from a concrete system, as follows:

Definition 4. *Given an LTS,* $\mathcal{M} = (S, Act, \to, s_0)$ *and a homomorphism* h, *defined as above, we say that the* Modal-*LTS defined by* $\widehat{\mathcal{M}} = (\widehat{S}, Act, \to_\Diamond, \to_\Box, \widehat{s}_0)$ *is the minimal abstraction by* h *(denoted by* $\widehat{\mathcal{M}} = min_h(\mathcal{M})$*) if and only if* $h(s_0) = \widehat{s}_0$ *and the following conditions hold:*

- $\widehat{s} \xrightarrow{a}_\Diamond \widehat{r} \iff \exists\, s, r, a.\, h(s) = \widehat{s} \wedge h(r) = \widehat{r} \wedge s \xrightarrow{a} r$
- $\widehat{s} \xrightarrow{a}_\Box \widehat{r} \iff \forall\, s. h(s) = \widehat{s} \Rightarrow (\exists\, r, a.\, h(r) = \widehat{r} \wedge s \xrightarrow{a} r)$

This definition gives the most accurate abstraction of a concrete system by using a given homomorphism h, in other words the one that preserves most information of the original system. Less precise abstractions would contain more *may* transitions and/or fewer *must* transitions. The relative precision of abstractions is formalized by the approximation operator $\sqsubseteq$, as follows:

Definition 5. *Given two* Modal-*LTSs*, $\mathcal{M} = (S, Act, \rightarrow_\diamond, \rightarrow_\Box, s_0)$ *and* $\mathcal{N} = (S, Act, \dashrightarrow_\diamond, \dashrightarrow_\Box, s_0)$ *built over the same sets of states and actions;* $\mathcal{N}$ *is an approximation of* $\mathcal{M}$*, denoted by* $\mathcal{M} \sqsubseteq \mathcal{N}$*, if the following conditions hold:*

- $s \xrightarrow{a}_\diamond r \implies s \overset{a}{\dashrightarrow}_\diamond r$
- $s \overset{a}{\dashrightarrow}_\Box r \implies s \xrightarrow{a}_\Box r$

These definitions were presented in [15]. Following the definitions we can see that the transition system of Figure 2 is a correct and minimal abstraction of the concrete *resettable* counter. Let us now present the new ideas to generate more expressive abstractions.

Definition 6. *Given an LTS,* $\mathcal{M} = (S, Act, \rightarrow, s_0)$ *and a mapping* $h : S \rightarrow \widehat{S}$*, we say that the* Accelerated Modal-*LTS,* $\widehat{\mathcal{M}}$ *defined by* $(\widehat{S}, Act, \rightarrow_\diamond, \rightarrow_\boxplus, \widehat{s}_0)$ *is the minimal abstraction by h (denoted by* $\widehat{\mathcal{M}} = min_h(\mathcal{M})$*) if and only if* $h(s_0) = \widehat{s}_0$ *and the following conditions hold:*

- $\widehat{s} \xrightarrow{a}_\diamond \widehat{r} \iff \exists\, s, r.\, h(s) = \widehat{s} \wedge h(r) = \widehat{r} \wedge s \xrightarrow{a} r$
- $\widehat{s} \xrightarrow{\sigma}_\boxplus \widehat{r} \iff (\forall\, s. h(s) = \widehat{s} \Rightarrow \exists\, r, a_0, ..., a_i.\, h(r) = \widehat{r} \wedge s \xrightarrow{a_0} ... \xrightarrow{a_i} r \wedge [a_0 \cdots a_i] \in [\![\sigma]\!])$

The condition on *may* transitions is as in definition 2, however the condition on *must* transitions changes. $[a_0 \cdots a_i] \in [\![\sigma]\!]$ means that the word $a_0 \cdots a_i$ belongs to the language generated by the regular expression σ. A transition $\widehat{s} \xrightarrow{\sigma}_\boxplus \widehat{r}$ means that for all s mapped to $\widehat{s}$ there exists an r mapped to $\widehat{r}$ such that we can go from s to r by a word contained in the language generated from σ.

This definition gives the most accurate abstraction of a concrete system for a given homomorphism h, in other words the one that preserves as much information as possible of the original system. Note that the minimal abstraction for *accelerated Modal*-LTS is infinite in general, due to the presence of regular expressions. In practice, one will compute a finite approximation of this. We now define our formal notion of approximation:

Definition 7. *Given two* Accelerated Modal-*LTSs*, $\mathcal{M} = (S, Act, \rightarrow_\diamond, \rightarrow_\boxplus, s_0)$ *and* $\mathcal{N} = (S, Act, \dashrightarrow_\diamond, \dashrightarrow_\boxplus, s_0)$ *built over the same sets of states and actions;* $\mathcal{N}$ *is an approximation of* $\mathcal{M}$*, denoted by* $\mathcal{M} \sqsubseteq \mathcal{N}$*, if the following conditions hold:*

- $s \xrightarrow{a}_\diamond r \implies s \overset{a}{\dashrightarrow}_\diamond r$
- $s \overset{\sigma}{\dashrightarrow}_\boxplus r \implies \exists\, \sigma_0, ..., \sigma_i.\, s \xrightarrow{\sigma_0}_\boxplus \cdots \xrightarrow{\sigma_i}_\boxplus r \wedge [\![\sigma_0 \cdots \sigma_i]\!] \subseteq [\![\sigma]\!]$

The relation $\sqsubseteq$ characterises the precision of the abstractions. $\mathcal{M}$ is more precise than $\mathcal{N}$ because it has less (or the same number of) *may*-transitions and more *accelerated must*-transitions or more precise ones. For example, considering only one *accelerated must*-transition $s \xrightarrow{\sigma}_\boxplus r$, we have $s \xrightarrow{a}_\boxplus r \sqsubseteq s \xrightarrow{a+}_\boxplus r \sqsubseteq s \xrightarrow{a*}_\boxplus r \sqsubseteq s \xrightarrow{a*|b}_\boxplus r \sqsubseteq s \not\rightarrow_\boxplus r$, where in the last case, we mean that there is not any *must* transition between s and r.

Another simple example, in which a transition abstracts away the intermediate states of a computation, would be $s \xrightarrow{a}_{\boxplus} t \xrightarrow{b}_{\boxplus} r \sqsubseteq s \xrightarrow{a \cdot b}_{\boxplus} r$.

Concluding this section, we have defined the relationship between a concrete system modeled by an LTS and an abstraction of it, modeled as an *accelerated Modal*-LTS. In particular, it can be checked that Figure 3 is an approximation (def. 7) of the minimal abstraction (def. 6) of Figure 1. It preserves the information that the counter can decrease to 0. This extra information will be used to infer the satisfaction of some liveness properties from the abstract model to the concrete. In the next section, we will present some results about the preservation of properties in this direction.

4 Logical Characterisation

We now investigate which properties can be inferred from abstract systems to concrete ones. For this purpose, we are going to use Propositional Dynamic Logic (PDL) which is a branching logic, in the style of HML [9] with regular expressions, less expressive than μ-calculus [18]. We consider three types of formulae, action (α), regular (β) and state formulae (φ), expressed by the following grammars:

$$\alpha ::= \mathsf{T} \mid \mathsf{F} \mid \neg\,\alpha \mid \alpha_1 \wedge \alpha_2 \mid \alpha_1 \vee \alpha_2 \mid a \qquad \beta ::= \alpha \mid \beta_1 \cdot \beta_2 \mid \beta_1 | \beta_2 \mid \beta*$$
$$\varphi ::= \mathsf{T} \mid \mathsf{F} \mid \neg\,\varphi \mid \varphi_1 \wedge \varphi_2 \mid \varphi_1 \vee \varphi_2 \mid [\beta]\varphi \mid \langle\beta\rangle\varphi$$

We have chosen this logic because of the fact that it is also built over regular formulae, which gives a cleaner relationship with the abstraction framework we have presented. We first describe informally the meaning of the formulae and then we will give the formal semantics: a stands for an action label, it matches transitions with the same action label. T matches all actions, $\neg\,\alpha$ matches all actions but the ones matched by α. F matches no action, it could have been expressed by $\neg\,\mathsf{T}$.

Regular formulae match sequences of actions; '$\cdot$' stands for the concatenation operator, '|' is the choice operator, '*' is the transitive and reflexive closure operator. Note that α is used to represent both a regular formula with only one action and an action formula.

The semantics of the state formulae is standard. $[\beta]\varphi$ holds in a state in which all continuations by sequences matching β end in a state satisfying φ. $\langle\beta\rangle\varphi$ holds in a state in which there exists at least one β sequence to a state satisfying φ.

The formal semantics of action formulae $[\![\alpha]\!]$, is as follows:

$$[\![\mathsf{T}]\!] = Act \qquad [\![a]\!] = \{a\} \qquad [\![\alpha_1 \wedge \alpha_2]\!] = [\![\alpha_1]\!] \cap [\![\alpha_2]\!]$$
$$[\![\mathsf{F}]\!] = \emptyset \qquad [\![\neg\alpha]\!] = Act \setminus [\![\alpha]\!] \qquad [\![\alpha_1 \vee \alpha_2]\!] = [\![\alpha_1]\!] \cup [\![\alpha_2]\!]$$

Now, we give the semantics for the regular formulae, $[\![\beta]\!]$:

$$[\![\alpha]\!] = \{[a] \mid a \in [\![\alpha]\!]\} \qquad [\![\beta_1|\beta_2]\!] = [\![\beta_1]\!] \cup [\![\beta_2]\!]$$
$$[\![\beta*]\!] = [\![\beta]\!]* \qquad [\![\beta_1 \cdot \beta_2]\!] = \{\, w_1 \cdot w_2 \mid w_1 \in [\![\beta_1]\!] \wedge w_2 \in [\![\beta_2]\!] \,\}$$

The interpretation of a state formula on a LTS returns a set of states, following the next rules:

$$[\![\mathsf{T}]\!] = S \qquad [\![\mathsf{F}]\!] = \emptyset \qquad [\![\neg\varphi]\!] = S \setminus [\![\varphi]\!]$$
$$[\![\varphi_1 \wedge \varphi_2]\!] = [\![\varphi_1]\!] \cap [\![\varphi_2]\!] \qquad [\![\varphi_1 \vee \varphi_2]\!] = [\![\varphi_1]\!] \cup [\![\varphi_2]\!]$$
$$[\![[\beta]\varphi]\!] = \{s \,|\, \forall r, a_0, ..., a_i.\, s \xrightarrow{a_0} \cdots \xrightarrow{a_i} r \wedge [a_0 \cdots a_i] \in [\![\beta]\!] \implies r \in [\![\varphi]\!]\}$$
$$[\![\langle\beta\rangle\varphi]\!] = \{s \,|\, \exists r, a_0, ..., a_i.\, s \xrightarrow{a_0} \cdots \xrightarrow{a_i} r \wedge [a_0 \cdots a_i] \in [\![\beta]\!] \wedge r \in [\![\varphi]\!]\}$$

We say that a state s satisfies a formula φ, denoted by $s \models \varphi$, if and only if $s \in [\![\varphi]\!]$. The notation $\mathcal{M}, s \models \varphi$ means that the state s satisfies the formula φ on $\mathcal{M}$. It is simple to see that we can use the classical dualities on state formulae: $\mathsf{T} = \neg\mathsf{F}$, $\varphi_1 \wedge \varphi_2 = \neg(\neg\varphi_1 \vee \neg\varphi_2)$ and $[\![[\beta]\varphi]\!] = [\![\neg\langle\beta\rangle\neg\varphi]\!]$.

We now define a three-valued semantics for *Accelerated Modal*-LTSs. This is analogous to [6], where the semantics of a state formula consists of two sets of states: A set of states that *necessarily* satisfy the formula and a set of states that *possibly* satisfy it. Thus, the semantics of the formulae is given by $[\![\varphi]\!] \in 2^S \times 2^S$ and the projections $[\![\varphi]\!]^{nec}$ and $[\![\varphi]\!]^{pos}$ give the first and the second component, respectively. First, we present the semantics for the *necessary* interpretation. We start with the state formulae in which basically the modality is pushed inwards in all the operators, and inverted for the negation:

$$[\![\mathsf{T}]\!]^{nec} = S \qquad [\![\mathsf{F}]\!]^{nec} = \emptyset \qquad [\![\neg\varphi]\!]^{nec} = S \setminus [\![\varphi]\!]^{pos}$$
$$[\![\varphi_1 \wedge \varphi_2]\!]^{nec} = [\![\varphi_1]\!]^{nec} \cap [\![\varphi_2]\!]^{nec} \qquad [\![\varphi_1 \vee \varphi_2]\!]^{nec} = [\![\varphi_1]\!]^{nec} \cup [\![\varphi_2]\!]^{nec}$$
$$[\![[\beta]\varphi]\!]^{nec} = \{s \,|\, \forall r, a_0, ..., a_i.\, s \xrightarrow{a_0}_{\Diamond} \cdots \xrightarrow{a_i}_{\Diamond} r \wedge [a_0 \cdots a_i] \in [\![\beta]\!] \Longrightarrow r \in [\![\varphi]\!]^{nec}\}$$
$$[\![\langle\beta\rangle\varphi]\!]^{nec} = \{s \,|\, \exists r, \sigma_0, ..., \sigma_i.\, s \xrightarrow{\sigma_0}_{\boxplus} \cdots \xrightarrow{\sigma_i}_{\boxplus} r \wedge [\![\sigma_0 \cdots \sigma_i]\!] \subseteq [\![\beta]\!] \wedge r \in [\![\varphi]\!]^{nec}\}$$

By the definition of negation if a system *necessarily* satisfies a negated property then it does not *possibly* satisfy the property (in positive form), i.e., $[\![\neg\varphi]\!]^{nec} = \neg[\![\varphi]\!]^{pos}$. The most interesting part of the semantics is the definition of the existential operator ($\langle\beta\rangle\varphi$). A given state satisfies an existential property if there exists a sequence of actions arriving at a suitable state r, such that the language generated by the concatenation of the action labels is contained in the language generated by the given β. The *necessary* semantics of the existential operator looks at *must*-transitions because it reasons about required behaviours, the executions that are guaranteed by the model. However the *necessary* semantics of the universal operator ($[\beta]\varphi$) considers *may* executions. This is because we have to check all possible continuations, which are represented by the *may*-transitions. The *possibly* semantics is dual, we just present it for the box and diamond operators (note the swap of the modalities):

$$[\![[\beta]\varphi]\!]^{pos} = \{s \,|\, \forall r, \sigma_0, ..., \sigma_i.\, s \xrightarrow{\sigma_0}_{\boxplus} \cdots \xrightarrow{\sigma_i}_{\boxplus} r \wedge [\![\sigma_0 \cdots \sigma_i]\!] \subseteq [\![\beta]\!] \Longrightarrow r \in [\![\varphi]\!]^{pos}\}$$
$$[\![\langle\beta\rangle\varphi]\!]^{pos} = \{s \,|\, \exists r, a_0, ..., a_i.\, s \xrightarrow{a_0}_{\Diamond} \cdots \xrightarrow{a_i}_{\Diamond} r \wedge [a_0 \cdots a_i] \in [\![\beta]\!] \wedge r \in [\![\varphi]\!]^{pos}\}$$

We say that a state s necessarily satisfies a formula φ, denoted by $s \models^{nec} \varphi$, if and only if $s \in [\![\varphi]\!]^{nec}$ and dually s possibly satisfies a formula φ, denoted by

$s \models^{pos} \varphi$, if and only if $s \in [\![\varphi]\!]^{pos}$. It is absolutely not trivial to see whether we can compute the semantics of a formula, the use of accelerations converts the model checking problem into a non-standard problem. In section 5, we propose an algorithm, showing that the PDL model checking problem on *Accelerated Modal*-LTS is decidable.

We remark that the *necessary* interpretation is consistent, i.e., we cannot prove at the same time that one formula and its negation are *necessarily* satisfied. Furthermore, the *possible* semantics is complete, we can either prove a property or its negation. We formally state these results.

Lemma 1. $[\![\varphi]\!]^{nec} \subseteq [\![\varphi]\!]^{pos}$

This lemma follows from the fact that, in *Accelerated Modal*-LTSs, every *accelerated must*-transition corresponds to a finite sequence of *may*-transitions (see Definition 3).

Lemma 2. *The* necessary *interpretation is consistent, i.e.,* $[\![\varphi \wedge \neg\varphi]\!]^{nec} = \emptyset$

Lemma 3. *The* possible *interpretation is complete, i.e.,* $[\![\varphi \vee \neg\varphi]\!]^{pos} = S$

These two properties follow trivially from the fact that $[\![\varphi]\!]^{nec} \subseteq [\![\varphi]\!]^{pos}$ and from the semantics of negation. Furthermore, the dualities between operators, presented above, are also satisfied by the three-valued semantics. We now present the preservation results; their proofs can be found in [19].

Theorem 1. *Given two* Accelerated Modal-*LTSs,* $\mathcal{M}$ *and* $\mathcal{N}$*, over the same sets of states and labels* S *and* Act*, with* $\mathcal{M} \sqsubseteq \mathcal{N}$ *for all* s *in* S *and for all formula* φ*, we have:*

$$\mathcal{N}, s \models^{nec} \varphi \implies \mathcal{M}, s \models^{nec} \varphi \quad \textit{and} \quad \mathcal{N}, s \not\models^{pos} \varphi \implies \mathcal{M}, s \not\models^{pos} \varphi$$

The theorem states that necessary properties can be inferred from approximations to more precise models and the other way around for possible properties. We can formalise a similar result for the relation between abstract systems and concrete systems:

Theorem 2. *Let* $\mathcal{M}$ *be the LTS,* $(S, Act, \rightarrow, s_0)$*,* h *be mapping between* S *and* $\widehat{S}$ *and let the* Accelerated Modal-*LTS,* $\widehat{\mathcal{M}}$ $(\widehat{S}, Act, \rightarrow_\diamond, \rightarrow_\boxplus, \widehat{S}_0)$ *be the minimal abstraction of* $(\widehat{\mathcal{M}} = min_h(\mathcal{M}))$*. Then for every* φ*, and for every* s *and* $\widehat{s}$ *such that* $h(s) = \widehat{s}$*, we have:*

$$\widehat{\mathcal{M}}, \widehat{s} \models^{nec} \varphi \implies \mathcal{M}, s \models \varphi \quad \textit{and} \quad \widehat{\mathcal{M}}, \widehat{s} \not\models^{pos} \varphi \implies \mathcal{M}, s \not\models \varphi$$

Theorem 1 defines the inference between abstractions with different precision, and Theorem 2 between a concrete system and its minimal abstraction. Together they can be used to infer properties from an abstract approximation to a concrete system. If we have to prove a property φ on a concrete system $\mathcal{M}$, it is enough that an abstract approximation $\widehat{\mathcal{M}}$ *necessarily* satisfies it. If we want to refute a property, we prove that $\widehat{\mathcal{M}}$ does not *possibly* satisfies it. If, however, $\widehat{\mathcal{M}}, \widehat{s} \not\models^{nec} \varphi$ and $\widehat{\mathcal{M}}, \widehat{s} \models^{pos} \varphi$ no conclusion on $\widehat{\mathcal{M}}$ can be drawn. This is inevitable, because abstraction may loose information.

4.1 Example

Let us consider again the example of the *resettable* counter and the following property, $[(\neg\text{"expire"})*]\langle \mathsf{T}*.\text{"expire"}\rangle \mathsf{T}$, which is read as: *after any sequence of actions different from* expire *there is a path that contains the action* expire. We can see that this is trivially satisfied by the concrete system presented in Figure 1. We can infer it from the abstract system of Figure 3. We recall that the universal modality is interpreted over *may* actions and the existential over *must* ones. Hence:

- From the initial state +, the states that *may* be reached by $(\neg\ \text{"expire"}*)$ are $\{+, 0\}$. Then,
- from + there is the path $\overset{\text{dec+}}{\longrightarrow}_{\boxplus}\overset{\text{expire}}{\longrightarrow}_{\boxplus}$ and $[\![\text{dec+}\cdot\text{expire}]\!] \subset [\![\mathsf{T}*\cdot\text{expire}]\!]$, therefore + satisfies $\langle\ \mathsf{T}*.\ \text{"expire"}\ \rangle\ \mathsf{T}$.
- From 0 we have the transition $\overset{\text{expire}}{\longrightarrow}_{\boxplus}$, and $[\![\text{expire}]\!] \subset [\![\mathsf{T}*\cdot\text{expire}]\!]$, therefore 0 also satisfies the $\langle\ \mathsf{T}*.\ \text{"expire"}\ \rangle\ \mathsf{T}$.
- Hence, the formula is *necessarily* satisfied in the abstraction and we can infer the satisfaction on the concrete system.

We remark that the formula cannot be proved using only the abstraction framework without *accelerations* (Figure 2) because there will not be a *must*-transition between + and 0. The next section is dedicated to describing a model checking algorithm that implements the semantics given above. It shows that the three-valued PDL model checking problem for *Accelerated Modal*-LTSs is still decidable.

5 Model Checking

Theorem 3. *The three-valued PDL model checking problem for* accelerated Modal-*LTSs is decidable.*

The rest of this section is devoted to the algorithm, and to an example of its application. Let a fixed *Accelerated Modal*-LTS $\mathcal{M}$ and a PDL formula φ be given. The model checking problem is solved by two functions *eval_nec* and *eval_pos* that given a formula compute a set of states. They work by analysing the subformula components of the original. They are derived from the semantics presented in section 4. We only present *eval_nec* which returns the set of states that *necessarily* satisfy a formula. *eval_pos* returns the set of states that *possibly* satisfy a formula and it is defined dually.

function *eval_nec*(φ) {
    if $\varphi = \mathsf{F}$ **then return** $\emptyset$;
    if $\varphi = \neg\varphi_1$ **then return** $S \setminus \mathit{eval\_pos}(\varphi_1)$;
    if $\varphi = \varphi_1 \vee \varphi_2$ **then return** $\mathit{eval\_nec}(\varphi_1) \cup \mathit{eval\_nec}(\varphi_2)$;
    if $\varphi = \langle\beta\rangle\varphi_1$ **then return** $\mathit{exists\_must}(\beta, \mathit{eval\_nec}(\varphi_1))$;
}

The rest of the cases ($\top$, $\wedge$ and $[]$) follow from the standard dualities. The auxiliary function *exists_must*(β, R) computes the part referring to the *accelerations.* It returns the set of states that can reach a state in R by performing a sequence of actions such that the language generated by the concatenation of the action labels is included in the language generated by β. The function that computes this is not trivial. We first give an algorithm to compute this function, and then provide an explanation for it. In the algorithm we use the following notation: Given an automaton B, $B_{(i,J)}$ denotes the automaton that is obtained from B by changing the initial state to i and the final states to J.

function *exists_must*(β, R) {
1 $B :=$ DFA(β); $b_0 :=$ START(B); $F :=$ FINAL(B);
2 **for all** σ **such that** $\exists\, s, r \in S.\, s \xrightarrow{\sigma}_{\boxplus} r$ **do** $R_\sigma := \{(i, J) \mid [\![\sigma]\!] \subseteq [\![B_{(i,J)}]\!]\}$
3 **for all** $s, r \in S$ **do** $R_{(s,r)} := \bigcup\{R_\sigma \mid s \xrightarrow{\sigma}_{\boxplus} r\}$
4 **while** $R_{(s,t)}$ is not yet stable for some s, t **do** {
$R_{(s,t)} := R_{(s,t)} \cup \{(h, J) \mid \exists\, r, I.\, (h, I) \in R_{(s,r)} \wedge \forall i \in I. (i, J) \in R_{(r,t)}\}$
5 **return** $\{s \mid \exists\, r \in R\, \exists\, (b_0, J) \in R_{(s,r)}.\, J \subseteq F\}$;
}

Let us see how the last algorithm works:

1. First, the algorithm computes a deterministic automaton (DFA) corresponding to the regular expression β. B denotes this automaton, b_0 its initial state and F the set of final states.
2. Then, for every regular expression σ of the *Modal* transition system, we compute R_σ which consists of a set of the pairs (i, J) in B, such that the language generated by σ is included in the language accepted by the automaton $B_{(i,J)}$. Note that if (i, J) is in R_σ then all pairs (i, J') with $J \subset J'$ are also in R_σ
3. In the third step, for every pair of states (s, r) of the transition system, we take the union of the sets associated to the transitions from s to r. That is, $(i, J) \in R_{(s,r)}$ implies that there exists a regular expression σ such that there is a transition from s to r labelled with σ, i.e, $s \xrightarrow{\sigma}_{\boxplus} r$, and the language of σ is included in the language accepted by $B_{(i,J)}$.
4. Then, for every pair of states, we compute the closure of the sets. The computation is done until the fixpoint is reached. If $(h, J) \in R_{(s,t)}$ then there exists a sequence of transitions from s to t, i.e., $s \xrightarrow{\sigma_0}_{\boxplus} \ldots \xrightarrow{\sigma_n}_{\boxplus} t$ such that the language of the concatenation of $\sigma_0, \ldots, \sigma_n$ is included in the language accepted by $B_{(h,J)}$.
5. Finally, the algorithm returns the states s that are related with a target state $r \in R$, such that the relation $R_{(s,r)}$ contains a pair (b_0, J) where b_0 is the initial state of B and J only contains final states of B. From $J \subseteq F$ follows that the language accepted by $B_{(b_0,J)}$ is included in the language accepted by B. And, by step 4, we see that there exists a sequence of regular expressions $\sigma_0, \ldots, \sigma_n$ such that there is a sequence of transitions from s to r labelled with

$\sigma_0, ..., \sigma_n$, i.e., $s \xrightarrow{\sigma_0}_{\boxplus} ... \xrightarrow{\sigma_n}_{\boxplus} r$ and the language of the concatenation of the σs is included in the language accepted by $B_{(b_0,J)}$, so also in the language accepted by B.

We can easily see that the function *exists_must* terminates, because for every pair (s, r) the relation $R_{(s,r)}$ will contain elements in $B \times \mathcal{P}(B)$, which is finite because B is a finite automaton. Furthermore, the fixpoint computation is monotonic which implies that the algorithm will finish.

The use of regular expressions adds more computational complexity to the *normal* model checking algorithm. The algorithm is exponential in the size of the automaton corresponding to β and the automata of the transitions, and in the size of the transition systems. Even though the complexity is very high, in practice the regular expressions that will appear will be rather simple, so we expect that it will not cause a significant slow down of the normal model checking algorithms.

To complete the model checking algorithm we have to provide also a definition for *eval_pos*, which is as similar to *eval_nec*, swapping the modalities as we did for the semantics in section 4.

5.1 Example

To end this section we include an example of the *exists_must* computation. Let us consider the transition system of the left part of Figure 4 (in which we only include *must* transitions). We want to compute *exists_must*(β, R) of it, with $\beta = a \cdot (b) * \cdot c$ and $R = \{t\}$. The first step is to transform β to a deterministic automaton (right part of Figure 4). From the DFA we can remove those edges that lead to states from which no accepting states can be reached anymore. This simplifies the example, without changing the final result.

- $R_{\sigma_0} = R_{\sigma_3} = R_{\sigma_4} = \{(h, \{i\}), (h, \{i, j\}), (h, \{i, h\}), (h, \{i, j, h\})\}$
- $R_{\sigma_1} = \{(i, \{j\}), (i, \{j, i\}), (i, \{j, h\}), (i, \{j, i, h\})\}$, $R_{\sigma_2} = \emptyset$ and
- $R_{\sigma_5} = \{(i, \{i\}), (i, \{i, j\}), (i, \{i, h\}), (i, \{i, j, h\})\}$.

Now, in step 3 we compute the relations between states of the transition system:

- $R_{(r,s)} = R_{\sigma_0}$, $R_{(s,t)} = R_{\sigma_1}$, $R_{(u,u)} = R_{\sigma_3} \cup R_{\sigma_4}$ and $R_{(u,t)} = R_{\sigma_5}$
- The sets for the rest of the pairs of states are empty.

We close the relations under concatenation, which adds:

- $R_{(r,t)} = \{(h, \{j\}), (h, \{j, i\}), (h, \{j, h\}), (h, \{j, i, h\})\}$
- $R_{(u,t)} = R_{\sigma_5} \cup \{(h, \{i\}), (h, \{j, i\}), (h, \{j, h\}), (h, \{j, i, h\})\}$

Finally, we see that $R_{(r,t)}$ contains the pair $(h, \{j\})$ which is the initial state of β and $\{j\} \subseteq F$. Therefore, there is a path from r to t for which the language of the concatenation of its labels is included in β. Hence, the result of the function will be $\{r\}$.

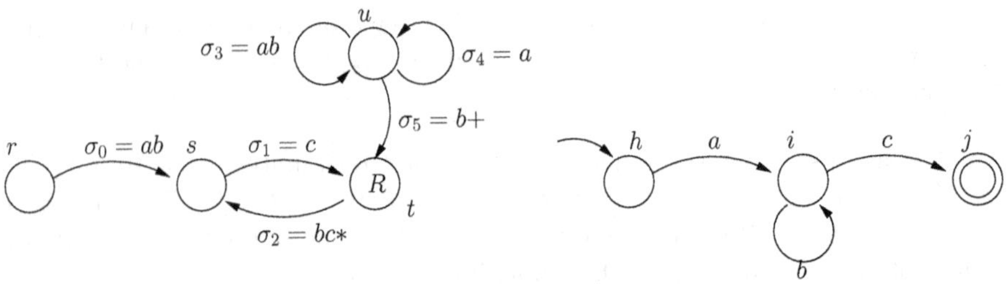

Fig. 4. *Accelerated Modal*-LTS (only *must*-part) and the DFA corresponding to β

6 Related Work

The problem of how to improve the expressiveness of abstractions has already been addressed by other authors. An interesting approach is, for example, the one proposed by Pnueli, in [10, 5]. His idea is to impose fairness constraints on the abstract system in order to remove undesirable behaviours. The fairness constraints are extracted from the knowledge we have of the concrete system. For example in the *resettable* counter example, we may know that the concrete system does not contain any infinite decreasing trace, hence the abstract one should not have it either. So, we can infer the following constraint: *"For any fair trace, if the transition* $+ \xrightarrow{dec} 0$ *is infinitely often enabled then it should be infinitely often taken."*

Any fair computation of the abstract system will not contain an infinite loop $+ \xrightarrow{dec} +$. This approach is valid to remove non-progressing traces (possible behaviours, represented by *may* traces in our framework). It has been used to infer properties coded in LTL by Pnueli and also recently by Bosnacki et al. [1]. In the latter approach the authors proved that in some specific cases, such as the counter abstraction, strong fairness constraints can be reduced to weak fairness which are more efficiently handled by model checkers. In our approach, we add necessary behaviours, so those approaches are independent of ours. Actually, both methods can be combined orthogonally, to even further make abstractions more precise.

The only other approach we have found to add more *must*-behaviour is based on adding hyper-transitions [17, 12]. It applies, for example, in cases of *if-then-else* constructions. If the condition is abstracted, we don't know the next state. So there will be two *may*-transitions, one $s \xrightarrow{a}_\diamond r$ and one $s \xrightarrow{b}_\diamond t$, but no *must*-transition. The idea of hyper-edges is to add a *must*-hyper-transition, starting from s, and pointing to both r and t. This indicates that there must be a transition, but we are not sure where it precisely ends. We can capture a similar effect by adding a step $s \xrightarrow{a|b}_\boxplus u$, where u is a state more abstract than r and t. Finally, we can even express much more complicated structures, such as nested loops, like in $s \xrightarrow{(b \cdot a* \cdot c)+}_\boxplus r$.

7 Future Work

We have described how to capture semantically a transition that represents a set of computations. A different problem is, given a specification of the system, how to add *sound accelerated*-transitions. We give here some ideas about this.

In some cases a transition of the type $s \xrightarrow{a*}_{\boxplus} r$ corresponds to a loop in the original specification. The transition expresses that the loop executes a number of actions a and then terminates. We do not know how many cycles it contains, but it ends at some point. Therefore to add such a transition to the abstract labelled transition system, we will have to prove termination of the concrete loop. Proving termination of sequential programs has been investigated for many years, we believe that many of the results of this field can be applied to our framework.

One of the common ways to prove termination is by checking that the computation progresses in a given well-founded domain. For example, in order to infer fairness constrains, Pnueli [10] uses a monitor process composed in parallel with the modelled system. The monitor controls the progress in the domain of the naturals. In some cases it is trivial to find the domain, for example for the decreasing counter, but this is not always the case.

In [19], we showed how our approach can be used to prove successful termination of a parametric Bounded Retransmission Protocol, see also [7, 4]. More experiments on realistic size case studies are needed to see which regular expressions typically arise in practice, and how our algorithm behaves on them. This will also provide feedback on how to optimize the algorithm. Furthermore, an interesting question is to determine the precise complexity of the three-valued PDL model checking problem on *accelerated Modal*-LTSs.

8 Conclusion

In this paper we introduced accelerated modal transitions. These capture sequences of required behaviour. They turned out to be useful in cases where usual modal LTSs are too imprecise, due to missing must-transitions. Besides defining *accelerated Modal*-LTSs, and suitable abstraction and approximation relations, we also showed that these relations preserve properties of three-valued PDL in the desired direction. As a consequence, after applying abstract interpretation, we are able to prove more liveness properties. Finally, we showed that the three-valued model checking problem for PDL on accelerated *Modal*-LTSs is decidable.

References

[1] D. Bosnacki, N. Ioustinova, and N. Sidorova. Using fairness to make abstractions work. In *Proc. of SPIN Model Checking and Software Verification*, volume 2989 of *LNCS*, pages 198–215. Springer, 2004.

[2] E. M. Clarke, O. Grumberg, and D. E. Long. Model checking and abstraction. *Journal of the ACM*, pages 343–354, 1992.

[3] D. Dams. *Abstract Interpretation and Partition Refinement for Model Checking.* PhD thesis, Eindhoven University of Technology, 1996.

[4] D. Dams and R. Gerth. The bounded retransmission protocol revisited. *ENTCS*, 9, 2000.

[5] Y. Fang, N. Piterman, A. Pnueli, and L. Zuck. Liveness with invisible ranking. In *Proc. of Verification Model Checking and Abstract Interpretation (VMCAI)*, volume 2937 of *LNCS*, pages 223–238. Springer, 2004.

[6] P. Godefroid, M. Huth, and R. Jagadeesan. Abstraction-based model checking using modal transition systems. In *Proc. of Concurrency Theory (CONCUR)*, volume 2154 of *LNCS*, pages 426–440. Springer, 2001.

[7] J. F. Groote and J.C. van de Pol. A bounded retransmission protocol for large data packets. In *Proc. of Algebraic Methodology and Software Technology (AMAST)*, volume 1101 of *LNCS*, pages 536–550. Springer, 1996.

[8] D. Harel, A. Pnueli, and J. Stavi. Propositional dynamic logic of context-free programs. *Foundations of Computer Science*, pages 310–321, 1981.

[9] M. Hennessey and R. Milner. On observing nondeterminism and concurrency. In *Proc. of International Conference on Automata, Languages and Programming (ICALP)*, volume 85 of *LNCS*, pages 295–309. Springer, 1980.

[10] Y. Kesten and A. Pnueli. Verifying liveness by augmented abstraction. In *Proc. of Computer Science Logic (CSL)*, LNCS 1683, pages 141–145. Springer, 1999.

[11] K. G. Larsen and B. Thomsen. A modal process logic. In *Proc. of Logic in Computer Science (LICS)*, pages 203–210. IEEE computer society, 1988.

[12] K. G. Larsen and L. Xinxin. Equation solving using modal transition systems. In *Proc. of Logic in Computer Science (LICS)*, pages 108–117. IEEE, 1990.

[13] C. Loiseaux, S. Graf, J. Sifakis, A. Bouajjani, and S. Bensalem. Property preserving abstractions for the verification of concurrent systems. *Formal Methods in System Design*, pages 11–44, 1995.

[14] A. Pnueli. Abstraction for liveness. In *Proc. of Verification Model Checking and Abstract Interpretation (VMCAI)*, LNCS 3385, pages 146–164. Springer, 2005.

[15] J.C. van de Pol and M. Valero Espada. Modal abstraction in μCRL. In *Proc. of Algebraic Methodology and Software Technology (AMAST)*, volume 3116 of *LNCS*, pages 409–425. Springer, 2004.

[16] D. Schmidt. Structure-preserving binary relations for program abstraction. In *Proc. of The Essence of Computation*, volume 2566 of *LNCS*, pages 245 – 268. Springer, 2002.

[17] S. Shoham and O. Grumberg. Monotonic abstraction-refinement for CTL. In *Proc. of Tools and Algorithms for the Construction and Analysis of Systems (TACAS)*, volume 2988 of *LNCS*, pages 546–560. Springer, 2004.

[18] C. Stirling. *Modal and Temporal Properties of Processes.* Texts in Computer Science. Springer, 2001.

[19] M. Valero Espada. *Modal Abstraction and Replication of Processes with Data.* PhD thesis, Free University Amsterdam, 2005.

A Compositional Semantics of Plan Revision in Intelligent Agents

M. Birna van Riemsdijk and John-Jules Ch. Meyer

Utrecht University, Department of Information and Computing Sciences
P.O. Box 80.089, 3508 TB Utrecht
The Netherlands
{birna, jj}@cs.uu.nl

Abstract. This paper revolves around the so-called plan revision rules of the agent programming language 3APL. These rules can be viewed as a generalization of procedures. This generalization however results in the semantics of programs of the 3APL language no longer being compositional. This gives rise to problems when trying to define a proof system for the language. In this paper we define a restricted version of plan revision rules which extends procedures, but which does have a compositional semantics, as we will formally show.

1 Introduction

An agent is commonly seen as an encapsulated computer system that is situated in some environment and that is capable of flexible, autonomous action in that environment in order to meet its design objectives [1]. Autonomy means that an agent encapsulates its state and makes decisions about what to do based on this state, without the direct intervention of humans or others. Agents are situated in some environment which can change during the execution of the agent. This requires *flexible* problem solving behavior, i.e., the agent should be able to respond adequately to changes in its environment. Programming flexible computing entities is not a trivial task. Consider for example a standard procedural language. The assumption in these languages is that the environment does not change while some procedure is executing. If problems do occur during the execution of a procedure, the program might throw an exception and terminate (see also [2]). This works well for many applications, but we need something more if change is the norm and not the exception.

A philosophical view that is well recognized in the AI literature is that rational behavior can be explained in terms of the concepts of *beliefs*, *goals* and *plans*. [3, 4, 5]. This view has been taken up within the AI community in the sense that it might be possible to *program* flexible, autonomous agents *using* these concepts. The idea is that an agent tries to fulfill its goals by selecting appropriate plans, depending on its beliefs about the world. Beliefs should thus represent the world or environment of the agent; the goals represent the state of the world the agent wants to realize and plans are the means to achieve these goals. When programming in terms of these concepts, beliefs can be compared

M. Johnson and V. Vene (Eds.): AMAST 2006, LNCS 4019, pp. 353–367, 2006.

to the program state, plans can be compared to statements, i.e., plans constitute the procedural part of the agent, and goals can be viewed as the (desired) post-conditions of executing the statement or plan. Through executing a plan, the world and therefore the beliefs reflecting the world will change and this execution should have the desired result, i.e., achievement of goals.

This view has been adopted by the designers of the agent programming language *3APL*[1] [6, 7], which is a well-known language in the agent programming community. The dynamic parts of a 3APL agent thus consist of a set of beliefs, a plan[2] and a set of goals. A plan can consist of sequences of so-called basic actions, which change the beliefs[3] if executed. To provide for the possibility of programming flexible behavior, so-called *plan revision* rules were added to the language. These rules can be compared to procedures in the sense that they have a head, which is comparable with the procedure name, and a body, which is a plan in the case of 3APL and a statement in the case of procedural languages.

The operational meaning of plan revision rules is similar to that of procedures: if the procedure name or head is encountered in a statement or plan, this name or head is replaced by the body of the procedure or rule, respectively (see [8] for the operational semantics of procedure calls). The difference however is that the head in a plan revision rule can be *any* plan (or statement) and not just a procedure name. In procedural languages it is furthermore usually assumed that procedure names are distinct. In 3APL however, it is possible that multiple rules are applicable at the same time. This provides for very general and flexible plan revision capabilities, which is a distinguishing feature of 3APL compared to other agent programming languages [9, 10, 11].

As argued, we consider these general plan revision capabilities to be an essential part of agenthood. The introduction of these capabilities now gives rise to interesting issues concerning the *semantics of plan execution*, which we will be concerned with in this paper.

The main issue which arises with the introduction of plan revision rules, is the issue of *compositionality* of semantics of plans. For standard procedural languages [8, Chapter 5], the semantics of statements is compositional, i.e., the semantics of a composed statement can be defined in terms of the semantics of the parts of which it is composed. The semantics of plans however, which can be viewed as the statements of 3APL, is *not* compositional. The reason for this lies in the presence of plan revision rules, which we will elaborate on in section 3.2.

The fact that the semantics of plans is not compositional, gives rise to problems when trying to reason about 3APL programs. A proof system for a programming language will typically contain rules by means of which properties of the entire program can be proven by proving properties of the parts of which the program is composed. Since the semantics of 3APL plans is not compositional, this is problematic in the case of 3APL. One way of trying to approach this

[1] 3APL is to be pronounced as "triple-a-p-l".

[2] In the original version this was a set of plans.

[3] A change in the environment is a possible "side effect" of the execution of a basic action.

problem is by defining a specialized logic for 3APL which tries to circumvent the issue, as was done in [12, 13]. The resulting logic is however non-standard and can be difficult to use, which will be explained in more detail in section 3.3.

The approach we take in this paper, is to try to *restrict* the allowed plan revision rules, such that the semantics of plans becomes compositional in some sense. It is not immediately obvious what kind of restriction would yield the desired result. In this paper, we propose such a restriction and prove that the semantics of plans in that case is compositional.

The outline of the paper is as follows. In section 2, we present the syntax and semantics of a simplified version of 3APL. It is important to note that we use a simplified version, in order to be able to focus on the issue of compositionality of plans. In particular, we do not include a model of the environment in the semantics, since this is not necessary for investigating the compositionality issue. In section 3 we elaborate on the issue of compositionality and explain why the semantics of full 3APL is not compositional. In section 4 we present our proposal for a restricted version of plan revision rules, and prove that the semantics of plans is compositional, given this restriction on plan revision rules. This paper aims to be a first step towards a compositional proof system for 3APL. Investigating automated theorem proving and providing accompanying tool support for this is left for future research.

2 3APL

2.1 Syntax

Below, we define belief bases and plans. A belief base is a set of propositional formulas. A plan is a sequence of basic actions. Basic actions can be executed, resulting in a change to the beliefs of the agent.

In the sequel, a language defined by inclusion shall be the smallest language containing the specified elements.

Definition 1 *(belief bases)***.** Assume a propositional language $\mathcal{L}$ with typical formula p and the connectives $\wedge$ and $\neg$ with the usual meaning. Then the set of belief bases Σ with typical element σ is defined to be $\wp(\mathcal{L})$.[4]

Definition 2 *(plans)***.** Assume that a set $\mathsf{BasicAction}$ with typical element a is given. The set of plans Plan with typical element π is then defined as follows.

$$\pi ::= a \mid \pi_1; \pi_2$$

We use ϵ to denote the empty plan and identify $\epsilon; \pi$ and $\pi; \epsilon$ with π.

Plan revision rules consist of a head π_h and a body π_b. Informally, an agent that has a plan π_h, can replace this plan by π_b when applying a plan revision rule of this form.

[4] $\wp(\mathcal{L})$ denotes the powerset of $\mathcal{L}$.

Definition 3 *(plan revision rules).* The set of plan revision rules $\mathcal{R}$ is defined as follows: $\mathcal{R} = \{\pi_h \rightsquigarrow \pi_b \mid \pi_h, \pi_b \in \mathsf{Plan}, \pi_h \neq \epsilon\}$.[5]

Take for example a plan $a; b$ where a and b are basic actions, and a plan revision rule $a; b \rightsquigarrow c$. The agent can then either execute the actions a and b one after the other, or it can apply the plan revision rule yielding a new plan c, which can in turn be executed.

Below, we provide the definition of a 3APL agent. The function $\mathcal{T}$, taking a basic action and a belief base and yielding a new belief base, is used to define how belief bases are updated when a basic action is executed.

Definition 4 *(3APL agent).* A 3APL agent $\mathcal{A}$ is a tuple $\langle \sigma_0, \pi_0, \mathsf{PR}, \mathcal{T} \rangle$ where $\sigma_0 \in \Sigma$, $\pi_0 \in \mathsf{Plan}$, $\mathsf{PR} \subseteq \mathcal{R}$ is a finite set of plan revision rules and $\mathcal{T} : (\mathsf{BasicAction} \times \Sigma) \to \Sigma$ is a partial function, expressing how belief bases are updated through basic action execution.

A plan and a belief base can together constitute a so-called configuration. During computation or execution of the agent, the elements in a configuration can change.

Definition 5 *(configuration).* Let Σ be the set of belief bases and let Plan be the set of plans. Then $\mathsf{Plan} \times \Sigma$ is the set of configurations of a 3APL agent. If $\langle \sigma_0, \pi_0, \mathsf{PR}, \mathcal{T} \rangle$ is an agent, then $\langle \pi_0, \sigma_0 \rangle$ is the initial configuration of the agent.

2.2 Semantics

The semantics of a programming language can be defined as a function taking a statement and a state, and yielding the set of states resulting from executing the initial statement in the initial state. In this way, a statement can be viewed as a transformation function on states. In 3APL, plans can be seen as statements and belief bases as states on which these plans operate. There are various ways of defining a semantic function and in this paper we are concerned with the so-called *operational* semantics (see for example De Bakker [8] for details on this subject).

The operational semantics of a language is usually defined using transition systems [14]. A transition system for a programming language consists of a set of axioms and derivation rules for deriving transitions for this language. A transition is a transformation of one configuration into another and it corresponds to a single computation step. Let $\mathcal{A}$ be a 3APL agent with a set of plan revision rules PR, belief update function $\mathcal{T}$, and let $\mathsf{BasicAction}$ be its set of basic actions. Below, we give the transition system $\mathsf{Trans}_{\mathcal{A}}$ for our simplified 3APL language,

[5] In [6], plan revision rules were defined to have a guard, i.e., rules were of the form $\pi_h \mid \phi \rightsquigarrow \pi_b$, where ϕ is a condition on the belief base. For a rule to be applicable, the guard should then hold. For technical convenience and because we want to focus on the plan revision aspect of these rules, we however leave out the guard in this paper.

which is based on the system given in [6]. This transition system is specific to agent $\mathcal{A}$.

There are two kinds of transitions, i.e., transitions describing the execution of basic actions and those describing the application of a plan revision rule. The transitions are labelled to denote the kind of transition. A basic action at the head of a plan can be executed in a configuration if the function $\mathcal{T}$ is defined for this action and the belief base in the configuration. The execution results in a change of belief base as specified through $\mathcal{T}$ and the action is removed from the plan.

Definition 6 *(action execution).* Let $a \in \mathsf{BasicAction}$.

$$\frac{\mathcal{T}(a, \sigma) = \sigma'}{\langle a; \pi, \sigma \rangle \rightarrow_{exec} \langle \pi, \sigma' \rangle}$$

A plan revision rule can be applied in a configuration if the head of the rule is equal to a prefix of the plan in the configuration. The application of the rule results in the revision of the plan, such that the prefix equal to the head of the rule is replaced by the plan in the body of the rule. A rule $a; b \rightsquigarrow c$ can for example be applied to the plan $a; b; c$, yielding the plan $c; c$. The belief base is not changed through plan revision.

Definition 7 *(rule application).* Let $\rho : \pi_h \rightsquigarrow \pi_b \in \mathsf{PR}$.

$$\langle \pi_h \bullet \pi, \sigma \rangle \rightarrow_{apply} \langle \pi_b \bullet \pi, \sigma \rangle$$

Using the transition system, individual transitions can be derived for a 3APL agent. These transitions can be put in sequel, yielding transition sequences, which are typically denoted by θ. From a transition sequence, one can obtain a *computation sequence* by removing the plan component of all configurations occurring in the transition sequence. In the following definitions, we formally define computation sequences and we specify the function yielding these sequences, given an initial configuration.

Definition 8 *(computation sequences).* The set Σ^+ of finite computation sequences is defined as $\{\sigma_1, \ldots, \sigma_i, \ldots, \sigma_n \mid \sigma_i \in \Sigma, 1 \leq i \leq n, n \in \mathbb{N}\}$.

Definition 9 *(function for calculating computation sequences).* Let $x_i \in \{exec, apply\}$ for $1 \leq i \leq m$. The function $\mathcal{C}^{\mathcal{A}} : (\mathsf{Plan} \times \Sigma) \rightarrow \wp(\Sigma^+)$ is then as defined below.

$$\mathcal{C}^{\mathcal{A}}(\pi, \sigma) = \{\sigma, \ldots, \sigma_m \in \Sigma^+ \mid \langle \pi, \sigma \rangle \rightarrow_{x_1} \ldots \rightarrow_{x_m} \langle \epsilon, \sigma_m \rangle \text{ is a finite sequence of transitions in } \mathsf{Trans}_{\mathcal{A}}\}.$$

Note that we only take into account successfully terminating transition sequences, i.e., those sequences ending in a configuration with an empty plan. Using the function defined above, we can now define the operational semantics of 3APL.

Definition 10 *(operational semantics).* Let $\kappa : \Sigma^+ \to \Sigma$ be a function yielding the last element of a finite computation sequence, extended to handle sets of computation sequences as follows, where I is some set of indices: $\kappa(\{\delta_i \mid i \in I\}) = \{\kappa(\delta_i) \mid i \in I\}$. The operational semantic function $\mathcal{O}^{\mathcal{A}} : \mathsf{Plan} \to (\Sigma \to \wp(\Sigma))$ is defined as follows:

$$\mathcal{O}^{\mathcal{A}}(\pi)(\sigma) = \kappa(\mathcal{C}^{\mathcal{A}}(\pi, \sigma)).$$

We will in the sequel omit the superscript $\mathcal{A}$ to functions as defined above, for reasons of presentation.

3 3APL and Non-compositionality

Before we go into discussing why the semantics of 3APL plans is not compositional, we consider compositionality of standard procedural languages.

3.1 Compositionality of Procedural Languages

The semantics of standard procedural languages such as described in [8, Chapter 5] are compositional. Informally, a semantics for a programming language is compositional if the semantics of a composed program can be defined in terms of the semantics of the parts of which it is composed. To be more specific, the meaning of a composed program $S_1; S_2$ should be definable in terms of the meaning of S_1 and S_2, for the semantics to be compositional.

A semantics can be defined directly in a compositional way, in which case the semantics is often termed a denotational semantics [8]. Alternatively, a semantics can be *defined* in a *non*-compositional way, such as an operational semantics defined using computation sequences, while it still *satisfies a compositionality property.* In this paper, we focus on the latter case. It turns out that the operational semantics for a procedural language such as discussed in [8, Chapter 5] satisfies such a compositionality property, while the operational semantics of 3APL of definition 10 does not. All results and definitions with respect to procedural languages which we refer to in section 3, can be found in [8, Chapter 5].

An operational semantics of a procedural language can be defined analogously to the operational semantics of 3APL of definition 10, where plans are statements and belief bases are states. Both operational semantics are defined in a non-compositional way, since they do not use the structure of the plan or statement to define its semantics. Nevertheless, the operational semantics of a procedural language does satisfy a compositionality property, i.e., the following holds: $\mathcal{O}(S_1; S_2)(\sigma) = \mathcal{O}(S_2)(\mathcal{O}(S_1)(\sigma))$, where S_1 and S_2 are statements. This property specifies that the set of states possibly resulting from the execution of a composed statement $S_1; S_2$ in σ is equal to the set of states resulting from the execution of S_2 in all states resulting from the execution of S_1 in σ.

3.2 Non-compositionality of 3APL

While the presented compositionality property is termed "natural" in [8, Chapter 5], it is *not* satisfied by the operational semantics of 3APL, i.e., it is not the case that $\mathcal{O}(\pi_1; \pi_2)(\sigma) = \mathcal{O}(\pi_2)(\mathcal{O}(\pi_1)(\sigma))$ always holds. The reason for this lies in the presence of plan revision rules. Take for example an agent with one plan revision rule $a; b \rightsquigarrow c$. Let σ_{ab} and σ_c be the belief bases resulting from the execution of actions a followed by b, and c in σ, respectively. We then have that $\mathcal{O}(a; b)(\sigma) = \{\sigma_{ab}, \sigma_c\}$, i.e., the agent can either execute the actions a and b one after the other, or it can apply the plan revision rule and then execute c.

If the semantics of 3APL plans would have been compositional, we would also have that $\mathcal{O}(b)(\mathcal{O}(a)(\sigma)) = \{\sigma_{ab}, \sigma_c\}$. This is however not the case, since $\mathcal{O}(b)(\mathcal{O}(a)(\sigma)) = \{\sigma_{ab}\}$.[6] This stems from the fact that if one "breaks" the composed plan $a; b$ in two, one can no longer apply the plan revision rule $a; b \rightsquigarrow c$, because this rule can only be applied if the composed plan $a; b$ is considered. The set of belief bases $\mathcal{O}(a)(\sigma)$ only contains those resulting from the execution of a. The action b is then executed on those belief bases, yielding $\mathcal{O}(b)(\mathcal{O}(a)(\sigma))$. The result thus does not contain σ_c.

3.3 Reasoning About 3APL

This non-compositionality property of 3APL plans gives rise to problems when trying to define a proof system for reasoning about 3APL plans. In standard procedural languages, the following proof rule is part of any Hoare logic for such a language [8], where p, p' and q are assertions.

$$\frac{\{p\}\ S_1\ \{p'\} \qquad \{p'\}\ S_2\ \{q\}}{\{p\}\ S_1; S_2\ \{q\}} \tag{3.1}$$

This rule specifies that one can reason about a composed program by proving properties of the parts of which it is composed. The soundness of this rule depends on the fact that $\mathcal{O}(S_1; S_2)(\sigma) = \mathcal{O}(S_2)(\mathcal{O}(S_1)(\sigma))$. Because this property does not hold for 3APL plans, a similar rule for 3APL would not be sound (see also the discussion in [13]). Nevertheless, one would still want to reason about composed 3APL plans.

In [13],[7] we have presented a specialized dynamic logic for this purpose. In that paper, we define a logic for reasoning about 3APL plans in which we can restrict the number of plan revision rule applications allowed to occur during the execution of the plan. Based on this logic, we define a logic for reasoning about 3APL plans in general. The resulting complete proof system however contains an infinitary proof rule, i.e., a rule with an infinite number of premises. In some cases, induction can be used to prove the premises of this rule. These induction

[6] Note that $\mathcal{O}(b)(\mathcal{O}(a)(\sigma)) \subseteq \mathcal{O}(a; b)(\sigma)$.

[7] Parts of [13] were published in [12].

proofs are however quite involved, and it is not yet clear whether these can somehow be automated, etc.

Another possible approach for reasoning about 3APL plans has been suggested in [15, 16]. In that paper, we define a denotational (i.e., compositional) semantics for a 3APL *meta*-language. This meta-language is relatively standard, as it is essentially a non-deterministic language with a `while` construct. It was suggested that it might by possible to reason about this meta-language, rather than about 3APL plans directly.

While the two discussed papers aim at reasoning about full 3APL, we take a different approach in this paper. Here, we investigate whether we can somehow *restrict* plan revision rules, such that the semantics of plans becomes compositional (in some sense). The idea is that given such a compositional semantics, it will be possible to come up with a more standard and easy to use proof system for 3APL.

4 Compositional 3APL

One obvious candidate for a restricted version of plan revision rules is the restriction to rules with an atomic head, i.e., to rules of the form $a \rightsquigarrow \pi$. These rules are very similar to procedures, apart from the fact that an action a could either be transformed using a plan revision rule, *or* executed directly. In contrast with actions, procedure variables cannot be executed, i.e., they can only be replaced by the body of a procedure. It is easy to see that a semantics for 3APL with only these plan revision rules would be compositional.

However, this kind of plan revision rules would capture very little of the general plan revision capabilities of the non-restricted rules. The challenge is thus to find a less restrictive kind of plan revision rules, which would still satisfy the desired compositionality property. Finding such a restricted kind of plan revision rules is non-trivial. We discuss the line of reasoning by which it can be obtained in section 4.1. In section 4.2, we present and explain the theorem that expresses that the proposed restriction on plan revision rules indeed establishes (some form of) compositionality. Finally, in section 5, we briefly address the issue of reasoning about 3APL with restricted plan revision rules, and point to directions for future research regarding this issue.

4.1 Restricted Plan Revision Rules

The restriction to plan revision rules that we propose is given in definition 11 below, and can be understood by trying to get to the essence of the compositionality problem arising from non-restricted plan revision rules.

First, we have to observe that the general kind of compositionality as specified in section 3.1 for procedural languages is in general not obtainable for 3APL, if the set of plan revision rules contains a rule with a non-atomic head. The property specifies that the semantics of a composed plan (or program) should be definable in terms of the parts of which it is composed. The property however does not specify how a composed plan should be broken down into parts. That is, for a plan to be compositional in the general sense, compositionality should hold, no matter how the plan is decomposed. Consider for example the plan $a; b; c$. It should then be the case that $\mathcal{O}(a; b; c)(\sigma) = \mathcal{O}(c)(\mathcal{O}(a; b)(\sigma)) = \mathcal{O}(b; c)(\mathcal{O}(a)(\sigma))$, i.e., the compositionality property should hold, no matter whether the plan is decomposed into $a; b$ and c, or a and $b; c$.

If a set of plan revision rules however contains a rule with a non-atomic head, it is always possible to come up with a plan (and belief base and belief update function) for which this property does not hold. This plan should contain the head of the plan revision rule. If the decomposition of the plan is then chosen such that it "breaks" this occurrence of the head of the rule in the plan, the compositionality property in general does not hold for this decomposition. This is because the plan revision rule can in that case not be applied when calculating the result of the operational semantic function. Consider for example the plan revision rule $a; b \rightsquigarrow c$ and the plan $a; b; c$. If the plan is decomposed into a and $b; c$, the rule cannot be applied and thus $\mathcal{O}(a; b; c)(\sigma) = \mathcal{O}(b; c)(\mathcal{O}(a)(\sigma))$ does not always hold.

The question is now which kind of compositionality *can* be obtained for 3APL. We have established that being allowed to decompose a composed plan into arbitrary parts for a definition of compositionality gives rise to problems in the case of 3APL. That is, the standard definition of compositionality will always be problematic if we want to consider plan revision rules with a non-atomic head. Since we want our restriction on plan revision rules to allow at least some form of non-atomicity (because otherwise we would essentially be considering procedures), we have to come up with another definition of compositionality if we want to make any progress.

The idea that we propose is essentially to take the operational meaning of a plan as the basis for a compositionality property. When executing a plan π, either the first action of π is executed, or an applicable plan revision rule is applied. In the first case, π has to be of the form $a; \pi_r$[8], and in the latter case of the form $\pi_h; \pi'_r$, given an applicable plan revision rule of the form $\pi_h \rightsquigarrow \pi_b$. Taking this into account, we are, broadly speaking, looking for a restriction to plan revision rules which allows us to decompose π into a and π_r, or π_h and π'_r. To be more specific, it should be possible to execute a and then consider π_r separately, or to apply the specified plan revision rule and then consider the body of the rule π_b and the rest of the plan, i.e., π'_r, separately. That is, we are after something like the following compositionality property:[9]

$$\mathcal{O}(\pi)(\sigma) = \mathcal{O}(\pi_r)(\mathcal{O}(a)(\sigma)) \cup \mathcal{O}(\pi'_r)(\mathcal{O}(\pi_b)(\sigma)). \tag{4.1}$$

In order to come up with a restriction on plan revision rules that gives us such a property, we have to understand why this property does not always hold in

[8] The subscript r here indicates that π_r is the *rest* of the plan π.

[9] The property that will be proven in section 4.2 differs slightly, as it takes into account the existence of multiple applicable plan revision rules.

the presence of non-restricted plan revision rules. Essentially, what this property specifies is that we can separate the semantics of certain prefixes of the plan π (i.e., a and π_h), from the semantics of the rest of π.

A case in which this is *not* possible, is the following. Consider a plan of the form $\pi_h; \pi'_h; \pi$, and plan revision rules of the form $\pi_h \rightsquigarrow \pi_b$ and $\pi_b; \pi'_h \rightsquigarrow \pi'_b$. We can apply the first rule to this plan, yielding $\pi_b; \pi'_h; \pi$. If the semantics of the plan would be compositional in the sense of (4.1), it should now be possible to consider the semantics of $\pi'_h; \pi$, i.e., the "rest" of the plan, separately. Given the second plan revision rule however, this is not possible: if we separate $\pi_b; \pi'_h; \pi$ into π_b and $\pi'_h; \pi$, we can no longer apply the second plan revision rule, whereas we *can* apply the rule if the plan is considered in its composed form. The semantics of the plan $\pi_h; \pi'_h; \pi$ is thus not compositional, given the two plan revision rules.

This argument is similar to the explanation of why the general notion of compositionality does not hold for 3APL. Contrary to the general case however, we can in the case of compositionality as defined in (4.1), specify a restriction to plan revision rules that prevents this problem from occurring. The restriction will thus allow us to consider the semantics of π'_r (see (4.1)) separately from the semantics of π_b, thereby establishing compositionality property (4.1).

As explained, if there is a plan revision rule of the form $\pi_h \rightsquigarrow \pi_b$, a plan revision rule with a head of the form $\pi_b; \pi'_h$ is problematic. A restriction one could thus consider, is the restriction that if there is a rule of the form $\pi_h \rightsquigarrow \pi_b$, there should not also be a rule of the form $\pi_b; \pi'_h \rightsquigarrow \pi'_b$, i.e., the body of a rule cannot be equal to the prefix of the head of another rule. This restriction however does not do the trick completely. The reason has to do with the fact that actions from a plan of the form $\pi_b; \pi'_h$ can be executed.

Consider for example a plan $a_1; a_2; b_1; b_2$ and plan revision rules $a_1; a_2 \rightsquigarrow c_1; c_2$ and $c_2; b_1 \rightsquigarrow c_3$. The head of the second rule does not have the form $c_1; c_2; \pi$, i.e., the body of the first rule is not equal to the prefix of the head of another rule. Therefore, according to the suggested restriction, this rule is allowed. We can apply the first rule to the plan, yielding $c_1; c_2; b_1; b_2$. If the compositionality property holds, we should now be able to consider the semantics of $b_1; b_2$ separately. Suppose the action c_1 is executed, resulting in the plan $c_2; b_1; b_2$. Considering the second plan revision rule, we observe that this rule is applicable to this plan. This is however only the case if we consider this plan in its composed form. If we separate the semantics of $b_1; b_2$ as specified by the compositionality property (4.1), we cannot apply the rule. Given the plan $a_1; a_2; b_1; b_2$ and the two plan revision rules, the compositionality property thus does not hold.

The solution to this problem is to adapt the suggested restriction which considers the body of a rule in relation with the prefix of the head of another rule, to a restriction which consider the *suffix* of the body of a rule in relation with the prefix of the head of another rule. The restriction should thus specify that the suffix of the body of a rule cannot be equal to the prefix of the head of another rule. Under that restriction, the second rule of the example discussed above would not be allowed, and the compositionality property (4.1) would hold.

This restriction on plan revision rules is specified formally below. The fact that under this restriction, the property (4.1) (or a slight variation thereof) holds, is formally shown in section 4.2.

Definition 11 *(restricted plan revision rules).* Let PR be a set of plan revision rules. Let suff be a function taking a plan and yielding all its suffixes, and let pref be a function taking a plan and yielding all its strict prefixes.[10] We say that PR is *restricted* iff the following holds:

$$\forall \rho \in \mathsf{PR} : (\rho : \pi_h \rightsquigarrow \pi_b) : \neg \exists \rho' \in \mathsf{PR} : (\rho' : \pi'_h \rightsquigarrow \pi'_b) : (\mathsf{suff}(\pi_b) \cap \mathsf{pref}(\pi'_h)) \neq \emptyset.$$

The fact that we define pref as yielding *strict* prefixes allows the suffix of the body of a plan revision rule to be exactly equal to the head of another rule. This does not violate the compositionality property, and it results in restricted plan revision rules being a superset of rules with an atomic head. Otherwise, a rule $b \rightsquigarrow c$, for example, would not be allowed if there is also a rule $a \rightsquigarrow a; b$, since b, i.e., the suffix of the latter rule, would then by definition be equal to the prefix of the head of the first rule.

4.2 Compositionality Theorem

The theorem expressing the compositionality property that holds for plans under a restricted set of plan revision rules, is given below. It is similar to property (4.1) specified in section 4.1, except that we take into account the existence of multiple applicable plan revision rules. A plan π can thus be decomposed into a and π_r (where π is of the form $a; \pi$), or into π_h^ρ and π_r^ρ (where π is of the form $\pi_h^\rho; \pi_r^\rho$) for any applicable plan revision rule ρ of the form $\pi_h^\rho \rightsquigarrow \pi_b^\rho$.

Theorem 1 *(compositionality of semantics of plans).* Let $\mathcal{A}$ be an agent with a restricted set of plan revision rules PR. Let ρ range over the set of rules from PR that are applicable to the plan π, and let π be of the form $\pi_h^\rho; \pi_r^\rho$ for an applicable rule ρ of the form $\pi_h^\rho \rightsquigarrow \pi_b^\rho$. Further, let a be the first action of π, i.e., let π be of the form $a; \pi_r$. We then have for all $\pi \neq \epsilon$ and σ:

$$\mathcal{O}(\pi)(\sigma) \quad = \quad \mathcal{O}(\pi_r)(\mathcal{O}(a)(\sigma)) \cup \bigcup_\rho \mathcal{O}(\pi_r^\rho)(\mathcal{O}(\pi_b^\rho)(\sigma)).$$

In order to prove this theorem, we use lemma 1 below. This lemma, broadly speaking, specifies that for a plan of the form $\pi_h; \pi$, the following is the case: after application of a plan revision rule of the form $\pi_h \rightsquigarrow \pi_b$, yielding the plan $\pi_b; \pi$, it will always be the case that π_b is executed entirely, before π is executed. Because of this, the semantics of π_b and of π can be considered separately, which is the core of our compositionality theorem.

[10] The plan a is for example a strict prefix of $a; b$, but the plan $a; b$ is not.

Lemma 1. Let $\mathcal{A}$ be an agent with a restricted set of plan revision rules PR, and let $\pi_h \rightsquigarrow \pi_b \in$ PR. We then have that any transition sequence $\langle \pi_b; \pi, \sigma \rangle \rightarrow \ldots \rightarrow \langle \epsilon, \sigma' \rangle$ has the form[11]

$$\langle \pi_b; \pi, \sigma \rangle \rightarrow \ldots \rightarrow \langle \pi, \sigma'' \rangle \rightarrow \ldots \rightarrow \langle \epsilon, \sigma' \rangle$$

such that each configuration in the first part of the sequence, i.e., in $\langle \pi_b; \pi, \sigma \rangle \rightarrow \ldots \rightarrow \langle \pi, \sigma'' \rangle = \theta$, has the form $\langle \pi_i; \pi, \sigma_i \rangle$. That is, π is always the suffix of the plan of the agent in each configuration of θ.

In the proof of this lemma, we use the notion of a plan π' *being suffix* in π with respect to some set of plan revision rules. A plan π' is suffix in π, if π is the suffix of π', i.e., if π' is of the form $\pi_{pre}; \pi$. Further, π_{pre} should be a concatenation of suffixes of the bodies of the relevant set of plan revision rules.

Definition 12 *(suffix in π).* Let PR be a set of plan revision rules. Let suf_i with $1 \leq i \leq n$ denote plans that are equal to the suffix of the body of a rule in PR, i.e., for each suf_i there is a rule in PR of the form $\pi_h \rightsquigarrow \pi_r; suf_i$. We say that *a plan π' is suffix in π* with respect to PR, iff π' is of the form $suf_1; \ldots; suf_n; \pi$, and the length of $suf_1; \ldots; suf_n$ is greater than 0.

The idea is that, given a plan of the form $\pi_b; \pi$ which is suffix in π by definition[12], this property is preserved until the plan is of the form π. If this is the case, we have that π is always the (strict) suffix of the plan of each configuration, until the plan equals π. We thus use the preservation of this property to prove lemma 1 (see below).

We need the fact that the part of the plan occurring before π is a sequence of suffixes, in order to prove that π is preserved as the suffix of the plan.[13] The reason is, that if this is the case, we know by the fact that our plan revision rules are restricted, that there cannot occur a rule application which transforms π, thereby violating our requirement that π remains the suffix of the plan of the agent, until the plan becomes equal to π. If a plan is of the form $suf_1; \ldots; suf_n; \pi$, where each suf_i denotes a plan that is equal to the suffix of the body of a plan revision rule, we know that any plan revision rule will only modify a prefix of suf_1, because the plan revision rules are restricted. There cannot be a rule with a head of the form $suf_1; \pi_h$, because this would violate the requirement of restricted plan revision rules.

Proof of lemma 1. Let $\mathcal{A}$ be an agent with a restricted set of plan revision rules PR. Let $\langle \pi_1, \sigma \rangle \rightarrow \langle \pi_2, \sigma' \rangle$ be a transition of $\mathcal{A}$. First, we show that if π_1 is suffix in π (with respect to PR), it has to be the case that π_2 is suffix in π, or that $\pi_2 = \pi$.

---

[11] In this lemma we omit the labels of transitions, for reasons of presentation.

[12] That is, if π_b is the body of a plan revision rule.

[13] Note that we use the term suffix to refer to suffixes of the plans of the bodies of plan revision rules, and to refer to the suffix of the plan in a configuration.

Assume that π_1 is suffix in π, i.e., let $\pi_1 = suf_1; \ldots ; suf_n; \pi$. If $\pi = \epsilon$, the result is immediate. Otherwise, the proof is as follows. A transition from $\langle \pi_1, \sigma \rangle$ results either from the execution of an action, or from the application of an applicable rule.

Let $suf_1 = a; suf_1'$. If action a is executed, π_2 is of the form $suf_1'; \ldots ; suf_n; \pi$. If $suf_1', \ldots , suf_n$ are ϵ, we have that $\pi_2 = \pi$. Otherwise, we have that π_2 is suffix in π.

Let $\rho : \pi_h \rightsquigarrow \pi_b$ be a rule from PR that is applicable to π_1. Then it must be the case that π_1 is of the form $\pi_h; \pi_r$. By the fact that PR is restricted, we have that there is not a rule ρ' of the form $suf_1; \pi' \rightsquigarrow \pi_b'$, i.e., such that suf_1, which is the suffix of the body of a rule, is the prefix of the head of ρ'. Given that ρ is applicable to π_1, it must thus be the case that π_h is a prefix of suf_1, i.e., that suf_1 is of the form $\pi_h; \pi''$. Applying ρ to π_1 thus yields a plan of the form $\pi_b; \pi''; suf_2; \ldots ; suf_n; \pi$. Since both π_b and π'' are suffixes of the bodies of rules in PR, we have that π_2 is suffix in π.

We have to show that any transition sequence θ of the form $\langle \pi_b; \pi, \sigma \rangle \to \ldots \to \langle \epsilon, \sigma' \rangle$ has a prefix θ' such that π is always a suffix of the plan in each configuration of θ'. Let π_2 be the plan of the second configuration of θ. We have that $\pi_b; \pi$ is suffix in π. Therefore, it must be the case that π_2 is also suffix in π, or that $\pi_2 = \pi$. In the latter case, we have the desired result. In the former case, we have that π is a suffix of π_2, in which case the first two configuration may form a prefix of θ'. Let π_3 be the plan of the third configuration of θ. If π_2 is suffix in π, it has to be the case that π_3 is suffix in π, or that $\pi_3 = \pi$. In the latter case, we are done. In the former case, the first three configurations may form a prefix of θ'. This line of reasoning can be continued. Since θ is a finite sequence, it has to be the case that at some point a configuration of the form $\langle \pi, \sigma'' \rangle$ is reached. This yields the desired result. □

Proof of theorem 1. We have to show the following:

$$\sigma' \in \mathcal{O}(\pi)(\sigma) \quad \Leftrightarrow \quad \sigma' \in \mathcal{O}(\pi_r)(\mathcal{O}(a)(\sigma)) \cup \bigcup_{\rho} \mathcal{O}(\pi_r^{\rho})(\mathcal{O}(\pi_b^{\rho})(\sigma)).$$

($\Leftarrow$) Follows in a straightforward way from the definitions.
($\Rightarrow$) Let n be the number of plan revision rules applicable to π, where $\pi_h^{\rho_i}$ and $\pi_b^{\rho_i}$ respectively denote the head and body of rule ρ_i. We then have to show:

$$\begin{aligned} \sigma' \in \mathcal{O}(\pi)(\sigma) \Rightarrow \sigma' &\in \mathcal{O}(\pi_r^{\rho_1})(\mathcal{O}(\pi_b^{\rho_1})(\sigma)) \text{ or} \\ &\vdots \\ \sigma' &\in \mathcal{O}(\pi_r^{\rho_n})(\mathcal{O}(\pi_b^{\rho_n})(\sigma)) \text{ or} \\ \sigma' &\in \mathcal{O}(\pi_r)(\mathcal{O}(a)(\sigma)). \end{aligned}$$

If $\sigma' \in \mathcal{O}(\pi)(\sigma)$, then there is a transition sequence of the form

$$\langle \pi, \sigma \rangle \to_x \ldots \to_x \langle \epsilon, \sigma' \rangle$$

i.e., if $\pi_h^\rho \rightsquigarrow \pi_b^\rho$ is an arbitrary rule ρ that is applicable to π, where $\pi = \pi_h^\rho; \pi_r^\rho$, there are transition sequences of the form

$$\langle \pi_h^\rho; \pi_r^\rho, \sigma \rangle \rightarrow_{apply} \langle \pi_b^\rho; \pi_r^\rho, \sigma \rangle \rightarrow_x \ldots \rightarrow_x \langle \epsilon, \sigma' \rangle \tag{4.2}$$

or, if $\pi = a; \pi_r$, of the form

$$\langle a; \pi_r, \sigma \rangle \rightarrow_{exec} \langle \pi_r, \sigma'' \rangle \rightarrow_x \ldots \rightarrow_x \langle \epsilon, \sigma' \rangle. \tag{4.3}$$

In case σ' has resulted from a transition sequence of form (4.2), we prove

$$\sigma' \in \mathcal{O}(\pi_r^\rho)(\mathcal{O}(\pi_b^\rho)(\sigma)). \tag{4.4}$$

In case σ' has resulted from a transition sequence of form (4.3), we prove

$$\sigma' \in \mathcal{O}(\pi_r)(\mathcal{O}(a)(\sigma)). \tag{4.5}$$

Assume σ' has resulted from a transition sequence of form (4.2). We then have to prove (4.4), i.e., we have to prove that there is a belief base $\sigma'' \in \mathcal{O}(\pi_b^\rho)(\sigma)$, such that $\sigma' \in \mathcal{O}(\pi_r^\rho)(\sigma'')$. That is, we have to prove that there are transition sequences of the form $\langle \pi_b^\rho, \sigma \rangle \rightarrow \ldots \rightarrow \langle \epsilon, \sigma'' \rangle$, and of the form $\langle \pi_r^\rho, \sigma'' \rangle \rightarrow \ldots \rightarrow \langle \epsilon, \sigma' \rangle$.

By definitions 6 and 7, we have that if $\langle \pi_1; \pi_2, \sigma \rangle \rightarrow \langle \pi_1'; \pi_2, \sigma' \rangle$ is a transition for arbitrary plans π_1 and π_2, then $\langle \pi_1, \sigma \rangle \rightarrow \langle \pi_1', \sigma' \rangle$ is also a transition. By lemma 1, we have that there is a prefix of (4.2) of the form $\langle \pi_h^\rho; \pi_r^\rho, \sigma \rangle \rightarrow \langle \pi_b^\rho; \pi_r^\rho, \sigma \rangle \rightarrow \ldots \rightarrow \langle \pi_r^\rho, \sigma'' \rangle$, such that the plan of each configuration in this sequence is of the form $\pi_i; \pi$. From this we can conclude the desired result, i.e., that there are transition sequences of the form $\langle \pi_b^\rho, \sigma \rangle \rightarrow \ldots \rightarrow \langle \epsilon, \sigma'' \rangle$, and of the form $\langle \pi_r^\rho, \sigma'' \rangle \rightarrow \ldots \rightarrow \langle \epsilon, \sigma' \rangle$.

Assume σ' has resulted from a transition sequence of form (4.3). Then proving (4.5) is analogous to proving (4.4), except that we do not need lemma 1. □

5 Conclusion and Future Work

As argued, an important reason for defining a variant of 3APL with a compositional semantics, is that it is more likely that it will be possible to come up with a more standard and easy to use proof system for such a language. A natural starting point for such an effort is the definition of a proof rule for sequential composition, analogous to rule (3.1), as specified below (we use the notation of theorem 1).

$$\frac{\{p\}\ a\ \{p'\} \quad \{p'\}\ \pi_r\ \{q\} \qquad \bigwedge_\rho(\{p\}\ \pi_b^\rho\ \{p'\} \text{ and } \{p'\}\ \pi_r^\rho\ \{q\})}{\{p\}\ \pi\ \{q\}} \tag{5.1}$$

The soundness proof of this rule is analogous to the soundness proof of rule (3.1) [8, Chapter 2], but using theorem 1 instead of $\mathcal{O}(S_1; S_2)(\sigma) = \mathcal{O}(S_2)(\mathcal{O}(S_1)(\sigma))$. A complete proof system for compositional 3APL would however also need an induction rule. We conjecture that it will be possible to define an analogue of Scott's induction rule [8, Chapter 5] which is used for proving properties of recursive procedures, for reasoning about plans in the context of plan revision rules. Investigating this is however left for future research.

References

1. Wooldridge, M.: Agent-based software engineering. IEEE Proceedings Software Engineering **144**(1) (1997) 26–37
2. Wooldridge, M., Ciancarini, P.: Agent-Oriented Software Engineering: The State of the Art. In Ciancarini, P., Wooldridge, M., eds.: First Int. Workshop on Agent-Oriented Software Engineering. Volume 1957. Springer-Verlag, Berlin (2001) 1–28
3. Bratman, M.E.: Intention, plans, and practical reason. Harvard University Press, Massachusetts (1987)
4. Rao, A.S., Georgeff, M.P.: Modeling rational agents within a BDI-architecture. In Allen, J., Fikes, R., Sandewall, E., eds.: Proceedings of the Second International Conference on Principles of Knowledge Representation and Reasoning (KR'91), Morgan Kaufmann (1991) 473–484
5. Cohen, P.R., Levesque, H.J.: Intention is choice with commitment. Artificial Intelligence **42** (1990) 213–261
6. Hindriks, K.V., de Boer, F.S., van der Hoek, W., Meyer, J.J.Ch.: Agent programming in 3APL. Int. J. of Autonomous Agents and Multi-Agent Systems **2**(4) (1999) 357–401
7. Dastani, M., van Riemsdijk, M.B., Dignum, F., Meyer, J.J.Ch.: A programming language for cognitive agents: goal directed 3APL. In: Programming multiagent systems, first international workshop (ProMAS'03). Volume 3067 of LNAI. Springer, Berlin (2004) 111–130
8. de Bakker, J.: Mathematical Theory of Program Correctness. Series in Computer Science. Prentice-Hall International, London (1980)
9. Giacomo, G.d., Lespérance, Y., Levesque, H.: *ConGolog*, a Concurrent Programming Language Based on the Situation Calculus. Artificial Intelligence **121**(1-2) (2000) 109–169
10. Rao, A.S.: AgentSpeak(L): BDI agents speak out in a logical computable language. In van der Velde, W., Perram, J., eds.: Agents Breaking Away (LNAI 1038), Springer-Verlag (1996) 42–55
11. Shoham, Y.: Agent-oriented programming. Artificial Intelligence **60** (1993) 51–92
12. van Riemsdijk, M.B., de Boer, F.S., Meyer, J.J.Ch.: Dynamic logic for plan revision in intelligent agents. In Leite, J.A., Torroni, P., eds.: Computational logic in multi-agent systems: fifth international workshop (CLIMA'04). Volume 3487 of LNAI. (2005) 16–32
13. van Riemsdijk, M.B., de Boer, F.S., Meyer, J.J.Ch.: Dynamic logic for plan revision in intelligent agents. Technical Report UU-CS-2005-013, Utrecht University, Institute of Information and Computing Sciences (2005) To appear in Journal of Logic and Computation.
14. Plotkin, G.D.: A Structural Approach to Operational Semantics. Technical Report DAIMI FN-19, University of Aarhus (1981)
15. van Riemsdijk, M.B., Meyer, J.J.Ch., de Boer, F.S.: Semantics of plan revision in intelligent agents. In Rattray, C., Maharaj, S., Shankland, C., eds.: Proceedings of the 10th International Conference on Algebraic Methodology And Software Technology (AMAST04). Volume 3116 of LNCS., Springer-Verlag (2004) 426–442
16. van Riemsdijk, M.B., Meyer, J.J.Ch., de Boer, F.S.: Semantics of plan revision in intelligent agents. Theoretical Computer Science **351**(2) (2006) 240–257 Special issue of Algebraic Methodology and Software Technology (AMAST'04).

ITP/OCL: A Rewriting-Based Validation Tool for UML+OCL Static Class Diagrams*

Manuel Clavel and Marina Egea

Universidad Complutense de Madrid, Spain
{clavel, marina_egea}@sip.ucm.es

Abstract. In this paper we present the ITP/OCL tool, a rewriting-based tool that supports automatic validation of UML class diagrams with respect to OCL constraints. Its implementation is directly based on the equational specification of UML+OCL class diagrams. It is written entirely in Maude making extensive use of its reflective capabilities. We also give notice of the Visual ITP/OCL, a Java graphical interface that can be used as a front-end for the ITP/OCL tool.

1 Introduction

The Unified Modeling Language (UML) [1] is a general-purpose visual modeling language that is used to specify, visualize, construct, and document the artifacts of a software system. The UML notation is largely based on diagrams. However, for certain aspects of a model, diagrams often do not provide the level of conciseness and expressiveness that a textual language can offer. The Object Constraint Language (OCL) [2] is a textual constraint language. OCL comes to provide help on precise information specification in UML models.

Validation and testing in software development have been recognized of key importance for long. There are many different approaches to validation: simulation, rapid prototyping, etc. We *validate* a model by checking whether its instances (also called "snapshots") fulfill the desired constraints. A number of CASE tools exist which facilitate drawing and documenting UML diagrams. However, there is little support for validating models during the design stage and generally no substantial support for constraints written in OCL. In this paper we present the ITP/OCL tool, a rewriting-based tool that supports automatic validation of UML class diagrams with respect to OCL constraints. The ITP/OCL implementation is directly based on the equational specification of UML+OCL class diagrams developed in [3, 4]. It is written entirely in Maude [5], making extensive use of its reflective capabilities to implement the user interface, thanks to which the tool's underlying equational semantics remains hidden to the user, who only must be familiar with the standard notions of UML diagrams and OCL constraints.

* Research supported by Spanish MEC Projects TIC2003-01000, TIN2005-09207-C03-03, and by Comunidad de Madrid Program S-0505/TIC/0407.

M. Johnson and V. Vene (Eds.): AMAST 2006, LNCS 4019, pp. 368–373, 2006.

2 UML+OCL Diagrams

The UML *static view* models concepts in the application domain as well as internal concepts invented as part of the implementation of an application. It does not describe the time-dependent behavior of the system, which is described in other views. Key elements in the static view are classes and their relationships, which can be of different kinds, including associations and generalizations. The static view is displayed in class diagrams.

Example 1. Consider the class diagram TRAINWAGON shown in Figure 1. It models an example from a railway context. A train may own wagons, and wagons may be connected to other wagons (their predecessor and successor wagons). Wagons can be either smoking or non-smoking.

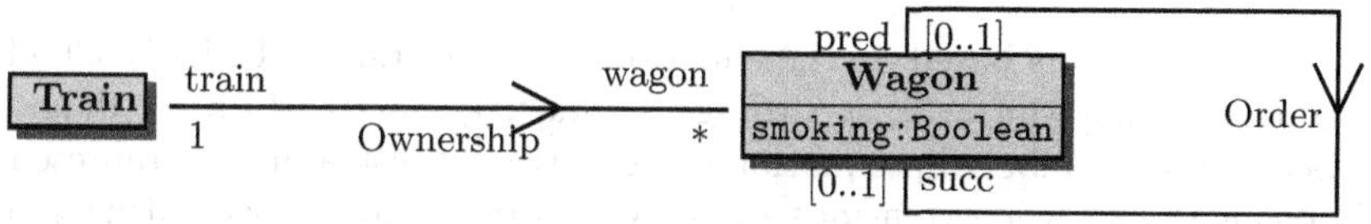

Fig. 1. The class diagram TRAINWAGON

A system may be in different states as it changes over time. An *object diagram* models the objects and links that represent the state of a system at a particular moment. An *object* is an instance of a class. A *link* is an instance of an association. An object diagram is primarily a tool for research and testing.

Example 2. Consider now the object diagram TRAINWAGON-1 shown in Figure 2. It describes a snapshot of the railway system modeled by the class diagram TRAINWAGON, although possibly an "undesired" one since it describes a train with two wagons linked in a cyclic way!

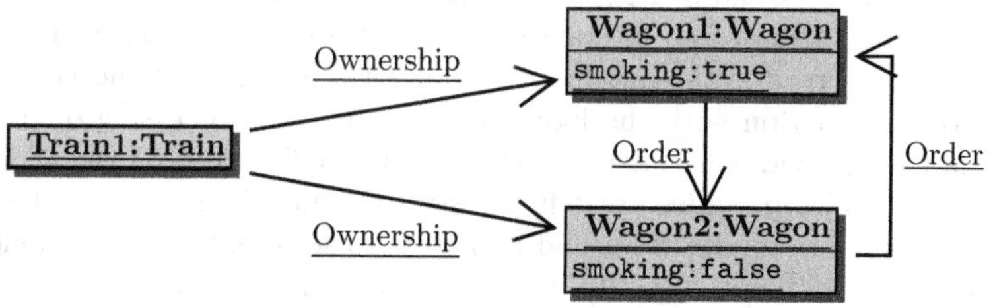

Fig. 2. The object diagram TRAINWAGON-1

OCL is a pure specification language on top of UML. It is a textual language with a notational style similar to common object oriented languages.

Example 3. Consider the following constraint over the class diagram TRAIN-WAGON: *There do not exist two different wagons directly linked to each other in a cyclic way.* This constraint can be expressed using OCL as the following invariant notInCyclicWay over the class diagram TRAINWAGON:

```
not ((Wagon.allInstances)
    →exists(w1:Wagon | (Wagon.allInstances) →exists(w2:Wagon |
            w1:Wagon <> w2:Wagon
            and (w1:Wagon.succ) →includes(w2:Wagon)
            and (w2:Wagon.succ) →includes(w1:Wagon)))) .
```

The object diagram TRAINWAGON-1 does not satisfy this constraint, since there exists two wagons, namely Wagon1 and Wagon2, such that the successors of Wagon1 include Wagon2, and the successors of Wagon2 include Wagon1.

3 The ITP/OCL Tool

The ITP/OCL tool is based on the equational specification of UML+OCL class diagrams developed in [3,4], according to which: i) class and object diagrams are specified as membership equational theories; ii) constraints are represented as Boolean terms over extensions of those theories; and iii) validating object diagrams with respect to constraints is reduced to checking whether the corresponding Boolean terms rewrite to true or false. The ITP/OCL tool is written entirely in Maude [5], a term-rewriting based programming language that implements membership equational logic (and rewriting logic). Maude is also a reflective programming language. This means, in particular, that both its parser and its rewriting engine are available to the programmer as built-in operations: we have taken advantage of the latter to implement the tool's OCL parser and of the former to implement the tool's UML+OCL rewriting-based validation engine. The implementation of the ITP/OCL tool comprises around 4,000 lines of Maude code. The latest version of the ITP/OCL tool, with the available documentation and a collection of examples (that includes class diagrams with enumeration classes, association classes, generalizations, attributes and query operations), can be found at `http://maude.sip.ucm.es/itp/ocl/`.

The implementation of an interactive tool in Maude comprises four different tasks: defining a read-eval-print loop; defining the syntax for the commands; defining the interaction with the loop; and defining the processing of the commands. Maude provides a generic input/output facility through its "loop objects" terms. It also provides great flexibility to define the syntax for the commands thanks to its mixfix front-end and to the use of *bubbles* (any nonempty list of Maude identifiers). Finally, the processing of the requests made to an interactive tool is defined in Maude by equations acting on the loop objects terms.

The ITP/OCL's commands can be grouped in four classes:

- *Commands that create a diagram.* They are defined by equations that add to the tool's database the module that, according with the tool's semantics, specifies an empty class (respectively, object) diagram. For example, to create a class diagram we use the command `(create-class-diagram` *CD* `.)`, where *CD* is the class diagram's name.

- *Commands that insert an element (class, attribute, association, and so on) in a diagram.* They are defined by equations that add to the module specifying the diagram in the tool's database the declarations (sorts, operators, memberships, equations) that, according with the tool's semantics, specify that the diagram has this element. E.g., to insert a class we use the command (`insert-class` *CD* : *C* .), where *CD* is the class diagram's name and *C* is the class' name.
- *Commands that state a constraint over a class diagram.* They are defined by equations that associate to the module specifying the class diagram in the tool's database the Boolean term that, according with the tool's semantics, represents this constraint. For instance, to state a *contextualized invariant* we use the command (`insert-invariant` *CD* : *C* : *INV* .), where *CD* is the class diagram's name, *C* is the contextual class' name, and *INV* is the invariant.
- *Commands that validate an object diagram.* They are defined by equations that check whether the Boolean terms representing the invariants reduce to true or false in the module that, according with the tool's semantics specifies the union of the class diagram, along with its invariants, and the object diagram. E.g., to check whether an object diagram validates the invariants stated over the class diagram of which it is an instance, we use the command (`check-invariants` *CD* : *OD* .), where *CD* is the class diagram's name and *OD* is the object diagram's name.

4 The Visual ITP/OCL Tool

The Visual ITP/OCL is a Java graphical interface for the ITP/OCL tool.[1] Events on the Visual ITP/OCL's worksheets and toolbars are transformed into ITP/OCL commands and are interpreted and executed in a Maude process running the ITP/OCL tool. In its current state:

- Diagrams can be graphically created in a similar way to other CASE tools, like Rational Rose [6] or Gentleware Poseidon UML [7]. By clicking, dragging, and dropping on the class diagram's (respectively, object diagram's) worksheet and toolbar, the user can interactively insert, modify, and delete classes and associations (respectively, objects and links.) Several class diagrams can be opened at the same time. Object diagrams are associated to their class diagrams. There is also a zooming facility to increase/decrease the scale. Finally, diagrams can also be saved (and loaded from) a MySQL database and can be exported as eps-files.
- Constraints on class diagrams (respectively, queries on object diagrams) are inserted through an OCL editor and parser. Constraints inserted in a class diagram can be automatically checked over specific object diagrams. Queries can also be automatically evaluated.

[1] The Visual ITP/OCL tool is being developed by F. Alcaraz, J. P. Gavela, and J. Arias as a Master's project.

5 Related and Future Work

A number of CASE tools exists which facilitate drawing and documenting UML diagrams [6, 7, 8]. However, there is little support for validating models during the design stage and generally no substantial support for constraints written in OCL. The USE tool [9] is, however, a significant exception. The USE tool expects as input a textual description of a model and its constraints. This textual description is then displayed in a graphical interface. Objects and links can be then graphically created by a drag and drop facility. In every system state, the constraints can be automatically checked. The USE tool also supports the validation of sequence diagrams by checking that, for each step in the sequence, the initial and the resulting diagrams satisfy, respectively, the pre- and post-conditions constraining the application of the corresponding (non-query) operation. This feature is not yet supported by the ITP/OCL tool. Notice, however, that validating sequence diagrams à la USE only requires the capability of checking constraints (the pre- and post-conditions) over object diagrams. We plan to add this feature in the next version of the ITP/OCL tool.

The ITP/OCL tool is based on the equational specification of UML+OCL class diagrams: validating invariants over (or evaluating OCL queries in) object diagrams is done by rewriting the corresponding term in the corresponding equational specification. This is clearly different from the USE tool's underlying semantics and its corresponding evaluation mechanism [10]. Finally, as suggested by various of our referees, we have tried a first comparison between both tools. In particular, Table 1 shows the time (in seconds) consumed the tools to validate a number of constraints over two object diagrams, TRAINWAGON-10x25 and TRAINWAGON-10x100, of the class diagram TRAINWAGON. In addition to the constraint notInCyclicWay introduced in Section 2, we have considered the following constraints:

- *All trains must own at least one wagon.*

  ```
  context Train inv atLeastOnewagon:
          self:Train.wagon size() ≥ 1.
  ```

- *A wagon and its successor wagon should belong to the same train.*

  ```
  context Wagon inv belongToTheSameTrain:
          self:Wagon.succ →notEmpty() implies
          self:Wagon.succ →forAll(w:Wagon | (w:Wagon.train = self:Wagon.train)).
  ```

- *All trains will have the same number of wagons.*

  ```
  context Train inv sameNumberOfWagons:
          Train.allInstances →forAll (t1:Train |
          (self:Train <> t1:Train implies
          (self:Train.wagon →size() = t1:Train.wagon →size()))).
  ```

The object diagram TRAINWAGON-10x25 contains 10 trains and 250 wagons, each train is linked to 25 different wagons, which are linked in the expected

Table 1. Validation times

	TRAINWAGON-10x25		TRAINWAGON-10x100	
	USE	ITP/OCL	USE	ITP/OCL
atLeastOnewagon	0.055s	0.076s	0.034s	0.116s
belongToTheSameTrain	0.207s	0.076s	0.970s	0.780s
sameNumberOfWagons	0.202s	0.120s	0.241s	0.936s
notInCyclicWay	45.745s	13.788s	2819.410s	233.710s

way; that is, each wagon has a predecessor and a successor, except for the first and the last wagon. The object diagram TRAINWAGON-10x100 contains 10 trains and 1000 wagons, each train is linked to 100 different wagons, which are also linked in the expected way. The object diagrams TRAINWAGON-10x25 and TRAINWAGON-10x100 satisfy indeed our four constraints. The validations have been carried out in a laptop computer with a 2GHz Pentium processor and 1 GB RAM. As expected, validating the constraint notInCyclicWay takes more time; essentially, it has to make, respectively, 250×250 and 1000×1000 comparisons. However, the time consumed by the USE tool is unexpectedly high. We have tried to obtain from the USE community an explanation for this fact (which may be simply due to our inexpert use of the tool) but we have not get an answer yet; as soon as we get it, we will publish it in the ITP/OCL web page.

References

1. Object Management Group: Unified Modeling Language Specification (2004) `http://www.uml.org`.
2. Object Management Group: Object Constraint Language Specification (2004) `http://www.omg.org`.
3. Egea, M.: ITP/OCL: a theorem prover-based tool for UML+OCL class diagrams. Master's thesis, Facultad de Informática, Universidad Complutense de Madrid (2005) `http://maude.sip.ucm.es/~marina/`.
4. Clavel, M., Egea, M.: Equational specifications of UML+OCL static class diagrams. `http://maude.sip.ucm.es/itp/~clavel` (2006)
5. Clavel, M., Durán, F., Eker, S., Lincoln, P., Martí-Oliet, N., Meseguer, J., Talcott, C.: Maude Manual (Version 2.2). (2005) SRI International, December 2005, `http://maude.cs.uiuc.edu`.
6. IBM: Rational Software (2006) `http://www-306.ibm.com/software/rational/`.
7. AG, G.: Poseidon Standard Edition (2006) `http://www.gentleware.com`.
8. Demuth, B., Löcher, S., Zschaler, S.: Structure of the Dresden OCL toolkit. Technical report, Technical University of Darmstadt, Germany (2004) Reviewed Conference Paper.
9. Richters, M.: The USE tool : A UML-based specification environment. (2001) `http://www.db.informatik.uni-bremen.de/projects/USE/`.
10. Gogolla, M., Richters, M., Bohling, J.: Tool support for validating UML and OCL models through automating snapshot generation. In: Proceedings of SAICSIT. (2003) 111–120

A Computational Group Theoretic Symmetry Reduction Package for the Spin Model Checker

Alastair F. Donaldson* and Alice Miller**

Department of Computing Science
University of Glasgow, 17 Lilybank Gardens
Glasgow, Scotland, G12 8QQ
{ally, alice}@dcs.gla.ac.uk

Abstract. Symmetry reduced model checking is hindered by two problems: how to identify state space symmetry when systems are not fully symmetric, and how to determine equivalence of states during search. We present TopSpin, a fully automatic symmetry reduction package for the Spin model checker. TopSpin uses the Gap computational algebra system to effectively detect state space symmetry from the associated Promela specification, and to choose an efficient symmetry reduction strategy by classifying automorphism groups as a disjoint/wreath product of subgroups. We present encouraging experimental results for a variety of Promela examples.

1 Introduction

Model checking concurrent systems comprised of replicated components can potentially be made easier by exploiting symmetries of a model of the system, induced by the replication. If such *component symmetries* can be identified before search then the model checking algorithm can be modified to consider a single state from each equivalence class of symmetric states. This results in reduced space requirements for verification by model checking.

However, symmetry reduction can only speed up model checking if an efficient procedure is available to determine whether or not a given state is equivalent to a previously reached state. A common approach to solving this problem for explicit state model checking is, given a total ordering on states and a symmetry group G, to convert a state s to $min[s]_G$—the *smallest* state in the equivalence class of s under G—before it is stored. Thus efficient algorithms are required to compute $min[s]_G$. This is the *constructive orbit problem*, which has been proved to be NP-hard [4]. Current implementations of symmetry reduction techniques for explicit state model checking, such as SymmSpin [2], are limited to dealing with full symmetry between components of a concurrent system—both symmetry detection and on-the-fly representative computation are easy for this special case.

* Supported by the Carnegie Trust for the Universities of Scotland.
** Partially funded by The Universitiy of Glasgow John Robertson Bequest, award number JR05/14.

M. Johnson and V. Vene (Eds.): AMAST 2006, LNCS 4019, pp. 374–380, 2006.

In previous work we proposed a framework for the automatic detection of arbitrary structural symmetry, with an implementation for the Promela specification language [6]. In this paper we present TopSPIN, a symmetry reduction package for the SPIN model checker which uses exact and approximate strategies for dealing with such arbitrary symmetries. The tool draws on theory and technology from computational group theory to efficiently compute equivalence class representatives. In particular, the GAP computational algebra system [10] is used both for symmetry detection, and for classifying an arbitrary group based on its structure as a direct/wreath product of basic subgroups, so that an appropriate symmetry reduction strategy may be chosen. For groups which cannot be classified in this way, TopSPIN uses an approximate symmetry reduction strategy based on hillclimbing local search, which is sub-optimal in terms of memory requirements but fast and safe. We present experimental results which demonstrate the effectiveness of our techniques. TopSPIN, together with Promela code for the specifications described in Sect. 5, can be found on our website [7]. Throughout the paper, we assume some basic knowledge of group theory.

2 Background and Notation

SPIN [11] is the bespoke model checker for the Promela specification language, and provides several reasoning mechanisms: assertion checking, acceptance and progress states and cycle detection, and satisfaction of temporal properties, expressed in linear temporal logic (LTL). SPIN translates each component defined in a Promela specification into a finite automaton and then computes the asynchronous interleaving product of these automata to obtain the global behavious of the concurrent system. This interleaving product is essentially a Kripke structure $\mathcal{M} = (S, s_o, R, L)$, where S is a finite set of states with initial state s_0, $R \subseteq S \times S$ a *total* transition relation, and $L : S \to 2^{AP}$ a labelling function. The set AP of atomic propositions refer to the values of local and global variables, and contents of buffered channels.

A bijection $\alpha : S \to S$ which satisfies, for all $(s,t) \in R$, $(\alpha(s), \alpha(t)) \in R$, is an *automorphism* or *symmetry* of $\mathcal{M}$, and all such symmetries form a group $Aut(\mathcal{M})$ under composition of mappings. If a subgroup G of $Aut(\mathcal{M})$ is known in advance then model checking can be performed over a *quotient* Kripke structure, $\mathcal{M}_G$, typically smaller than the original [12]. Kripke structure automorphisms induced by symmetry between components of the concurrent system, i.e. bijections of the component index set which give rise to automorphisms when lifted to act component-wise on states, are called *component symmetries* [9]. In this work we restrict our attention to component symmetries. If $G \leq Aut(\mathcal{M})$ and $s \in S$, then $[s]_G = \{\alpha(s) : \alpha \in G\}$ is the *orbit* of s under G.

3 An Overview of TopSPIN

In order to check properties of a Promela specification, SPIN first converts the specification into a C source file, `pan.c`, which is then compiled into an executable verifier. The state space thus generated is then searched. If the property being checked is proved to be false, a counterexample is given. TopSpin follows the approach used by the SymmSpin symmetry reduction package [2], where `pan.c` is generated as usual by Spin, and then converted to a new file, `sympan.c`, which includes algorithms for symmetry reduction. With TopSpin because we allow for arbitrary system topologies, symmetry must be detected before `sympan.c` can be generated. This is illustrated in Fig. 1.

First, the *static channel diagram* (SCD) of the Promela specification is extracted by the SymmExtractor tool [6]. The SCD is a graphical representation of potential communication between components of the specification. The group of symmetries of the SCD, $Aut(SCD)$, is computed using the *saucy* tool [5], which we have extended to handle directed graphs. The generators of $Aut(SCD)$ are checked against the Promela specification for validity (an assurance that they induce symmetries of the underlying state space). TopSpin uses Gap to compute, from the set of valid generators, the largest group $G \leq Aut(SCD)$ which can be safely used for symmetry-reduced model checking. Gap is then used to classify the structure of G in order to choose an efficient symmetry reduction strategy. The chosen strategy is merged with `pan.c` to form `sympan.c`, which can be compiled and executed as usual.

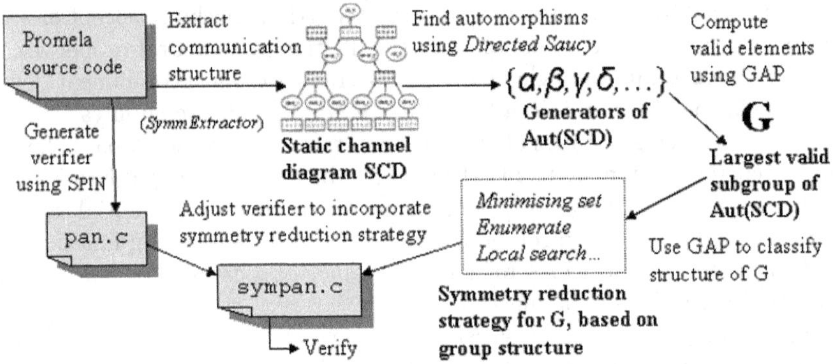

Fig. 1. The symmetry reduction process

4 Symmetry Reduction Strategies

We refer to processes and channels of a Promela specification as *components*, and restrict our attention to Promela specifications with a fixed number of components. Throughout, we assume that $G \leq S_n$ is a nontrivial symmetry group for a Promela specification consisting of n components. In this section we outline various strategies which TopSpin uses to compute $min[s]_G$ for a state s and an *arbitrary* group G. An appropriate strategy for G is chosen based on analysis of the structure of G before search. Note that, during verification, the C function `memcmp` provides a total ordering on states.

4.1 The Strategies

Enumeration. If G is a relatively small group ($|G| < 100$ say) then for a state s, $min[s]_G$ can be computed by *enumerating* the elements of G, and returning $min\{\alpha(s) : \alpha \in G\}$. TopSPIN implements this approach with two optimisations, applied simultaneously. As the operation of applying a transposition to a state is less expensive than that of applying an arbitrary permutation, a group element α is expressed as a product of transpositions and $\alpha(s)$ is computed by applying these transpositions to s in order. TopSPIN uses a *stabiliser chain* to enumerate the elements of G. Given a stabiliser chain $G = G^{(1)} \geq G^{(2)} \geq \cdots \geq G^{(k)} = \{id\}$ for some $k > 1$, every element of G can be uniquely expressed as a product $u_{k-1}u_{k-2}\ldots u_1$, where, for $1 \leq i < k$, u_i is a representative of a coset of $G^{(i+1)}$ in $G^{(i)}$ [3]. Thus each $\alpha \in G$ need not be applied to a state s from scratch: partial images of s under the coset representatives may be re-used.

Minimising Sets. Using terminology from [9], a group H is said to be *nice* if there is a small set $X \subseteq H$ such that $t = min[s]_H$ iff $\alpha(t) \geq t\ \forall \alpha \in X$. If H is nice with respect to a subset X then we call X a *minimising set* for H. Given a minimising set X for G, the element $min[s]_G$ can be computed by setting $t = s$, and applying elements of X to t until a fixpoint is reached. TopSPIN uses this symmetry reduction strategy in cases where G is isomorphic to a fully symmetric group S_m, for some $m \leq n$, which simultaneously permutes several disjoint subsets of $\{1, 2, \ldots, n\}$. (Such groups occur commonly in practice, e.g. a set of processes may have associated channels, so that any permutation of the processes must also permute the associated channels.) In this case, let $\alpha_{i,j}$ denote the permutation which simultaneously transposes the ith and jth elements of each subset. If G is generated by the set $X = \{\sigma_{j,k} : 1 \leq j < k \leq m\}$ then it can be shown that X is a minimising set for G, and $|X|$ is quadratic in n even though G may be very large. If TopSPIN detects that G is isomorphic to S_m for some $m \leq n$ then it attempts to construct a minimising set of the above form.

Disjoint Products. If G is the disjoint product of subgroups $H_1, H_2, \ldots, H_k$ for some $k > 1$ then $min[s]_G = min[\ldots min[min[s]_{H_1}]_{H_2} \cdots]_{H_k}$ [4]. TopSPIN constructs an equivalence relation on the generators of G to detect whether G is a disjoint product. The approach is very efficient, but not complete—it does not guarantee detection of the finest decomposition of G as a disjoint product. However, we have found it to work well in practice.

Wreath Products. If G is a wreath product $H \wr K$ of two subgroups H and K then G contains r copies of H for some $r \geq 1$, denoted $H_1, H_2, \ldots, H_r$, which each permute elements within a distinct "block" of components of the specification, and K permutes the blocks. In this case, it can be shown that $min[s]_G = min[min[\ldots min[min[s]_{H_1}]_{H_2} \cdots]_{H_r}]_K$ [4]. If reduction strategies can be found for H and K, then analogous strategies to that for H can easily be obtained for each H_i, and $min[s]_G$ can be computed by applying the strategy for each H_i, followed by the strategy for K. To efficiently detect a wreath product decomposition for G, TopSPIN identifies candidate blocks by using GAP to compute non-trivial *block systems* for G. Corresponding groups H and K are derived for the candidate blocks, and a check is made to see whether or not G is the wreath product of these groups.

Local Search. If none of the above strategies are applicable then, since enumeration is very expensive, it may be infeasible to compute $min[s]_G$. TopSPIN implements an *approximate* symmetry reduction strategy based on hillclimbing local search using the group generators, which does not guarantee unique representatives, but is safe to use when model checking as it guarantees storage of at least one state per equivalence class. Though not as space-efficient as enumeration, this strategy can work considerably faster.

4.2 Choosing a Reduction Strategy

TopSPIN uses a top-down recursive algorithm to choose a symmetry reduction strategy for an arbitrary group G with respect to a set of n components. If G is isomorphic to a cyclic group and $|G| \leq n$, or to a dihedral group and $|G| \leq 2n$, then the enumeration strategy is selected. If $|G|$ is isomorphic to the group S_m for some $m \leq n$ then TopSPIN attempts to construct a minimising set for G of the form described above, so that the minimising set strategy can be chosen. If G can be shown to decompose as a product of subgroups then a composite strategy is obtained by choosing a strategy for each subgroup. Otherwise, the local search strategy is chosen. In order to compare strategies it is possible to select the strategy used (rather than let TopSPIN choose the most efficient).

5 Experimental Results

Table 1 gives experimental results applying our techniques to three families of Promela specifications. For each specification, we give the number of model states without symmetry reduction (**orig**), with full symmetry reduction (**red**), and using the strategy chosen by TopSPIN (**best**). If the latter two are equal, '=' appears for the TopSPIN strategy. The use of state compression, provided by SPIN, is indicated by the number of states in italics. For each strategy (**basic** for enumeration without the optimisations described in Sect. 4.1, **enum** for optimised enumeration, and **best** for the strategy chosen by TopSPIN), and when symmetry reduction is not applied (**orig**), we give the time taken for verification (in seconds). Verification attempts which exceeded available resources, or did not terminate within 5 hours, are indicated by '-'. All experiments were performed on a PC with a 2.4GHz Intel Xeon processor, 3Gb of available main memory, running SPIN version 4.2.3. The first family of specifications model flow of control in a three-tiered architecture consisting of a database, a layer of p servers, and a layer of pq clients, where q clients are connected to each server (a D-S-C system). Here models exhibit wreath product symmetry: there is full symmetry between the q clients in each block, and the blocks of clients, with their associated servers, are interchangeable. A configuration with p servers and q clients per server is

Table 1. Experimental results for various configurations of the three-tiered (D-S-C), resource allocator (R-C) and hypercube (HC) specifications

system	config.	states orig	time orig	$\|G\|$	states red	time basic	time enum	states best	time best
D-S-C	2/3	103105	5	72	2656	7	4	=	2
D-S-C	2/4	1.1×10^6	37	1152	5012	276	108	=	2
D-S-C	3/3	*2.54×10⁷*	4156	1296	50396	4228	1689	=	19
D-S-C	3/4	-	-	82944	130348	-	-	=	104
R-C	3,3	16768	0.2	36	1501	0.9	0.3	=	0.1
R-C	4,4	199018	2	576	3826	57	19	=	0.4
R-C	5,5	2.2×10^6	42	14400	8212	4358	1234	=	2
R-C	4,4,4	2.39×10^7	1587	13824	84377	-	12029	=	17
R-C	5,5,5	-	-	1728000	254091	-	-	=	115
HC	3d	13181	0.3	48	308	0.6	0.3	468	0.2
HC	4d	380537	18	384	1240	58	34	6986	13
HC	5d	9.6×10^6	2965	3840	3907	7442	5241	90442	946

denoted p/q. The second family of specifications model a resource allocator process which controls access to a resource by a competing set of prioritised clients (an R-C system). Models of these specifications exhibit disjoint product symmetry: there is full symmetry between each set of clients with the same priority level. A configuration with p_i clients of priority level i is denoted $p_1, p_2, \ldots, p_k$, where k is the number of priority levels. Finally, we consider specifications modelling message routing in an n-dimensional hypercube network (an HC system). The symmetry group here is isomorphic to the group of geometrical symmetries of an n-dimensional hypercube, which cannot be decomposed as a disjoint or wreath product of subgroups, and thus must be handled using either the *enumeration* or *local search* strategies. An n-dimensional hypercube specification is denoted nd. For all specifications, we verify deadlock freedom, and check the satisfaction of basic safety properties expressed using assertions.

In all cases, the basic enumeration strategy is significantly slower than the optimised enumeration strategy, which is in turn slower than the strategies chosen by TopSPIN. For hypercube configurations, TopSPIN chooses the local search strategy, which requires storage of more states than the enumeration strategy, but still results in a greatly reduced state space.

6 Related and Future Work

The SymmSpin symmetry reduction package avoids the problem of automatic symmetry detection by requiring symmetries to be specified using *scalarsets*, an approach proposed in [12]. Scalarsets can only specify full symmetry between identical components, thus the three-tiered architecture and hypercube examples of Sect. 5 could not be handled by SymmSpin. Multiple scalarset types could be used to specify symmetry between clients with the same priority level in the resource allocator example, but the automatic approach to symmetry detection provided by TopSPIN is clearly preferrable.

Automatic symmetry detection by static channel diagram analysis is similar to an approach for deriving symmetry in a shared variable model of comunication [4]. However, this approach is not directly applicable to the specification language of a mainstream model checker such as SPIN. Certain classes of groups for which orbit representatives can be efficiently computed are also presented in [4]. We extend this work by providing techniques to automatically determine whether a group belongs to one of these classes.

Future work includes extending TopSPIN to allow symmetry-reduced verification of LTL properties under weak fairness, as described in [1]. This will involve combining strategies for representative computation with the nested depth first search algorithm employed by SPIN [11]. The notion of *virtual* symmetry is suggested in [8] to deal with systems which are "almost" symmetric. The symmetry detection techniques which TopSPIN uses could potentially be extended to handle virtual symmetry, allowing state-space reductions for examples with less symmetry than those which we present.

References

1. D. Bosnacki. A light-weight algorithm for model checking with symmetry reduction and weak fairness. In *SPIN'03*, LNCS 2648, pages 89–103. Springer, 2003.
2. D. Bosnacki, D. Dams, and L. Holenderski. Symmetric spin. *International Journal on Software Tools for Technology Transfer*, 4(1):65–80, 2002.
3. G. Butler. *Fundamental Algorithms for Permutation Groups*, volume 559 of *LNCS*. Springer-Verlag, 1991.
4. E.M. Clarke, E.A. Emerson, S. Jha, and A.P. Sistla. Symmetry reductions in model checking. In *CAV'98*, LNCS 1427, pages 147–158. Springer, 1998.
5. P.T. Darga, M.H. Liffiton, K.A. Sakallah, and I.L. Markov. Exploiting structure in symmetry detection for CNF. In *DAC'04*, pages 530–534. ACM Press, 2004.
6. A. F. Donaldson and A. Miller. Automatic symmetry detection for model checking using computational group theory. In *FM'05*, LNCS 3582, pages 418–496. Springer, 2005.
7. A. F. Donaldson and A. Miller. TopSPIN Website: `http://www.dcs.gla.ac.uk/people/personal/ally/topspin/`.
8. E.A. Emerson, J. Havlicek, and R.J. Trefler. Virtual symmetry reduction. In *LICS'00*, pages 121–131. IEEE Computer Society Press, 2000.
9. E.A. Emerson and T. Wahl. Dynamic symmetry reduction. In *TACAS'05*, LNCS 3440, pages 382–396. Springer, 2005.
10. The Gap Group. *GAP–Groups, Algorithms, and Programming, Version 4.4*; 2006. `http://www.gap-system.org`.
11. G.J. Holzmann *The SPIN model checker: primer and reference manual.* Addison Wesley, 2003.
12. C. Ip and D. Dill. Better verification through symmetry. *Formal Methods in System Design*, 9:41–75, 1996.

Using Category Theory as a Basis for a Heterogeneous Data Source Search Meta-engine: The Prométhée Framework

Paul-Christophe Varoutas[1,2,⋆], Philippe Rizand[2], and Alain Livartowski[1]

[1] Institut Curie, Service d'Information Médicale. 25 rue d'Ulm, 75005 Paris, France
[2] Institut Curie, Direction des Systèmes d'Information et Informatique. 25 rue d'Ulm, 75005 Paris, France
paul-christophe.varoutas@curie.fr
http://www.curie.fr

Abstract. It is generally acknowledged that integration of large-scale information systems is a challenging problem. A domain that particularly encounters this problem currently is healthcare, where information systems tend to be heterogeneous and in constant evolution. A particular need for health professionals is the ability to ask medical questions across a heterogeneous information system, then visualise and analyse the results in a synthetic and coherent manner. In this paper, we present a case study of a heterogeneous data source search meta-engine framework, based on category theory. This framework addresses the problem of cross-interrogation of heterogeneous data sources, such as relational database management systems, documentary database systems, or collections of documents. It additionally attempts to address the problem of constant evolution of such information systems. The framework has been successfully applied to the biomedical data of the medical information system at the Institut Curie, a major French cancer care centre. Different aspects of this work are illustrated, such as the mathematical foundations of the Prométhée framework, the methodology used for its implementation, and the impact that Prométhée has encountered since its deployment in a hospital environment.

1 Introduction

Integration of large-scale information systems is a topic well recognized as both challenging and important. The integration problem particularly arises in current health information systems. In hospitals, the difficulty originates not only from the heterogeneity of the specialized medical systems and equipment, but also from the constant change of the information system. Indeed, the continuous progress in medicine and science, the evolution of medical procedures which affect the hospital's organization, and the constant evolution of underlying technologies lead to a constant evolution of the information systems.

Heterogeneous data source integration is a complex problem which has various levels and aspects, such as cross-aggregation, -interoperability, -quality control,

⋆ Contact author.

M. Johnson and V. Vene (Eds.): AMAST 2006, LNCS 4019, pp. 381–387, 2006.

and -interrogation of data. Of particular interest to health professionals is the need to formulate medical or health-related questions, ask them across a heterogeneous medical information system, and finally visualise and analyse the results in a synthetic and coherent way.

The Institut Curie, a cancer care centre in Paris, France, has recently succeeded in conceiving, specifying, implementing and deploying a solution that enables the cross-interrogation of its medical information system by its authorized healthcare professionals.

In this paper, we present this short case study of a framework, based on algebraic methodological concepts, that addresses both the problems of cross-interrogation of a large-scale heterogeneous medical information system, and of adaptation to constant evolutions of the information system.

2 The Concept of Prométhée

Prométhée is a framework that enables the deployment, within an entity, corporation or institution (the term "agency" will be used from now on), of a fully functional infocentre that enables the aggregation, cross-interrogation, visualization and simple statistical analysis of a large-scale heterogeneous information system.

Prométhée attempts to directly address the complexity of both the heterogeneity and constant evolution of large-scale information systems. It operates without requiring modification or adaptation of the different data sources it interconnects. It is specifically designed to minimize the maintenance tasks needed when data sources are modified or replaced.

As constantly evolving technology is one of the major problems to be addressed by such a framework, we support the opinion that technology by itself cannot be the basis for a sustainable solution to the problem. This is the reason why Prométhée does not rely on technologies, but rather on a theoretic model which is based on algebraic concepts.

The theoretic data model of Prométhée mainly relies on category theory, graph theory and set theory. We have made no theoretic advances in these fields; instead we have combined and applied them to address the problem of integration and cross-interrogation of large-scale heterogeneous information systems.

In order to deploy Prométhée within an agency, the first step consists in defining one or more ontology diagrams containing the major entities manipulated during the agency's activity, as well as possible transformations between these entities. These ontology models are created using a specialized user interface. They are stored in the algebraic model of Prométhée, which is based on category theory (figure 1). The use of category theory to support ontologies in information systems integration is treated in some detail in [3], [4].

Once the ontological diagram has been defined, data sources (or subsets of sources) can then be easily attached at any entity (domain) of the diagram (category). Data sources thus become entities (domains) themselves within the diagram.

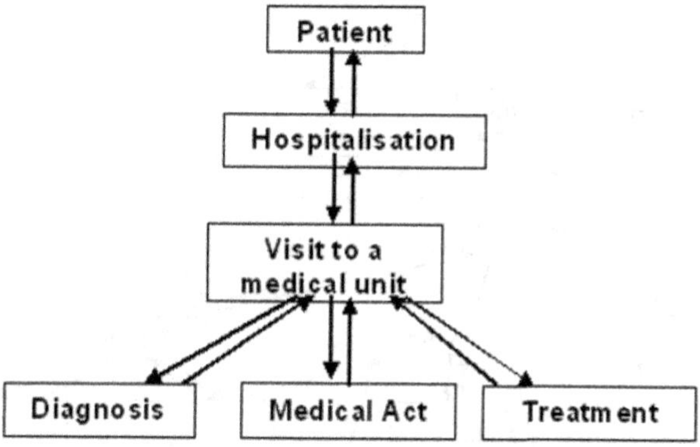

Fig. 1. Simplified diagram illustrating a subset of the Prométhée category model as applied to the hospital information system of the Institut Curie. The ontological diagram can be viewed as a category, the entities as domains, and the arrows as morphisms between different domains.

Data sources that are attachable to the model in our functional prototype include: Relational Database Management Systems (such as Oracle, MySQL, PostgreSQL and Microsoft SQL Server), Documentary Management Systems (such as Lotus Notes), HTML (web) content, XML content, and collections of files (such Microsoft Word or Adobe PDF documents).

In the following example drawn from the medical information system of the Institut Curie, data sources attached to the ontological diagram include the patient identity server, the hospitalisation, radiology and surgery reports documentary systems, the PMSI system (this system, present in all French hospitals, reports the hospital's activity to the French Ministry of Health), the e-prescription system, and the chemotherapy database system (figure 2).

In order to ask a medical question across the information system via Prométhée (figure 3), an authorized professional uses the framework's web user interface. He first selects one or more data sources, and then formulates his question by filling in one or several web forms. When a query is submitted, it is first processed by the query translation module. This module translates the web form-formulated question into computer search engine and into human language. The first makes the query understandable by a particular search engine, the latter enables the user to verify the correct formulation of his question. The query, frequently consisting of several subqueries, is then addressed to the meta-engine, which dispatches the subqueries to one or more different search engines. The engines either query the data sources directly, or query specific indexes generated and maintained by Prométhée. They then collect the subquery results. The engines then address their sets of results to the meta-engine. These sets are processed by the meta-engine's algebraic model, via consecutive set operations and entity transformations using the morphisms defined in the meta-engine's category theory model. The final results of the query are obtained, and rendered to the user via the user interface.

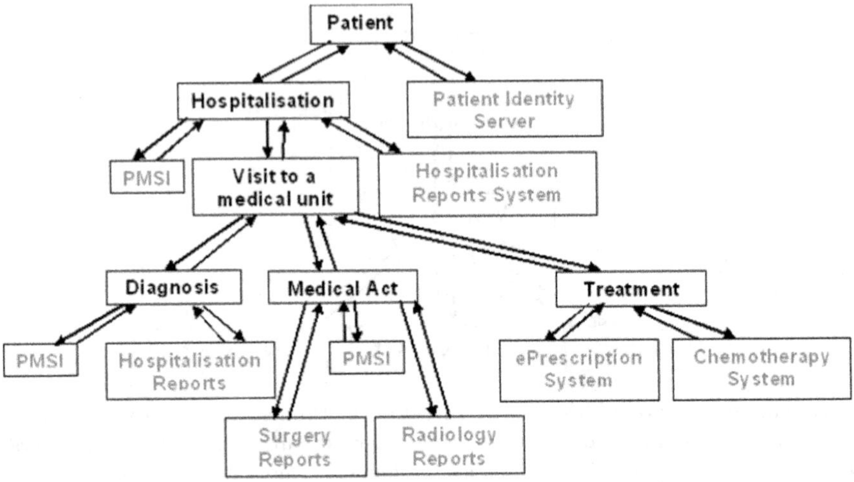

Fig. 2. Attaching heterogeneous data sources to the category model. Note that different subsets of the same system can be attached to different domains of the model.

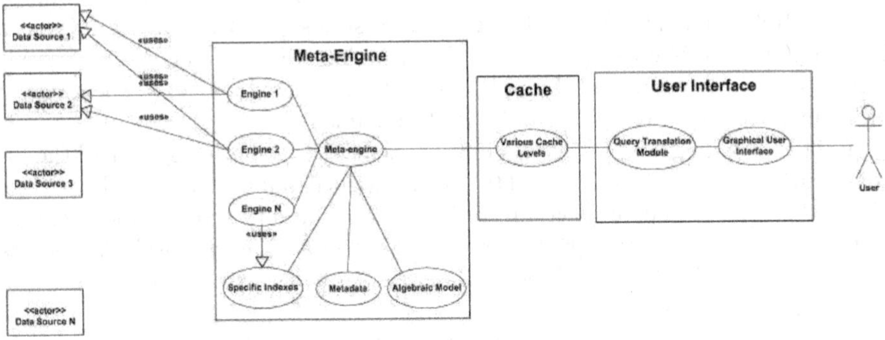

Fig. 3. Prométhée simplified use case diagram

3 Software Design Methodology

The Prométhée concept has been translated into a framework specification, using design pattern methodology. This specification has been implemented into a software development kit (Prométhée SDK), which is a set of coordinated objects, components and services. It is a generic toolkit, designed to address various aspects of heterogeneous data source integration, cross-interrogation, and cross-interrogation related problems.

The Prométhée SDK has been used to derive various applications, services and tools. Some examples are a web application that enables to design the integrated model of a heterogeneous information system, a web application which enables users to cross-query the information system, or a cross-heterogeneous data source quality control system that generates nightly quality reports.

For the framework's specification and implementation, we are using various software design paradigms. Using methodologies based on experimentation, quick prototyping and testing, we selected the most appropriate paradigms to accomplish specific tasks and/or roles within the Prométhée framework, in terms of design, maintainability, and evolutivity. For example, object-oriented and component-oriented programming are used for the user interfaces, aspect-oriented programming is used to provide authorization and audit capabilities; functional programming is heavily used within the algebraic model. At the highest design level, subsystems are encapsulated using object-oriented methodology and their interactions are formalized using design pattern methodology.

4 Applying the Concept: The Institut Curie's Medical Infocentre

The Institut Curie is the conjunction of a hospital and a research centre focused on cancer. It is a private, non-profit foundation accredited as a public service since 1921, and constitutes a "reference site" in the French health care system. The Institut Curie's originality relies on its double mission of treatment of and research on cancer, as well as the continuity between its various activities, from fundamental research to clinical research to healthcare.

The Institut Curie hospital specialises in breast cancer, paediatric tumours, ocular tumours, and sarcomas. It is one of the first fully-computerized cancer care centres in Europe, and relies on the sum of 60 distinct ICT systems. Each one of these systems ensures the coherency, quality, traceability, availability and completeness of one or more specific medical or medico-technical activities.

Prométhée currently interconnects 30 of these systems. Integrated information covers patient identity data, tumour identity data, medical reports (hospitalisation, surgery, consultation, etc), clinical (biostatistics) data, pathology data (reports, tissue images, marker/antibody analysis results, etc), radiology data, radiotherapy data, specimen banks (tumour, serum, haematology, pharmacology, and genetics banks). From the content-type point of view, this covers structured, free-text and multimedia data.

5 Impact

In 2005, the European Commission launched its eHealth Impact Study [2], with the objective to develop a generic, context-adaptable assessment and evaluation method for eHealth applications and services, and obtain reliable evidence on the positive (economic and other) impacts of eHeath systems used in real-life medical or health situations, by applying the method in depth at ten sites.

Together with Elios, the Institut Curie's electronic patient record, Prométhée has been selected to feature within a list of 100 good practice examples from all EU Member States and across all important Information Technology (IT) for Health application fields. Furthermore, our Institute has been selected as one of the ten European sites where an in-depth impact study was conducted [5].

This study shows that, since its first deployment in 2002, the Prométhée medical infocentre is increasingly used by the healthcare professionals of our Institute (figure 4).

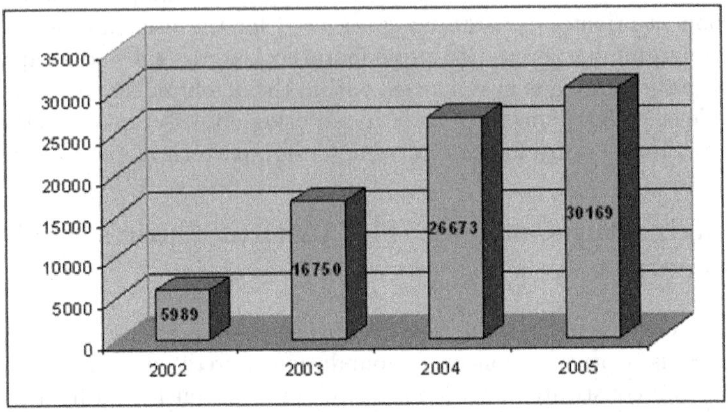

Fig. 4. Usage statistics of Prométhée since 2002. Each use in the chart represents a cross-interrogation, a 1- or 2-variable statistical analysis, or a data export of query results. Note that the leap observed between 2002 and 2003 corresponds to a major version release of Prométhée.

Additionally, the study analyses the progressive use of the tool in different and increasingly complex cognitive contexts of use: activity reporting, support to medical training or course preparation, support to the management of human-specimen bank systems where the management needs to be based on criteria present in independent information systems, and evaluation of medical practices [1,6].

Finally, an interesting conclusion of this study is that, after 4 years of deployment, Prométhée matures towards an operational steering tool to support the reorganisation of medical processes, progressively leading to major organizational impact as well as improved economic efficiency. Thus we could speculate that, after succeeding to adapt to numerous evolutions of the hospital's needs and organization, Prométhée starts to contribute itself to this evolution.

These results are in compliance with our initial goal to design a tool that is generic and flexible enough to enable its use in particular cognitive contexts, which were not envisaged or formalized during the tool's design phase.

6 Future Developments

As the framework will continuously be used in the safety-critical hospital environment of the Institut Curie, we currently seek to enhance and further develop the Prométhée meta-engine in terms of algebraic methodological tools and mathematical correctness.

The Prométhée meta-engine, as a service, provides a framework upon which various new applications can be built. Our current efforts are oriented towards cross-heterogeneous data source specialized views, quality control, and event-triggered notification systems.

7 Conclusion

Prométhée probably defines a novel family of heterogeneous data source search meta-engine tools, based on category theory. Applied to the hospital environment of the Institut Curie, it proves to be an ideal framework to support the emerging needs of biomedical informatics and health informatics.

While the tool is not directly used for healthcare delivery, it is used in a safety-critical environment, and in increasingly complex situations. Thus, the theoretic foundations for this type of systems must be further formalised. We seek to participate in this effort by contributing our pragmatic approach and prototypes used in a real-life medical environment.

References

1. C. Buron and A. Livartowski. Evaluation médico-économique de la macrobiopsie assistée par aspiration (Mammotome®) comparée à la biopsie chirurgicale dans la prise en charge des lésions infracliniques du sein. Study conducted through 16 French healthcare centres and coordinated by the Institut Curie for the DHOS (Direction of Hospitalisation and Organisation of Healthcare), French Ministry of Health, 2005.
2. eHealth IMPACT. Study on Economic and Productivity Impact of eHealth, commissioned by the European Commission. DG Information Society and Media. http://www.ehealth-impact.org
3. M. Johnson and C.N.G. Dampney. On category theory as a (meta)ontology for information systems research. In Formal Ontology in Information Systems edited by Chris Welty and Barry Smith, 59–69, 2001, ACM Press.
4. M. Johnson and C.N.G. Dampney. Enterprise information systems: specifying the links among project data models using category theory. Proceedings of the International Conference on Enterprise Information Systems, 619–626, 2001.
5. P.-C. Varoutas, A. Livartowski, M. Jarossay, B. Sigal-Zafrani, F. Gros, M. Cosquer, P. Rizand, T. Jones, D. Ambroise and V. Stroetmann. eHealth Impact : Institut Curie - Elios and Prométhée. Proceedings of the eHealth 2006 High Level Conference and Exhibition (in press). http://www.ehealthconference2006.org
6. B. Sigal-Zafrani, K. Müller, C. El Khoury, P.-C. Varoutas, C. Buron, A. Vincent-Salomon, S. Alran, A. Livartowski, R.J. Salmon, S. Neuenschwander. Large core vacuum assisted biopsy improves the management of patients presenting breast microcalcifications: analysis of 1009 cases. British Journal of Surgery (submitted).

Author Index

Lecture Notes in Computer Science

For information about Vols. 1–3969

please contact your bookseller or Springer

Vol. 4067: D. Thomas (Ed.), ECOOP 2006 – Object-Oriented Programming. XIV, 527 pages. 2006.

Vol. 4063: I. Gorton, G.T. Heineman, I. Crnkovic, H.W. Schmidt, J.A. Stafford, C.A. Szyperski, K. Wallnau (Eds.), Component-Based Software Engineering. XI, 394 pages. 2006.

Vol. 4060: K. Futatsugi, J.-P. Jouannaud, J. Meseguer (Eds.), Algebra, Meaning and Computation. XXXVIII, 643 pages. 2006.

Vol. 4059: L. Arge, R. Freivalds (Eds.), Algorithm Theory – SWAT 2006. XII, 436 pages. 2006.

Vol. 4058: L.M. Batten, R. Safavi-Naini (Eds.), Information Security and Privacy. XII, 446 pages. 2006.

Vol. 4057: J.P. W. Pluim, B. Likar, F.A. Gerritsen (Eds.), Biomedical Image Registration. XII, 324 pages. 2006.

Vol. 4056: P. Flocchini, L. Gąsieniec (Eds.), Structural Information and Communication Complexity. X, 357 pages. 2006.

Vol. 4055: J. Lee, J. Shim, S.-g. Lee, C. Bussler, S. Shim (Eds.), Data Engineering Issues in E-Commerce and Services. IX, 290 pages. 2006.

Vol. 4054: A. Horváth, M. Telek (Eds.), Formal Methods and Stochastic Models for Performance Evaluation. VIII, 239 pages. 2006.

Vol. 4053: M. Ikeda, K.D. Ashley, T.-W. Chan (Eds.), Intelligent Tutoring Systems. XXVI, 821 pages. 2006.

Vol. 4048: L. Goble, J.-J.C.. Meyer (Eds.), Deontic Logic and Artificial Normative Systems. X, 273 pages. 2006. (Sublibrary LNAI).

Vol. 4046: S.M. Astley, M. Brady, C. Rose, R. Zwiggelaar (Eds.), Digital Mammography. XVI, 654 pages. 2006.

Vol. 4045: D. Barker-Plummer, R. Cox, N. Swoboda (Eds.), Diagrammatic Representation and Inference. XII, 301 pages. 2006. (Sublibrary LNAI).

Vol. 4044: P. Abrahamsson, M. Marchesi, G. Succi (Eds.), Extreme Programming and Agile Processes in Software Engineering. XII, 230 pages. 2006.

Vol. 4043: A.S. Atzeni, A. Lioy (Eds.), Public Key Infrastructure. XI, 261 pages. 2006.

Vol. 4041: S.-W. Cheng, C.K. Poon (Eds.), Algorithmic Aspects in Information and Management. XI, 395 pages. 2006.

Vol. 4040: R. Reulke, U. Eckardt, B. Flach, U. Knauer, K. Polthier (Eds.), Combinatorial Image Analysis. XII, 482 pages. 2006.

Vol. 4039: M. Morisio (Ed.), Reuse of Off-the-Shelf Components. XIII, 444 pages. 2006.

Vol. 4038: P. Ciancarini, H. Wiklicky (Eds.), Coordination Models and Languages. VIII, 299 pages. 2006.

Vol. 4037: R. Gorrieri, H. Wehrheim (Eds.), Formal Methods for Open Object-Based Distributed Systems. XVII, 474 pages. 2006.

Vol. 4036: O. H. Ibarra, Z. Dang (Eds.), Developments in Language Theory. XII, 456 pages. 2006.

Vol. 4035: H.-P. Seidel, T. Nishita, Q. Peng (Eds.), Advances in Computer Graphics. XX, 771 pages. 2006.

Vol. 4034: J. Münch, M. Vierimaa (Eds.), Product-Focused Software Process Improvement. XVII, 474 pages. 2006.

Vol. 4033: B. Stiller, P. Reichl, B. Tuffin (Eds.), Performability Has its Price. X, 103 pages. 2006.

Vol. 4031: M. Ali, R. Dapoigny (Eds.), Innovations in Applied Artificial Intelligence. XXIII, 1353 pages. 2006. (Sublibrary LNAI).

Vol. 4027: H.L. Larsen, G. Pasi, D. Ortiz-Arroyo, T. Andreasen, H. Christiansen (Eds.), Flexible Query Answering Systems. XVIII, 714 pages. 2006. (Sublibrary LNAI).

Vol. 4026: P.B. Gibbons, T. Abdelzaher, J. Aspnes, R. Rao (Eds.), Distributed Computing in Sensor Systems. XIV, 566 pages. 2006.

Vol. 4025: F. Eliassen, A. Montresor (Eds.), Distributed Applications and Interoperable Systems. XI, 355 pages. 2006.

Vol. 4024: S. Donatelli, P. S. Thiagarajan (Eds.), Petri Nets and Other Models of Concurrency - ICATPN 2006. XI, 441 pages. 2006.

Vol. 4021: E. André, L. Dybkjær, W. Minker, H. Neumann, M. Weber (Eds.), Perception and Interactive Technologies. XI, 217 pages. 2006. (Sublibrary LNAI).

Vol. 4020: A. Bredenfeld, A. Jacoff, I. Noda, Y. Takahashi (Eds.), RoboCup 2005: Robot Soccer World Cup IX. XVII, 727 pages. 2006. (Sublibrary LNAI).

Vol. 4019: M. Johnson, V. Vene (Eds.), Algebraic Methodology and Software Technology. XI, 389 pages. 2006.

Vol. 4018: V. Wade, H. Ashman, B. Smyth (Eds.), Adaptive Hypermedia and Adaptive Web-Based Systems. XVI, 474 pages. 2006.

Vol. 4016: J.X. Yu, M. Kitsuregawa, H.V. Leong (Eds.), Advances in Web-Age Information Management. XVII, 606 pages. 2006.

Vol. 4014: T. Uustalu (Ed.), Mathematics of Program Construction. X, 455 pages. 2006.

Vol. 4013: L. Lamontagne, M. Marchand (Eds.), Advances in Artificial Intelligence. XIII, 564 pages. 2006. (Sublibrary LNAI).

Vol. 4012: T. Washio, A. Sakurai, K. Nakajima, H. Takeda, S. Tojo, M. Yokoo (Eds.), New Frontiers in Artificial Intelligence. XIII, 484 pages. 2006. (Sublibrary LNAI).

Vol. 4011: Y. Sure, J. Domingue (Eds.), The Semantic Web: Research and Applications. XIX, 726 pages. 2006.

Vol. 4010: S. Dunne, B. Stoddart (Eds.), Unifying Theories of Programming. VIII, 257 pages. 2006.

Vol. 4009: M. Lewenstein, G. Valiente (Eds.), Combinatorial Pattern Matching. XII, 414 pages. 2006.

Vol. 4007: C. Àlvarez, M. Serna (Eds.), Experimental Algorithms. XI, 329 pages. 2006.

Vol. 4006: L.M. Pinho, M. González Harbour (Eds.), Reliable Software Technologies – Ada-Europe 2006. XII, 241 pages. 2006.

Vol. 4005: G. Lugosi, H.U. Simon (Eds.), Learning Theory. XI, 656 pages. 2006. (Sublibrary LNAI).

Vol. 4004: S. Vaudenay (Ed.), Advances in Cryptology - EUROCRYPT 2006. XIV, 613 pages. 2006.

Vol. 4003: Y. Koucheryavy, J. Harju, V.B. Iversen (Eds.), Next Generation Teletraffic and Wired/Wireless Advanced Networking. XVI, 582 pages. 2006.

Vol. 4001: E. Dubois, K. Pohl (Eds.), Advanced Information Systems Engineering. XVI, 560 pages. 2006.

Vol. 3999: C. Kop, G. Fliedl, H.C. Mayr, E. Métais (Eds.), Natural Language Processing and Information Systems. XIII, 227 pages. 2006.

Vol. 3998: T. Calamoneri, I. Finocchi, G.F. Italiano (Eds.), Algorithms and Complexity. XII, 394 pages. 2006.

Vol. 3997: W. Grieskamp, C. Weise (Eds.), Formal Approaches to Software Testing. XII, 219 pages. 2006.

Vol. 3996: A. Keller, J.-P. Martin-Flatin (Eds.), Self-Managed Networks, Systems, and Services. X, 185 pages. 2006.

Vol. 3995: G. Müller (Ed.), Emerging Trends in Information and Communication Security. XX, 524 pages. 2006.

Vol. 3994: V.N. Alexandrov, G.D. van Albada, P.M.A. Sloot, J. Dongarra (Eds.), Computational Science – ICCS 2006, Part IV. XXXV, 1096 pages. 2006.

Vol. 3993: V.N. Alexandrov, G.D. van Albada, P.M.A. Sloot, J. Dongarra (Eds.), Computational Science – ICCS 2006, Part III. XXXVI, 1136 pages. 2006.

Vol. 3992: V.N. Alexandrov, G.D. van Albada, P.M.A. Sloot, J. Dongarra (Eds.), Computational Science – ICCS 2006, Part II. XXXV, 1122 pages. 2006.

Vol. 3991: V.N. Alexandrov, G.D. van Albada, P.M.A. Sloot, J. Dongarra (Eds.), Computational Science – ICCS 2006, Part I. LXXXI, 1096 pages. 2006.

Vol. 3990: J. C. Beck, B.M. Smith (Eds.), Integration of AI and OR Techniques in Constraint Programming for Combinatorial Optimization Problems. X, 301 pages. 2006.

Vol. 3989: J. Zhou, M. Yung, F. Bao, Applied Cryptography and Network Security. XIV, 488 pages. 2006.

Vol. 3988: A. Beckmann, U. Berger, B. Löwe, J.V. Tucker (Eds.), Logical Apporaches to Computational Barriers. XV, 608 pages. 2006.

Vol. 3987: M. Hazas, J. Krumm, T. Strang (Eds.), Location- and Context-Awareness. X, 289 pages. 2006.

Vol. 3986: K. Stølen, W.H. Winsborough, F. Martinelli, F. Massacci (Eds.), Trust Management. XIV, 474 pages. 2006.

Vol. 3984: M. Gavrilova, O. Gervasi, V. Kumar, C.J. K. Tan, D. Taniar, A. Laganà, Y. Mun, H. Choo (Eds.), Computational Science and Its Applications - ICCSA 2006, Part V. XXV, 1045 pages. 2006.

Vol. 3983: M. Gavrilova, O. Gervasi, V. Kumar, C.J. K. Tan, D. Taniar, A. Laganà, Y. Mun, H. Choo (Eds.), Computational Science and Its Applications - ICCSA 2006, Part IV. XXVI, 1191 pages. 2006.

Vol. 3982: M. Gavrilova, O. Gervasi, V. Kumar, C.J. K. Tan, D. Taniar, A. Laganà, Y. Mun, H. Choo (Eds.), Computational Science and Its Applications - ICCSA 2006, Part III. XXV, 1243 pages. 2006.

Vol. 3981: M. Gavrilova, O. Gervasi, V. Kumar, C.J. K. Tan, D. Taniar, A. Laganà, Y. Mun, H. Choo (Eds.), Computational Science and Its Applications - ICCSA 2006, Part II. XXVI, 1255 pages. 2006.

Vol. 3980: M. Gavrilova, O. Gervasi, V. Kumar, C.J. K. Tan, D. Taniar, A. Laganà, Y. Mun, H. Choo (Eds.), Computational Science and Its Applications - ICCSA 2006, Part I. LXXV, 1199 pages. 2006.

Vol. 3979: T.S. Huang, N. Sebe, M.S. Lew, V. Pavlović, M. Kölsch, A. Galata, B. Kisačanin (Eds.), Computer Vision in Human-Computer Interaction. XII, 121 pages. 2006.

Vol. 3978: B. Hnich, M. Carlsson, F. Fages, F. Rossi (Eds.), Recent Advances in Constraints. VIII, 179 pages. 2006. (Sublibrary LNAI).

Vol. 3977: N. Fuhr, M. Lalmas, S. Malik, G. Kazai (Eds.), Advances in XML Information Retrieval and Evaluation. XII, 556 pages. 2006.

Vol. 3976: F. Boavida, T. Plagemann, B. Stiller, C. Westphal, E. Monteiro (Eds.), NETWORKING 2006. Networking Technologies, Services, and Protocols; Performance of Computer and Communication Networks; Mobile and Wireless Communications Systems. XXVI, 1276 pages. 2006.

Vol. 3975: S. Mehrotra, D.D. Zeng, H. Chen, B. Thuraisingham, F.-Y. Wang (Eds.), Intelligence and Security Informatics. XXII, 772 pages. 2006.

Vol. 3973: J. Wang, Z. Yi, J.M. Zurada, B.-L. Lu, H. Yin (Eds.), Advances in Neural Networks - ISNN 2006, Part III. XXIX, 1402 pages. 2006.

Vol. 3972: J. Wang, Z. Yi, J.M. Zurada, B.-L. Lu, H. Yin (Eds.), Advances in Neural Networks - ISNN 2006, Part II. XXVII, 1444 pages. 2006.

Vol. 3971: J. Wang, Z. Yi, J.M. Zurada, B.-L. Lu, H. Yin (Eds.), Advances in Neural Networks - ISNN 2006, Part I. LXVII, 1442 pages. 2006.

Vol. 3970: T. Braun, G. Carle, S. Fahmy, Y. Koucheryavy (Eds.), Wired/Wireless Internet Communications. XIV, 350 pages. 2006.